Sustainability in Plant and Crop Protection

Series editor

Aurelio Ciancio, Sezione di Bari, Consiglio Nazionale delle Ricerche Istituto per la Protezione delle Piante, Bari, Italy

More information about this series at http://www.springer.com/series/13031

Rosa H. Manzanilla-López
Luis V. Lopez-Llorca
Editors

Perspectives in Sustainable Nematode Management Through *Pochonia chlamydosporia* Applications for Root and Rhizosphere Health

Springer

Editors
Rosa H. Manzanilla-López
Centro de Desarrollo de Productos Bióticos (Visiting Professor)
Instituto Politécnico Nacional
Yautepec, Morelos, Mexico

Luis V. Lopez-Llorca
Department of Marine Sciences and Applied Biology
University of Alicante
Alicante, Spain

Sustainability in Plant and Crop Protection
ISBN 978-3-319-59222-0 ISBN 978-3-319-59224-4 (eBook)
DOI 10.1007/978-3-319-59224-4

Library of Congress Control Number: 2017947835

Printed on acid-free paper

This Springer imprint is published by Springer Nature
The registered company is Springer International Publishing AG
The registered company address is: Gewerbestrasse 11, 6330 Cham, Switzerland

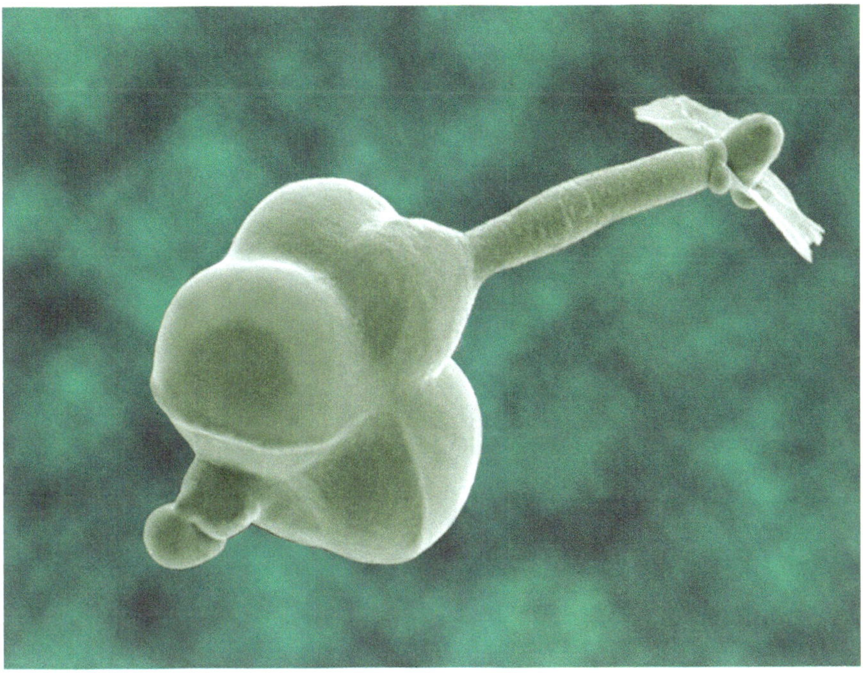

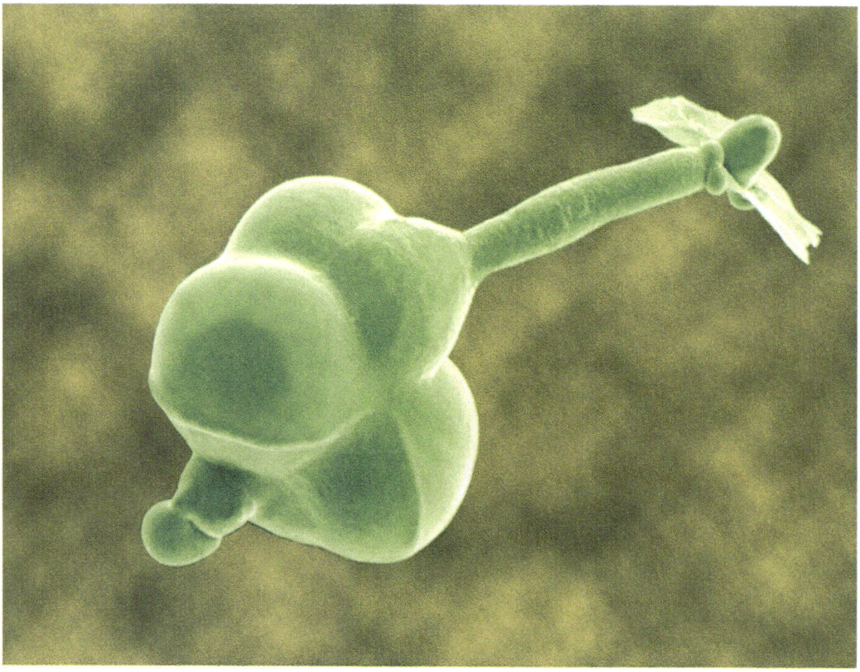

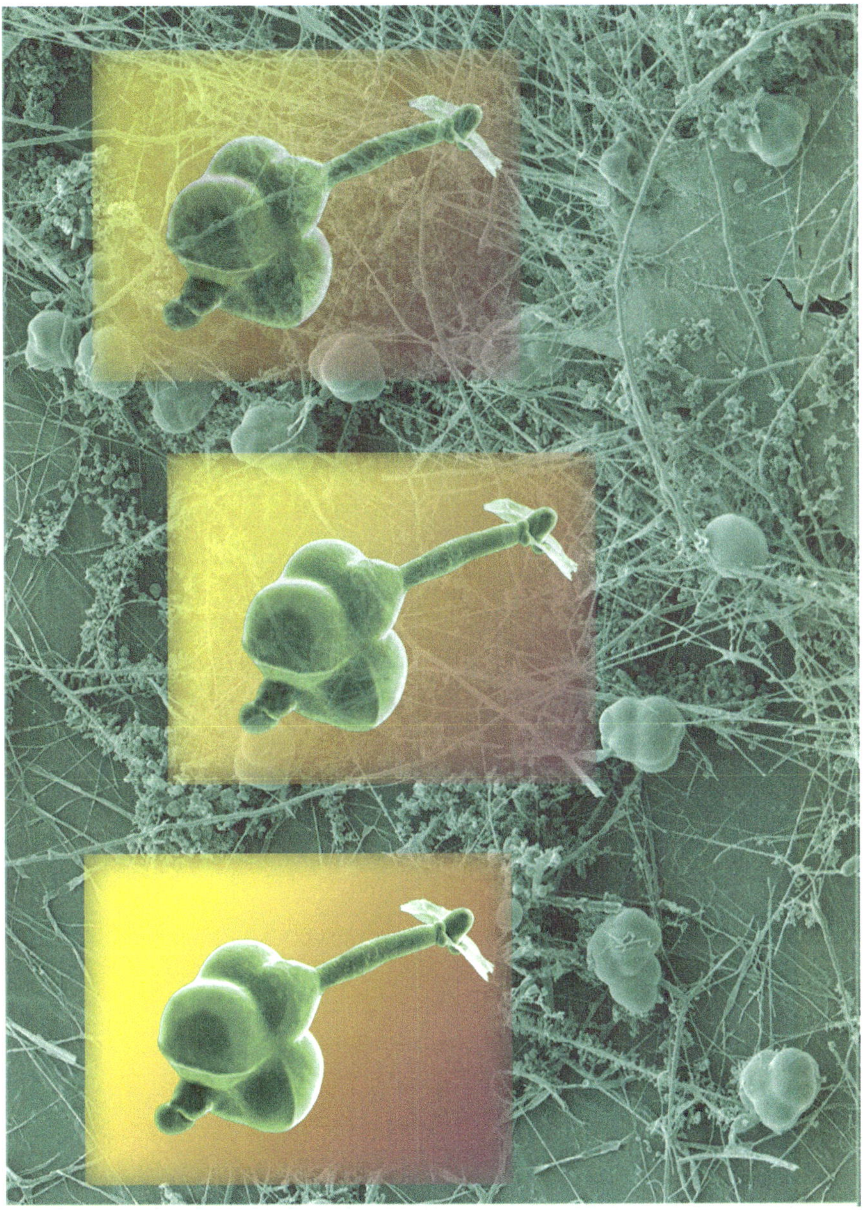

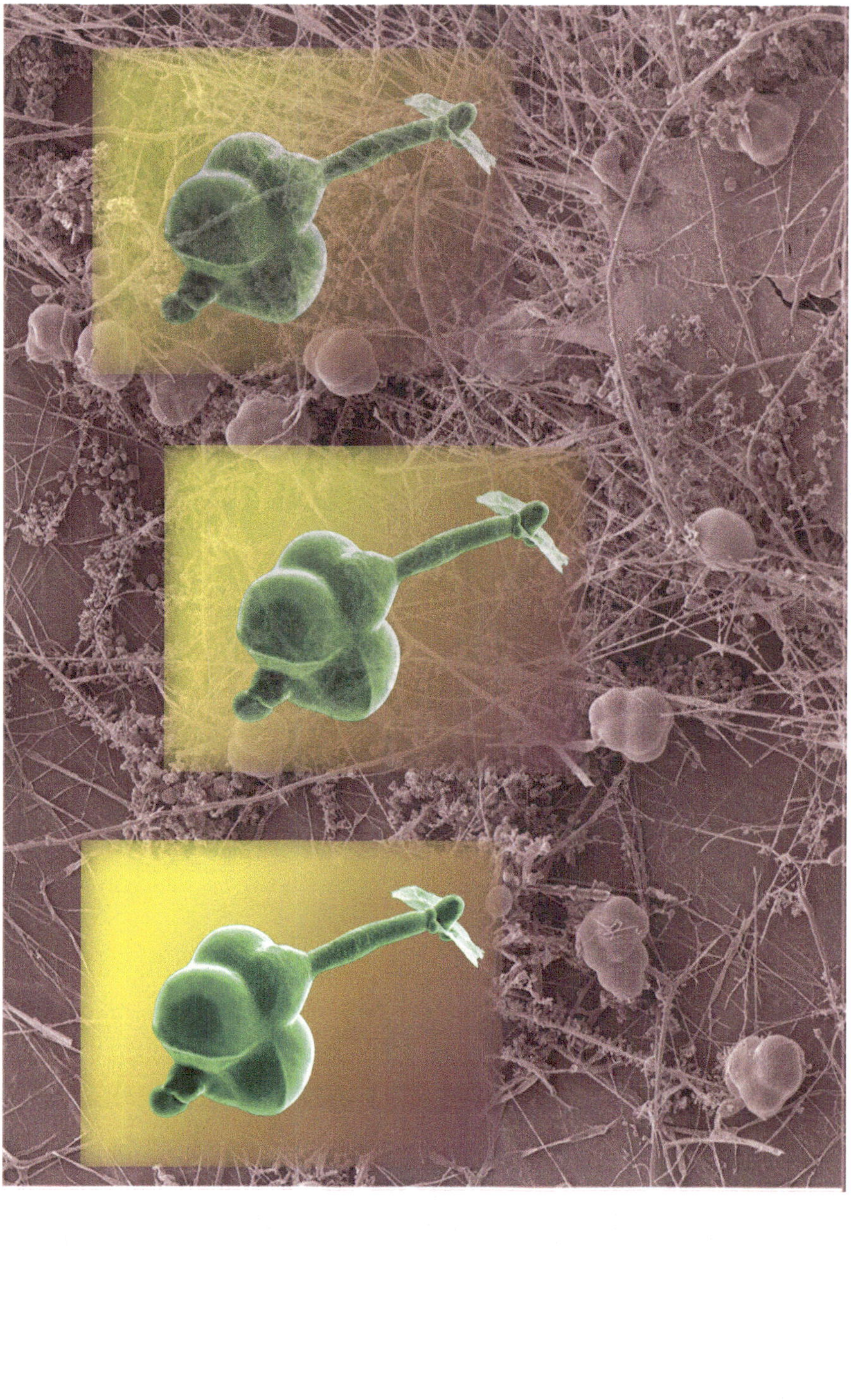

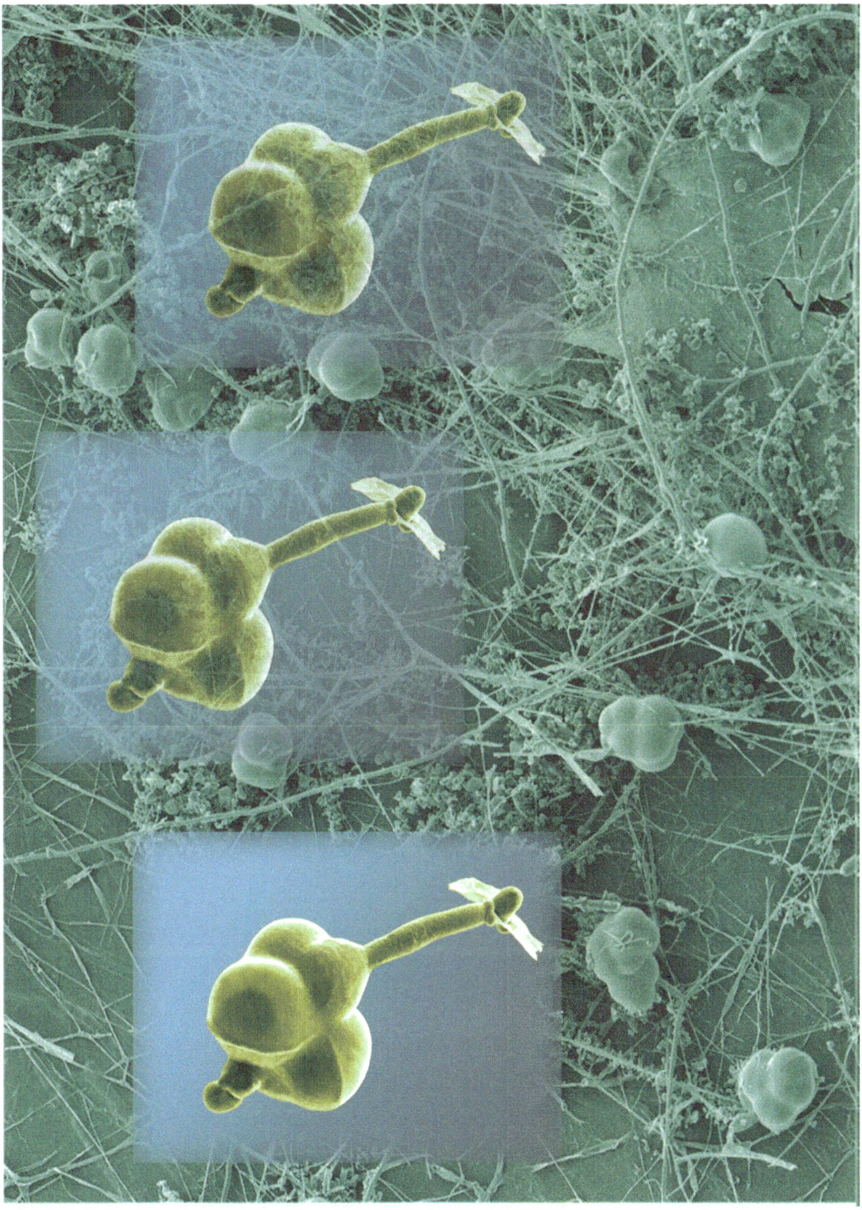

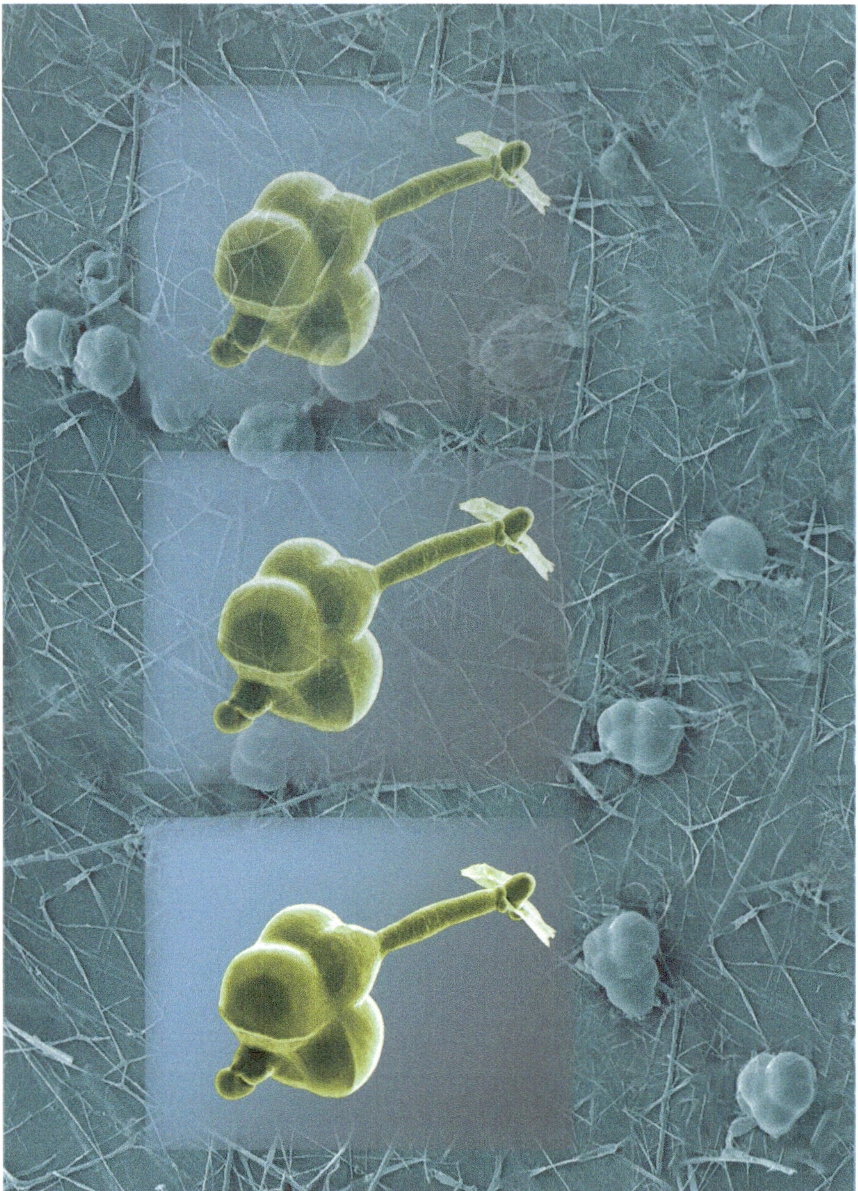

Figure legend

Cryo-SEM of a chlamydospore of *Pochonia chlamydosporia* isolate Pc60. Image design Jean Devonshire (Original image Copyright Rothamsted Research Ltd.)

Foreword

Plant-parasitic nematodes pose a serious and continuing threat to crop production worldwide. The diseases they cause are difficult to diagnose and often overlooked by comparison with other biotic constraints such as insect pests, microbial pathogens and weeds. Nonetheless, their impact, either alone or as members of disease complexes, is substantial, with losses running into many billions of dollars. In resource-poor regions, such losses can have a direct impact on food security.

Management of plant-parasitic nematodes also poses major challenges. Natural resistance is present in some crops, but may not be durable or effective against all strains of a nematode. In other crop species, few natural sources of genetic resistance have been found, and engineering via biotechnological approaches may be the only potential solution. Most plant-parasitic nematodes live in soil or association with plant roots and are difficult to target with chemicals. Many of the most effective nematicides originally available for use have now been withdrawn due to their more general biocidal effects, and concerns over their environmental impact. Alternative sustainable strategies for nematode control are urgently needed. Biological approaches, exploiting the natural enemies of these pests, are therefore of particular interest.

The topic of this book starts with a detective story, as pioneering scientists searched for a biological explanation for the observation that on some hosts in certain soils the severity of infestation by nematodes is naturally suppressed. Amongst the suspects was a fungus with the capability to attack some stages in the nematode life cycle, in particular female worms and eggs. The fungus, originally named *Verticillium chlamydosporium*, was later reclassified as a single species in the related genus *Pochonia*. But this was not a simple case. The main suspect, *P. chlamydosporia,* turned out to have several identities, as a dormant survivor, soil saprophyte, facultative parasite, and root endophyte, with the ability to switch between these different states. To exploit the fungus as a biological control agent required more research and a greater understanding of its ecology and interactions with plant roots and nematode parasites.

The editors of this book have assembled an international team of authors expert in all aspects of the biology of *P. chlamydosporia* and its use as a biocontrol agent in horticulture and agriculture. The chapters cover a wide range of topics, from laboratory culture and identification, through physiology and biochemistry, to state of the art studies of the recently assembled genome sequence, proteomics and secondary metabolism. Examples of current and potential applications in the control of both plant and animal parasitic nematodes are covered, as well as recent findings on the beneficial effects on plant development of colonisation as an endophyte. While the focus is on a single fungal taxon, research on *P. chlamydosporia* has raised questions of much wider biological significance in soil ecology, fungal biology, and host-pathogen interactions. The genetic and molecular basis of the multi-trophic lifestyle is a recurrent theme relevant to other fungi that undergo physiological transitions at different stages of their life cycle. Improved understanding of the factors regulating such processes should aid effective production, formulation and delivery of the fungus as a bio-inoculant.

The impetus for all this scientific investigation is the prospect of developing the fungus as an important ally in the sustainable management of parasitic nematodes. Much progress has been made, especially in the local production and use of the fungus, but challenges remain in achieving consistently high levels of control, and scaling up application to broad-acre staple crops. There are no quick fixes in the business of sustainable crop protection. In the longer term, however, the knowledge and experience gained should be invaluable in developing integrated pest management systems that will stand the test of time.

Former Head of Plant Pathology and Microbiology John A. Lucas,
Rothamsted Research
Herts, UK

Series Preface

This is the third Volume of the Series *Sustainability in Plant and Crop Protection* (SUPP). Its specific focus is on the beneficial hyphomycete *Pochonia chlamydosporia*, treated and examined in relation to its biology and potential for crop protection and sustainable management. The Volume considers the fungus biochemistry, taxonomy, ecology and application, providing actualised data on its genome, gene expression, and metabolism. It has been produced thanks to the endless efforts and enthusiastic endeavours of the Editors, Dr Rosa Manzanilla Lopez and Prof. Luis Vicente Lopez Llorca, and all the contributing Authors.

Management of nematode pests is a challenging task, due to the widespread distribution of species that attack the most important food and industrial crops or livestock, the peculiar ecology of parasitism including the environments in which nematodes live and multiply, and the complexity of the rhizosphere interactions. Although the use of synthetic pesticides against parasitic nematodes is still applied in many agroecosystems, concerns about the environment and human safety issues progressively increased the demand and use of organic products and technologies. Several studies and research efforts thus focused on more sustainable control practices, oriented towards non-chemical management approaches. Biological control and plant-microorganisms interactions are now established research fields within plant protection and nematology disciplines. However, detailed knowledge on the communities of microorganisms and their interactions are fundamental, finalized at switching from pesticide-based to information-based, and more resource-conservative, sustainable approaches.

Studies on soil microorganisms such as *P. chlamydosporia* are, therefore, welcome. This fungus has been the object of intense research work by taxonomists and nematologists since the early 70's, through laboratory and field-based studies, including the pioneering research work developed by Brian Kerry at Rothamsted Research. Brian first identified and reported the potential of *P. chlamydosporia* in cyst nematode management in the cereal crops in Northern Europe, and led a fruitful research activity in this field for many years, until his premature loss.

Many of the researchers involved in this Volume had the priviledge of his friendship and worked with him, following up with many studies with innovative and advanced research approaches, including the exploration of molecular and genomic dimensions. It is hence with interest that the reader may look at his Volume. It provides comprehensive reviews of many aspects of the *P. chlamydosporia* interactions with plants and soil, both as an endophyte and biological control agent, which are presented, described and illustrated in several chapters.

The production of data on the ecology, taxonomy and genetics of biological control agents, antagonists and root endophytes is indeed a challenge for the years to come. This is true because of the large number of species present in soil, in their vast majority still unclassified, and of the many interactions taking place on the root and soil interfaces.

This Volume provides a first comprehensive contribution to key aspects of a single species, within the context of its support to sustainable farming. The information provided is impressive, flanked by a rich bibliography, updating actual knowledge through the Authors' personal experience and work. It is desirable that it could be followed by many other similar treatises, dealing with the functions and services produced by other members of the soil microbial communities. Scientific data will provide us a clearer view on underground complex food webs, on their role in sustaining crops health and, ultimately, food production.

Aurelio Ciancio

Preface

The idea of preparing a book written by world experts on *Pochonia chlamydosporia* had been maturing for several years due to an increasing interest in *Pochonia chlamydosporia* (Fungi: *Ascomycota*), one of the most promising microorganisms for the biological control of plant-parasitic nematodes such as cyst (*Heterodera* spp., *Globodera* spp.), root-knot (*Meloidogyne* spp.) and false root-knot nematodes (*Nacobbus aberrans sensu lato*). This book is the result of colleagues willing to share their expertise and knowledge with others working on *Pochonia*. We hope that you may enjoy and find the knowledge compiled between the covers of this book useful.

The book contains 18 chapters divided into six sections, starting with a historical background chapter, followed by 16 chapters, each contributed by experts, concerning those key aspects necessary to work with this biocontrol agent in a multidisciplinary treatise. Topics covered include systematics, biology, nematode-fungus interactions, nematode management strategies, secondary metabolites and other methods including more novel research areas such as molecular, *-omics*, plant growth enhancement and endophytic abilities of *P. chlamydosporia*. The final chapter deals with the future perspectives of *P. chlamydosporia* research.

The book is aimed at researchers, professionals, practitioners and graduate or undergraduate students involved in both basic and applied studies, as well as in activities dealing with biocontrol agents of plant-parasitic nematodes, helminths, crop protection and integrated pest management. It concentrates on fundamental aspects and on the evolution of nematophagous fungi.

The focus of the book is based on *P. chlamydosporia* varieties. The authors and editors have tried to put together most of the information available on the fungus up to March 2017. Scientific names of fungi, as they appear cited in the book, have been revised using the *Index Fungorum* (http://www.indexfungorum.org/names/names.asp) to provide their updated, valid names. Readers wishing to know more on the research made on other biological control agents of nematodes are encouraged to consult specialised books on the subject, such as those written by Stirling (1991, 2014).

Finally, we would like to express our sincere gratitude to our families, authors and friends and colleagues, whose enthusiasm, support and hard work have made the completion of this task possible and also for sharing their images with us to illustrate the book.

Harpenden, Herts, UK
Alicante, Spain
April 2017

Rosa H. Manzanilla-López
Luis V. Lopez-Llorca

References

Stirling, G. R. (1991). *Biological control of plant parasitic nematodes: Progress, problems and prospects*. Wallingford: CABI Publishing.

Stirling, G. R. (2014). *Biological control of plant-parasitic nematodes: soil ecosystem management in sustainable agriculture* (2nd ed). Wallingford: CABI.

Acknowledgments

Rosa H. Manzanilla-López would like to thank the authorities and colleagues of the Centro de Desarrollo de Productos Bióticos (CEPROBI) of the National Polytechnic Institute of Mexico for their continued support to her as an invited professor participating in various academic and research activities, including the production of this book.

Luis V. Lopez-Llorca wishes to thank Prof. N. Talbot (University of Exeter, UK) for his support during a sabbatical stay at his laboratory, when most of his editorial work was carried out. Thanks are also due to the Spanish Ministry of Economy and Competitiveness for grant AGL 2015-66833-R, which covered the sabbatical stay.

We also thank Dr. David J. Hunt for his guidance on matters of editorial style and his invaluable help in editing and revising the text. Thanks are also due to our Springer editors, Dr. Aurelio Ciancio (Editor in Chief), Mariska van der Stigchel and Zuzana Bernhart.

Contents

Contributors

Isabel Abrantes Centre for Functional Ecology (CFE), University of Coimbra, Coimbra, Portugal

Thalita S. Avelar Monteiro Universidade Federal de Viçosa, Campus de Rio Paranaíba, Instituto de Ciências Agrárias, Rio Paranaíba, MG, Brazil

Matthew A. Back Centre for Integrated Pest Management, Harper Adams University, Newport, Shropshire, UK

Xinyue Cheng College of Life Sciences, Beijing Normal University, Beijing, China

Aurelio Ciancio Istituto per la Protezione Sostenibile delle Piante, CNR, Bari, Italy

Mariantonietta Colagiero Istituto per la Protezione Sostenibile delle Piante, CNR, Bari, Italy

Rosane H.C. Curtis Bionemax UK Ltd, Rothamsted Centre for Research and Enterprise, Harpenden, Herts, UK

Jean Devonshire Formerly at Rothamsted Research, Harpenden, Herts, UK

Nuria Escudero Laboratory of Plant Pathology, Multidisciplinary Institute for Environmental Studies (MIES) Ramón Margalef, Department of Marine Sciences and Applied Biology, University of Alicante, Alicante, Spain

Department of Agri-Food Engineering and Biotechnology, Universitat Politècnica de Catalunya, Catalunya, Spain

Ivânia Esteves CFE-Centre for Functional Ecology, Department of Life Sciences, Calçada Martim de Freitas, University of Coimbra, Coimbra, Portugal

Harry C. Evans CAB InternationalEgham, Surrey, UK

Ken Evans Formerly at Rothamsted Research, Harpenden, Herts, UK

Francisco Franco-Navarro Programa de Fitopatología, Colegio de Posgraduados-Campus Montecillo, Texcoco, Estado de México, Mexico

Sonia Gómez-Vidal Laboratory of Plant Pathology, Multidisciplinary Institute for Environmental Studies (MIES) Ramón Margalef, Department of Marine Sciences and Applied Biology, University of Alicante, Alicante, Spain

Research Technical Services, University of Alicante, Alicante, Spain

Leandro Grassi de Freitas Departamento de Fitopatología, Universidade Federal de Viçosa, Minas Gerais, Brazil

Leopoldo Hidalgo-Díaz Dirección de Sanidad Vegetal, Centro Nacional de Sanidad Agropecuaria, San José de las Lajas, Mayabeque, Cuba

Ricardo Holgado NIBIO, Norsk institutt for bioøkonomi /Norwegian Institute of Bioeconomy Research, Ås, Akershus, Norway

Paul M. Kirk Royal Botanic Gardens, Kew, Richmond, Surrey, UK

Runmao Lin College of Life Sciences, Beijing Normal University, Beijing, China

Institute of Vegetables and Flowers, Chinese Academy of Agricultural Sciences, Beijing, China

Everaldo A. Lopes Instituto de Ciências Agrárias, Campus de Rio Paranaíba, Universidade Federal de Viçosa, Rio Paranaíba, Minas Gerais, Brazil

Luis V. Lopez-Llorca Laboratory of Plant Pathology, Multidisciplinary Institute for Environmental Studies (MIES) Ramón Margalef, Department of Marine Sciences and Applied Biology, University of Alicante, Alicante, Spain

Federico Lopez-Moya Laboratory of Plant Pathology, Multidisciplinary Institute for Environment Studies (MIES) Ramón Margalef. Department of Marine Sciences and Applied Biology, University of Alicante, Alicante, Spain

Rosa H. Manzanilla-López Formerly at Rothamsted Research, Harpenden, UK

Centro de Desarrollo de Productos Bióticos (Visiting Professor), Yautepec, MOR, Mexico

Frutos Marhuenda-Egea Multidisciplinary Institute for Environmental Studies (MIES) Ramón Margalef, Department of Marine Sciences and Applied Biology, University of Alicante, Alicante, Spain

Department of Agrochemistry and Biochemistry, University of Alicante, Alicante, Spain

Laboratory of Plant Pathology, Department of Marine Sciences and Applied Biology, University of Alicante, Alicante, Spain

María Gabriela Medina-Canales Escuela Nacional de Ciencias Biológicas del Instituto Politécnico Nacional, Ciudad de México, Mexico

Pedro Mendoza de Gives National Centre for Disciplinary Research in Veterinary Parasitology (CENID-PAVET-INIFAP-MEXICO), Jiutepec, MOR, Mexico

Rosa Navarrete-Maya UNIGRAS (Unidad de Investigación en Granos y Semillas), Facultad de Estudios Superiores Cuautitlán, Universidad Nacional Autónoma de México, Cuautitlán Izcalli, Mexico

Xue-Mei Niu State Key Laboratory for Conservation and Utilization of Bio-resources and Key Laboratory for Microbial Resources of the Ministry of Education, Yunnan University, Kunming, People's Republic of China

Isabella Pentimone Istituto per la Protezione Sostenibile delle Piante, CNR, Bari, Italy

Fábio Ribeiro Braga Department of Veterinary, Federal University of Viçosa, University Campus s/n, Viçosa, MG, Brazil

Aída V. Rodríguez-Tovar Escuela Nacional de Ciencias Biológicas del Instituto Politécnico Nacional, Ciudad de México, Mexico

Laura Rosso Istituto per la Protezione Sostenibile delle Piante, CNR, Bari, Italy

Danuta Sosnowska Institute of Plant Protection – National Research Institute, Poznan, Poland

Maria Clara Vieira dos Santos Centre for Functional Ecology (CFE), University of Coimbra, Coimbra, Portugal

Bingyan Xie Institute of Vegetables and Flowers, Chinese Academy of Agricultural Sciences, Beijing, China

Part I
Introduction and Systematics

Chapter 1
Introduction (Historical and Overview)

Ken Evans, Rosa H. Manzanilla-López, and Luis V. Lopez-Llorca

Abstract Biological control is an alternative to chemical control of plant-parasitic nematodes. This is largely due to public demand for biologically-based and environment-friendly management options for safer pest control. Such demands have had an important impact on biological control research expansion and funding. However, the development of any strain of a biological control agent for nematode control requires many years of research, experimentation, validation and safe-use tests before the biological control agent becomes available to farmers or is further developed by industry as a commercial biopesticide or bionematicide. Biological control potential can be unconstrained when biological control agents are used in combination with compatible integrated pest management tactics, which may include some chemical products and other biological control agent-based products that are currently available on the biopesticide market. This chapter presents part of the history behind some of the initial studies that help to illustrate the scientific work carried out by the many scientists who laid the foundations and helped to develop *Pochonia chlamydosporia* as a viable, sustainable alternative to chemical control in the integrated management of plant-parasitic nematodes.

K. Evans
Formerly at Rothamsted Research, 1 Birch Way, AL5 5TP Harpenden, Herts, UK
e-mail: kenavis.evans@tiscali.co.uk

R.H. Manzanilla-López (✉)
Centro de Desarrollo de Productos Bióticos (Visiting Professor),
Yautepec, Morelos, C.P. 62731, Mexico
e-mail: rosa.manzanilla@gmail.com

L.V. Lopez-Llorca
Laboratory of Plant Pathology, Multidisciplinary Institute for Environmental Studies (MIES)
Ramón Margalef, Department of Marine Sciences and Applied Biology, University of
Alicante, Alicante, Spain
e-mail: luislopezllorca1@gmail.com

R.H. Manzanilla-López, L.V. Lopez-Llorca (eds.), *Perspectives in Sustainable Nematode Management Through* Pochonia chlamydosporia *Applications for Root and Rhizosphere Health*, Sustainability in Plant and Crop Protection,
DOI 10.1007/978-3-319-59224-4_1

1.1 Introduction

Pochonia chlamydosporia is one of the most studied biological control agents (BCA) of major nematode pests such as cyst nematodes (*Globodera* spp., *Heterodera* spp.) and root-knot nematodes (*Meloidogyne* spp.). The number of publications on this fungus is increasing continuously, not only as a result of the demand for research on new environmentally friendly strategies to manage plant-parasitic nematodes (PPN), but also due to the availability of new molecular technologies and sophisticated equipment at more accessible costs. Researchers worldwide have taken advantage of such factors to explore and understand this BCA, not only at the macro level but also at the molecular level. This chapter presents part of the history behind the development of this microorganism as a sustainable alternative to chemical management of PPN as one of the available tactics of Integrated Pest Management (IPM). The present account is largely based on UK experience but may also apply to other countries.

The nematology research carried out at Rothamsted Research (RRes) in the UK, including biological control of PPN, during a period that spanned more than six decades (1948 to 2014), has been reviewed (Evans and Manzanilla-López 2017). Part of that history is presented in this chapter. Many references to seminal papers on *P. chlamydosporia* that were produced during this period at Rothamsted and outside the UK can be found within the chapters in this book, covering the basic, pre-molecular beginnings through to the molecular, high tech and '*-omics*' studies currently in vogue.

1.2 The Beginnings (Pre-molecular)

From the late 1990s and well into the new century, increasing worldwide pressure was put on growers to reduce their reliance on chemical nematicides. The availability of fewer chemicals registered for such use, strict supermarket protocols and extended harvest intervals (i.e., the minimum time permitted between chemical application and crop harvest) were some of the factors responsible for this pressure. In response, the then department of Entomology and Nematology of Rothamsted became involved in a number of collaborative projects looking into alternative technologies as components of IPM.

In 1950, the Rothamsted Nematology department reported observation of hyperparasites on *Ditylenchus dipsaci*, mentioning the genera *Harposporium* and *Arthrobotrys*. Interest in these relationships was taken up, thus paving the way for future biological control research. The idea of using one species to limit the numbers of a second species that causes economic damage (biological control) has been around for a long time, and it was natural to apply this concept to the control of PPN. Terry Williams reported, in 1972, that formalin treatments in experiments from 1964 and onwards increased the numbers of *Heterodera avenae* females found on cereal roots and thereby caused a large increase in soil populations of the nematodes. A fungus was suspected of infecting the females and the improved nematode reproduction was put down to control of the fungus by formalin. In 1973, Brian Kerry found an *Entomophthora*-like fungus (ELF) on *H. avenae* females at Rothamsted and proposed that it was at least partly responsible for the natural control of this nematode. Despite continuous cropping with cereals, the number of *H. avenae* rarely exceeds 10–20/g soil in the UK. In 1974, Kerry found the ELF and *Verticillium chlamydosporium* (= *Pochonia chlamydosporia*[1]) (71%) and *Cylindrocarpon destructans* (= *Ilyonectria destructans*) (52%) in many of the examined field populations of *H. avenae*. In 1975, Kerry confirmed that the ELF killed *H. avenae* females whilst *V. chlamydosporium* parasitized eggs (Chaps. 3 and 4). All *Heterodera* species tested were infected by the ELF (later on described as *Nematophthora gynophila*) and *V. chlamydosporium* but the potato cyst nematodes (*Globodera* spp.) were only infected by *V. chlamydosporium*. Kerry went on to point out that the cereal monoculture practised after World War II had provided the opportunity for the fungi to reach an equilibrium with *H. avenae* that held the nematode population density below the economic damage threshold. The realisation that this equilibrium existed between the nematode and its parasites and that soils can become suppressive to nematodes was the stimulus for Brian Kerry to go on to develop an understanding and subsequently exploit this relationship.

Finding soil suppressiveness to PPN coincided with the withdrawal of some nematicides from the market due to their toxicity to humans (Kerry 1991). Surveys showed *N. gynophila* to occur in fields in Denmark, Poland, Holland, California and Tennessee (USA), while *V. chlamydosporium* occurred in Denmark, Holland and California (see also Pyrowolakis et al. 2002; Westphal and Becker 1999, 2000, 2001). Morgan-Jones et al. (1981, a, b) encountered *V. chlamydosporium* parasitizing *Meloidogyne arenaria* females and cysts of *Heterodera glycines* (see also Gintis et al. 1982, 1983; Godoy et al. 1982; Morgan-Jones et al. 1983). Some specialization enabled these opportunistic fungi to exploit a unique ecological niche. Their frequency of occurrence among cyst nematode populations indicated a capacity to compete successfully in agricultural soils (Rodríguez-Kábana et al. 1984; Morgan-Jones and Rodriguez-Kabana 1987).

The biological control (BC) approach later followed at Rothamsted would largely correspond to that defined by Stirling in 1991: "A reduction of nematode populations which is accomplished through the action of living organisms other than nematode-resistant host plants, which occurs naturally or through the manipulation of the environment or the introduction of antagonists".

[1] *Verticillium chlamydosporium* is a synonym of *Pochonia chlamydosporia*, the current valid name of the fungus.

1.3 Honing Skills and Harmonizing Protocols

In 1976, the 'pathogens of nematodes subgroup', also known as the 'Working Group on the integrated Control of Soil Pests', was formed as part of the International Organization for Biological and Integrated Control of Noxious Animals and Plants/ West Palaearctic Regional Section (IOBC/WPRS) (Kerry 1991). At that time, some of the research activities were focused on the control of PPN by the application to soil of selected fungal agents, including those that had already been developed commercially. However, the levels of control were variable and too little was known about these fungi to obtain predictable results (Kerry 1991). Biological control research was not only concentrated on targeting PPN but also on the use of nematode-trapping fungi for the control of animal parasitic nematodes (e.g. in Denmark and France), as some fungal isolates had shown considerable potential as control agents of nematodes that infect ruminants (Kerry 1991), a line of research that will extend in future to other latitudes (see Chap. 17).

There was a time (1970s–1980s) when the training of mycologists was moving from morphology and taxonomy as main subjects to include fungal biology. In 1977, the publication of the book *The nematode-destroying fungi* by G.L. Barron showed that nematode-destroying fungi were taxonomically diverse, as shown by morphological adaptations to the nematophagous habit. However, little was known about the morphogenesis of trapping/infection devices and the physiology of these fungi (Barron 1977). The book also included methods and techniques to recover and culture predatory and nematode-parasitic fungi.

Research on biological control of nematodes had also begun to move from being a predominantly observational to a more experimental science (Kerry 1991). Several laboratories from different countries in Europe (e.g., Belgium, England, France, Germany, The Netherlands) were already involved by 1988 in the development of nematophagous fungi or products containing them for the control of nematodes, and in the development of a range of methods for the isolation, selection and testing of nematophagous fungi. These methods were brought together and made available to the public in a special IOBC bulletin (Kerry and Crump 1991). However, at the time it was acknowledged that techniques still needed to be standardised (Kerry 1991). Therefore, fundamental knowledge was required and research was done on several aspects of the biology, ecology, epidemiology, mode of action and suitable methods of study of the fungi. In 1991, much of the information that had become available on biological control was put together by Graham Stirling in a book (Stirling 1991) on biological control of plant-parasitic nematodes, an important source of information that became a bench mark, and which was followed by a second edition in 2014 (Stirling 2014).

The nematophagous fungus *Verticillium chlamydosporium* was considered easier to culture and handle than other fungi isolated from UK fields. Attention became focused on tests with formulations of carrier materials for application to soil and mutants were selected for their ability to tolerate carbendazim (see Chap. 11) so that this fungicide could be incorporated into inoculum granules and thereby allow the chosen isolate to multiply after application to soil, competing fungi being selectively held back. Various media were tested as carriers for such introductions but could not provide the population densities of 1000 propagules per g of soil that *V. chlamydosporium* was found to reach naturally. Another concern was the possibility of loss of infectivity after repeated subculturing, so original isolates were carefully preserved (see Chap. 12). As a result, more than 100 isolates of the fungus were held in culture at RRes for long term storage.

The publication of a set of improved and standardised methods and protocols was achieved in 2002 thanks to the many contributions made by scientists to the *Manual for research on* Verticillium chlamydosporium, *a potential biological control for root-knot nematodes* (Kerry and Bourne 2002). Many of these protocols are still in use worldwide (see Chap. 12).

1.4 Understanding the Fungus

Although fungi such as *N. gynophila* had been shown to have a significant impact on the multiplication of *H. avenae*, the fungus biology, environmental requirements and mode of nutrition hindered its use as a BCA (Stirling 1991). Progress was also made on understanding the infection process in cereal cyst nematodes (CCN). Nematode defences to fungal infection such as egg shell thickness (Lopez-Llorca and Robertson 1992a) and cross-linking of nematode proteins were identified (Lopez-Llorca and Fry 1989). A collaboration with Mr. D. Claugher from The Natural History Museum, London, led to the study of *Pochonia* appressoria differentiation. This research discovered adhesive production by the fungus infecting nematode eggs (Lopez-Llorca and Claugher 1990). This was possible using high resolution (Field Emission) Scanning Electron Microscopy. A serine protease (P32) from *Pochonia rubescens* [= *Metapochonia rubescens*], isolated from CCN suppressive soils in Scotland, was purified and characterized (Lopez-Llorca 1990). It was later immunolocalized in appressoria of the fungus infecting nematode eggs (Lopez-Llorca and Robertson 1992b). This provided the first evidence of the involvement of *Pochonia* spp. extracellular enzymes in nematode egg parasitism (see Chap. 4).

Verticillium chlamydosporium (= *Pochonia chlamydosporia*) grows saprophytically in soil and will survive when no nematodes are available if it is added to soil together with an energy source. When commercially available preparations of parasitic fungi were stringently tested, they failed to control *Meloidogyne incognita*, one of the most important PPN species in the world. Therefore, one of the principal targets of the fungus studies became the root-knot nematodes *Meloidogyne* spp. and isolates able to control them were selected in RRes from the early 1990s. Key features to improve fungus performance focused on selection of isolates capable of: i) colonizing the rhizosphere of plants; ii) production *in vitro* of chlamydospores; and iii) infecting nematode eggs (Kerry 2000). Chlamydospores produced in large numbers on solid media could be added as aqueous suspensions, thereby rapidly establishing the fungus. However, the method of mass culturing of *P. chlamydosporia* also affected the subsequent survival and proliferation of the fungus (see Chap. 11). Therefore, work continued on its culturing and formulation followed by a very productive collaboration with the Centro Nacional de Sanidad Agropecuaria (CENSA) in Cuba. Local isolates of the fungus were identified in Cuba and a pilot production plant was set up to produce a commercially viable preparation known as KlamiC®, launched commercially in 2006 (see Chap. 15).

1.5 *Pochonia chlamydosporia* and Compatible IPM Strategies

The ability of *P. chlamydosporia* to compete with other nematophagous and soil fungi was also tested *in vitro*. In a study to assess potential interactions between nematophagous fungi and their responses to chemical pesticides for potato cyst nematode control, *P. chlamydosporia* was the most effective fungus at inhibiting growth of saprophytic fungi and was the least susceptible to being inhibited by saprophytic fungi. It was also the most tolerant to nematicides. Some potato growers were applying two control measures for potato cyst nematodes at this time, a fumigant in the autumn followed by a granular nematicide in the spring at a cost of ~£900/ha. Separate applications of two biological control agents could therefore be feasible and cost-effective (see Jacobs et al. 2003; Tobin et al. 2008 and Chap. 16). Nowadays, the use of the fungus in combination with other compatible strategies ranges from field to glasshouse crops (Verdejo-Lucas et al. 2002; Giné et al. 2016; Sellitto et al. 2016).

1.6 *Pochonia* as a Bionematicide Product

The many years spent in developing a bionematicide product that could be used in commercial management programmes for nematodes came to fruition at a time when many chemicals for the control of nematodes were being banned around the world due to environmental concerns. *Pochonia chlamydosporia* has been developed as a commercial bionematicide in several countries, including Cuba, Brazil, China, and Italy (see Chap. 15). The combined use of poor nematode host plants and the soil application of the fungus may be a useful strategy for the management of RKN as it may result in greater control by the fungus in those plants that are poor hosts for the nematode but which support extensive rhizosphere fungal growth. Increased dependence of the fungus on nematodes as a source of nutrition has been associated with a decrease in saprotrophic competitiveness, which is often negatively correlated with parasitic competence. Establishment and reproduction of the fungus, following application of an adequate initial inoculum was, under proper management conditions, able to achieve levels found in naturally suppressive soils and eventually produced long-term effective control of nematodes (Kerry 2000).

1.7 Cooperation Links and Training

The approach to *Pochonia chlamydosporia* research in RRes was strongly collaborative, with an important training element that brought together nematologists, molecular biologists, fungal physiologists, natural chemists, mathematical modellers and soil microbial molecular ecologists at Rothamsted (Fig. 1.1) and in UK universities (Cambridge, Cranfield, Lancaster, Leeds, Nottingham, Reading, Sheffield), and from overseas (Cuba, Kenya, Italy, Malawi, Mexico, Pakistan, Poland, Portugal, Spain, Tanzania, Uganda, Zimbabwe) (Fig. 1.2). Studies carried out included morphology, biology, culturing conditions, formulations and application of the fungus (Chaps. 12 and 15), ecology, and tri-trophic and multi-trophic interactions. The latter included the discovery of the endophytic behaviour of the fungus and associated plant growth, development, yield and defence promotion. Physiology of the parasitism, nutrition, compatibility of the fungus with other root-knot and cyst nematode control strategies, secondary metabolites with chemical structures partly similar to phomalactone (Chap. 7), and molecular studies (see Chaps. 4 and 13) were also conducted.

Fig. 1.1 Rothamsted Nematology Department (1988). Front row (*from left to right*): Janet Cowland, Jack Beane, Marie-Louise Rodgers, Corinna Flynn, David Crump, Keith Davies. Back row (*from left to right*): Mike Robinson, Mike Russell, Fiona De Leij, Ken Evans, Frans De Leij, David Cooper, Alan Nicholls, John Feil, Alan Whitehead, Roland Perry, Robin Webb, Paul Burrows, Marj Leijdens (courtesy of K. Evans)

Fig. 1.2 Maceió (Brazil) ONTA-SBN Meeting 2009. Front row (*from left to right*): Lucy Conceição, Isabel Abrantes, Ivânia Esteves. Back row (*from left to right*): Rosa H. Manzanilla-López, Alejandro Tovar-Soto, Myrian Tigano, Leopoldo Hidalgo-Díaz, Maria J. Cunha, Clara Vieira dos Santos (courtesy of R.H. Manzanilla-López)

1.8 Knowledge Transfer

Some limitations to uptake of fungal control agents in peri-urban and urban vegetable production in Latin America and Africa included an incomplete knowledge of the parameters affecting transmission, a lack of field and farm-scale evaluation, and quality control problems in production, formulation and application. The EU sponsored project MiCoSPa (2002–2006) led by Judith Pell (Rothamsted) made significant advances in the fundamental understanding of the transmission process, effective mass production, and practical use in the field of beneficial entomopathogenic and nematophagous fungi based on stakeholder requirements (Fig. 1.3).

Socio-economic surveys of vegetable growers and biological control companies were designed and completed to identify restrictions and enable processes for the uptake of microbial control agents in Cuba and Mexico. Information was collected from more than 100 farmers in Cuba who were surveyed on such parameters as farm size, main crops and pest control strategies used, the surveys proving useful in identifying farming practices that were complementary to microbial control. Evaluation of inundative strategies included those focused on inundative application of *P. chla-*

Fig. 1.3 MiCoSPa Project (2005). Meeting with farmers in Mexico (courtesy of F. Franco-Navarro)

mydosporia in crop rotations for *M. incognita* control in Cuba and *Nacobbus aberrans* control in Mexico. Such experiments provided field efficacy data for these strategies (see Chap. 15).

Companies that produced BCA from different regions in Mexico and Cuba were also questioned on their products, production methods, and quality control protocols. Commercial production methods and Good Laboratory Practice and Standard Operating Procedures were developed and adapted for uptake by producers of microbial control agents (Hidalgo-Díaz 2004; Chap. 15). Laboratory and field studies allowed significant advances in the fundamental understanding of the transmission process of beneficial fungi at the population, whole organism and molecular scales, and were used to develop epidemiological models describing the processes and predicting the impact of perturbations in the nematode population. Field trials evaluated the use of BCA under different strategies and the results were provided to local farmers and producers of microbial control agents. Numerous scientists were trained in all participating countries. In addition, the technology from the project was transferred to local farmers and to those involved in the commercial production of BCA in Cuba and Mexico. Training in production methods developed in Cuba was also transferred to Brazil, Honduras, Kenya, India, and Italy. Furthermore, some of the collaborations established are still extant (Judith Pell pers. comm.; see also Chap. 15).

1.9 Molecular Studies

The use of molecular tools to study genetic characterisation, diversity and parasitism started in the mid-1990s at RRes with work by Tariq But and Rudy Segers. Immunolocalization of the serine protease VCP1 at the penetration site indicated that it degraded the vitelline membrane on the surface of the egg shell and exposed the chitin layer, thus playing an important role in the infection process (see Sect. 1.4). Its expression was induced by the nematode host and depletion of nutrients *in vitro*. The enzyme was repressed by glucose and easily metabolized N sources and induced by nitrate and insoluble proteins. It was considered that polymorphism differences observed in VCP1 could contribute to host preference (see Segers et al. 1995; Morton et al. 2003 and Chaps. 3, 4, 6, 13).

Penny R. Hirsch and her team of microbiologists at RRes developed the use of fingerprinting protocols, the design of specific primers to discriminate between *P. chlamydosporia* var. *chlamydosporia* and *P. chlamydosporia* var. *catenulata*, quantitative PCR primers, study of VCP1 polymorphisms, and RNA expression studies (see Chaps. 5 and 13). PCR-based DNA fingerprinting techniques provided a rapid means to examine the genetic variation of fungal isolates from various geographical regions and nematode hosts and also to monitor biocontrol agents after their release into soil. The development of KlamiC® more or less coincided with the re-classification of the fungus, which was subsequently known as *Pochonia chlamydosporia* (see Chap. 2). By the time of the release, diagnostic bands had been identified for the species in polymerase chain reaction (PCR) using specific primers. DNA fingerprinting using PCR with arbitrary primers showed that isolates (of which Rothamsted was holding more than 400 by this time) tended to group on the basis of their original host and, to a lesser extent, on their geographic origin (see Chap. 13).

Molecular tools have been used increasingly in identification and taxonomy studies of *Pochonia* (*Clavicipitaceae*). In recent years there have been important advances in *Clavicipitaceae* taxonomy, including resolving the differences between *Verticillium* and *Pochonia* through ITS sequence molecular analysis (Chaps. 2 and 13). *Pochonia chlamydosporia* has at least five different varieties and some former species of *Pochonia* are now placed in the genus *Metapochonia* (see Chap. 2).

1.10 Genomics

Genomics aims to study a living organism's genome, taking into account its primary structure in order to understand its organization and gene functions, as well as trying to establish its phylogenetic relationships with other organisms (Finetti-Sialer and Manzanilla-López 2011). The advent of new technologies has opened an era for *Pochonia* studies on gene expression (Transcriptomics, Chap. 5), protein synthesis and interactions (Proteomics, Chap. 6), secondary metabolites (Metabolomics,

Chap. 8), whole genome studies (Genomics, Chaps. 9 and 10), and the possibility of assessing the DNA biodiversity at any community level (Metagenomics, Chap. 13).

Although the genome annotation of isolate Pc10 could not be completed due to the closure of nematology at RRes in 2011 (Evans and Manzanilla-López 2017), *Pochonia chlamydosporia* var. *chlamydosporia* isolate Pc10 (IMI 331547 / CBS 101244) had been sequenced in collaboration between RRes and The Genome Analysis Centre (TGAC, BBSRC sequencing facility, UK) obtaining a ~ 45 Mb genome estimate (E. Ward et al. unpub. data; Manzanilla-López et al. 2013). Meanwhile, researchers in Spain carried out the sequencing, gene validation (using RNA-seq of barley colonized endophytically by the fungus), and annotation of the genome (~41 Mb) of the Spanish *P. chlamydosporia* isolate 123 (Larriba et al. 2014; Chap. 10). The mitogenome or mitochondrial genome of *P. chlamydosporia* strain 170 (~25,615 bp) was sequenced in China (Lin et al. 2015; Chap. 9).

1.11 Molecular Discoveries and Integrated Pest Management Approaches

Invasion of host tissues by *P. chlamydosporia* via appressoria formation was thought to be a nutritional response with low C:N ratios influencing egg pathogenicity (Chaps. 3, 4 and 5). The change from the saprophytic to the parasitic phase of the fungus was therefore thought to be related to nutrients (including C and N) that were either released by plants into their rhizosphere or available in nematode eggs. The C:N ratio is an important issue since incorporation into the soil via organic manures of nutrients that are of easy access to the fungus reduces the parasitic activity of the fungus. Hence, there is a need to assess the effects of adding different organic amendments to check their compatibility with the fungus without reducing the control it exerts (Chaps. 4 and 15).

At the molecular level, an analysis of serine protease enzyme production, as related to nutrients available for the fungus in the presence or absence of nematode eggs, found that the enzyme VCP1, a serine protease involved in the early stages of egg penetration, was repressed by glucose and stimulated by ammonium chloride, but there was also evidence of an increase in VCP1 mRNA when glucose and eggs were present, with VCP1 expression higher in more alkaline pH conditions (Ward et al. 2012).

1.12 Endophytism

For a long time it was considered that *P. chlamydosporia* remained localized in the more peripheral layers of root tissues and did not colonize the vascular system, a feature that differentiated *P. chlamydosporia* from plant pathogenic species of

Verticillium. Subsequent studies by researchers in Spain discovered the endophytic behaviour of *P. chlamydosporia* in both monocotyledonous and dicotyledonous crop plants (Bordallo et al. 2002). This behaviour of the fungus was confirmed through genetic transformation of isolates with the green fluorescent protein (GFP) gene by Maciá-Vicente et al. (2009) using an *Agrobacterium tumefasciens*-mediated transformation protocol. This allowed laser confocal microscopy analysis of plant responses and patterns of root colonization in tomato (Escudero and Lopez-Llorca 2012). The discovery of the endophytic root colonization by egg-parasitic fungi such as *Pochonia* has implications for the infection process of eggs and females of plant endoparasitic nematodes, which may occur not only via external root colonization of females and nematode eggs external to the root but also through internal colonization (Chap. 4), thus opening the way for new application methods of the fungus to the plant and soil (Lopez-Llorca et al. 2008; Larriba et al. 2014). Some isolates of *P. chlamydosporia* promote root growth and reduce flowering time of tomato (Zavala-Gonzalez et al. 2015). Recently, endophytism studies have taken advantage of *Arabidopsis thaliana* as a model system to understand this mutualistic phase of the fungus as *P. chlamydosporia* readily colonizes *A. thaliana*. Fungal inoculation reduces flowering time, measured by the number of days of rosette leaves produced at bolting stage and leaves produced at the time of appearance of the first open flowers, stimulating plant growth and increasing seed/plant yield (Zavala-Gonzalez et al. 2017). Jasmonate signalling is an important factor modulating *A. thaliana* colonization by the fungus (Zavala-Gonzalez et al. 2017).

1.13 Inducing Plant Defences

The *Pochonia* research group of Spain led by Luis Lopez-Llorca (www.fungalinteractions.org) in collaboration with Prof. Sivasithamparam from the University of Western Australia, found that *P. chlamydosporia* can reduce the colonisation, root damage and associated stress (peroxidase activity) in wheat roots caused by the fungal root pathogen *Gaeumannomyces graminis* var. *tritici* (Monfort et al. 2005). Lopez-Llorca and his research team later found that induction of genes involved in stress response (e.g., heat-shock proteins), biosynthesis of hormones, and plant immunity were enriched in barley roots colonised endophytically by *P. chlamydosporia* (Larriba et al. 2015). This demonstrates another role of the fungus in inducing plant growth and defences. There are two recognized types of induced resistance: systemic acquired resistance and induced systemic resistance, which are differentiated by their signal transduction pathways (Timper 2014). Research carried out in Portugal and the UK on combining the fungus with jasmonic acid treatments has shown promising results in controlling RKN (Chap. 14; see also Sect. 1.12).

1.14 Future Perspectives

The search for isolates that parasitize PPN females and eggs continues worldwide (Chap. 12) with the aim of taking advantage of natural biological control agents for biopesticide development. However, research on the fungus has evolved from being purely focused on applied nematode management to other, more novel, research areas such as molecular and *-omics* technologies, which are being carried out mainly in China, Spain and Italy, and other areas, such as secondary metabolites with nematicidal or drug potential development, which are being investigated in China. Transcriptomics, proteomics, metabolomics, genomics and other *-omics* approaches are currently applied to the study of *P. chlamydosporia* (Chaps. 5, 6, 8, 9 and 10). Genome sequencing and informative technology tools will support analyses that may range from a broad, biodiverse community down to single organisms. In the study of fungal parasitism, such approaches supply information of great value in terms of host or parasite biology, providing opportunity for detailed comparisons between different organisms, especially when few genetic or molecular data are available (Finneti-Sialer and Manzanilla-López 2011). There is also increasing evidence relating to the role of chitosan as a modulator of *P. chlamydosporia* gene expression and pathogenicity (Escudero et al. 2016).

1.15 Concluding Remarks

Biological control is complex and requires in-depth knowledge of the biology and ecology of the biological control agent, crop and target host. One of the practical aims of the research led by B.R. Kerry (Fig. 1.4) was how to develop and increase suppressiveness to PPN through inoculation and establishment of the fungus in the soil and plant rhizosphere. The strategy should be accompanied by a careful selection of fungal isolates according to nematode biotype, host plant, and population levels of PPN. Equally important is the need to provide the best conditions to enhance fungus performance as a nematode parasite through habitat manipulations, thereby providing favourable conditions that are either lacking in the habitat or present at inadequate levels (Barbosa 1998). This kind of strategy promoted a number of studies on the effect of adding compost and manures to modify C:N ratios in order to retain activity or even enhance the parasitic phase of the fungus (see Chaps. 15 and 16). With this purpose in mind, experiments were carried out in laboratory, glasshouse and field plots in countries such as Cuba and Mexico (peri-urban agriculture) and Kenya. Results of the research were also intended to be made available to small and medium farmers, thus making possible the inclusion of biological control in the IPM of RKN (Gowen 2002, 2005). Therefore, substrates other than rice (and also cheaper) were tested to find alternatives for production of the fungus in an affordable manner in parts of Africa and Latin America (Pérez-Rodríguez et al. 2007). The development of a reliable production method of the fungus has not

Fig. 1.4 Brian R. Kerry (Copyright Rothamsted Research Ltd.)

remained limited to cottage industries as larger scale production is now carried out (see Chaps. 9 and 14). Successful industrial production and workable formulations have been achieved in Brazil, China, Cuba and Italy (Manzanilla-López et al. 2013).

Secondary metabolite production from *P. chlamydosporia* strains (and those from other *Pochonia* spp.) may vary across geographical and smaller spatial scales. It would therefore appear important to explore further the role of environmental variation in the production of secondary metabolites by *P. chlamydosporia* (Chap. 7).

A better understanding of the fungus-nematode-plant association and interaction at different molecular levels, i.e., from the whole genome (*-omics*) to the single gene (molecular), will improve and drive management programmes and assist in the development of alternative methods, including biological control, to manage major plant-parasitic nematode species.

Acknowledgements The information of the MiCoSPa project outputs was kindly provided by Dr. Judith Pell.

References

Barbosa, P. (1998). Agroecosystems and conservation biological control. In P. Barbosa (Ed.), *Conservative Biological control* (pp. 39–54). San Diego: Academic.

Barron, G. L. (1977). *The nematode-destroying fungi, Topics in Mycobiology No. 1*. Guelph: Canadian Biological Publications.

Bordallo, J. J., Lopez-Llorca, L. V., Salinas, J., et al. (2002). Colonization of plant roots by egg-parasitic and nematode-trapping fungi. *The New Phytologist, 154*, 491–499.

Escudero, N., & Lopez-Llorca, L. V. (2012). Effects of plant growth and root-knot nematode infection of an endophytic GFP transformant on the nematophagous fungus *Pochonia chlamydosporia. Symbiosis, 57*, 33–42.

Escudero, N., Ferreira, S. R., Lopez-Moya, F., et al. (2016). Chitosan enhances parasitism of *Meloidogyne javanica* eggs by the nematophagous fungus *Pochonia chlamydosporia. Fungal Biology, 120*, 572–585.

Evans, K., & Manzanilla-López, R. (2017). The history of the nematology Department at Rothamsted. *The Annals of Applied Biology, 170*, 4–44.

Finetti-Sialer, M. M., & Manzanilla-López, R. H. (2011). Exploiting "-omics" and molecular approaches in plant nematology research. In R. Rodríguez-Herrera, C. N. Aguilar, J. K. Simpson-Williamson, et al. (Eds.), *Phytopathology in the Omics Era, Transworld Research Network* (pp. 39–68). Thiruvananthapuram: Research Signpost.

Giné, A., Bonmati, M., Sarro, A., et al. (2016). Natural occurrence of fungal egg parasites of root-knot nematodes, *Meloidogyne* spp. in organic and integrated vegetable production systems in Spain. *BioControl, 58*, 407–416.

Gintis, B. O. G., Morgan-Jones, G., & Rodríguez-Kábana, R. (1982). Mycoflora of young cysts of *Heterodera glycines* in North Carolina soils. *Nematropica, 12*, 295–303.

Gintis, B. O., Morgan-Jones, G., & Rodríguez-Kábana, R. (1983). Fungi associated with several developmental stages of *Heterodera glycines* from Alabama field soil. *Nematropica, 13*, 181–200.

Godoy, G., Rodríguez-Kábana, R., & Morgan-Jones, G. (1982). Fungal parasites of *Meloidogyne arenaria* eggs in an Alabama soil. A mycological survey and greenhouse studies. *Nematropica, 13*, 201–213.

Gowen, S. R. (2002). *Integrated management of root-knot nematodes on vegetables in Kenya R 7472 (Za 0324)*. Final technical report (1 October 1999–30 September 2002). DFID R7472 Crop Protection Programme. Reading, UK, p. 52.

Gowen, S. R. (2005). *Promotion of sustainable approaches for the management of root-knot nematodes of vegetables in Kenya (R8296 (ZA 0568))*. Crop protection programme. Final Report, DFID, p. 39.

Hidalgo-Díaz, L. (2004) Registration of biological pesticides in Cuba. In M. N. Wabule, P. N. Ngaruiga, & F. K. Kimmins et al. (Eds.), *Registration for biological control agents in Kenya, Proceedings of the Pest Control Products Board/Kenya/Agricultural Research Institute/ Department for International Development Crop Protection Programme Workshop*, Nakuru, Kenya, 14–16 May 2003. KARI/PCPB, Nairobi, Kenya, and Natural Resources International Ltd., Aylesford, UK, pp. 117–133

Jacobs, H., Gray, S. N., & Crump, D. H. (2003). Interactions between nematophagous fungi and consequences for their potential as biological agents for the control of potato cyst nematodes. *Mycological Research, 107*, 47–56.

Kerry, B. R. (1991). Preface. In B. R. Kerry, & D. H. Crump (Eds.), *Methods for studying nematophagous fungi. Working group "Integrated control of soil pests"* (pp. I–II). IOBC/WPRS/ Bulletin XIV/2.

Kerry, B. R. (2000). Rhizosphere interactions and the exploitation of microbial agents for the biological control of plant-parasitic nematodes. *Annual Review of Phytopathology, 38*, 423–441.

Kerry, B. R., & Bourne, J. (2002). *A manual for research on Verticillium chlamydosporium a potential biocontrol agent for root-knot nematodes*. Gent: IOBC/WPRS.

Kerry B. R., & Crump D. H. (1991). Methods for studying nematophagous fungi. IOBC/WPRS Bulletin XIV (2).

Larriba, E., Jaime, M. D. L. A., Carbonell-Caballero, J., et al. (2014). Sequencing and functional analysis of the genome of a nematode egg-parasitic fungus, *Pochonia chlamydosporia. Fungal Genetics and Biology, 65*, 69–80.

Larriba, E., Jaime, M. D. L. A., Nislow, C., et al. (2015). Endophytic colonization of barley (*Hordeum vulgare*) roots by the nematophagous fungus *Pochonia chlamydosporia* reveals plant growth promotion and a general defense and stress transcriptomic response. *Journal of Plant Research*. doi:10.1007/s10265-015-0731-x.

Lin, R., Liu, C., Shen, B., et al. (2015). Analysis of the complete mitochondrial genome of *Pochonia chlamydosporia* suggests a close relationship to the invertebrate-pathogenic fungi in Hypocreales. *BMC Microbiology, 15*, 5.

Lopez-Llorca, L. V. (1990). Purification and properties of extracellular proteases produced by the nematophagous fungus *Verticillium suchlasporium*. *Canadian Journal of Microbiology, 36*, 530–537.

Lopez-Llorca, L. V., & Claugher, D. (1990). Appressoria of the nematophagous fungus *Verticillium suchlasporium*. *Micron and Microscopica Acta, 21*, 125–130.

Lopez-Llorca, L. V., & Fry, S. C. (1989). Dityrosine and tetratyrosine, potential croos-links in proteins of plant parasitic nematodes. *Nematologica, 35*, 165–179.

Lopez-Llorca, L. V., & Robertson, W. M. (1992a). Ultrastructure of infection of cyst nematode eggs by the nematophagous fungus *Verticillium suchlasporium*. *Nematologica, 39*, 65–74.

Lopez-Llorca, L. V., & Robertson, W. (1992b). Immunocytochemical localization of a 32 kDa protease from the nematophagous fungus *Verticillium suchlasporium* in infected nematode eggs. *Experimental Mycology, 16*, 261–267.

Lopez-Llorca, L. V., Maciá-Vicente, J. G., & Jansson, H. B. (2008). Mode of action and interactions of nematophagous fungi. In A. Ciancio & K. G. Mukerji (Eds.), *Integrated management and biocontrol of vegetable and grains crops Nematodes* (pp. 51–76). Heidelberg: Springer.

Maciá-Vicente, J. G., Jansson, H. B., Talbot, N. J., et al. (2009). Real-time PCR quantification and live-cell imaging of endophytic colonization of barley (*Hordeum vulgare*) roots by *Fusarium equiseti* and *Pochonia chlamydosporia*. *The New Phytologist, 182*, 213–228.

Manzanilla-López, R. H., Esteves, I., Finetti-Sialer, M. M., et al. (2013). *Pochonia chlamydosporia*: Advances and challenges to improve its performance as a biological control agent of sedentary endo-parasitic nematodes. *Journal of Nematology, 45*, 1–7.

Monfort, E., Lopez-Llorca, L. V., Jansson, H. B., et al. (2005). Colonisation of seminal roots of wheat and barley by egg-parasitic nematophagous fungi and their effects on *Gaeumannomyces graminis* var. *tritici* and development of root-rot. *Soil Biology and Biochemistry, 37*, 129–1235.

Morgan-Jones, G., & Rodríguez-Kábana, R. (1981). Fungi associated with cysts of *Heterodera glycines* in an Alabama soil. *Nematropica, 11*, 69–74.

Morgan-Jones, G., & Rodriguez-Kabana, R. (1987). Fungal biocontrol for the management of nematodes. In J. A. Veech & D. W. Dickson (Eds.), *Vistas on Nematology: A commemoration of the twenty-fifth anniversary of the Society of Nematologists* (pp. 94–99). De Leon Springs: E.O. Painter Printing.

Morgan-Jones, G., Godoy, G., & Rodríguez-Kábana, R. (1981a). *Verticillium chlamydosporium*, fungal parasite of *Meloidogyne arenaria* females. *Nematropica, 11*, 115–119.

Morgan-Jones, G., Gintis, B. O., & Rodríguez-Kábana, R. (1981b). Fungal colonization of *Heterodera glycines* cysts in Arkansas, Florida, Mississippi and Missouri soils. *Nematropica, 11*, 155–163.

Morgan-Jones, G., White, J. F., & Rodríguez-Kábana, R. (1983). Phytonematode pathology; ultrastructural studies. I. Parasitism of *Meloidogyne arenaria* eggs by *Verticillium chlamydosporium*. *Nematropica, 13*, 245–260.

Morton, C. O., Hirsch, P. R., Peberdy, J. P., et al. (2003). Cloning of and genetic variation in protease VCP1 from the nematophagous fungus *Pochonia chlamydosporia*. *Mycological Research, 107*, 38–46.

Pérez-Rodríguez, I., Doroteo-Mendoza, A., Franco-Navarro, F., et al. (2007). Isolates of *Pochonia chlamydosporia* var. *chlamydosporia* from Mexico as potential biological control agents of *Nacobbus aberrans*. *Nematropica, 37*, 127–134.

Pyrowolakis, A., Westphal, A., Sikora, R. A., et al. (2002). Identification of root-knot nematode suppressive soils. *Applied Soil Ecology, 19*, 51–56.

Rodríguez-Kábana, R., Morgan-Jones, G., Godoy, G., et al. (1984). Effectiveness of species of *Gliocladium*, *Paecilomyces* and *Verticillium* for control of *Meloidogyne arenaria* in field soil. *Nematropica, 14*, 155–170.

Segers, R., Butt, T. M., Keen, J. N., et al. (1995). The subtilisins of the invertebrate mycopathogens *Verticillium chlamydosporium* and *Metarhizium anisopliae* are serologically and functionally related. *FEMS Microbiology Letters, 126*, 227–232.

Sellitto, V. M., Curto, G., Dallavalle, E., et al. (2016). Effect of *Pochonia chlamydosporia*-based formulates on the regulation of root-knot nematodes and plant growth response. *Frontiers in Life Science, 9*(3), 177–181. http:dx.doi.org/101080/21553769.2016.1193827.

Stirling, G. R. (1991). *Biological control of plant parasitic nematodes: Progress, problems and prospects*. Wallingford: CABI Publishing.

Stirling, G. R. (2014). *Biological control of plant-parasitic nematodes: Soil ecosystem management in sustainable agriculture* (2nd ed.). Croydon: CABI.

Timper, P. (2014). Conserving and enhancing biological control of nematodes. *Journal of Nematology, 46*, 75–89.

Tobin, J. D., Haydock, P. P. J., Hare, M. C., et al. (2008). The compatibility of the fungicide azoxistrobin with *Pochonia chlamydosporia*, a biological control agent for potato cyst nematodes (*Globodera* spp.) *The Annals of Applied Biology, 152*, 301–305.

Verdejo-Lucas, S., Ornat, C., Sorribas, F. J., & Stchiegel, A. (2002). Species of root-knot nematodes and fungal egg parasites recovered from vegetables in Almería and Barcelona, Spain. *Journal of Nematology, 34*, 405–408.

Ward, E., Kerry, B. R., Manzanilla-López, R. H., et al. (2012). The *Pochonia chlamydosporia* serine protease gene *vcp1* is subject to regulation by carbon, nitrogen and pH: Implications for nematode biocontrol. *PLoS One, 7*(4), e35657.

Westphal, A., & Becker, J. O. (1999). Biological suppression and natural population decline of *Heterodera schachtii* in a California field. *Phytopathology, 89*, 434–440.

Westphal, A., & Becker, J. O. (2000). Transfer of biological soil suppressiveness against *Heterodera schachtii*. *Phytopathology, 90*, 401–406.

Westphal, A., & Becker, J. O. (2001). Components of soil suppressiveness against *Heterodera schachtii*. *Soil Biology and Biochemistry, 33*, 9–16.

Zavala-Gonzalez, E. A., Escudero, N., Lopez-Moya, F., et al. (2015). Some isolates of the nematophagous fungus *Pochonia chlamydosporia* promote root growth and reduce flowering time of tomato. *The Annals of Applied Biology, 166*, 472–483.

Zavala-González, E. A., Rodríguez-Cazorla, E., Escudero, N., et al. (2017). *Arabidopsis thaliana* root colonization by the nematophagous fungus *Pochonia chlamydosporia* is modulated by jasmonate signalling and leads to accelerated flowering and improved yield. *The New Phytologist, 213*, 351–364.

Chapter 2
Systematics of *Pochonia*

Harry C. Evans and Paul M. Kirk

Abstract The history of the genus *Pochonia* (*Clavicipitaceae*, *Hypocreales*) is charted from the pre- to post-molecular era. The relatively recent use of more sophisticated molecular techniques – as exemplified by multigene phylogenetics – has enabled the unravelling of this convoluted taxonomic tale and there is now light at the end of the tunnel, especially with the realisation of the 'One fungus, One name' (1F, 1 N) initiative. The present status of the genus – and its near relatives on nematode hosts – is discussed. Currently, only one species, *Pochonia chlamydosporia*, is recognised within the genus, with five varieties being delimited, two of which – var. *chlamydosporia* and var. *catenulata* – produce sexual morphs on alternate hosts: molluscs (snail eggs) and insects (beetle larvae), respectively. Species originally assigned to *Pochonia* are now accommodated in the genus *Metapochonia*. *Pochonia chlamydosporia* is a multitrophic species, parasitic on, or pathogenic to, both invertebrate and nematode hosts, as well as being a plant symbiont. Finally, recommendations are made for future research: it is expected that the varieties of *P. chlamydosporia* will be elevated to species rank, whilst it is predicted that many more species will be identified within the genus as intensive and more focused, selective sampling is undertaken, particularly of soils and plant roots in tropical forest ecosystems.

2.1 Introduction

Systematics is often viewed as a 'dry' subject of little interest and relevance to non-taxonomists, no more so than in mycology with the added confusion between, and the multiplicity of, names for the asexual and sexual states or morphs of the same

H.C. Evans (✉)
CAB International, Egham, Surrey TW20 9TY, UK
e-mail: h.evans@cabi.org

P.M. Kirk
Royal Botanic Gardens, Kew, Richmond, Surrey TW9 3AB, UK
e-mail: p.kirk@kew.org

R.H. Manzanilla-López, L.V. Lopez-Llorca (eds.), *Perspectives in Sustainable Nematode Management Through* Pochonia chlamydosporia *Applications for Root and Rhizosphere Health*, Sustainability in Plant and Crop Protection,
DOI 10.1007/978-3-319-59224-4_2

organism. The order *Hypocreales* – and, in particular, the genus *Pochonia* Bat. & O.M. Fonseca – is a prime example of seemingly baffling nomenclatural changes, but which in fact demonstrates that far from being passive, systematic mycology is constantly on the move and can even be said to have reached supersonic speeds with the now routine use of the latest molecular techniques to advance and clarify fungal phylogeny. Exemplified, perhaps, by the fact that mycology was an integral part of botany – whereas, nowadays it would 'phylogenetically' fall within zoology, if such disciplines were still included in the university curriculum – whilst some of the infamous 'fungal plagues' of history are now consigned to Kingdom *Straminipila* (*Chromista*) – allied to the algae – rather than to the *Animalia*-related Kingdom *Fungi* (Kirk et al. 2008).

Thus, mycological systematics can be said to be in a state of flux, especially with the recent revision of and changes made to Article 59 in the ICN or *International Code of Nomenclature for algae, fungi, and plants* (McNeill 2012). The original Article allowed for the naming of both sexual and asexual states of many fungal species with unique Latin binomials, known as the dual system of fungal nomenclature. The ICN now states that "for a taxon of non-lichen-forming *Ascomycota* and *Basidiomycota* [all names] compete for priority", regardless of their particular morph. Thus, each species of fungus should now have only one scientific name in accordance with the other groups of organisms governed by the ICN, with the scientific name based on the principle of priority. However, this is not so straightforward in mycology since situations exist in which applying a strict principle of priority does not always contribute to nomenclatural stability, the raison d'être for the Code. This has put a major onus on mycologists, who must now decide on a unified name for the various possible names existing for different states or morphs of the same species of fungus (Hibbett and Taylor 2013). Specialist groups or committees have already begun to propose names which should be protected or suppressed within the *Hypocreales* in accordance with the 'one fungus one name' policy (Kepler et al. 2014; Quandt et al. 2014) and others are in progress. Historically, the general rule has been to give priority to the generic name typified by the sexual morph (teleomorph).

However, further splitting of 'sexual' genera may become advisable in some cases rather than simply assimilating multiple, disparate asexual genera within one teleomorphic genus. This situation is exemplified by the recently introduced teleomorphic genus *Metacordyceps* G.H. Sung et al. (*Hypocreales*: *Clavicipitaceae*; Sung et al. 2007a; see also Kepler et al. 2012), whose asexual morphs or anamorphs include the much older, globally distributed and commercially important entomopathogenic species of the genera *Metarhizium* Sorokin 1883 and *Nomuraea* Maubl. 1903, as well as *Pochonia* Bat. & O.M. Fonseca 1965; note that the citations for authors of fungal names have been standardised throughout the text, in accordance with Kirk and Ansell (1992). As Gams et al. (2012) have concluded: "That these anamorph genera are phylogenetically and ecologically distinct may justify the preferential use of the anamorph names. Nevertheless, the younger teleomorph name *Metacordyceps* also carries relevant information". In the light of these recent events, the convoluted history of the genus *Pochonia* – and, in particular, of the 'flagship' species *P. chlamydosporia* – is now unravelled and the current systematic status is discussed.

2.2 Historical: Pre-molecular

The first comprehensive scientific description of the fungus eventually referred to as *P. chlamydosporia* was by Goddard (1913) who isolated it from garden soil in Ann Arbor (Michigan, USA) during a study of the soil mycobiota and recognised it as a new species within the asexual (Hyphomycete) genus *Verticillium*, based on the "verticillate, conidial fructification" (see Fig. 2.1A), in which the sporogenous structures (phialides) are produced in whorls or verticils. The full diagnosis is presented below:

> *Mycelium orbicular, spreading into a thick mat with little zonation, at first white, later cream, and finally ochroleucous to ochraceous forming a firm crust; surface powdery with age; reverse color yellow (flavus) to orange; no coloration of the medium. Hyphae branched, septate, hyaline, 2–3.3 μm broad. Conidiophores upright, branched verticillately; end branches 15–30 μm long, in whorls of 3 or 4, tapering to a knobbed end which bears a single spore. Conidia oval, hyaline, 2.2–3.5 μm in diameter, easily falling away. Chlamydospore formation especially common; chlamydospores multicellular, 4–9-celled, with granular contents, thick- walled, globular, 10–25 μm in diameter when mature, slightly lobed, borne on short side branches which are 15–30 μm long, very persistent.*

However, he also acknowledged that the production of purported chlamydospores is the main diagnostic feature of the fungus and these spores were said to form a creamy or ochraceous powder on the surface of older colonies (Goddard 1913). From his description of *Verticillium chlamydosporium* Goddard, as well as the accompanying illustrations (Fig. 2.1A), there is little doubt as to the identity of the fungus. Nevertheless, his observation of the conidia (asexual dispersal spores or mitospores) being formed singly and "easily falling away", is at odds with the reality of the spores clumped in mucoid heads. There is also some ambiguity about the use of the term chlamydospore which is defined as: "an asexual 1-celled spore (primarily for perennation, not dissemination) originating endogenously and singly within part of a pre-existing cell, by the contraction of the protoplast" (Kirk et al. 2008). Clearly, this is not the form or the function of the 'chlamydospores' in this fungus.

The next description of the fungus was by Petch (1939) in Ceylon (now Sri Lanka), who isolated it from egg masses of the giant African land snail, *Achatina fulica*, and placed it in the genus *Stemphyliopsis*, as *S. ovorum* Petch, with the original Latin diagnosis:

> *Mycelio tenui, albo, farinoso; hyphis regularibus, septatis, hyalinis, 1.5μ diam.; conidiophoris lateralibus, simplicibus, unisporis, ad 10μ alt., 1μ diam., apice leniter incrassates, juxta apicem septatis; conidiis terminalibus, 3–6 cellulosis, globosis, 9μ diam., vel ovoideis, ad 16 × 12μ, hyalinis, levibus, cellulis cuneatis, exterior convexis, parietibus 2–3μ crassis.*

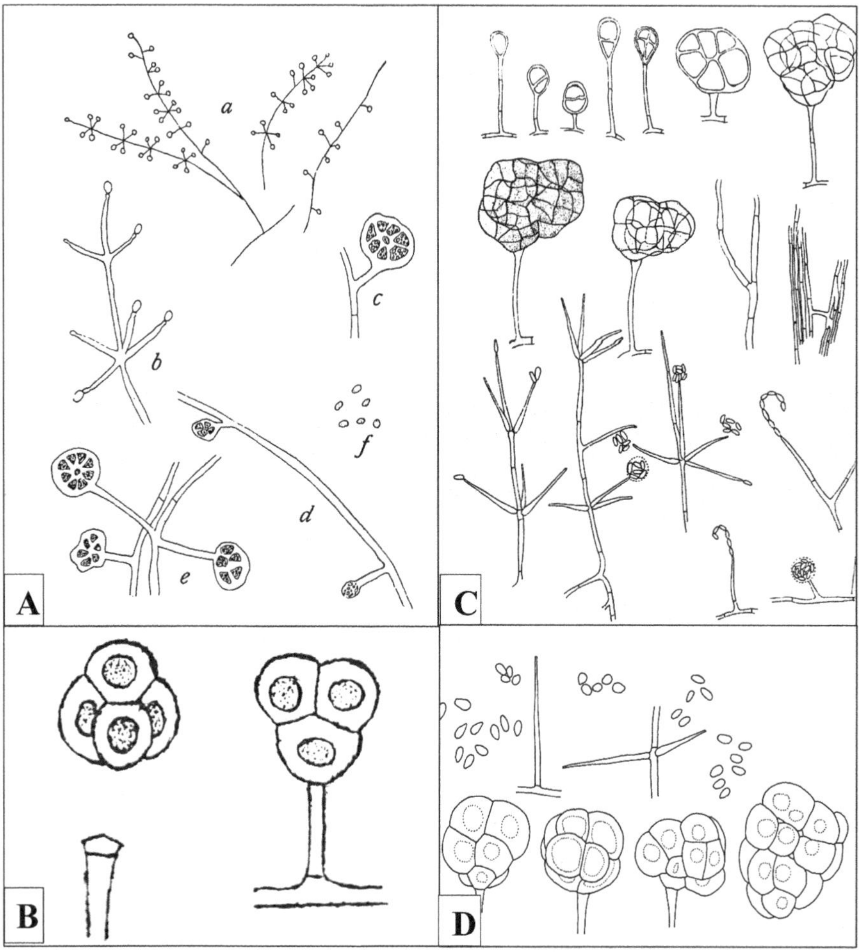

Fig. 2.1 Original illustrations of *Pochonia chlamydosporia*. (**A**) Type of *Verticillium chlamydosporium* from Goddard (1913); (**B**) Type of *Stemphyliopsis ovorum* from Petch (1939); (**C**) Type of *Pochonia humicola* from Batista and Fonseca (1965); (**D**) Neotype of *Pochonia chlamydosporia* from Zare et al. (2001)

Post-1935, it became obligatory to give the description of a new fungal species in Latin for the name to be validly published, and Petch gives a succinct account of the 'chlamydospore' stage – accompanied by a simple illustration (see Fig. 2.1B) – which he described as a compound conidium or dispersal spore, rather than a chlamydospore, which detaches readily from a short pedicel. However, the verticillate conidial stage was not included in the description and was either overlooked or overgrown by the 'chlamydospores'. There are some disparities with Goddard's description: smaller spores, 9–16 μm compared to 10–25 μm; fewer cells, 3–6 compared to 4–9; and, the easily detached rather than on "very persistent" side branches

(Goddard 1913), which could question the synonymy. Fortunately, the single collection on which Petch based his description was deposited in the Fungarium at the Royal Botanic gardens, Kew (K(M) 156,674) – although designating a holotype did not become obligatory until 1958, a single collection is always the holotype – and, later, this was examined by Barron and Onions (1966) who, on finding the characteristic verticillate stage, identified it as *Verticillium chlamydosporium*. Of particular relevance about this Sri Lankan specimen is that it appears to be the first record of the fungus as a pathogen of invertebrates. Petch (1939) had investigated and isolated the fungus from eggs of the snail – which "interfered with breeding experiments during investigations into the life history of that pest" – but had not considered the potential importance of the pathogen as a biocontrol agent of this highly invasive alien mollusc. And, in fact, he even delayed formalising the new species "which is not an entomogenous fungus…but its description has been awaiting publication for more than ten years" (Petch 1939). The giant African land snail has since become a major pest in many parts of the tropics (CABI 2014), where it had been introduced deliberately as a protein source (Van der Weijden et al. 2007) or accidentally, and is now known more correctly under the name *Lissachatina fulica*. Interestingly, *S. ovorum* is still recognised in MycoBank as a distinct species:

> MycoBank 273177 - *Stemphyliopsis ovorum* Petch, n.sp., Genus: *Stemphyliopsis*, Species: *ovorum*, Country: Sri Lanka, Collection Year: 1927, Associated Organism: *Achatina fulica* (giant African land snail), Substratum: on (snail) eggs. Type Status: Holotype of *Stemphyliopsis ovorum* Petch 1939, Collector: T. Petch.

The next description of the fungus was over 25 years later, when Batista and Fonseca (1965) introduced the genus *Pochonia* to accommodate the new species *P. humicola* – "uma curiosa entidade fungica" [a strange fungal entity] – isolated from soils in the States of Paraíba, Pernambuco and Rio Grande do Norte in north-east Brazil, and which they considered to be an integral part of the soil microbiota. The genus was named in honour of Prof. J. Pochon (Pasteur Institute, France) and was recognised as being polymorphic (pleomorphic) since it has two asexual stages, designated as the *Verticillium*-like, micro-conidial state and the macro-conidial or dictyospore state (Fig. 2.1C). This is the first use of the term dictyospore – "divided by intersecting septa in more than one plane" (Kirk et al. 2008), "septa running both transversely and longitudinally" (Seifert et al. 2011) – to describe these structures.

Shortly afterwards, Barron and Onions (1966) investigated cultures and fungarium material deposited in the IMI collections at the Commonwealth Mycological Institute (now CABI) under the name *Verticillium chlamydosporium*, and compared them to cultures deposited under the generic names *Diheterospora*, *Paecilomyces* and *Stemphyliopsis*. As reported above, they concluded that *Stemphyliopsis ovorum* is a later synonym, and that the type species of the genus *Diheterospora*, *D. heterospora* Kamyschko, is also conspecific with *V. chlamydosporium*. The new genus was proposed by Kamyschko (1962) to accommodate two species, isolated from soil in Russia, with multicellular chlamydospores – termed aleuriospores by Barron and Onions (1966) – but each with different conidial or phialospore (spores produced from phialides) states: *Verticillium* in *D. heterospora* and *Paecilomyces* in *D.*

catenulata. In addition, although type material could not be obtained for study, Barron and Onions (1966) considered that these taxa had been validly published and that, from the illustrations and description, there was no doubt that the type species of the genus is the same as that described much earlier by Goddard (1913). They further concluded that the aleuriospore state – "a thick-walled terminal spore formed from the blown out end of a sporogenous cell" (Barron 1968) – which was described as muriform – "longitudinally and transversely septate" (Barron 1968) – is the dominant and consistent spore form and, therefore, that *V. chlamydosporium* should be transferred to *Diheterospora*, as *D. chlamydosporia* (Goddard) G.L. Barron & Onions. They also noted that, although these fungi have been isolated from soil over a wide geographic area, their true role may be as parasites of snail eggs: "an interesting ecological speciality" (Barron and Onions 1966). None of the records, thus far, had been from nematodes.

Subsequently, Barron (1968) included the genus *Diheterospora* in his book on 'The genera of Hyphomycetes from soil', but, unfortunately, the type species was designated erroneously as *D. chlamydosporia* (Kamyschko) Barron & Onions, and with the generic diagnosis:

> *Aleuriospore state conspicuous: aleuriospores large, muriform, at first thin-walled and hyaline, later becoming yellowish or golden brown and very thick-walled, smooth, borne on short pedicels arising more or less at right angles to the vegetative hyphae; accessory phialide state of the* Verticillium *or* Paecilomyces *types also produced.*

An aleuriospore, however, has subsequently been considered to be a 'confused term' and should be regarded as obsolete (Kirk et al. 2008); although Seifert et al. (2011) have retained the term to describe "solitary, thallic conidia; usually with thickened and darkened walls". Once again, none of the aforementioned definitions of aleuriospores correspond to the dominant spores produced by *V. chlamydosporium*. Barron (1968) was the first to recognise – based on the illustrations (see Fig. 2.1D) and description (Batista and Fonseca 1965) – that *Pochonia humicola* is identical and, therefore, that this name should be regarded as a synonym. Presumably, in the earlier publication by Barron (Barron and Onions 1966), there had been no opportunity to examine the paper by Batista and Fonseca (1965).

However, shortly after, Gams (1971) re-assessed the genus *Verticillium* Nees and reduced the genus *Diheterospora*, as well as *Pochonia*, to synonymy; establishing the new section *Verticillium* sect. *Prostrata* to accommodate the entomopathogenic fungi with verticillate asexual states, to which *V. chlamydosporium* was re-assigned, along with *Diheterospora catenulata* under the new combination *Verticillium catenulatum* (Kamyschko ex G.L. Barron & Onions) W. Gams. He also included another synonym, *Dictyoarthrinopsis kelleyi* Dominik & Majchrowicz, for a fungus isolated from soils in Guinea, in which the diagnostic dictyospores were described and illustrated but not the verticillate state (Dominik and Majchrowicz 1966). Interestingly, at this stage, all of the records examined by Gams (1971) of both *V. chlamydosporium* (five isolates) and *V. catenulatum* (eight isolates) were from soil – predominantly in Europe – with no suggestion that they were parasitic on nematodes. The term chosen by Gams (1971) for the previously designated chlamydospore or aleuriospore state was dictyochlamydospore: "a non-deciduous multicelled chlamydospore composed of an outer wall separable from the walls of the component cells" (Kirk et al. 2008).

Nevertheless, Barron (1980a, 1985) continued to use the name *Diheterospora* and described new endoparasites of rotifers within the genus. Gams (1988), in his revision of nematophagous species of *Verticillium*, revisited the controversy and rejected the separation of *Verticillium* and *Diheterospora* based on the presence of dictyochlamydospores; arguing that "these structures can occur in quite variable densities and can be completely absent in some cases" and, moreover, that the phialidic states "of these fungi are quite indistinguishable from those of other *Verticillium* species" (Gams 1988). He did accept, however, that the rotifer parasites were distinct from the *Verticillium* species parasitic on, and isolated from, nematodes and merited further study. By now, of course, the nematophagous habit of *V. chlamydosporium* had been well established – following studies in Europe (Bursnall and Tribe 1974; Tribe 1977; Kerry 1980) and later in North America (Morgan-Jones et al. 1981) – rather than simply being a soil saprobe.

In his key to the nematophagous species of *Verticillium* sect. *Prostrata*, Gams (1988) separated *V. chlamydosporium* from the other taxa based on the abundance of yellowish, very thick-walled dictyochlamydospores in the aerial mycelium and recognised two varieties: *V. chlamydosporium* Goddard var. *chlamydosporium* and *V. chlamydosporium* var. *catenulatum* (Kamyschko ex Barron & Onions) W. Gams. He concluded that separation at the species level was no longer warranted – in contrast to previous taxonomic treatments (Barron and Onions 1966; Gams 1971) – because the cultures and dichtyochlamydospore are more or less identical, and that "the distinction between conidial chains and heads has limited weight" (Gams 1988). Another nematophagous, dichtyochlamydospore-producing fungus was also recognised and delimited as a separate taxon – *V. suchlasporium* W. Gams & Dackman – in which these spores are far less numerous and submerged in the medium rather than aerial. The specific epithet reflects this characteristic – being a contraction of '*subchlamydosporium*' (fide Gams 1988) – and *V. suchlasporium* was considered to be the commonest taxon isolated from cysts of *Heterodera* in northern Europe. Nevertheless, this separation was not supported by subsequent studies on secreted enzyme activities and RFLP (restriction fragment length polymorphism) analysis (Carder et al. 1993). When Gams (1971) proposed the new section *Verticillium* sect. *Prostrata*, he recognised the uniqueness of the verticillate species accommodated within it compared to other *Verticillium* species. However, because of the simplistic generic diagnostic features – phialides in whorls – there were no tools available to permit further separation of this obviously polyphyletic genus. Later, Jun et al. (1991) attempted to resolve the taxonomy of the genus *Verticillium* through an integrated approach employing morphological, physiological and biochemical characters. They concluded that the plant pathogenic species, such as *V. alboatrum*, were significantly different from the species from *Verticillium* sect. *Prostrata* and that – from an UPGMA dendrogram based on over 40 physiological and morphological characters – six cluster groups could be separated. However, an isolate of *V. chlamydosporium* from mollusc eggs remained ungrouped and sat well apart from other ungrouped isolates. With the routine use of more sophisticated molecular techniques in the last 20 years, generic concepts have been refined and phylogenetic relationships are now being resolved.

2.3 Historical: Post-molecular

2.3.1 Asexual Morph

When Zare et al. (2000) used ITS sequences they were able to distinguish three major clades (B, C and D) within *Verticillium* sect. *Prostrata*, which could be subdivided further (D1–3) employing ribosomal DNA sequences (Sung et al. 2001). The commonest and most distinctive species of group D2, *V. chlamydosporium*, was found to be remote from the other taxa – based on both ITS and LSU+SSU sequences of rDNA – making the group paraphyletic. However, Gams and Zare (2001) elected to classify all the D2 species in *Pochonia*, since this was considered to be the oldest generic name (Batista and Fonseca 1965). *Diheterospora* pre-dated *Pochonia* but, because no type was designated by Kamyschko (1962), the name was not validly published (Art. 40.1, ICN) – it was validated later by Barron and Onions (1966) – and they chose not to propose the name for conservation. In addition, Zare et al. (2001) also argued that the name *Pochonia* was simpler, as well as nomenclaturally correct, and, therefore, preferable to *Diheterospora* – the change was also supported since "the dictyochlamydospores are subordinate in importance" (Gams and Zare 2001). As previously discussed, Gams (1971, 1988) did not recognise *Pochonia* as a distinct genus because separation was based only on the presence or absence of dictyochlamydospores, which is an unstable or variable character, and reduced it to synonymy with *Verticillium*. Following further molecular analyses of small and large rDNA subunits, the 5.8S and ITS regions (Gams and Zare 2001), the concept of the genus *Verticillium* was restricted to plant pathogenic and plant-associated species. Subsequently, it was proposed to conserve the generic name – with a conserved type *Verticillium dahliae* Kleb. (Gams et al. 2005) – and, finally, the new family *Plectosphaerellaceae* W. Gams, Summerb. & Zare 2007 (Zare et al. 2007) was erected to accommodate *Verticillium* and related genera. This is a sister family to *Glomerellaceae*, both of which are of uncertain taxonomic position (*incertae sedis*), but far removed from the *Clavicipitaceae* in the *Hypocreales* (Zhang et al. 2006).

The generic description of *Pochonia*, as envisaged and modified by Zare et al. (2001), is thus:

> *Colonies rather fast-growing, reaching 15–40 mm diam in 10 days. Conidiophores usually prostrate and little differentiated from the vegetative hyphae, but sometimes erect and differentiated. Conidiogenous cells phialides, aculeate, verticillate or solitary. Conidia adhering in globose heads or chains, subglobose, ellipsoidal to rod-shaped, isodiametric-polyhedric, or falcate with blunt ends. Dictyochlamydospores produced on the surface of the colony or submerged in the agar, three-dimensional, pluricellular, thick-walled structures, usually formed on short stalks.*

The type species is cited as *Pochonia humicola* [= *Verticillium chlamydosporium* var. *chlamydosporium*]. However, in proposing the new combination, *Pochonia chlamydosporia* (Goddard) Zare & W. Gams var. *chlamydosporia*, *P. humicola* was not listed in the synonyms by Zare and Gams (2001), whilst it appeared in the list compiled by Zare et al. (2001). And, to compound the confusion, although the type of *P. humicola* from Brazil had been available and examined in both studies (Gams 1988; Zare et al. 2001), it was decided to select a neotype, based on an isolate from soil in Germany, which is not permitted under the rules in the ICN. Goddard (1913) had not designated a type in the original protologue although this was not mandatory for valid publication at this time. Moreover, there are still doubts about the correct citation for the genus type – due to confusion between nomenclature and taxonomy – since, in consecutive book chapters, it is given initially as *P. humicola* (Gams and Zare 2003) and then as *P. chlamydosporia* (Hodge 2003).

In their revision, six species of *Pochonia* were accepted and distinguished on the basis of conidial shape – all with dictyochlamydospores, "or at least some irregularly swollen hyphae" (Gams and Zare 2001) – and were keyed out: *P. chlamydosporia*, *P. suchlasporia* (W. Gams & Dackman) Zare & W. Gams, *P. rubescens* Zare, W. Gams & López-Llorca, *P. gonioides* (Drechsler) Zare & W. Gams, *P. bulbillosa* (W. Gams & Malla) Zare & W. Gams, and *P. microbactrospora* W. Gams & Zare, with two varieties being recognised in both *P. chlamydosporia* and *P. suchlasporia* (Zare et al. 2001). Taxa of the genus were said to be mainly parasites of nematode cysts, although only the first four species had been associated with or isolated from nematode hosts. In a subsequent review of the clavicipitaceous fungi parasitizing nematodes, Gams and Zare (2003) modified the key, but still retained the same species structure. However, they added the rider that, in the molecular analyses, *P. chlamydosporia* proved to be heterogeneous and concluded that: "This species is not very closely related to the remaining taxa of the genus", as suggested previously by Sung et al. (2001). This heterogeneity of, or variation between, isolates of *P. chlamydosporia* had first been highlighted by Kerry et al. (1986) and Irving and Kerry (1986), following *in vitro* growth studies, and was confirmed later by Arora et al. (1996), using PCR-based molecular analysis. Apart from producing dictyochlamydospores, species of *Pochonia* were also distinguished from those of the closely related genus *Lecanicillium* Zare & Gams by the cyanophilic nature – readily absorbing blue stains such as cotton blue – of the conidia (Zare and Gams 2001, 2004).

Subsequently, Zare and Gams (2007) added another species, *P. globispora* Zare & W. Gams, distinguished by its globose spores in berry-like heads and apparent absence of dictyochlamydospores. In addition, it is a root endophyte, rather than a nematode parasite, although they acknowledged that more studies were needed and that a nematophagous habit may be established. The cyanophilic phialides and the very high ITS sequence similarities (99%) with *P. chlamydosporia* placed it in the genus *Pochonia* rather than *Lecanicillium*. In fact, it was concluded that the ITS sequences of *P. globispora* establish a link between *P. chlamydosporia* and the other species in the genus, which is well illustrated in the neighbour-joining tree based on ITS1-5.8S-ITS2 (Zare and Gams 2007). At the same time, Sung et al. (2007a, b) also added a further species, *P. parasitica* (G.L. Barron) G.H. Sung,

J.M. Sung, Hywel-Jones & Spatafora – based on a fungus originally isolated from rotifers (Barron 1980a, b) – since it clustered with other *Pochonia* species in the phylogenetic tree. This tree examined phylogenetic relationships between more than 160 taxa in the *Clavicipitaceae* (Lindau) Earle ex Rogerson *sensu stricto*, and related families in the *Hypocreales* – based on Bayesian analyses of five genes (nrSSU, nrLSU, tef1, rpb1 and rpb2) – and separated the *Clavicipitaceae* into three clades. Species of *Pochonia*, including isolates of *P. chlamydosporia* from mollusc and nematode hosts, clustered in clade A within the subclade *Cordyceps taii* along with *Metarhizium*, although it was concluded from the analysis that *Pochonia*, unlike *Metarhizium*, is not a monophyletic lineage.

During surveys in Japan for novel fungal metabolites, Nonaka et al. (2013) targeted soil fungi associated with plant roots and, of the 118 *Verticillium*-like isolates, 50 were identified as *Pochonia*, based on morphology and multigene (SSU, LSU, tef, rpb1 and rpb2) phylogenetics. All the strains of *P. chlamydosporia* formed a strongly supported monophyletic clade which was further divided into four subclades corresponding to: *P. chlamydosporia* var. *chlamydosporia*, *P. chlamydosporia* var. *catenulata*, *P. chlamydosporia* var. *ellipsospora* Nonaka, Kaifuchi & Masuma and *P. chlamydosporia* var. *spinulospora* Nonaka, Kaifuchi & Masuma. As the varietal epithets indicate, the latter two varieties have a radically different micro-conidial morphology to the original and much older nematophagous varieties; although both produce similar dictyochlamydospores, albeit significantly larger in *P. chlamydosporia* var. *spinulospora*. A new species, *P. boninensis* Nonaka, Kaifuchi & Masuma – with irregularly swollen hyphae rather than dictyochlamydospores – clustered in a separate clade. Even more recently, an additional new variety – *P. chlamydosporia* var. *mexicana* – has been recognised, isolated from eggs of *Meloidogyne* in Mexico (Medina-Canales et al. 2014). In the phylogenetic tree presented – based on the ITS1-5.8S-ITS2 region – isolates of this new variety clustered in a separate subgroup from the much older nematophagous varieties, separated by the subgroup defined by *P. bulbillosa*, the whole being interpreted as a monophyletic group or clade. This is somewhat at odds with the findings of Nonaka et al. (2013) who placed *P. bulbillosa* in a distinct and distant clade and concluded that "the inclusion of new *Pochonia* taxa from Japan failed to reconstruct the genus as monophyletic"; confirming the results obtained by Sung et al. (2007a, b). In addition, because this new variety lacked a description and designation of a type, it was not validly published. Shortly afterwards, however, this error was recognised and rectified, when a holotype was designated (Medina-Canales et al. 2015).

2.3.2 Sexual Morph

More confusion was added to the 'nomenclatural mix' when a sexual stage (sexual morph or teleomorph), *Cordyceps chlamydosporia* H.C. Evans, was described from and associated with *P. chlamydosporia* on mollusc eggs from tropical forests in Ecuador (both western and eastern) and Amazonian Brazil (Zare et al. 2001; see Figs 2.2 and 2.3). The designated holotype, with abundant dictyochlamydospores

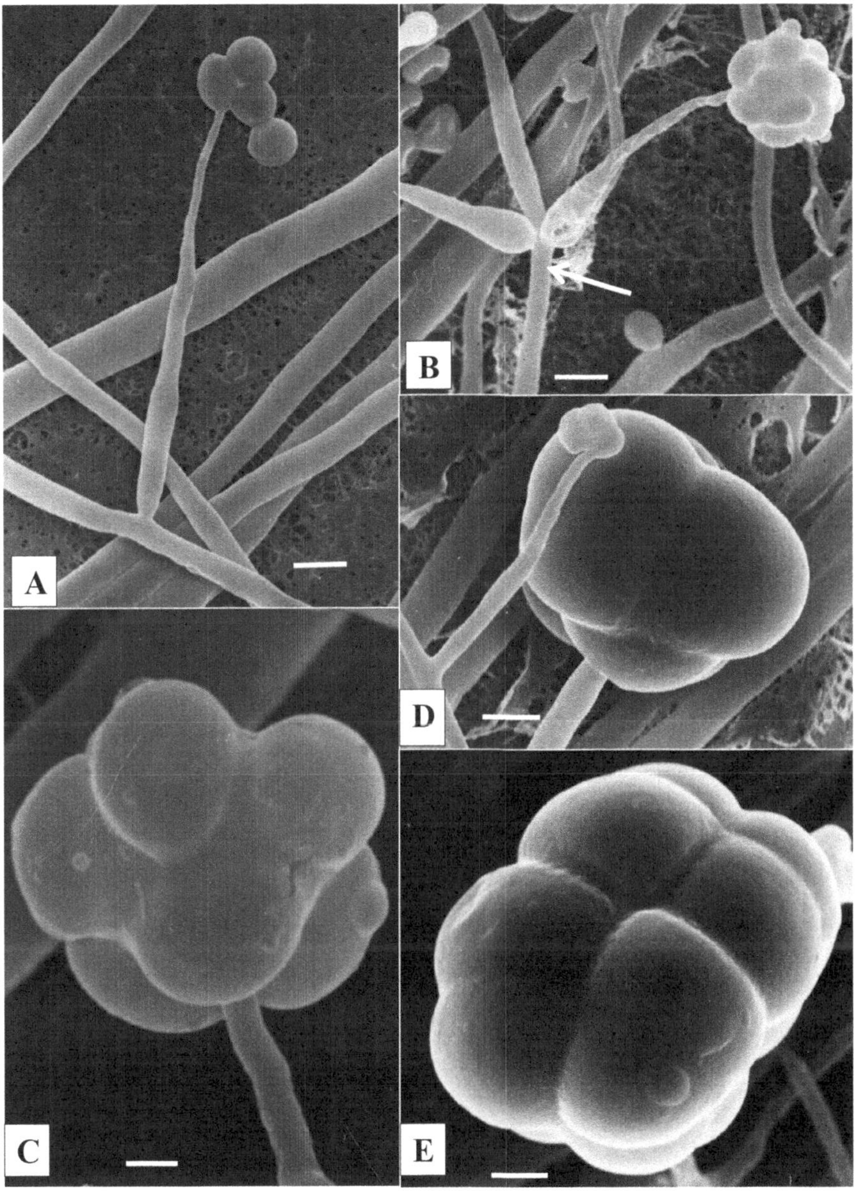

Fig. 2.2 SEM of *Pochonia chlamydosporia* isolated from mollusc eggs from Amazonian forest, Rondônia, Brazil. (**A**) Solitary phialide with terminal group of conidia; (**B**) Verticil of phialides (*arrow*) with conidia in a mucoid head; (**C**) Close up of mucoid spore head; (**D**) Developing dictyochlamydospore, note solitary phialide on left; (**E**) Mature dictyochlamydospore (Bars: **A**, **B** = 4 μm; **C** = 1.5 μm; **D**, **E** = 3 μm)

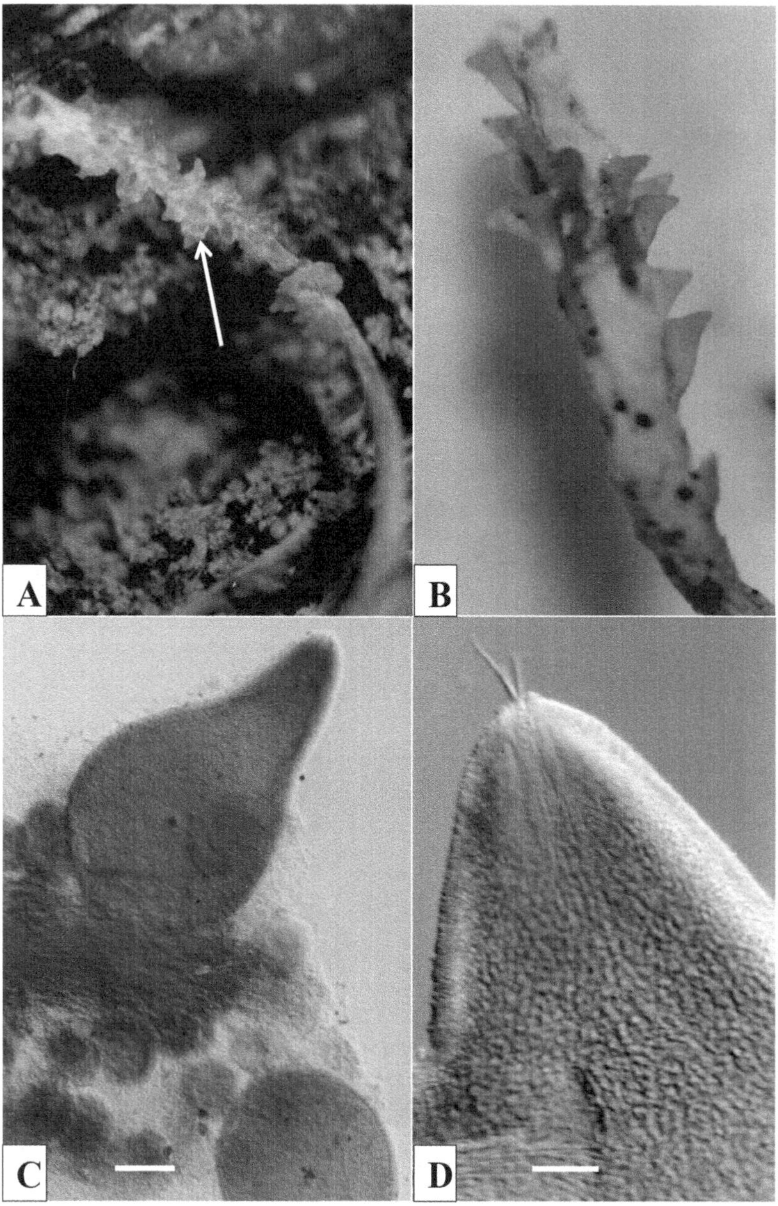

Fig. 2.3 *Metacordyceps chlamydosporia*, type on mollusc eggs from tropical forest, western Ecuador. (**A**) Sexual structure (*clava or ascostroma*) emerging from egg mass buried in soil, bearing rows of erumpent or superficial perithecia (*arrow*); (**B**) Close up of above, showing flask-shaped perithecia; (**C**) Light microscope image of perithecium at tip of clava; (**D**) Close up of neck region, with two cylindrical sexual spores (ascospores) emerging from apical pore of ostiole (Bars: **C** = 90 μm; **D** = 25 μm)

covering the egg masses – as well as forming on well-defined aerial structures (synnemata) – provided circumstantial evidence of a sexual-asexual connection, which was later confirmed by isolation of the asexual morph from ascospores. The following diagnosis was given:

> *Mycelium white to yellow densely covering the egg mass. Stromata single, unbranched, white to pale orange, narrow cylindrical, 12–16 × 0.5–1.25 mm, widening towards the tip. Fertile area terminal, 2.0–3.5 mm in length, white to pale orange. Perithecia crowded, initially immersed in the mycelium, becoming erumpent and superficial; smooth, pale yellow, occasionally brownish orange, broadly flask-shaped, (500–) 600–650 × 250-300 μm, with a prominent ostiole and a long neck region, 150–200 × 80–120 μm, lined with periphyses. Asci 8-spored, cylindrical, (160–) 200–500 × (2–) 3–4 (4.5) μm, with a prominent cap. Ascospores hyaline, filiform, two fascicles of four, somewhat distended in the ascus, septa indistinct, not fragmenting into part-spores, 150–350 × 1.0 μm.*

DNA sequences of *Cordyceps chlamydosporia* from these mollusc isolates were not included in the phylogenetic tree constructed during this study (Zare et al. 2001), but isolates from mollusc and nematode hosts were subsequently compared in a more comprehensive five-gene phylogenetic classification (Sung et al. 2007a, b) and the sequences found to be identical. These clustered in a clade containing species of *Metarhizium* within the *Clavicipitaceae sensu stricto* for which sexual morphs had been assigned previously to the genus *Cordyceps*. However, since *Cordyceps sensu stricto* clustered in a separate clade – within the new family *Cordycipitaceae* Kreisel ex G.H. Sung, J.M. Sung, Hywel-Jones & Spatafora – the new genus *Metacordyceps* was established, not only to accommodate species with *Metarhizium* asexual morphs, but also *P. chlamydosporia*, resulting in the publication of the new combination *Metacordyceps chlamydosporia* (H.C. Evans) G.H. Sung, J.M. Sung, Hywel-Jones & Spatafora (Sung et al. 2007a, b). With the creation of this new teleomorph-typified genus, the question of name priority quickly became an issue – not only for the well-known and long established genus *Metarhizium*, but also for *Pochonia* – as the lobbying increased for the One fungus, One name (1F, 1 N) initiative, this concern being highlighted by Gams et al. (2012).

Metacordyceps chlamydosporia was included in a systematic study of the entomopathogenic genus *Torrubiella* using multi-gene phylogeny, and this mollusc isolate fell within the *Clavicipitaceae sensu stricto*, although it clustered with a species of *Rotiferophthora* rather than with another species of *Pochonia*, *P. gonioides* (Johnson et al. 2013). Similarly, Kepler et al. (2013) included *M. chlamydosporia* from a mollusc host and *P. chlamydosporia* from a nematode host, as well as four other species of *Pochonia*, in another phylogenetic study of clavicipitioid genera and the *P. chlamydosporia* isolates grouped together next to species of *Rotiferophthora* and *Metarhizium*, but somewhat distant from the other species of *Pochonia*.

This distinction between the type of *Pochonia* and other species assigned to the genus had been noted earlier in a phylogenetic study of the genus *Metacordyceps* (Kepler et al. 2012). A core *Metacordyceps* clade was identified – with pigmented ascostromata and immersed or buried ascomata – in which the asexual morphs pertain predominantly to species of *Metarhizium*, and which could readily be separated from the other species in the clade with non-pigmented ascostromata and superficial or strongly erumpent ascomata, as typified by *Metacordyceps chlamydosporia*. As noted previously, in all recent multi-gene phylogenetic studies *M. chlamydosporia* clusters nearer to species of *Rotiferophthora* – parasites of bdelloid rotifers with similarly prominent but flattened, two-dimensional dictyochlamydospores – than to other species of *Pochonia*. These taxa group in a clade basal to the rest of the *Metacordyceps* clade for which a sexual morph has never been associated. Following on from this study, Kepler et al. (2014) employed a molecular dataset of the protein-coding genes – beta-tubulin (BTUB), largest and second largest RNA Polymerase II subunit (RPB1 and RPB2) and translation elongation factor 1-alpha (TEF) – in order to clarify relationships between *Metarhizium* and related taxa, including *Pochonia*. In a far reaching decision – for research communities in general, and for biocontrol practitioners in particular – it was proposed that *Metarhizium* should take priority rather than the sexual morph, *Metacordyceps*, which was made in order "to limit disruption and to maintain the utility of currently recognised names" (Kepler et al. 2014). *Metacordyceps* has now been subsumed as a synonym within the emended genus *Metarhizium* Sorokin. Moreover, as long suspected, it was found that the genus *Pochonia* is polyphyletic and forms two strongly supported but significantly different clades. Based on this evidence, it was proposed that *Pochonia* should be restricted to those taxa forming a monophyletic clade with the type, *P. chlamydosporia*. Most of the other *Pochonia*-like species grouped in a clade that sub-tended the rest of the non-outgroup taxa and these are now accommodated in the new genus *Metapochonia* Kepler, S.A. Rehner & Humber (Kepler et al. 2014). Nevertheless, it was also conceded that this is not the final word since the systematics of *Pochonia* is still in a state of transition and it was predicted that, with more comprehensive sampling and molecular data, there will probably be sufficient support to elevate the varieties of *P. chlamydosporia* to species status. This continues a trend not to recognise infraspecific taxa.

2.4 Present Taxonomic Status: One Fungus, One Name (1F, 1N)

This section summarises the current situation regarding the *Pochonia*-like fungi associated with plant-parasitic nematodes. However, there are several issues concerning the correct terminology to be used for certain critical morphological features, specifically the spore stages. As discussed previously, various words have been used to describe the macrospore stage in species of *Pochonia*, from chlamydospore or dictyochlamydospore to aleuriospore or dictyospore: the former names imply non-dispersed, usually non-infective, resting structures; the latter names are used for asexual dispersal spores or conidia. Campbell and Griffiths (1975) used SEM to study the development of these spores and concluded that they are not aleuriospores and that the multicellular structure would be better classified as a dictyochlamydospore. However, this terminology is confusing since it implies that the spore function is exclusively for perennation and not for dispersal. It is obvious from the morphology that these aerial spores are liberated freely and have a dispersal function. In fact, they are usually the dominant spore form and the propagule of choice in mass production. Therefore, by definition, this spore would be classified as a type of conidium. In the latest taxonomic treatment of *Pochonia* (Seifert et al. 2011), it is described as "A dimorphic anamorph genus for species with verticillium-like conidiophores and often dictychlamydospores". The spore states are described, thus: A-anamorph [asexual morph] – ameroconidia (aseptate spores), hyaline, slimy; B-anamorph, hyaline, stalked dictyochlamydospore. In the glossary, a chlamydospore is defined as "thick-walled, resistant; persistent, not being liberated by a dehiscence mechanism" (Seifert et al. 2011). Clearly, there is considerable ambiguity, and it is suggested here that a working compromise would be to adopt the dimorphic terminology for the two asexual morphs: A-anamorph (ameroconidium); B-anamorph (dictyoconidium); the latter being defined as a "conidium with septa running both transversally and longitudinally" (Seifert et al. 2011).

Pochonia Bat. & O.M. Fonseca, Publ Inst Micol Recife 462: 4, 1965

Clavicipitaceae*, *Hypocreales*, *Hypocreomycetidae*, *Sordariomycetes*, *Pezizomycotina*, *Ascomycota*, *Fungi

= *Diheterospora* Kamyschko, Bot Mater Otd Sporov Rast Bot Inst Komarova Akad Nauk USSR 15: 137, 1962 (*nomen invalidum*) ex G.L. Barron & Onions, Can J Bot 44: 866, 1966

Type Species: Pochonia humicola Bat. & O.M. Fonseca 1965 = *P. chlamydosporia* (Goddard) Zare & W. Gams (Zare et al. 2001).

The following varieties are recognised:

Pochonia chlamydosporia (Goddard) Zare & W Gams var. ***chlamydosporia***, Nova Hedwigia 72: 334, 2001

= *Verticillium chlamydosporium* Goddard, Bot Gaz 56: 275, 1913
= *Diheterospora chlamydosporia* (Goddard) G.L. Barron & Onions, Can J Bot 44: 866, 1966
= *Stemphyliopsis ovorum* Petch, Trans Br Mycol Soc 23: 146, 1939
= *Diheterospora heterospora* Kamyschko, Bot Mater Otd Sporov Rast Bot Inst Komarova Akad Nauk USSR 15: 138, 1962 (*nomen invalidum*) ex G.L. Barron & Onions, Can J Bot 44: 866, 1966
= *Pochonia humicola* Bat. & O.M. Fonseca, Publ Inst Micol Recife 462: 5,1965
= *Dictyoarthrinopsis kelleyi* Dominik & Majchr., Mycopath Mycol Appl 28: 210, 1966
= *Cordyceps chlamydosporia* H.C. Evans, Nova Hedwigia 72: 59, 2001
= *Metacordyceps chlamydosporia* (H.C. Evans) G.H. Sung, J.M. Sung, Hywel-Jones & Spatafora, Stud Mycol 57: 35, 2007

Cosmopolitan: isolated directly from soil, and as a parasite on nematode eggs, cysts and egg masses within galls (*Heterodera*, *Meloidogyne*, *Globodera*), as well as on eggs of molluscs – including the giant African snail (*Achatina fulica*) – from the rhizosphere, and as a root endophyte from dicots (Barron and Onions 1966; Lopez-Llorca and Boag 1993a; Zare et al. 2001; Zare and Gams 2003a).

Zare and Gams (2003a) also listed it from insect hosts (Aphididae), and from tapeworm (*Ascaris*) eggs, but these are considered to be dubious records or from laboratory infection experiments.

Pochonia chlamydosporia var. ***catenulata*** (Kamyschko ex G.L. Barron & Onions) Zare & W. Gams, Nova Hedwigia 72: 334, 2001

= *Diheterospora catenulata* Kamyschko, Bot Mater Otd Sporov Rast Bot Inst Komarova Akad Nauk USSR 15: 138, 1962 (*nomen invalidum*) ex G.L. Barron & Onions, Can J Bot 44: 868, 1966

= *Verticillium catenulatum* (Kamyschko ex G.L. Barron & Onions) W. Gams, Neth J Pl Pathol 94: 134, 1988

Cosmopolitan: isolated from soil, eggs of *Heterodera*, slug eggs (*Agriolimax agrestis*) and as an entomopathogen associated with a *Cordyceps* sexual morph on a beetle larva (Barron and Onions 1966; Gams 1988; Zare et al. 2001; Zare and Gams 2003a).

The phialidic asexual morph (A-anamorph) was considered by Barron and Onions (1966) as pertaining to the genus *Paecilomyces* rather than to *Verticillium*, because of the common occurrence of primary and secondary verticils and the dry spores in chains rather than in mucoid heads, and separated it from *V. chlamydosporium* at the species level within the genus *Diheterospora*, and based on the aleuriospore asexual morph (= B-anamorph or dictyospore). Subsequently, as noted earlier, Gams (1971, 1988) and Zare et al. (2001) rejected this generic and species concept and chose to classify these taxa at the varietal rank within *Verticillium* and then in *Pochonia*, with the dictyochlamydospores becoming a secondary character. However, as well as differences in conidiogenesis, the phialospores can also be distinguished morphologically; those of *P. chlamydosporia* var. *catenulata* being subglobose to limoniform, compared to the ovoid to short cylindric or ellipsoidal and slightly larger spores of *P. chlamydosporia* var. *chlamydosporia*.

Pochonia chlamydosporia var. ***ellipsospora*** Nonaka, Kaifuchi & Masuma, Mycologia 105: 1207, 2013

Pochonia chlamydosporia var. ***spinulospora*** Nonaka, Kaifuchi & Masuma, Mycologia 105: 1209, 2013

Japan: isolated from soil.

As described by Nonaka et al. (2013), in *P. chlamydosporia* var. *ellipsospora* the predominantly ellipsoidal phialospores are produced in small heads, contrasting strongly with those of *P. chlamydosporia* var. *spinulosa* which are globose to subglobose, as well as markedly ornamented or spiny, and produced in chains.

Pochonia chlamydosporia var. ***mexicana*** M.G. Medina-Canales, A.V. Rodríguez-Tovar, R.H. Manzanilla-López, G. Zúñiga, A. Tovar-Soto, Index Fungorum 239: 1, 2015

Mexico: isolated from soil, rhizosphere and plant roots, as well as from galls of *Meloidogyne* (Medina-Canales et al. 2014). Isolates were shown to be parasitic on eggs of various root-knot nematodes (*Meloidogyne* spp.) and were considered to have biological control potential.

The following is an abridged description from the protologue (Medina-Canales et al. 2015): Phialides solitary. Conidia of different size and shape; elongate conidia of two sizes, the smaller (1 ± 0.25 × 0.3 ± 0.09 μm), termed microconidia, and the larger (3.2 ± 0.83 × 0.9 ± 0.26 μm); and globose conidia (2.4 ± 0.55 × 1.5 ± 0.37 μm). Colonies white becoming beige with pale yellow to orange reverse. A description of the 'chlamydospores' was lacking in this protologue, although SEM images of these appeared in the original publication (Medina-Canales et al. 2014). It is obvious from the description of the phialospores that this taxon is morphologically, as well as molecularly distinct from the type species of *Pochonia*.

Metapochonia Kepler, S.A. Rehner & Humber, Mycologia 106: 820, 2014

Verticillium-like, producing conidia on slender awl-shaped phialides, solitary or in whorls; dictyochlamydospores often present, but scarce and typically submerged in the substrate; phylogenetic position near to but outside the true *Pochonia* clade. Unlike *Pochonia*, the macrospores are true resting spores with no other function other than perennation; having no form of dehiscence and being buried. In fact, their relative scarcity and simple structure suggests that they are primitive compared to the dictyospores of *Pochonia* which may have evolved from them to function as the dominant aerial, dispersal, infective spore and to play a key role in the life cycle.

Type Species: Pochonia suchlasporia (W. Gams & Dackman) Zare & W. Gams, Nova Hedwigia 73: 67, 2001.

According to Kepler et al. (2014), the genus "Encompasses at least seven species", however, only those associated with nematodes are included here.

Metapochonia gonoides (Drechsler) Kepler, S.A. Rehner & Humber, Mycologia 106: 820, 2014

= *Acrostalagmus goniodes* Drechsler, J Washington Acad Sci 32: 347, 1942
= *Verticillium goniodes* [as *gonioides*] (Drechsler) W. Gams & Stalpers, Neth J Pl Pathol 94:
143, 1988
= *Pochonia gonoides* [as *gonioides*] (Drechsler) Zare & W. Gams, Nova Hedwigia 73: 72, 2001

Temperate: isolated from plants and leaf litter; originally described from cysts and eggs of *Heterodera*, as well as an endoparasite of *Rhabditis*, and from free-living nematodes, such as *Bunonema* (Gams 1988; Zare et al. 2001).

Metapochonia rubescens (Zare, W. Gams & López-Llorca) Kepler, S.A. Rehner & Humber, Mycologia 106: 820, 2014

= *Pochonia rubescens* Zare, W.Gams & López-Llorca, Nova Hedwigia 73: 69, 2001

Temperate: isolated from soil; from cysts and eggs of *Heterodera*, culture filtrates have been shown to be nematicidal to *Globodera rostochiensis* (Lopez-Llorca and Boag 1993b).

Metapochonia suchlasporia (W. Gams & Dackman) Kepler, S.A. Rehner & Humber, Mycologia 106: 820, 2014

= *Verticillium suchlasporia* W. Gams & Dackman, Neth J Pl Pathol 94: 136, 1988
= *Pochonia suchlasporia* (W. Gams & Dackman) Zare & W. Gams, Nova Hedwigia 73: 67, 2001

Temperate: isolated from soil; from cysts and eggs of *Heterodera* and *Globodera* (Gams 1988; Zare et al. 2001; Zare and Gams 2003b).

The latter authors distinguished two varieties within this species, *Metapochonia suchlasporia* var. *suchlasporia* and *M. suchlasporia* var. *catenata*, with almost the same morphological characters – conidia produced in heads as opposed to chains – used to separate *Pochonia chlamydosporia* var. *chlamydosporia* and *P. chlamydosporia* var. *catenulata*.

2.5 Future Taxonomic Status?

As emphasised in the introduction, fungal systematics is in a state of flux, no more so than in the *Hypocreales* to which many of the fungal pathogens or parasites of invertebrates belong: a still largely untapped source of potential biological control agents and novel metabolites. The following areas are identified where future actions would be needed to clarify the taxonomic status of *Pochonia chlamydosporia*, and its nematode-parasitising relatives, as well as its biology and relationship with other groups within the *Hypocreales*:

- It is obvious from Sect. 4, that the differences – both morphological and molecular – between the five varieties of *P. chlamydosporia* justify their elevation to species rank, as happened in the genus *Metarhizium* (Bischoff et al. 2009; Kepler et al. 2014). This trend has also been adopted for hypocrealean pathogens of ants which previously had been afforded varietal status within ant complexes (Evans and Samson 1982, 1984), but latterly – with the inclusion of more characters, including spore germination patterns, and the advent of molecular phylogenetics – species can readily been delimited (Evans et al. 2011; Araújo et al. 2015). Indeed, there is now the probability that these *Ophiocordyceps*-ant complexes contain numerous, even hundreds of 'hidden' species, rather than the simplistic view of one over-arching species with many varieties, which could be comparable to the situation in the *Pochonia*-nematode complex. More robust sampling and phlylogenetic analyses could reveal an equivalent richness of species, especially in the tropics, as shown by recent in-depth studies of the soil mycobiota in Japan and Mexico (Nonaka et al. 2013; Medina-Canales et al. 2014). Indeed, this richness and potential megadiversity of the genus *Pochonia* has been exemplified recently by the isolation of a species from a *Myrmica* ant in the UK – originally listed as *P. chlamydosporia* (Evans et al. 2010) – which, following a detailed morphological analysis, was found to produce four spore types; including dictyospores and conidia with prominent ornamentation, as well as true thick-walled, aseptate resting spores (H.C. Evans, unpublished).

- The association of *P. chlamydosporia* var. *chlamydosporia* with hosts as diverse as plant-parasitic nematodes and molluscs – and the production of a well-defined sexual morph on the latter – is intriguing, as well as puzzling. Moreover, this link has been supported consistently by phylogenetic analyses, leading to the suggestion that such a connection points to novel heteroxenous – having more than one host – life cycles within these fungi (Hodge 2003). Indeed, there is an example of such a life cycle with the discovery of a new species of *Harposporium* on an arboreal coleopteran larva, which produced compound synnemata together with a synanamorph in the genus *Hirsutella* (= *Ophiocordyceps*), with the former prominent within the bark and the latter emerging onto the surface (Evans and Whitehead 2005). Until recently, *Harposporium* was considered to be predominantly a nematophagous genus – parasitic on saprophagous nematode genera in which the characteristic crescent-shaped or lunate conidia are ingested and lodge in the digestive tract and thereby initiate infection – and there are several other examples of this potential heteroxenous habit (Hodge 2003; Zare and Gams 2003a, b). It was posited that the *Harposporium* conidia are deposited between the inner bark and infect free-living nematodes feeding on the detritus, whilst the *Hirsutella* conidia infect beetle larvae crawling over the bark (Evans and Whitehead 2005). The *Ophiocordyceps* sexual morph with forcibly discharged aerial spores (ascospores) would be eventually produced on the mummified cadaver. This is mirrored in *Pochonia chlamydosporia* var. *chlamydosporia* on mollusc eggs in the tropics: the conidia of the A-asexual morph may be exclusively to infect the snail eggs, whilst those of the B-asexual morph (dictyospores) may function to infect soil nematodes, as well as to serve as survival structures. Given the size and complexity of the sexual morph, there is no way that this stage can ever form on nematode hosts and hence genetic variation within the species is dependent on the 'intermediate' mollusc (or insect) host. The same occurs with *P. chlamydosporia* var. *catenulata* which has been isolated on several occasions from ascospores of a robust *Cordyceps*-like species on beetle larvae in tropical forests; producing typical A-asexual and B-asexual morphs in culture (H.C. Evans, unpublished). Isolates have been included in phylogenetic analyses (Zare and Gams 2001; Zare and Gams 2007), and proved to be indistinguishable from strains isolated from temperate soil. A recent analysis of the mitochondrial genome of *P. chlamydosporia* has confirmed this close relationship with the invertebrate-pathogenic fungi in the *Hypocreales* and they form a monophyletic group sharing a common ancestor (Lin et al. 2015). Further surveys in the tropics should reveal similar links with other varieties/species in the *P. chlamydosporia* complex producing sexual morphs on insect and mollusc hosts.
- The ability of *P. chlamydosporia* to live as a true or coevolved endophyte within plant roots is now being recognised; offering intriguing biological control pos-

sibilities, as well as indicating a potential additional role as a plant growth promoter (Bordallo et al. 2002; Monfort et al. 2005; Larriba et al. 2014). The multitrophic habit of this fungus means it functions both as a pathogen of invertebrates (insect, nematodes and molluscs) and as a plant symbiont, which it shares with its close relatives in the genus *Metarhizium* (Sasan and Bidochka 2012), and may even extend to movement of nitrogen from the soil into the plant (Behie and Bidochka 2014). Most genera of the entomopathogenic *Hypocreales* show a high tolerance to fungicides, such as dodine (Beilharz et al. 1982), and both *P. chlamydosporia* and *Metapochonia suchlasporia* have also demonstrated a degree of tolerance to copper (Bååth 1991). During experiments with fungicide-based selective media to isolate species of *Metarhizium* from root systems of mature coffee trees in Brazil, the plates were dominated by *P. chlamydosporia sensu lato*; however, in previous studies with coffee soil, the plates were dominated by *Metarhizium* (H.C. Evans, unpublished). It appears highly likely, therefore, that *P. chlamydosporia* is a common resident of plant roots, especially in the tropics, and future surveys should concentrate on plant root systems – as well as soils – in tropical forest ecosystems in order to measure the true diversity within the *P. chlamydosporia* complex.

References

Araújo, J. P. M., Evans, H. C., Geiser, D. M., et al. (2015). Unravelling the diversity behind the *Ophiocordyceps unilateralis* (*Ophiocordycipitaceae*) complex: Three new species of zombie-ant fungi from the Brazilian Amazon. *Phytotaxa, 220*, 224–238.

Arora, D. K., Hirsch, P. R., & Kerry, B. R. (1996). PCR-based molecular discrimination of *Verticillium chlamydosporium* isolates. *Mycological Research, 100*, 801–809.

Bååth, E. (1991). Tolerance of copper by entomogenous fungi and the use of copper-amended media for isolation of entomogenous fungi from soil. *Mycological Research, 95*, 1140–1142.

Barron, G. L. (1968). *The genera of Hyphomycetes from soil*. Baltimore: Williams & Wilkins.

Barron, G. L. (1980a). Fungal parasites of rotifers: Two new verticillate endoparasites with aerial conidiophores. *Canadian Journal of Botany, 58*, 432–438.

Barron, G. L. (1980b). Fungal parasites of rotifers: A new *Tolypocladium* with underwater conidiation. *Canadian Journal of Botany, 58*, 439–442.

Barron, G. L. (1985). Fungal parasites of bdelloid rotifers: *Diheterospora*. *Canadian Journal of Botany, 63*, 211–222.

Barron, G. L., & Onions, A. H. S. (1966). *Verticillium chlamydosporium* and its relationships to *Diheterospora, Stemphyliopsis*, and *Paecilomyces*. *Canadian Journal of Botany, 44*, 861–869.

Batista, A. C., & Fonseca, O. M. (1965). *Pochonia humicola* n. gen. e n. sp., uma curiosa entidade fungica dos solos do Nordeste do Brasil. *Publicacões do Instituto de Micologia, Universidade do Recife, 462*, 1–11.

Behie, S. W., & Bidochka, M. J. (2014). Ubiquity of insect-derived nitrogen transfer to plants by endophytic insect-pathogenic fungi: An additional branch of the soil nitrogen cycle. *Applied and Environmental Microbiology, 80*, 1553–15560.

Beilharz, V. C., Parberry, D. G., & Swart, H. J. (1982). Dodine: A selective agent for certain soil fungi. *Transactions of the British Mycological Society, 79*, 505–511.

Bordallo, J. J., Lopez-Llorca, L., Jansson, H. B., et al. (2002). Colonization of plant roots by egg-parasitic and nematode-trapping fungi. *The New Phytologist, 154*, 491–499.

Bursnall, L. A., & Tribe, H. T. (1974). Fungal parasitism in cysts of *Heterodera*. II. Egg parasites of *H. schacthtii*. *Transactions of the British Mycological Society, 62*, 595–601.
CABI. (2014). *Lissachatina fulica* (giant African land snail). *CABI Invasive Species Compendium*. Wallingford: CAB International.
Campbell, W. P., & Griffiths, D. A. (1975). The development of thick-walled, multicellular, aerial spores in *Diheterospora chlamydosporia* (= *Verticillium chlamydosporium*). *Canadian Journal of Microbiology, 21*, 963–971.
Carder, J. H., Segers, R., Butt, T. M., et al. (1993). Taxonomy of the nematophagous fungi *Verticillium chlamydosporium* and *V. suchlasporium* based on secreted enzyme activities and RFLP analysis. *Journal of Invertebrate Pathology, 62*, 178–184.
Dominik, T., & Majchrowicz, I. (1966). Some new species of fungi from the soil of Conakry. *Mycopathologia et Mycologia Applicata, 28*, 209–219.
Evans, H. C., & Samson, R. A. (1982). *Cordyceps* species and their anamorphs pathogenic on ants (Formicidae) in tropical forest ecosystems. I. The *Cephalotes* (Myrmicinae) complex. *Transactions of the British Mycological Society, 79*, 431–453.
Evans, H. C., & Samson, R. A. (1984). *Cordyceps* species and their anamorphs pathogenic on ants (Formicidae) in tropical forest ecosystems. II. The *Camponotus* (Formicinae) complex. *Transactions of the British Mycological Society, 82*, 127–150.
Evans, H. C., & Whitehead, P. F. (2005). Entomogenous fungi of arboreal Coleoptera from Worcestershire, England, including the new species *Harposporium bredonense*. *Mycological Progress, 4*, 91–99.
Evans, H. C., Groden, E., & Bischoff, J. F. (2010). New fungal pathogens of the red ant, *Myrmica rubra*, from the UK and implications for ant invasions in the USA. *Fungal Biology, 114*, 451–466.
Evans, H. C., Elliott, S. L., & Hughes, D. P. (2011). Hidden diversity behind the zombie-ant fungus *Ophiocordyceps unilateralis*: Four new species described from carpenter ants in Minas Gerais, Brazil. *PloS One, 6*, e17024.
Gams, W. (1971). *Cephalosporium-artige Schimmelpilze (Hyphomycetes)*. Stuttgart: Fischer.
Gams, W. (1988). A contribution to the knowledge of nematophagous species of *Verticillium*. *Netherlands Journal of Plant Pathology, 94*, 123–148.
Gams, W., & Zare, R. (2001). A revision of *Verticillium* section *Prostrata*. III. Generic classification. *Nova Hedwigia, 72*, 329–337.
Gams, W., & Zare, R. (2003). A taxonomic review of the clavicipitaceous anamorphs parasitizing nematodes and other microinvertebrates. In J. F. White, C. W. Bacon, N. L. Hywel-Jones, & J. W. Spatafora (Eds.), *Clavicipitalean fungi: Evolutionary biology, chemistry, biocontrol and cultural impacts* (pp. 17–73). New York: Marcel Dekker.
Gams, W., Zare, R., & Summerbell, R. C. (2005). Proposal to conserve the generic name *Verticillium* (anamorphic Ascomycetes) with a conserved type. *Taxon, 54*, 179.
Gams, W., Humber, R. A., Jaklitsch, W., et al. (2012). Minimizing the chaos following the loss of Article 59: Suggestions for a discussion. *Mycotaxon, 119*, 495–507.
Goddard, N. H. (1913). Can fungi living in agricultural soil assimilate free nitrogen? *Botanical Gazette, 56*, 249–305.
Hibbett, D. S., & Taylor, J. W. (2013). Fungal systematics: Is a new age of enlightenment at hand? *Nature Reviews Microbiology, 11*, 129–133.
Hodge, K. T. (2003). Clavicipitaceous anamorphs. In J. F. White, C. W. Bacon, N. L. Hywel-Jones, & J. W. Spatafora (Eds.), *Clavicipitalean fungi: Evolutionary biology, chemistry, biocontrol and cultural impacts* (pp. 75–123). New York: Marcel Dekker.
Irving, F., & Kerry, B. R. (1986). Variation between strains of the nematophagous fungus, *Verticillium chlamydosporium*. *Nematologica, 32*, 474–485.
Johnson, D., Sung, G. H., Hywel Jones, N. L., et al. (2013). Systematics and evolution of the genus *Torrubiella* (*Hypocreales*, *Ascomycota*). *Mycological Research, 113*, 279–289.
Jun, Y., Bridge, P. D., & Evans, H. C. (1991). An integrated approach to the taxonomy of the genus *Verticillium*. *Journal of General Microbiology, 137*, 1437–1444.

Kamyschko, O. P. (1962). De Monilialibus terrestribus novis notula. *Botanicheskie Materialy Otdela Sporovyh Rastenji Botanicheskogo Instituti Imeni V.L. Komarova Akademie Nauk SSSR, 15*, 137–141.
Kepler, R. M., Humber, R. A., Bischoff, J. F., et al. (2014). Clarification of generic and species boundaries for *Metarhizium* and related fungi through multigene phylogenetics. *Mycologia, 106*, 811–829.
Kepler, R. M., Sung, G.-H., Ban, S., et al. (2012). New teleomorph combinations in the entomopathogenic genus *Metacordyceps*. *Mycologia, 104*, 182–197.
Kepler, R. M., Ban, S., Nakagiri, A., et al. (2013). The phylogenetic placement of hypocrealean insect pathogens in the genus *Polycephalomyces*: An application of one fungus one name. *Fungal Biology, 117*, 611–622.
Kerry, B. R. (1980). Biocontrol: Fungal parasites of female cyst nematodes. *Journal of Nematology, 12*, 253–259.
Kerry, B. R., Irving, F., & Hornsey, J. C. (1986). Variation between strains of the nematophagous fungus, *Verticillium chlamydosporium* Goddard. I. Factors Affecting Growth *in vitro*. *Nematologica, 32*, 461–473.
Kirk, P. M., & Ansell, A. E. (1992). *Authors of fungal names*. Wallingford: CAB International.
Kirk, P. M., Cannon, P. F., Minter, D. W., et al. (2008). *Dictionary of the fungi* (10th ed.). Wallingford: CAB International.
Larriba, R., Jaime, M. D. L. A., Carbonell-Caballero, J., et al. (2014). Sequencing and functional analysis of the genome of a nematode egg-parasitic fungus, *Pochonia chlamydosporia*. *Fungal Genetics and Biology, 65*, 69–80.
Lin, R., Liu, C., Shen, B., et al. (2015). Analysis of the complete mitochondrial genome of *Pochonia chlamydosporia* suggests a close relationship to the invertebrate-pathogenic fungi in Hypocreales. *BMC Microbiology, 15*, 5.
Lopez-Llorca, L. V., & Boag, B. (1993a). A contribution to the ecology of fungi associated with females and cysts of *Heterodera avenae* in eastern Scotland. *Nematologia Mediterranea, 21*, 187–187.
Lopez-Llorca, L. V., & Boag, B. (1993b). Biological properties of a new pigment produced by the nematophagous fungus *Verticillium suchlasporium*. *Nematologia Mediterranea, 21*, 143–149.
McNeill, J. (2012). Guidelines for requests for binding decisions on application of the Code. *Taxon, 61*, 477–478.
Medina-Canales, M. G., Rodríguez-Tovar, A. V., Manzanilla-López, R. H., et al. (2014). Identification and molecular characterization of new Mexican isolates of *Pochonia chlamydosporia* for the management of *Meloidogyne* spp. *Biocontrol Science and Technology, 24*, 1–21.
Medina-Canales, M. G., Rodríguez-Tovar, A. V., Manzanilla-López, R. H., et al. (2015). *Index Fungorum, 239*, 1.
Monfort, E., Lopez-Llorca, L. V., Jansson, H. B., et al. (2005). Colonization of seminal roots of wheat and barley by egg-parasitic nematophagous fungi and their effect on *Gaeumannomyces graminis* var. *tritici* and development of root-rot. *Soil Biology and Biochemistry, 37*, 1229–1235.
Morgan-Jones, G., Godoy, G., & Rodríguez-Kábana, R. (1981). *Verticillium chlamydosporium*, fungal parasite of *Meloidogyne arenaria* females. *Nematropica, 11*, 115–120.
Nonaka, K., Omura, S., Masuma, R., et al. (2013). Three new *Pochonia* taxa (Clavicipitaceae) from soils in Japan. *Mycologia, 105*, 1202–1218.
Petch, T. (1939). Notes on entomogenous fungi. *Transactions of the British Mycological Society, 23*, 127–148.
Quandt, C. A., Kepler, R. M., Gams, W., et al. (2014). Phylogenetic-based nomenclatural proposal for *Ophiocordycipitaceae* (*Hypocreales*) with new combinations in *Tolypocladium*. *IMA Fungus, 5*, 121–134.
Sasan, R. K., & Bidochka, M. J. (2012). The insect-pathogenic fungus *Metarhizium robertsii* (Clavicipitaceae) is also an endophyte that stimulates plant root development. *American Journal of Botany, 99*, 101–107.
Seifert, K. A., Morgan-Jones, G., Gams, W., et al. (2011). *The genera of Hyphomycetes*. Utrecht: CBS-KNAW Fungal Biodiversity Centre.

Sung, G.-H., Spatafora, J. W., Zare, R., et al. (2001). A revision of *Verticillium* sect. *Prostrata*. II. Phylogenetic analyses of SSU and LSU nuclear rDNA sequences from anamorphs and teleomorphs of the Clavicipitaceae. *Nova Hedwigia, 72*, 29–46.
Sung, G.-H., Hywel-Jones, N. L., Sung, J.-M., et al. (2007a). Phylogenetic classification of *Cordyceps* and the clavicipitaceous fungi. *Studies in Mycology, 57*, 5–59.
Sung, G.-H., Sung, J.-M., Hywel-Jones, N. L., et al. (2007b). A multi-gene phylogeny of Clavicipitaceae (Ascomycota, Fungi): Identification of localized incongruence using a combinational bootstrap approach. *Molecular Phylogenetics and Evolution, 44*, 1204–1223.
Tribe, H. T. (1977). Pathology of cyst nematodes. *Biological Reviews, 52*, 477–507.
Van der Weijden, W. J., Leewis, R., & Bol, P. (2007). *Biological globalisation*. Utrecht: KNNV Publishing.
Zare, R., & Gams, W. (2001). A revision of *Verticillium* section *Prostrata*. IV. The genera *Lecanicillium* and *Simplicillium* gen. nov. *Nova Hedwigia, 73*, 1–50.
Zare, R., & Gams, W. (2003a). *Pochonia chlamydosporia*. *IMI Descriptions of Fungi and Bacteria, 1569*, 1–4.
Zare, R., & Gams, W. (2003b). *Pochonia suchlasporia*. *IMI Descriptions of Fungi and Bacteria, 1570*, 1–2.
Zare, R., & Gams, W. (2004). A monograph of *Verticillium* section *Prostrata*. *Rostaniha, 3*, 1–188.
Zare, R., & Gams, W. (2007). *Pochonia globispora* sp. nov. *Nova Hedwigia, 84*, 421–428.
Zare, R., Gams, W., & Culham, A. (2000). A revision of *Verticillium* sect. *Prostrata*. I. Phylogenetic studies using ITS sequences. *Nova Hedwigia, 71*, 465–480.
Zare, R., Gams, W., & Evans, H. C. (2001). A revision of *Verticillium* section *Prostrata*. V. The genus *Pochonia*, with notes on *Rotiferophthora*. *Nova Hedwigia, 73*, 51–86.
Zare, R., Gams, W., Starink-Willemse, M., et al. (2007). *Gibellulopsis*, a suitable genus for *Verticillium nigrescens*, and *Muscillium*, a new genus for *V. theobromae*. *Nova Hedwigia, 85*, 463–489.
Zhang, N., Castlebury, L. A., Miller, A. N., et al. (2006). An overview of the systematics of the Sordariomycetes based on a four-gene phylogeny. *Mycologia, 98*, 1076–1087.

Part II
Bionomics

Chapter 3
Biology and Management of *Pochonia chlamydosporia* and Plant-Parasitic Nematodes

Rosa H. Manzanilla-López, Ivânia Esteves, and Jean Devonshire

Abstract The nematophagous fungus *Pochonia chlamydosporia* (*Clavicipitaceae*) is a facultative parasite of major plant-parasitic nematodes pests such as cyst (*Globodera* spp., *Heterodera* spp.), root-knot (*Meloidogyne* spp.), false root-knot (*Nacobbus* spp.) and reniform (*Rotylenchulus reniformis*) nematodes. The potential of *P. chlamydosporia* as a biological control agent and biopesticide has been the subject of numerous studies aimed at understanding the micro-ecological conditions that allow the fungus to thrive in the soil and rhizosphere environments. *Pochonia* survives in soil in the absence of plant and nematode hosts and can also behave as an endophyte. Research evidence points to a physiological 'switch' from the saprophytic to the parasitic stage that is triggered by nutrition. The basic biology of the fungus and sedentary endoparasitic plant nematodes is reviewed to provide insights into the fungus multitrophic behaviour, as well as its importance as a biocontrol agent within an integrated pest management approach.

R.H. Manzanilla-López (✉)
Centro de Desarrollo de Productos Bióticos (Visiting Professor),
Yautepec, Morelos, C.P. 62731, Mexico
e-mail: rosa.manzanilla@gmail.com

I. Esteves
CFE-Centre for Functional Ecology, Department of Life Sciences, Calçada Martim de Freitas, University of Coimbra, P3000-456 Coimbra, Portugal
e-mail: iesteves@cu.pt

J. Devonshire
Formerly at Rothamsted Research, 1A Grove Avenue, Harpenden AL5 1EQ, Herts, UK
e-mail: jean.devonshire@gmail.com

R.H. Manzanilla-López, L.V. Lopez-Llorca (eds.), *Perspectives in Sustainable Nematode Management Through* Pochonia chlamydosporia *Applications for Root and Rhizosphere Health*, Sustainability in Plant and Crop Protection,
DOI 10.1007/978-3-319-59224-4_3

3.1 Introduction

Large-scale estimate of crop loss due to plant-parasitic nematodes (PPN) probably exceed US$ 100 billion annually (Sasser and Freckman 1987). Losses may be even higher at present costs but PPN interactions with other disease-causing agents has made it difficult accurately to measure their impact on yield. Plant-parasitic nematodes are often regarded as capable of producing a single, recognizable disease, but may also facilitate the entry and establishment of other microorganisms, including plant pathogens. Often the nematode partner is affected, either to its benefit or disadvantage, leading to 'interrelationships' and 'interactions' with viruses, fungi and bacteria (Manzanilla-López and Hunt 2008). *Pochonia chlamydosporia* and its varieties are facultative parasites of eggs of potato cyst (*Globodera pallida, G. rostochiensis*), cereal cyst (*Heterodera avenae*), beet cyst (*H. schachtii*), soybean cyst (*H. glycines*), root-knot (*Meloidogyne* spp.), false root-knot (*Nacobbus aberrans sensu lato*) and reniform (*Rotylenchulus reniformis*) nematodes. Other potential hosts include helminth eggs and eggs of animal-parasitic nematodes in which the ovicidal activity of the fungus has been demonstrated experimentally (see Chap. 17).

Among other factors, the phasing out of nematicides and the increasing need for environment-friendly alternatives that are compatible with integrated pest management (IPM) and integrated biology soil management approaches (Stirling 2014) have encouraged the screening and testing of numerous isolates of the fungus in order to find those that can work as potential biocontrol agents against PPN (see Chap. 12). However, only a few strains of *P. chlamydosporia* are commercially available (see Chap. 15). In this chapter, we will review the biology of the fungus, selected ecological factors within the tritrophic (plant, fungus and nematode) and multitrophic interactions, and how they may affect the effectiveness of two of the most studied varieties of the fungus: *P. chlamydosporia* var. *chlamydosporia* and *P. chlamydosporia* var. *catenulata* as biological control agents (BCA) and their inclusion as part of a nematode IPM programme (see Chaps. 15 and 16). Relevant information on other species that formerly belonged to *Pochonia*, but which are now placed in *Metapochonia* (see Chap. 2), is also included.

3.2 Classification

Pochonia chlamydosporia has been isolated worldwide from soils as fungal spores and as a parasite of eggs of various invertebrates. The genus *Pochonia* Bat. & O.M. Fonseca 1965 is included in the order *Hypocreales* (*Clavicipitaceae*) and closely related to *Verticillium*. Molecular studies have allowed the recognition of at least five *Pochonia chlamydosporia* varieties: *P. chlamydosporia* var. *chlamydosporia, P. chlamydosporia.* var. *catenulata, P. chlamydosporia* var. *ellipsospora, P. chlamydosporia* var. *mexicana* and *P. chlamydosporia* var. *spinulospora.* Classification of former *Pochonia* spp. that are now placed within the genus *Metapochonia* and teleomorphs of the fungus (i.e., *Cordyceps chlamydosporia, Metacordyceps chlamydosporia*) have been revised by Evans and Kirk in Chap. 2.

3.3 Morphology

Pochonia colonies on agar are white or whitish to ochre yellow and cottony. It is a relatively slow growing fungus achieving up to 0.3 cm per day according to incubating temperature (Kerry 1991; Clyde 1993). Conidiophores are prostrate and little differentiated from vegetative hyphae or erect and differentiated, verticillate or solitary. Conidia adhere to form globose mucoid spore heads or chains (Gams 1988). Chlamydospores (dictyochlamydospores) are variable in shape, formed on short stalks, or less differentiated and without a stalk. Chlamydospores are commonly referred to as the resting structure of the fungus although the term 'dictyochlamydospores' is more accurate (see Chap. 2). Perithecia (subglobose or flask-like ostiolate ascomas; Kirk et al. 2008) are known to be formed only in the sexual phase of the fungus occurring in mollusc eggs (see Chap. 2).

Sporulation may be controlled by genetic switches triggered by environmental conditions – spores (conidia and chlamydospores) germinate in soil giving rise to propagules (Clyde 1993). The individual conidium has one nucleus (Fig. 3.1). However, during germination the conidium can give the impression of having two nuclei. The germinating conidia produce one germinating tube that can be followed by a second. The germination tube grows and gives rise to hyphae (Fig. 3.2). Hyphae of *P. chlamydosporia* var. *chlamydosporia* isolates form loops, anastomosing hyphae, hyphae webs, and other structures when grown in agar cultures (Fig. 3.3A–F). An individual hypha can twist and intertwine with other hyphae to form a sort of filamentous aggregation of hyphae that may also anastomose (Fig. 3.3B). Hyphal anastomosis (i.e., fusion between different branches of the same or different hyphae) is an important feature in intra-hyphal communication, translocation of water, nutrients and colony homeostasis (Hickey et al. 2002). *Pochonia chlamydosporia* var. *chlamydosporia*

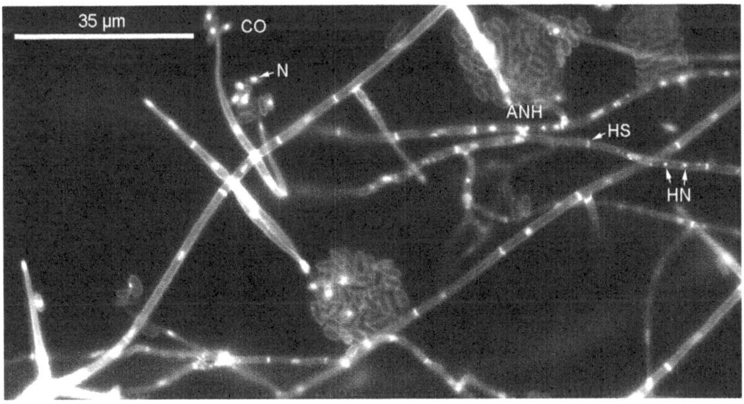

Fig. 3.1 *Pochonia chlamydosporia* var. *chlamydosporia* conidia, conidiophores and hyphae stained with fluorescence stain HOE 33258. conidia (CO), conidium nucleus (N; *arrow*), anastomosing hyphae (ANH), hypha septum (HS; *arrow*), multinucleate (HN; *arrow*) hypha cells (Copyright Rothamsted Research Ltd.)

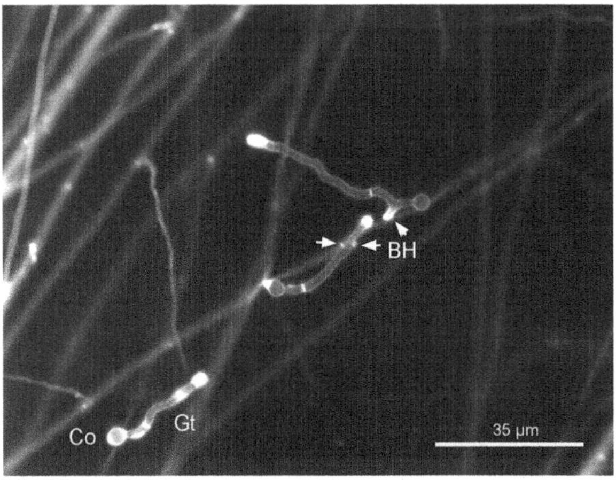

Fig. 3.2 *Pochonia chlamydosporia* conidium (Co), germination tube (Gt) and budding hyphae (BH) (*arrows*). Fluorescence stain HOE 33258 (Copyright Rothamsted Research Ltd.)

hyphal cells are multinucleate as shown when stained with HOE 33258 (Fig. 3.1). Appressoria are formed by hyphae in contact with the epidermal cell walls to penetrate the root epidermis cells and nematode eggs (Lopez-Llorca et al. 2002a, b).

The mycelial biomass growing in liquid cultures can aggregate to form pellets (Sykes 1994; Ward et al. 2012; see Chap. 11, Sect. 11.4). Mycelia pellets (Fig. 3.4A) are similar to 'microsclerotia' of the entomopathogenic fungus *Metarhizium anisopliae* (Jackson and Jaronski 2009). Some isolates of *P. chlamydosporia* also produce microsclerotia-like structures formed by aggregates of hyphae and conidia (Fig. 3.4B Manzanilla-López unpublished data) when cultured in a basal salts liquid media supplemented with 81 g/l of glucose and 9.0 g/l casamino acids, trace elements, and vitamins according to the method described by Jackson and Jaronski (2009); similar structures were occasionally found in old cultures of the fungus. The capacity to produce microsclerotia has been reported recently in the nematophagous fungus *Purpureocillium lilacinum* (Song et al. 2016). However, production of microsclerotia-like structures by different isolates of *P. chlamydosporia* var. *chlamydosporia* requires further study.

Crystals were observed when isolates Pc60 (UK), Pc190 (Kenya), Pc280 (Jersey), and Pc309 (Zimbabwe) of *P. chlamydosporia* from the Rothamsted culture collection were grown in glass slides covered with a thin layer of agar composed of Czapek Dox Broth and Daishin agar (see Chap. 11, Sect. 11.4) and kept in a humid chamber for 3–4 days at 26 °C (Fig. 3.5A) (Manzanilla-López unpublished information). Crystals have also been observed on the surface of colonized eggs (Fig. 3.5B). Energy dispersive spectrometry (EDS) of crystals of isolate Pc190 indicates the presence of calcium in the crystals (Fig. 3.5C, D). The presence of calcium, oxygen and carbon indicate calcium oxalate (CaC_2O_4) composition. The octahedral crystals (dihydrate form) and rhomboid 'picket fence' crystals (monohydrate form) of calcium oxalate are observed in light and scanning electron microscopy of Pc190 (Fig. 3.5A). A flat surface exhibited in the monohydrate form makes it suitable for analysis in an SEM using EDS. Calcium oxalate crystals have been identified using SME and EDS in other fungi such as *Lecanicillium fungicola* (see Santana Nunes et al. 2017).

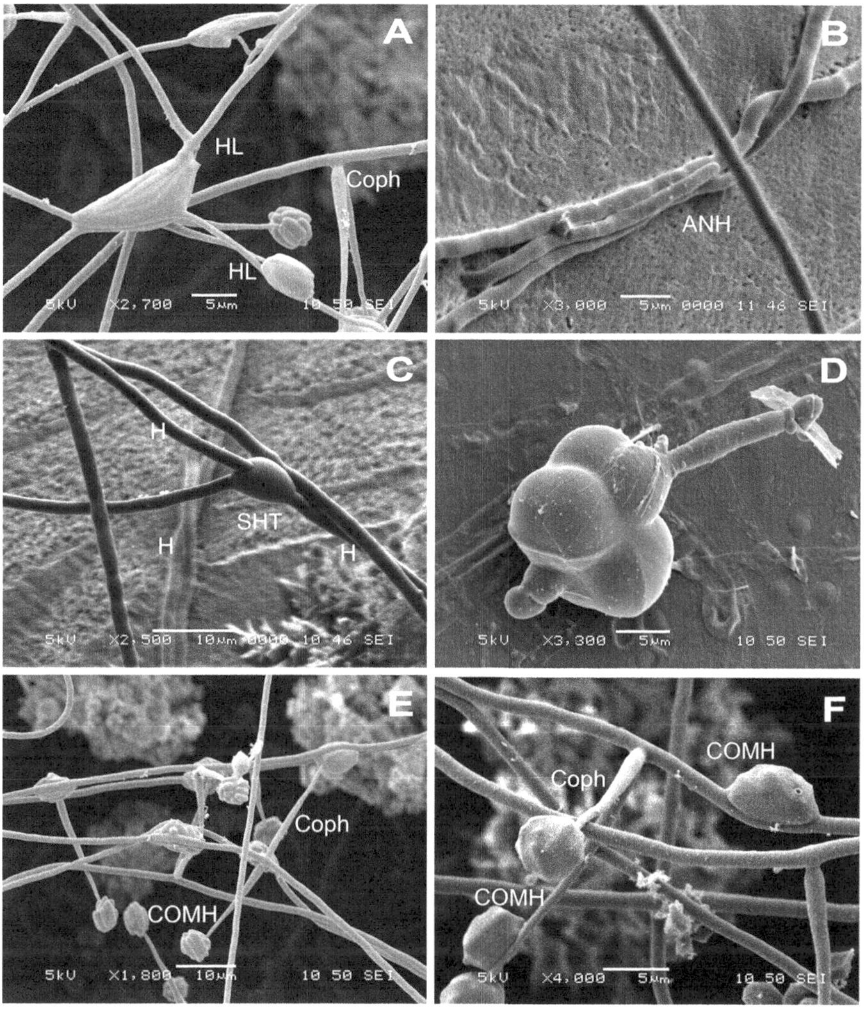

Fig. 3.3 *Pochonia chlamydosporia* var. *chlamydosporia* hyphae and mucoid heads of conidia (Cryo SEM micrographs). (**A**) Hyphal loops and conidiophores (phialides); (**B**) Hyphae intertwined and anastomosed; (**C**) Two hyphae branching out from the swollen tip of a hypha; (**D**) Chlamydospore; (**E**) Mucoid spore heads supported by conidiophores. The mucoid spore heads can be pushed between hyphae (*right corner*) or become stuck on the surface of other hyphae; (**F**) Note the hole (*scar*) produced on a mucoid spore head as a result of the detachment of the conidiophore tip. Anastomosing hyphae (ANH); conidiophores (Coph); conidia (CO); conidia mucoid head (COMH); hyphae (H); hyphal loops (HL); swollen hyphal tip (SHT) (Copyright Rothamsted Research Ltd.)

The absence of crystals has been considered one of the features that can distinguish *Pochonia* from *Lecanicillium* species (Zare and Gams 2007). Crystals have been reported for *Pochonia microbactrospora* (= *Metapochonia microbactrospora*)

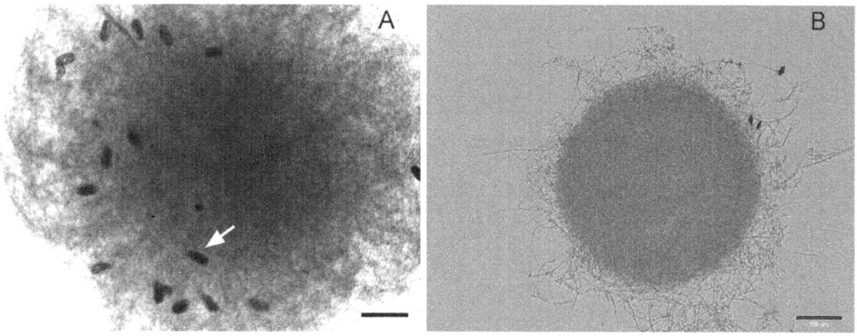

Fig. 3.4 *Pochonia chlamydosporia* var. *chlamydosporia* mycelia pellets and 'microsclerotia-like' structures produced in liquid culture. (**A**) Mycelia pellet (spherical mycelia aggregate) with *Meloidogyne incognita* eggs (arrow). Eggs were added to the fungus culture media and incubated at 28 °C for 1–2 days (Ward et al. 2012); (**B**) 'Micro-sclerotia-like' structures produced in liquid media (Jackson and Jaronski 2009) (Copyright Rothamsted Research Ltd.)

but not for *P. chlamydosporia* (Zare et al. 2001). The presence/absence of crystals in *P. chlamydosporia* varieties also deserves further investigation.

3.3.1 Virulence of Fungal Propagules to Nematodes

According to the *Dictionary of the Fungi* (Kirk et al. 2008) a pathogen is "a parasite able to cause disease in a particular host or range of hosts" and virulence refers to "the degree or measure of pathogenicity of the parasite". The fungus can infect nematodes with conidia and hyphae but the pathogenicity to nematodes of propagules of *P. chlamydosporia*, produced under different nutritional conditions, has not been fully assessed (Mo et al. 2005; see also Chaps. 11 and 12). The fungus does not produce chlamydospores in liquid culture but conidia on phialides. Although some isolates do not produce chlamydospores in solid media, this does not indicate that they are never produced. It is possible that isolates can produce chlamydospores given the appropriate environmental stimulus; for example, *P. chlamydosporia* isolate I49 produced chlamydospores ('spores') *in vitro* only after 7 months of incubation at the optimum growth temperature (Clyde 1993). Chlamydospores have sufficient food reserves to enable the fungus to establish without the addition of other energy sources (Kerry et al. 1993; Mauchline et al. 2002). They are one of the commonest inocula sources for adding the fungus to soil to control PPN.

3.4 Trophic Interactions

The biology of *P. chlamydosporia* is more complex than previously thought, the fungus being potentially able to display multitrophic (multimodal) and symbiotic abilities according to available substrata, hosts, and environmental conditions (see

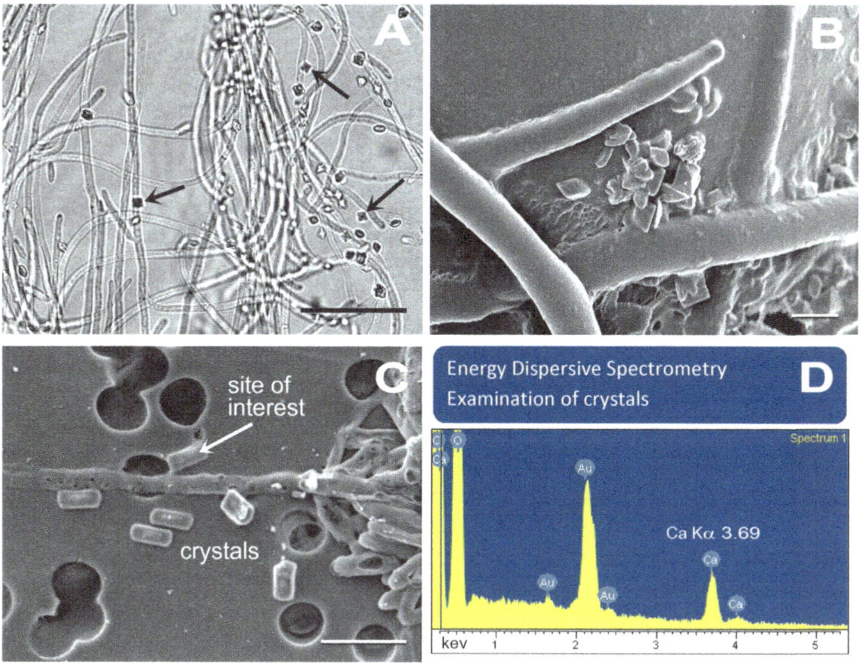

Fig. 3.5 *Pochonia chlamydosporia* var. *chlamydosporia* crystals. (**A**) Light microscopy indicating octahedral crystals (*arrows*); (**B**) SEM Cluster of octahedral crystals adjacent to *P. chlamydosporia* hyphae on the surface of a colonized *Meloidogyne incognita* egg. The fungus was cultured on a nuclepore filter (Millipore, 10 mm diam., 3 µm pore) coated with one drop of Czapek Dox broth agar medium; (**C**) SEM with rhomboid crystals adjacent to hypha of *Pochonia chlamydosporia* cultured on a nuclepore filter and indicating the Energy Dispersive Spectrometry (EDS) site of examination. Spectrum obtained from site of interest indicated in SME micrograph of crystals; (**D**) Energy dispersive spectrometry of crystals of *P. chlamydosporia* var. *chlamydosporia* isolate Pc190 of the fungus. Calcium (Ca) is identified in the crystals. The gold (Au) peaks arise from the coating the sample received during preparation. Scale bar: A 35 µm; B 2 µm; C 5 µm (Copyright Rothamsted Research Ltd.)

Chap. 2). Fungi adapted to different mutualism and parasitism interactions, such as *Pochonia*, are likely to have developed different strategies to overcome their host's defence (Stadler et al. 2003; see Chaps. 5 and 7). The parasitic phase of *P. chlamydosporia* has not been so extensively studied in invertebrate hosts such as molluscs and insects, and there are currently very few data available for this aspect (see Chap. 2). By contrast the tritrophic interaction between the fungus, the plant and the nematode is amongst the most studied. *Pochonia* can survive in its saprophytic stage in the absence of both the plant and nematode hosts (see Chap. 4). As a soil saprotroph, the fungus is able to compete effectively with other soil microflora (Kerry 1988). Kerry et al. (1993) have noted that, once established in soil, factors that affect rhizosphere colonization are likely to determine the efficacy of the fungus. Screening of isolates as potential BCA has also shown that isolates can behave as good or poor saprophytes, rhizosphere colonizers, and parasites (see Chap. 12). Some isolates of

P. chlamydosporia grow rapidly while others grow poorly in different soils following application, as shown by root-knot nematode biotype Pc392 which had a higher growth rate than potato cyst nematode biotype Pc280 (Manzanilla-López et al. 2011a). The phase of the fungus as a rhizosphere colonizer, and/or as an endophyte, occurs through interaction with plant roots (see Chap. 5). *Pochonia* can colonize the epidermis of some species of *Gramineae* and *Solanaceae*, but they do not invade the vascular cylinder in comparison to plant pathogenic fungi. Research has shown that the fungus also displays an endophytic behaviour (Maciá-Vicente et al. 2009a). This endophytic ability needs to be added to the BCA screening process of isolates.

3.4.1 Rhizosphere Colonization

Hyphal growth can be a limited means of spread for the fungus but it may be more important in the rhizosphere than in surrounding soil (Clyde 1993). Early establishment on host crops may give *P. chlamydosporia* an advantage over competitive rhizosphere microflora; it may also maximise infective potential by being present at high densities when nematode females first appear (Clyde 1993). Furthermore, some *Pochonia* spp. isolates have shown an endophytic behaviour that may be beneficial to the host plant's defence against other soil borne pathogens (Bordallo et al. 2002; Maciá-Vicente et al. 2009a, b; see also Chap. 5).

Fungal root colonization produces intra- and inter-cellular hyphal nets. Root surface colonization is higher than internal colonization (Fig. 3.6A, B), the latter reaching as far as the inner cortical cells of the root tissues. Root colonization also shows differences between mono- and dicotyledonous plants. In barley roots (*Hordeum vulgare*, monocotyledon) intercellular hyphae form 'steps' along the cell walls; hyphal penetration occurs through the longitudinal cell walls of the root (Lopez-Llorca et al. 2002a, b). Hyphal loops, which have being associated with nutrients absorption, also occur within root cells in both epidermal and cortical cells. Nuclei of epidermal cells have been sometimes found to be colonised by the fungus (Lopez-Llorca et al. 2002b). Chlamydospores occur in the rhizoplane (Fig. 3.6A) but they are sometimes produced within epidermal cells. On tomato roots (*Solanum lycopersicum*, dicotyledon), conidia and chlamydospores can be produced 1 month after inoculation (Bordallo et al. 2002). Lignitubers ensheath hyphae in tomato roots but not in barley, a feature that may partly explain the different extent of root colonization and host defence mechanisms acting in both plant hosts, including stress (Bordallo et al. 2002). The histochemistry of colonized roots has shown deposits of lignin and phenolics in papillae and cell walls related to host defence reactions (Bordallo et al. 2002). Guaiacol peroxidase is associated with many important biosynthetic processes and defence against abiotic and biotic stresses (Sharma et al. 2012). Guaiacol peroxidase (GPO) activity was significantly lower in barley seedlings inoculated with *Gaeumannomyces graminis* combined with *P. chlamydosporia* or *Lecanicillium lecanii* isolate 4629 than in seedlings inoculated with *G. graminis* alone. Hence, reduction in peroxidase production was considered stress-decrease related (Monfort et al. 2005).

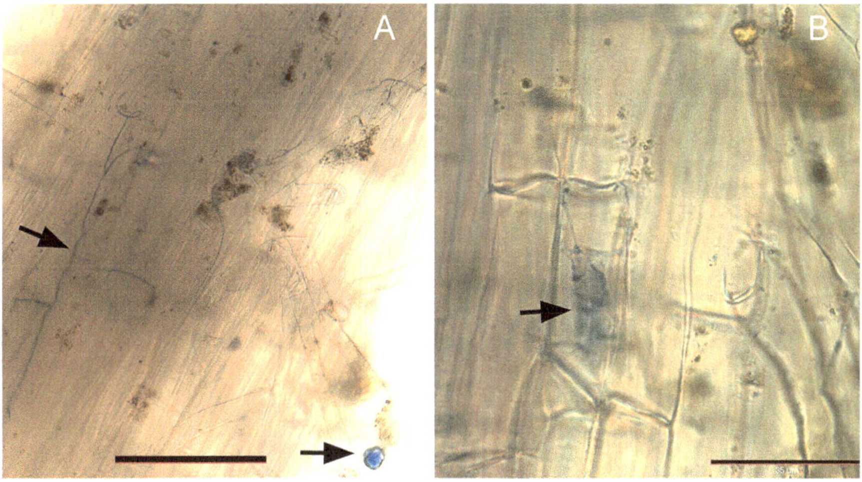

Fig. 3.6 Rhizosphere colonization of wheat by *Pochonia chlamydosporia* var. *chlamydosporia*. (**A**) Hypha (*left arrow*) and chlamydospore (*right arrow*); (**B**) Hyphae inside an epidermal root cell (*arrow*). Scale bar: (**A**) 10 μm; (**B**) 35 μm (After Manzanilla-López et al. 2011a. Copyright Rothamsted Research Ltd.)

Monitoring of root colonization by Green Fluorescent Protein GFP-expressing transformants has provided further evidence of a plant response against endophyte colonization in barley roots (Maciá-Vicente et al. 2009a). Laser scanning confocal microscopy showed that 3 days after inoculation (dai), transformant Pc123gfp formed discrete hyphae aggregates close to root hairs on the root surface. Penetration of the epidermis cells was through appressorium-like structures, which elicited papilla formation to prevent fungal penetration. Hyphal death resulted from the host response but living hyphae developed in other parts of the root to escape the defence reaction. It has been suggested that long-term endophytic root colonization depends on reaching an equilibrium between the plant defence mechanisms and the fungal rhizosphere competence of *Pochonia* (See Chap. 4). Root endophytic performance by biocontrol fungi is an important factor for control of root pathogens, through phenomena such as escape from competition with soil microbiota, enhancement of plant growth and modulation of plant defences (see Chap. 14).

3.4.2 Crop Growth Enhancement

Increased growth occurred with a single application of a native isolate of *P. chlamydosporia* (5×10^3 chlamydospore/g soil) to reduce the numbers of *R. reniformis* on cotton roots and soil under glasshouse conditions (Wang et al. 2005). Longer roots and shoots are produced in potato plants inoculated with *P. chlamydosporia* var. *chlamydosporia* in the absence of nematodes. However, a negative effect occurred on potato plant growth when the fungus and *M. incognita* were inoculated simultaneously on plants (Manzanilla-López et al. 2011a). Other species, such as *Pochonia* [*Metapochonia*] *rubescens*, apparently induce changes in the host plant that increase root length of barley seedlings (Monfort et al. 2005; Lopez-Llorca et al. 2008).

Although the parasitic effect of the fungus on eggs of sedentary endoparasitic plant nematodes is better known, there is no information about how the control of migratory plant-parasitic nematodes is affected by the fungus (see Chaps. 4 and 15). It has been proposed that, since the fungus is a root endophyte, it can affect migratory endo-parasitic nematodes such as *Pratylenchus* spp. and *Radopholus similis* by producing a toxin or by inducing resistance in plants such as banana (Freitas et al. 2009). Combination of the fungus with treatments of jasmonic acid, a plant signalling molecule that mediates plant defence responses (Cheng-Gang and Chuan-Chao 2012), has shown promising results in controlling root-knot nematodes (see Chap. 14).

3.5 Plant-Parasitic Nematodes and *Pochonia chlamydosporia* Parasitism

Different evolutionary strategies within the phylum Nematoda encompass a wide range of trophic habits, including plant parasitism. Endoparasites (migratory or sedentary) conduct most of their vital processes inside the roots (except during the dispersive stages), their mobility and feeding habits resulting in death of cells and disruption of root function. Migration through roots characterizes the first level of obligate endoparasitism (*Pratylenchus*, *Radopholus*), the next stage being sessile endoparasitism. The development of females that induce trophic nurse cells and galls allowed the sedentary habit to develop free of the necrotic reaction caused by the migratory habit, the second-stage juveniles (J2) and males being the only mobile stages. Such sedentary females may be transformed into a tough cyst containing J2 within eggs (*Heterodera* spp., *Globodera* spp.), or the females may be non-cyst-forming (*Meloidogyne* spp.) but protected within a gall formed from proliferation of the cortical root cells. Nurse cells include uninucleate giant cells (*Meloidodera floridensis*, *Rotylenchulus macrosoma*); multinucleate giant cells (*Meloidogyne*); or multinucleate syncytia (*Globodera* spp., *Heterodera* spp., *R. reniformis*, *Nacobbus* spp.) (Baldwin et al. 2004; Jones et al. 2013).

The rectal glands of *Meloidogyne* females produce a gelatinous matrix where eggs are laid down forming an egg mass. *Nacobbus aberrans s.l.* also lays eggs in a gelatinous matrix but no glands have been identified (Manzanilla-López et al. 1998). *Rotylenchus reniformis* gelatinous matrix is secreted by uterine glands (Jones et al. 2013) and by the excretory system in *Tylenchulus semipenetrans* (another host for *Pochonia* sp.). The gelatinous matrix contains glycoproteins (Sharon and Spiegel 1993; Sharon et al. 1993; Manzanilla-López et al. 1998; Agudelo et al. 2004). The egg mass and components of the egg shell are important for the survival of the developing embryo and the fully formed juvenile stages within the egg. The egg mass provides 'a first line of defence' against hostile soil elements, predators, parasites, low oxygen levels and desiccation (Evans and Perry 2009).

The body wall of *Meloidogyne* spp., which protects the female from the outside environment, has three major layers: the cuticle, the hypodermis and somatic muscles. The non-cellular, elastic, cuticle is secreted by the hypodermis and covers the entire body and all the openings (Eisenback and Hunt 2009). The external cuticular or epicuticle is covered in many species by a surface coat material composed of protein, carbohydrates and lipids (Spiegel and McClure 1995). The ability to shed and renew the surface coat continuously may help the nematode avoid recognition and the induction of resistance in the host plant (Curtis et al. 2009; see also Chap. 14). The parasitic relation of *Pochonia* spp. with nematode hosts has a nutritional basis (see Chap. 4). Both fungus and nematode share the habitat provided by the host plant and different nutrients can be obtained by the fungus, not only from the body and eggs of the nematode, but from other substrates such as the gelatinous matrix, nematode secretions and metabolites produced in root exudation as a result of the plant-nematode interaction. Differences in native microhabitats and niches occupied by fungal isolates, e.g., nematode females, nematode eggs, or soil, can be related to production of different secondary metabolites and inhibitory activities displayed by isolates (see Chap. 7).

3.5.1 Cyst Nematode Infection

The fungus may get into the cysts (female) through the natural body openings and/or cuticle. Infection of *H. avenae* immature females and cysts is followed by internal body colonization and subsequent destruction of the body contents leaving only resting spores (Lopez-Llorca and Duncan 1991). Cyst colonization of *H. schachtii* has shown that *V. chlamydosporium* (= *P. chlamydosporia*) infection may be age-related, primarily occurring in young females (<8 days post-emergence). Infected females tan prematurely, their body wall shrivels and the whole female may even disintegrate. Fungal colonization in 60% of white females was usually confined to the vulva (Clyde 1993). The mucilaginous drop at the vulva, produced by cells of the uterine wall (MacKintosh 1960), is often colonised by the fungus and eggs extruded into this mucilage are vulnerable to fungal parasitism (Fig. 3.7). The possibility that the fungus may have also direct access via the nematode mouth which is embedded in the host root tissue deserves further investigation since the endophytic behaviour of the fungus has been demonstrated and the fungus has been observed, for example, within *M. incognita* giant cells (see Chap. 4).

The root cap mucigel surrounding the root tip may also be exploited by the fungus as a nutrient source. Clyde (1993) found that rhizosphere-competent isolates were most abundant in the region of the root tip whereas hyphae were sparse or absent on older roots; isolated propagules along the root also could infect females as they emerged (Clyde 1993). However, in the case of *P. chlamydosporia* stable transformant Pc123gfp, root colonization was low when close to the root cap but

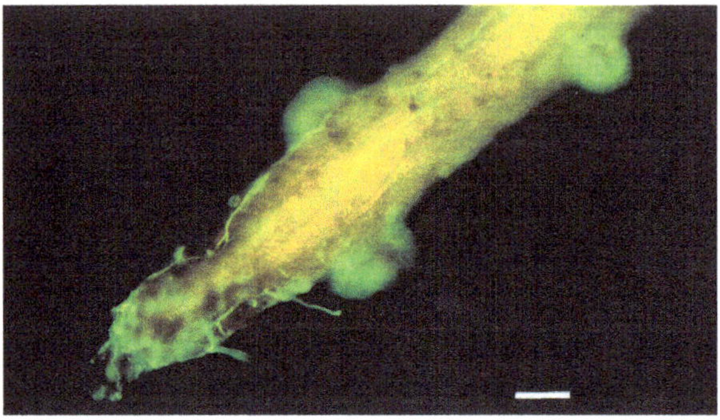

Fig. 3.7 Root cap colonization by *Pochonia chlamydosporia* var. *chlamydosporia* isolate I33 as seen with conventional epifluorescence. Scale bar 70 μm (Image from Clyde 1993; courtesy of H. Atkinson, Leeds University, UK)

increased away from the root cap forming a patchy hyphal network (Escudero and Lopez-Llorca 2012).

3.5.2 Meloidogyne *spp. Fungal Infection and Plant Host Interaction*

Plants and their rhizodeposits are an important source of C and N for soil microbiota and to maintain some entomopathogenic and nematophagous fungi in the saprophytic stage within soil (Bruck 2010). Root exudates include a variety of low molecular weight compounds that may influence nematode behaviour at the rhizoplane (Curtis et al. 2009). Most root exudation from young plants and healthy plants come from the root cap region. Second-stage juveniles of *Meloidogyne* spp. are attracted to growing root tips and invade the roots through the elongation zone behind the root tip. Root hairs and lateral roots also produce exudates as do areas where tissue damage, perhaps caused by J2 penetration, cause leakage of cell contents (Curtis et al. 2009).

3.5.3 Root Colonization and Infection of Nematode Females and Eggs

Differences in cyst and root-knot nematodes biology and their adaptations to plant parasitism may affect fungal infection (Morgan-Jones et al. 1983; Lopez-Llorca and Duncan 1991). Although fecundity can be reduced by fungal infection, differences could be due not only to a fungal effect, but to other factors such as host plant status of the target nematode, rate of nematode development on the host, environment temperature, and position on roots of females or developing juveniles (Clyde 1993). As a parasite of eggs, colonization of the root surface by the fungus has been linked to egg production and also to changes in root exudation due to nematode infection of the roots (Bourne and Kerry 1999).

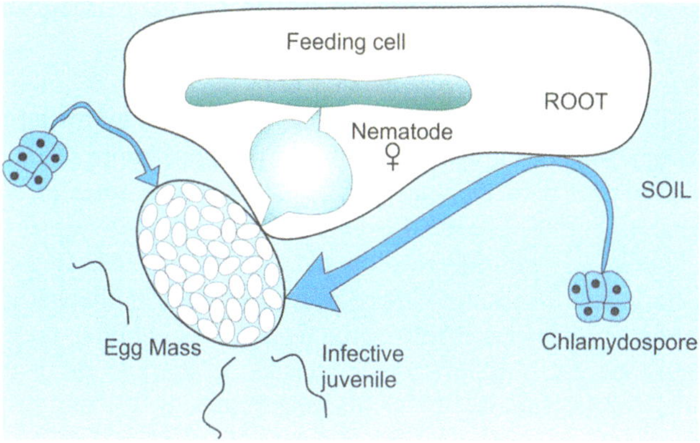

Fig. 3.8 Diagram of *Pochonia chlamydosporia* fungal colonization of root-knot female and egg mass laid outside the roots. Chlamydospores are the resting stage of the fungus in soil (Copyright Rothamsted Research Ltd.)

Most studies of egg infection by *Pochonia* spp. have been done on cyst nematodes (Lopez-Llorca and Claugher, 1990; Lopez-Llorca and Robertson, 1992) and only a few on infection of root-knot species (RKN) species (Morgan-Jones et al. 1983; Manzanilla-López et al. 2014) or *N. aberrans s.l.* (Flores-Camacho et al. 2007). Little is known of the infection process in *R. reniformis* (Wang et al. 2005). Differences in biology and embryogenesis time between nematode hosts can affect egg parasitism. Root-knot nematode eggs are laid inside a gelatinous matrix that is pushed outside the root galls leaving the egg-mass exposed in the rhizosphere (Fig. 3.8). Eggs within the gelatinous matrix of *Meloidogyne* spp. and *Nacobbus* spp. can be found in different stages of embryogenesis, e.g., from segmentation to fully developed J2. The embryogenesis period varies according to nematode species but is shorter at higher temperatures. In contrast to RKN, cyst nematode eggs are not directly exposed to the rhizosphere since they remain inside the encysted female until they develop into J2. Therefore, it is considered that the time that the fungus has to infect and kill RKN eggs is shorter than that for cyst nematode eggs (Kerry 1997). However, other anatomical and physiological factors affect the infection process, e.g., egg shell thickness, nutrient availability for the fungus, and enzymatic activity within fungal isolates (see Chap. 4). For example, conidia *in vitro* infection tests of eggs of root-knot and cyst nematodes in a yeast extract liquid medium has shown differences in percentage of egg parasitism within 48 h depending on the biotype and isolate of cyst or root-knot nematode (Esteves 2007; Ward et al. 2012; Lumbuano-Nyoni et al. 2015).

Newly formed cysts of *H. avenae* exposed to fungal infection contain a mixture of healthy eggs, eggs containing hyphae, hatched J2, and/or only hyphae (Lopez-Llorca and Duncan 1991). A similar variation in egg and fungal colonization occurs in RKN eggs (Fig. 3.9A–D). It is generally accepted that eggs in the earlier stages of embryogenesis are more susceptible to infection than those containing J2 (Lopez-Llorca and Duncan 1991; Kerry 2000) but all stages of egg development (Fig. 3.9E) can be attacked (Kerry and Crump 1977; Lopez-Llorca and Duncan 1988, 1991) as also shown in RKN eggs containing viable J2, which were susceptible to parasitism (Morgan-Jones et al. 1983). Egg location within the egg mass (more external or internal) and gelatinous matrix (Fig. 3.9F), egg mass age, and fungal biotype may all affect susceptibility of the egg to infection (Manzanilla-López et al. 2014).

Hyphae in contact with the egg (Fig. 3.9B) can form a net of anastomosing hyphae around the egg (Fig. 3.9C). Enzymes secreted *in situ* by the fungus weaken the egg shell before vegetative hyphae can penetrate (Fig. 3.9D). Egg shell penetration occurs via an appressorium followed by a penetration hypha that perforates the egg shell and forms an infection bulb (see Chap. 4). Penetration of the eggs can also occur through the germ tube of conidia attached to the outside of the egg shell (Lopez-Llorca and Robertson 1992). Infection bulbs develop into trophic hyphae that destroy the egg contents and degrade the egg shell (Fig. 3.10) from the inside (Lopez-Llorca and Robertson 1992). The change, or 'switch', from the saprophytic to the parasitic phase of the fungus may be related to nutrients. Adhesion, penetration and colonization processes involving serine proteases VCP1, P32, SCP1, chitinase and other enzymes are reviewed in Chap. 4, and proteomics in Chap. 6.

3.6 Host Preference

Fungal isolates are often referred to as biotypes according to the nematode hosts from which they were originally isolated: root-knot nematode (RKN biotype), potato cyst nematode (PCN biotype), cereal cyst (CN biotype). Biotypes can show a host preference, a feature that is related to differences in egg parasitism associated with specific amino acid polymorphisms in the fungal serine protease VCP1, at least for root-knot and cyst nematodes (Morton et al. 2003a, b). Serine proteases enzymes are putative determinants of pathogenicity or virulence (Butt et al. 1998). VCP1 is repressed by glucose and stimulated by ammonium chloride and there is evidence for an increase in VCP1 mRNA when glucose and eggs are present (Ward et al. 2012). Gene expression profiles in *P. chlamydosporia* have shown significant transcriptional reprogramming between the 'switch' from saprophytic to parasitic phase in two isolates of RKN and PCN biotypes in which some genes were induced/expressed or repressed, thereby suggesting a concerted regulation, especially when activated after exposure to eggs (Rosso et al. 2011; see also Chap. 5).

3.6.1 Virulence and Saprophytic Fitness Cost

Increased dependence of the fungus on nematodes as a source of nutrition has been associated with a decrease in saprotrophic competitiveness (Siddiqui et al. 2009). A negative correlation was shown between population densities in soil of different *P. chlamydosporia* biotypes and their parasitism of *M. hapla* and *G. pallida* eggs in

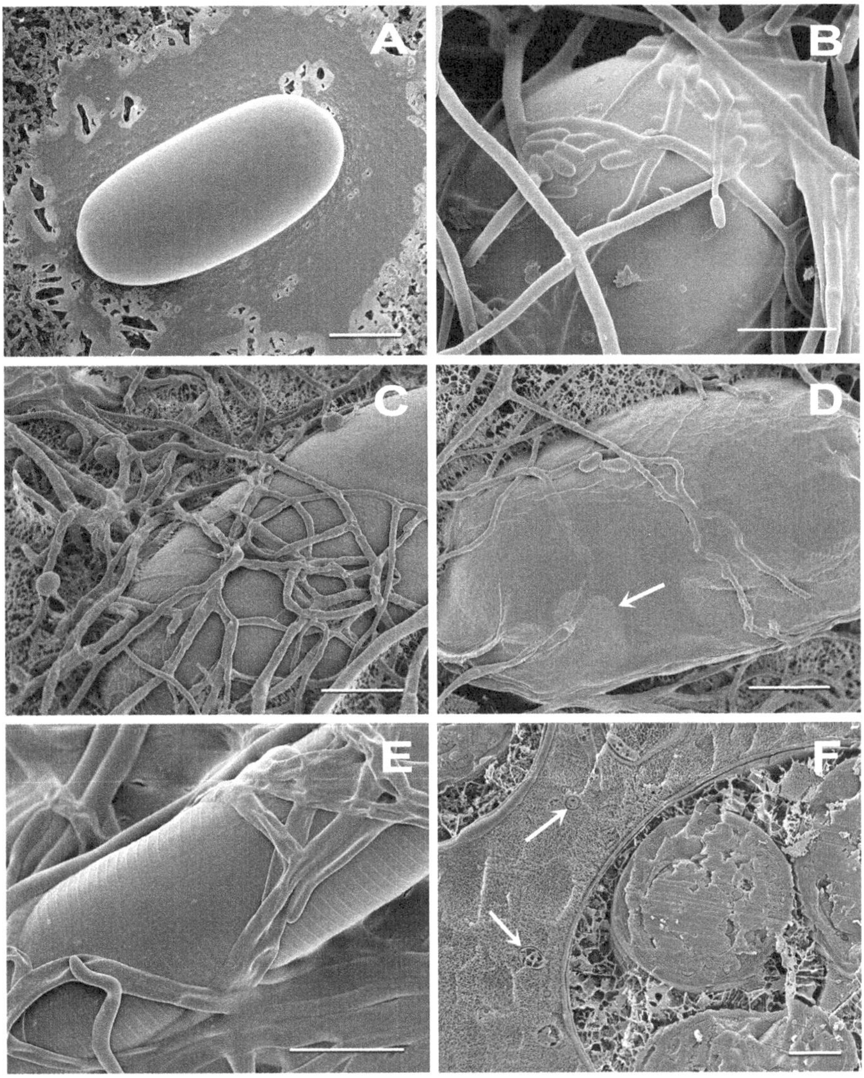

Fig. 3.9 *Pochonia chlamydosporia* Cryo-SEM. (**A**) *Meloidogyne incognita* egg displaying a clean, smooth surface before inoculation of the fungus; (**B**) Egg surface showing mycelium and conidia of the fungus after inoculation; (**C**) Net of hyphae on the egg surface; (**D**) Micrograph showing VCP1 erosion of the egg surface (*arrow*); (**E**) Hyphae on the J2 cuticle surface; (**F**) Egg mass gelatinous matrix and eggs colonized by the fungus (*arrows*). Scale bars: (**A**) 20 μm; (**B**–**E**) 10 μm; (**F**) 5 μm (J. Devonshire; **B** and **F** after Manzanilla-López et al. 2014. Copyright Rothamsted Research Ltd.)

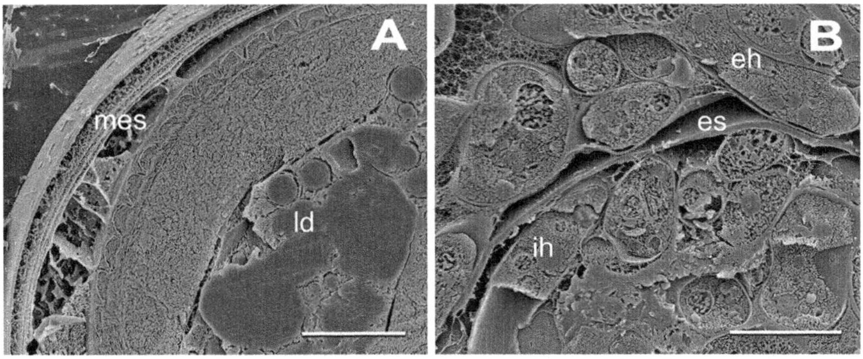

Fig. 3.10 Cryoplaned SEM micrographs of *Pochonia chlamydosporia* showing colonization of *Meloidogyne incognita* eggs. (**A**) Before fungus inoculation; (**B**) Egg colonized by the fungus showing no recognizable internal organs of the nematode embryo. Egg shell (es); external hyphae (eh); internal hyphae (ih); lipid deposits of the nematode embryo (ld); multi-layered egg shell (mes). Scale bars: 4 µm (After Manzanilla-López et al. 2014. Copyright Rothamsted Research Ltd.)

tests using compatible biotypes. The cyst nematode biotype was the most virulent against both nematode eggs, but was a less effective soil and rhizosphere colonizer in comparison to other isolates tested, thereby suggesting that virulence may have a fitness cost (Siddiqui et al. 2009).

3.6.2 Nematicidal and Nematostatic Effect of the Fungus on Juvenile Stages

Many fungi produce nematicidal or nematostatic compounds that can help regulate nematode densities in soil (Shamim et al. 2012). It is possible that long association and coevolution has resulted in native fungal strains better adapted to their host and that such strains may occur in the geographic area of origin and domestication of the crop and/or nematode (Meyer et al. 2004). To obtain a sample of these native fungal strains, fungi were isolated from soybean cyst nematode (SCN) eggs collected in Chinese agricultural fields. Filtrates of culture broths of isolates were screened for effects on SCN and the RKN, *M. incognita*. Compounds effects were inhibitory or stimulatory to egg hatch and affected J2 mobility. *Pochonia chlamydosporia* isolate L247 filtrates strongly stimulated SCN hatch (162%) but inhibited *M. incognita* hatch (Meyer et al. 2004). The nematicidal action of the fungal filtrate of *P. chlamydosporia* has been reported to produce ca 13% J2 mortality in *M. javanica* (Mukhtar and Pervaz 2003) and *H. schachtii* (Ayatollahy et al. 2008). In another study, *P. chlamydosporia* culture filtrate caused 60% mortality in *M. javanica* J2 after 48 h exposure to 1:10 filtrate dilution, or 73% mortality if undiluted (Shamim et al. 2012). The red pigment produced by *Pochonia* [*Metapochonia*] *suchlasporia*

(= *Verticillium suchlasporium*) can reduce the mobility of the J2 of *G. rostochiensis* (Lopez-Llorca and Boag 1993). The effect of filtrates and pigments may be partly due to the compounds and secondary metabolites found in the filtrates (see Chap. 7). There are few reports on the effect of the fungus on the juvenile stages of animal-parasitic nematodes (see Chap. 17).

3.7 Habitat and Niche

Pochonia chlamydosporia has multitrophic and symbiotic abilities: soil saprotroph, rhizosphere colonizer, root endophyte, facultative parasite of invertebrates, and hyperparasite of other fungi. When released in soil, *P. chlamydosporia* is exposed to fluctuating abiotic factors which can have a great impact on establishment in the field and compromise its efficacy in controlling nematodes. Some of these factors include soil pH, temperature, soil type, fungistatic activity (reviewed in Chap. 4), and water potential (Chap. 11). In general, the mean numbers of colony forming units (cfu) that proliferate from chlamydospores are greater in organic than in mineral soils (Kerry et al. 1993). Establishment and reproduction of the fungus following an adequate initial level of fungal inoculum applied to soil can build up levels found in naturally suppressive soils (i.e. 2–8 × 10^3 cfu/g soil) under proper management conditions (Kerry et al. 1993), and eventually produce long-term efficacy in the control of nematodes. However, density dependent mechanisms may operate (Clyde 1993; Mauchline et al. 2002). The combination of the nematode interaction with the host plant (i.e., resistant, susceptible, tolerant) will determine nematode reproduction, as well as the extent of damage and yield loss caused to the host. Plants that are good hosts for *Pochonia chlamydosporia* can be good or bad (poor) hosts for the nematode. Crops and plant species that can support more than 200 colony forming units (cfu)/cm^2 of root are considered good hosts, among them: beans (*Phaseolus vulgaris*), cabbage (*Brassica oleracea* var. *capitata*), *Crotalaria* spp., kale (*Brassica oleracea* var. *sabellica*), pigeonpea (*Cajanus cajan*), potato (*Solanum tuberosum*), pumpkin (*Cucurbita pepo*), and tomato (De Leij et al. 1992a; Bourne et al. 1996). In general, fungal control can be higher in plants that are poor hosts for the nematode but which also support extensive rhizosphere fungal growth. A relatively poor host for the nematode produces small galls that results in most egg masses being exposed in the rhizosphere, thus reducing the potential 'escape' of fungal infection by those eggs retained, and hence protected, inside the large galls formed on good hosts (Atkins et al. 2003a).

3.7.1 *Compatibility and Competition Between Isolates*

Different *P. chlamydosporia* strains have been fingerprinted and identified using ERIC PCR (Enterobacterial Repetitive Intergenic Consensus) showing that under experimental conditions they can occur together in the same niche and may compete with each other within the rhizosphere, although their capabilities as root colonizers or egg parasites can vary when they are applied singly or in combination (Mauchline et al. 2004; Manzanilla-López et al. 2009). However, the question remains as to whether diversity of isolates can be maintained in shared habitats (i.e., soil, rhizosphere, eggs) with other native/non-native populations when non-native populations of the fungus are introduced to the soil. Preliminary studies on compatibility or anastomosis of selected isolates of *P. chlamydosporia* biotypes (from *Meloidogyne* sp., *Heterodera* sp. and *Globodera* sp.) to assess vegetative or heterokaryon compatibility suggested that biotypes isolated from different hosts did not anastomose but further studies are required to confirm incompatibility (Manzanilla-López et al. 2011b).

3.8 Multitrophic Interactions

The extent of the tritrophic interaction between plant, nematode and fungus and how it influences other rhizosphere microorganisms and soil organisms through multitrophic interactions is largely unknown. The fungus does not affect the abundance of other nematodes and microbes on different plant species (Tahseen et al. 2005). However, it is likely that changes to root morphology and root exudation alter the composition of the saprophytic microflora (O'Flaherty et al. 2003).

Pochonia chlamydosporia (= *V. chlamydosporium*) is a mycoparasite of several species of *Puccinia* (rust fungi), and an antagonistic to soil borne fungal plant pathogens such as *Pythium aphanidermatum* (Lumsden et al. 1982), *Fusarium oxysporum* and *Rhizoctonia solani* [= *Thanatephorus cucumeris*] (Leinhos and Buchenauer 1992; Jacobs et al. 2003; Monfort et al. 2005). *Pochonia chlamydosporia* also has antagonistic effects against *Phytophthora capsici* when applied as a root dip (Sutherland and Papavizas 1991; see also Chap. 7).

Different species of nematophagous fungi, such as *Paecilomyces lilacinus* (= *Purpureocillium lilacinum*) and *Monographella cucumerina*, can occur with *Pochonia* in nematode infested soils (Gaspard et al. 1990; Olivares-Bernabeu and Lopez-Llorca 2002; Atkins et al. 2004; Manzanilla-López et al. 2009). Puertas et al. (2006) tested the compatibility of *P. chlamydosporia* var. *catenulata* with *Rhizobium* sp., *Trichoderma harzianum* and *Glomus clarum* (= *Rhizophagus clarus*) to control *M. incognita*, the parasitic activity of *P. chlamydosporia* var. *catenulata* being enhanced in combined applications with the other microbia (see Chap. 15).

Pre-colonization of *H. glycines* cysts by *Chaeotomium cochliodes* (= *Chaetomium globosum*), a saprophytic or weakly parasitic fungus, reduced parasitism of eggs in cysts that were subsequently treated with *Verticillium chlamydosporium* (= *Pochonia chlamydosporia*). However, in another soil test egg parasitism was not affected by the pre-colonizing fungi *Cylindrocarpon destructans*, *F. oxysporum* and *F. solani*, but was reduced by *Mortierella* sp., *Pyrenochaeta terrestris* (= *Setophoma terrestris*), and *C. cochliodes* (Chen and Chen 2003). It was unclear whether the cyst precolonizer fungus: i) prevented other fungi from entering the cysts, or ii) inhibited infection of eggs after they entered the cysts, but an inhibitory effect on the growth of post-colonizer parasitic fungi was considered since the presence of some fungi may affect other fungi through nutritional depletion, space occupation, and toxic effects (Chen and Chen 2003).

In a study to assess *P. chlamydosporia* potential interactions and compatibility with the nematophagous fungi *P. lilacinum* (= *Paecilomyces lilacinus*), *Monographella cucumerina* (= *Plectosphaerella cucumerina*) and the soil fungi *Rhizoctonia solani* [= *Thanatephorus cucumeris*], *Chaetomium globosum*, *Fusarium oxysporum*, *Penicillium bilaiaie* and *T. harzianum*, showed that *Pochonia chlamydosporia* radial growth was the least susceptible to growth inhibition by the other fungi (Jacobs et al. 2003). Application of *P. chlamydosporia* and *Hirsutella rhossiliensis*, alone or in combination did not reduce final population densities of *M. hapla* on lettuce (Viaene and Abawi 2000). Other interaction studies have included nematode trapping fungi (Campos and Campos 1997; D'Angieri and Campos 1997; Stirling and Smith 1998). Compatibility studies between *Pochonia* and other BCA microbia are becoming increasingly important, especially in the case of formulations including more than one BCA (see Chap. 4).

3.8.1 Rhizosphere Bacteria

Pasteuria penetrans and *P. chlamydosporia* (= *Verticillium chlamydosporium*) tend to complement each other providing up to 92% control of *M. incognita* (De Leij et al. 1992b). Pseudomonad bacteria can cause growth inhibition of *Verticillium* [*Metapochonia*] *suchlasporium* (Lopez-Llorca and Boag 1990). *Pochonia chlamydosporia* germination and mycelia growth can be inhibited by volatile compounds produced by different genera of bacteria (Zou et al. 2007). Mycostasis by rhizosphere bacteria may act to protect young females from parasitism by nematophagous fungi egg parasites (Lopez-Llorca and Boag 1990). On the other hand, *Verticillium chlamydosporium* (= *P. chlamydosporia*) has shown antibacterial activity (Filipello-Marchisio 1976).

The effect of *M. incognita*, in comparison to that of *P. chlamydosporia* (Pc10) and the nematicide aldicarb (carbamate), on the size and diversity of the rhizosphere population of heterotrophic bacteria in cabbage and tomato, showed that there was little difference between the treatments in the number of cultivable bacteria and fungi in the rhizospheres (O'Flaherty et al. 2003). Tomato roots infested with nematodes supported a different bacterial population from roots on which infestation is controlled either by fungus or by nematicide. Aldicarb and isolate Vc10 (= Pc10) gave a degree of control similar to that provided by the nematicide, neither treatment having a distinct impact on the rhizosphere bacteria compared to that of the nematode. Other interactions between *P. chlamydosporia* and soil microorganisms are revised in Chaps. 4 and 15.

3.8.2 Microflora of the Gelatinous Matrix

The microflora of the gelatinous matrix of the egg mass (i.e., eggs embedded in a gelatinous matrix) has been largely ignored as a factor influencing egg parasites of nematodes (Papert and Kok 2000). One of the gelatinous matrix functions is egg protection through antimicrobial activity (Orion and Kritzman 1991; Niblack and Karr 1994). The bacteria populations of egg masses can be much greater than that of rhizosphere soil, which may indicate a protection role against antagonists, including *P. chlamydosporia* (Papert and Kok 2000; Kok et al. 2001). The resident microbial population of egg masses is apparently adapted to the antimicrobial properties of the gelatinous matrix compounds as more than 50% of the bacteria species isolated from gelatinous matrix are unidentified species that may be more strictly associated with the egg mass niche (Kok et al. 2001).

The egg masses also contain fungi that have been associated with nematode eggs and cysts (Rodríguez-Kábana and Morgan-Jones 1988; Crump 1991; Kok et al. 2001; Meyer et al. 2004). Some fungi, such as *T. harzianum*, have shown a strong reaction against *P. chlamydosporia* (Kok et al. 2001) although this negative effect can apparently be overcome (Puertas et al. 2006; see Chap. 15). The distinct behaviours reported for interactions between *T. harzianum* and *P. chlamydosporia* strains serve to exemplify the need that *T. harzianum*-, nematophagous fungi- or bacteria-based products should be assessed for their effect on other egg parasitic fungi when formulated or applied together with *P. chlamydosporia* (see also Chap. 15). The inhibitory nature of the egg mass microflora should be considered in programmes to develop biological control of *Meloidogyne* spp. with nematode egg parasites (Kok et al. 2001).

3.8.3 Soil Microarthropods

Nematodes have a diverse range of natural enemies but biological control has been dominated by research on microbial antagonists (Stirling 2014). Collembola and mites are the most widely studied predators of nematodes. However, these groups are often neglected when screening for BCA although these generalist predators dominate the higher trophic levels in the soil food web and provide a background level of nematode suppression in all soils, including plant-parasitic nematodes (Sánchez-Moreno et al. 2009). The effect of microarthropod foraging on soil fungi, such as *Pochonia* and other nematophagous and entomopathogenic related genera, remains to be investigated. Fungi that have been cultivated from oribatid mites or identified in DNA extracts obtained from the mites include: *Trichoderma koningii* isolated from *Oribatella quadricornuta*, *Verticillium* sp. from *Oribatula tibialis*, and *Beauveria bassiana* and *Plectosphaerella cucumerina* from *Paradamaeus clavipes*. *Beauveria* has often been found associated with oribatid mites and other soil microarthropods – the fungus is able to degrade chitin of the living, and later that of dead, animals (Renker et al. 2005).

3.9 Epidemiology of Biological Control Agents

The development of models incorporating multiple variables is critical to get a better understanding of the factors that operate together to affect fungal BCA efficacy (Jaronsky 2010). However, there is limited knowledge of the epidemiology and quantitative relations of the fungus with its hosts. It has been proposed that levels of fungal parasitism increase as inoculum densities increase and that *P. chlamydosporia* may be adapted to take advantages of the exponential increase in food supply that becomes available with second and subsequent generations of the RKN populations and increased nematode levels resulting from susceptible crops being grown in consecutive growing seasons. The reproductive capacity of PPN is strongly influenced by the host and temperature, both of which may affect the number of nematode generations in a single crop season. The nematode population structure is pyramidal with relatively few of the large number of eggs and J2 at the base of the pyramid developing into egg-laying adults (Stirling 1991).

Modelling fungal attack and the effect of parasitic fungi on numbers of *H. avenae* has been done for *Nematophthora gynophila* (Perry 1978). Epidemiological models are not yet available for *P. chlamydosporia*, although there have been attempts to model the parasitic activity of the fungus using the Lotka-Volterra system of equations (Ana Puertas unpublished data). The assumptions for the model for *N. gynophila* (Perry 1978) were that: i) the two stages of infection (resting and infective spores) do not overlap in time; ii) spores are randomly distributed in soil; iii) distances between females are large compared with the distances over which spores can infect; iv) ability of resting spores to infect is independent of that of infected spores; v) the number of resting spores produced by females is the same whether infection was by resting spores or not; and vi) all infected females and their eggs are killed. Most of the above assumptions can apply to *P. chlamydosporia* epidemics and illustrate the complexity faced when producing epidemiological models of fungal attack of eggs and female nematodes.

Fungal transmission rate depend upon host threshold and parasite population densities (Anderson and May 1981). A nematode host threshold refers to the density of hosts required to insure that the pathogen does not become locally extinct (Jaffee et al. 1992). Results from a long term field study to evaluate the effect of *P. chlamydosporia* and other biocontrol agents on slow decline of black pepper (*Piper nigrum*), emphasised the importance of the density-dependant multiplication of RKN to suppress nematode populations through time following application of BCA alone or in combination (Eapen et al. 2009). Egg mortality caused by fungal parasitism can effectively reduce the primary inoculum, or pool of eggs, available in the soil to produce the next generation of J2 to infect the plant. However, BCA removal of surplus individuals does not necessarily reduce the number of nematodes in roots (Stirling 1991).

The number of nematode generations should be accounted for when assessing fungus-plant-nematode interactions in a particular crop since increasing nematode population and root galling cannot be prevented in the short term by the fungus, although multiplication of the nematode can be significantly reduced. Viaene and

Abawi (2000) reported that the percentage of *M. hapla* egg masses colonized by *P. chlamydosporia* (= *Verticillium chlamydosporium*) increased with increasing fungus application levels. However, in *M. hapla* infested soils at population levels equivalent to the threshold damage level (8 eggs/cm^3 soil), no difference in colonization could be detected when 5000 or 10,000 chlamydospores were added to soils.

Host plant susceptibility is likely to be a critical factor determining incidence of infection of females as it affects nematode development, and the fungus will have an advantage on plants where egg laying is prolonged (Clyde 1993). Presence of cyst females on roots may result in increased colonisation, particularly by rhizosphere competent isolates. However, this is not always the case for rhizosphere competent isolates. *Pochonia chlamydosporia* isolates I33 or I77 are rhizosphere competent but were not detected by cfu counts of roots with *H. schachtii*, possibly because the fungus was at levels close to the carrying capacity of the environment.

Population densities and epidemiology of the introduced BCA should be tested and monitored to ensure that it has survived through the period that activity against the nematode target is required. This can be achieved by using a combination of classic and molecular methods (see Chap. 13) that also enable the detection of molecularly fingerprinted specific isolates of fungi (Atkins et al. 2003b; Sidiqqui et al. 2009). Population dynamics and epidemiological models for some plant-parasitic nematodes have been produced (Cristóbal-Alejo et al. 2006; Schomaker and Been 2013) and can potentially be integrated with epidemiology models for the fungus so as to help application timing and BCA management.

3.10 Nematophagous and Entomopathogenic Multitrophic Mode of Action

Phylogenetic analysis of 18S ribosomal DNA sequences indicates a close relationship between arthropods, nematodes, and all other moulting phyla which are included in the clade Ecdysozoa (Aguinaldo et al. 1997; Dunn et al. 2008). Nematodes and insects share the presence of chitin in their cuticle (insects) or egg shell (nematodes). The nematode egg shell middle layer has a chitin microfibril core with a collagen-like protein coat (Perry 2002).

Nematophagous and entomopathogenic fungi belong to a wide range of taxa of parasites or facultative parasites of invertebrates which have similar infection patterns and enzymes that help to penetrate and colonize the host. Proteases of *P. chlamydosporia*, *Pochonia* [*Metapochonia*] *rubescens* and *M. anisopliae* are serologically and functionally related (Segers et al. 1995). The proteases involved in the infection processes are extracellular subtilisin-type serine-proteases class II, capable of digesting the protein component of nematode eggs (Figs. 3.9 and 3.10) and insect cuticle and are usually under induction-derepression-repression control (Butt et al. 1998; Segers et al. 1999; Ward et al. 2012). Most invertebrate pathogen fungal strains, including *P. chlamydosporia*, secrete a single subtilisin but some strains produce multiple isoforms that potentially provide the fungus with the tools to hydrolyse additional substrates, possibly increasing the potential host range and/or enabling them to exploit both a parasitic and saprophytic mode of nutrition (Butt et al. 1998). The role of hydrolytic enzymes of *P. chlamydosporia* in the infection process is becoming a subject for genetic engineering; for example, a chitinase-encoding gene *pcchi44* was cloned from *P. chlamydosporia* and over-expressed in *Escherichia coli* BL21. The recombinant chitinase PCCHI44 showed potent antagonistic effect against eggs of *M. incognita* and the silk worm, *Bombyx mori* (Mi et al. 2010).

The similarities in enzymatic capacities and infection mechanisms of entomopathogenic and nematophagous fungi may indicate that both groups are phylogenetically related (see Chaps. 2 and 13). Plant pathogenic fungi produce enzymes that degrade cellulose, xylan or cutin, but do not show chitinolytic or proteolytic activity towards elastine and mucine, whereas entomopathogenic fungi are able to degrade both these and a wider array of proteins (Lopez-Llorca and Jansson 2007). Recent studies have confirmed that the invertebrate-pathogenic fungi and the plant pathogens form a monophyletic group sharing a recent common ancestor (see Lin et al. 2015 and Chap. 9). The multitrophic or multimodal behaviour of these fungi can broad in practical terms, their commonest use as biocontrol agents (i.e., parasites) including their plant symbiont activities, such as modulation of plant defences and crop growth enhancement (Lopez-Llorca and Jansson 2007; see also Chaps. 2, 4, 14 and 15).

3.11 Biological Control and Integrated Pest Management of Nematodes

To combine a BCA with other IPM strategies is important to understand how a BCA interacts with the resident microbiota, including non-target pathogens, chemical pesticides and botanical extracts (Vilchis-Martínez et al. 2013). The fungus can be tolerant to some fungicides (see Jacobs et al. 2003; Tobin et al. 2008 and Chap. 16) and herbicides, e.g., Sencorex (metribuzin), Basagran (bentazone), and Stomp (pendimethalin), which produce no significant effects on the growth of *P. chlamydosporia* (Keith G Davies and Ivânia Esteves unpublished data).

The combined use of poor nematode hosts and the soil application of selected fungus biotypes may be a useful strategy for the management of RKN and PCN (see Chaps. 15 and 16). Effectiveness of *Pochonia* for controlling RKN has been tested in different crops (see also Chap. 15) and plastic-house vegetable production (Verdejo-Lucas et al. 2003; van Damme et al. 2005; Giné et al. 2016). Inclusion of BCA within the preferred (or acceptable) cropping system in combination with other compatible pest control strategies, such as crop rotation, cover crops, addition of organic amendments, biofumigation, plant resistance to nematodes, and other IPM tactics, is becoming a more common and effective practice. At present, the fungus is mainly applied as chlamydospores on colonised substrates, but dose and application methods can vary greatly depending on the formulation of the fungus (see Chaps. 15 and 16), either alone or in combination with other microbes (see Sellitto et al. 2016).

3.12 Future Perspectives

Pochonia chlamydosporia is an important enemy of nematodes in suppressive soils, hence its potential use to control nematode pests. Bio-management strategies for controlling nematodes should incorporate the use of *P. chlamydosporia* based upon a deep understanding of the biology of the fungus, and the careful selection and combination of the fungal isolate biotype with selected cultivars of host plants that are less susceptible or resistant to the nematode and which support extensive growth of the fungus in their rhizosphere. Hosts that could support fungal growth in the rhizosphere and/or an endophytic behaviour will also need to be considered. From the ecological perspective it is important to understand how a BCA interacts with the resident microbiota (e.g., competition, predation and parasitism by other microorganisms), including non-target pathogens, and other potential biocontrol agents, as well as their response to chemical or non-chemical pesticides. More research is needed on the multitrophic abilities of the fungus, including the effect or function that soil isolates may have on the sustainability of agroecosystems. There is also a need to build up infrastructure capability to follow the establishment, dynamics and, if possible, the epidemiology of the fungus in the soil and rhizosphere using molecular tools once an isolate is introduced for biocontrol purposes. Genomics, proteomics, transcriptomics and other *–omics* are opening new research opportunities for the fungus (see Chaps. 5, 6, 8, 9 and 10). Sequencing of fungal isolates and characterization of structure and function of the genome is expected to reveal new genes and mechanisms involved in the host-parasite relationship that lead to a more efficient and effective use of this biocontrol agent. Biological control as part of IPM of PPN is increasingly becoming part of a more holistic approach of integrated soil biology management (Stirling 2014), an approach capable of delivering a sustained crop yield while preserving soil health.

Acknowledgements The authors acknowledge Penny R. Hirsch (Rothamsted Research) and Elaine Ward, (formerly at Rothamsted Research), DFID grant BB/F003994/1 and Defra Link Project LK0966. Light microscopy and SEM studies were performed at Bioimaging Rothamsted Research facilities.

References

Agudelo, P., Robbins, R. T., Stewart, J. M. D., et al. (2004). Glycoproteins in the gelatinous matrix of *Rotylenchulus reniformis*. *Nematropica, 34*, 229–234.

Aguinaldo, A. M., Turbeville, J. M., Linford, L. S., et al. (1997). Evidence for a clade of nematodes, arthropods and other moulting animals. *Nature, 387*, 489–493.

Anderson, R. M., & May, R. M. (1981). The population dynamics of microparasites and their invertebrate host. *Philosophical Transactions of the Royal Society B, 291*, 451–524.

Atkins, S. D., Hidalgo-Díaz, L., Kalisz, H., et al. (2003a). Development of a new management strategy for the control of root-knot nematodes (*Meloidogyne* spp.) in organic vegetable production. *Pest Management Science, 59*, 183–189.

Atkins, S. D., Hidalgo-Díaz, L., Clark, I. M., et al. (2003b). Approaches for monitoring the release of *P. chlamydosporia* var. *catenulata*, a biological control agent of root-knot nematodes. *Mycological Research, 107*, 206–212.

Atkins, S. D., Sosnowska, D., Evans, V. J., et al. (2004). Investigation of three nematophagous fungi in two potato cyst nematode suppressive soils. *Multitrophic Interactions in Soil and Integrated Control IOBC wprs Bulletin, 27*, 1–8.

Ayatollahy, E., Fatemy, S., & Etebarian, H. R. (2008). Potential for biological control of *Heterodera schachtii* by *Pochonia chlamydosporia* var. *chlamydosporia* on sugar beet. *Biocontrol Science and Technology, 18*, 157–167.

Baldwin, J. G., Nadler, S. A., & Adams, B. J. (2004). Evolution of plant parasitism among nematodes. *Annual Review of Phytopathology, 42*, 83–105.

Bordallo, J. J., Lopez-Llorca, L. V., Hansson, H. B., et al. (2002). Colonization of plant roots by egg-parasitic and nematode trapping-fungi. *The New Phytologist, 154*, 491–499.

Bourne, J. M., & Kerry, B. R. (1999). Effect of the host plant on the efficacy of *Verticillium chlamydosporium* as a biological control agent of root-knot nematodes at different nematode densities and fungal application rates. *Soil Biology and Biochemistry, 31*, 75–84.

Bourne, J. M., Kerry, B. R., & De Leij, F. A. A. M. (1996). The importance of the host plant on the interaction between root-knot nematodes (*Meloidogyne* spp.) and the nematophagous fungus *Verticillium chlamydosporium* Goddard. *Biocontrol Science and Technology, 6*, 539–548.

Bruck, D. J. (2010). Fungal entomopathogens in the rhizosphere. *BioControl, 55*, 103–112.

Butt, T. M., Segers, R., Leal, S. C., et al. (1998). Variation in the subtilisins of fungal pathogens of insects and nematodes. In P. D. Bridge, Y. Couteaudier, & J. M. Clarkson (Eds.), *Molecular variability of fungal pathogens* (pp. 149–169). Wallingford: CABI International.

Campos, H. D., & Campos, V. P. (1997). Efeito da época e forma de aplicação dos fungus *Arthrobototrys conocides*, *Arthobotrys musiformis*, *Paecilomyces lilacinus* e *Verticillium chlamydosporium* no controle de *Meloidogyne exigua* do cafeeiro. *Fitopatologia Brasileira, 22*, 361–365.

Chen, S. Y., & Chen, F. J. (2003). Fungal parasitism of *Heterodera glycines* eggs as influenced by egg age and pre-colonization of cysts by other fungi. *Journal of Nematology, 35*, 271–277.

Cheng-Gang, R., & Chuan-Chao, D. (2012). Jasmonic acid is involved in the signaling pathway for fungal endophyte-induced volatile oil accumulation of *Atractylodes lancea* plantlets. *BMC Plant Biology, 12*, 128. doi:10.1186/1471-2229-12-128.

Clyde, J. M. F. (1993). The cyst nematode pathogen *Verticillium chlamydosporium*. PhD Thesis, The University of Leeds, Department of Pure and Applied Biology.

Cristóbal-Alejo, J., Mora-Aguilera, G., Manzanilla-López, R. H., et al. (2006). Epidemiology and integrated control of *Nacobbus aberrans* on tomato (*Lycopersicon esculentum* mill.) in Mexico. *Nematology, 8*, 727–737.

Crump, D. H. (1991). Estimation of suppressiveness and isolation of fungal parasites of cyst nematodes. In B. R. Kerry & D. G. Crump (Eds.), *Methods for studying nematophagous fungi* (IOBC / WPRS Bulletin, XIV (2), pp. 18–22). Wageningen: International Union of Biological Sciences.

Curtis, H. C. R., Robinson, F., & Perry, R. N. (2009). Hatch and host location. In R. N. Perry, M. Moens, & J. L. Starr (Eds.), *Root-knot nematodes* (pp. 139–162). Wallingford: CABI.

D'Angieri, C. N. F., & Campos, V. P. (1997). Control de Meloidogyne javanica em Jaborandi (*Pilocarpus microphyllus*) com *Arthrobotrys conoides*, *Paecilomyces lilacinus e Verticillium chlamydosporia*. *Nematologia Brasileira, 21*, 23–30.

De Leij, F. A. A. M., Kerry, B. R., & Dennehy, J. A. (1992a). The effect of fungal application rate and nematode density on the effectiveness of *Verticillium chlamydosporium* as a biological control agent for *Meloidogyne incognia*. *Nematologica, 38*, 112–122.

De Leij, F. A. A. M., Davies, K. G., & Kerry, B. R. (1992b). The use of *Verticillium chlamydosporium* and *Pasteuria penetrans* alone and in combination to control *Meloidogyne incognita* on tomato plants. *Fundamental and Applied Nematology, 15*, 235–242.

Dunn, C. W., Hejnol, A., Matus, D. Q., et al. (2008). Broad phylogenomic sampling improves resolution of the animal tree of life. *Nature, 452*, 745–749.

Eapen, J. A., Beena, B., & Ramana, K. V. (2009). Field evaluation of *Trichoderma harzianum*, *Pochonia chlamydosporia* and *Pasteuria penetrans* in a root knot nematode infested black pepper (*Piper nigrum* L.) garden in India. *Journal of Plantation Crops, 37*, 196–200.

Eisenback, J. D., & Hunt, D. J. (2009). General morphology. In R. N. Perry, M. Moens, & J. L. Starr (Eds.), *Root-knot nematodes* (pp. 18–54). Wallingford: CABI.

Escudero, N., & Lopez-Llorca, L. V. (2012). Effects on plant growth and root-knot nematode infection of an endophytic GFP transformant of the nematophagous fungus *Pochonia chlamydosporia*. *Symbiosis, 57*, 33–42.

Esteves, I. (2007). *Factors affecting the performance of Pochonia chlamydosporia as a biological control agent for nematodes*. PhD Thesis, Cranfield University.

Evans, A. F., & Perry, R. N. (2009). Survival mechanisms. In R. N. Perry, M. Moens, & J. L. Starr (Eds.), *Root-knot nematodes* (pp. 201–222). Wallingford: CABI.

Filipello-Marchisio, V. (1976). Sull'attivita antibiótica di *Diheterospora chlamydosporia* e di *Oidiodendron truncatum*. *Allionia, 21*, 67–71.

Flores-Camacho, R., Manzanilla-López, R. H., Cid del Prado-Vera, I., et al. (2007). Control of *Nacobbus aberrans* (Thorne, 1935) Thorne y Allen, 1944 with *Pochonia chlamydosporia* (= *Verticillium chlamydosporium*) (Goddard) Zare and W. Gams. *Revista Mexicana de Fitopatología, 25*, 26–34.

Freitas, L. G., Dallemole-Giaretta, R., Ferraz, S., et al. (2009). Controle biológico de nematoides: Estudo de casos. In L. Zambolim & M. C. Picanço (Eds.), *Controle biológico de pragas e doenças: exemplos práticos* (pp. 41–82). Viçosa: UFV/DFP.

Gams, W. (1988). A contribution to the knowledge of nematophagous species of *Verticillium*. *Netherlands Journal of Plant Pathology, 94*, 123–148.

Gaspard, J. T., Jaffee, B. A., & Ferris, H. (1990). *Meloidogyne incognita* survival in soil infested with *Paecilomyces lilacinus* and *Verticillium chlamydosporium*. *Journal of Nematology, 22*, 176–181.

Giné, A., Carrasquilla, M., Martínez-Alonso, M., et al. (2016). Characterization of soil suppressiveness to root-knot nematodes in organic horticulture in plastic greenhouse. *Frontiers in Plant Science*. doi:10.3389/fpls.2016.00164.

Hickey, P. C., Jacobson, N. D., & Glass, L. N. (2002). Live-cell imaging of vegetative hyphal fusion in *Neurospora crassa*. *Fungal Genetics and Biology, 37*, 109–119.

Jackson, M., & Jaronski, S. T. (2009). Production of microsclerotia of the fungal entomopathogen *Metarhizium anisopliae* and their potential use as a biocontrol agent for soil-inhabiting insects. *Mycological Research, 113*, 842–850.

Jacobs, H., Gray, S. N., & Crump, D. H. (2003). Interactions between nematophagous fungi and consequences for their potential as biological agents for the control of potato cyst nematodes. *Mycological Research, 107*, 47–56.

Jaffee, B. A., Philips, R., Muldoon, A., et al. (1992). Density-dependant host-pathogen dynamics in soil microcosms. *Ecology, 73*, 495–506.

Jaronski, S. T. (2010). Ecological factors in the inundative use of fungal entomopathogens. *BioControl, 55*, 159–185.

Jones, J. T., Haegeman, A., Danchin, E. G. T., et al. (2013). Top 10 plant-parasitic nematodes. *Molecular Plant Pathology*. doi:10.1111/mpp.12057.

Kerry, B. R. (1988) *Two microorganisms for the control of plant parasitic nematodes*. Proceedings of the Brighton Crop Protection Conference-Pests and diseases, pp. 603–607.

Kerry, B. R. (1991). Methods for studying the growth and survival of the nematophagous fungus, *Verticillium chlamydosporium* Goddard, in soil. *IOBC/WPRS Bulletin, 14*, 34–38.

Kerry, B. R. (1997). Biological control of nematodes: Prospects and opportunities. *FAO corporate document repository. Plant nematode problems and their control in the Near East Region* (FAO Plant Production and Protection Paper, 144). http://www.fao.docrep/V9978E00.htm. Accessed 20 Oct 2016

Kerry, B. R. (2000). Rhizosphere interactions and the exploitation of microbial agents for the biological control of plant-parasitic nematodes. *Annual Review of Phytopathology, 38*, 423–441.
Kerry, B. R., & Crump, D. H. (1977). Observations on fungal parasites of females and eggs of the cereal cyst-nematode, *Heterodera avenae*, and other cyst nematodes. *Nematologica, 23*, 193–201.
Kerry, B. R., Kirkwood, I. A., De Leij, F. A. A. M., et al. (1993). Growth and survival of *Verticillium chlamydosporium* Goddard, a parasite of nematodes in soil. *Biocontrol Science and Technology, 3*, 355–365.
Kirk, P. M., Cannon, P. F., Minter, D. W., et al. (2008). *Dictionary of the fungi* (10th ed.). Wallingford: CAB International.
Kok, C. J., Papert, A., & Hok-A-Hin, C. H. (2001). Microflora of *Meloidogyne* egg masses: Species composition, population density and effect on the biocontrol agent *Verticillium chlamydosporium* (Goddard). *Nematology, 3*, 729–734.
Leinhos, G. M. E., & Buchenauer, H. (1992). Inhibition of rust diseases of cereals by metabolic products of *Verticillium chlamydosporium. Journal of Phytopathology, 136*, 177–193.
Lin, R., Liu, C., Shen, B., et al. (2015). Analysis of the complete mitochondrial genome of *Pochonia chlamydosporia* suggests a close relationship to the invertebrate-pathogenic fungi in Hypocreales. *BMC Microbiology, 15*, 5.
Lopez-Llorca, L. A., & Boag, B. (1990). Inhibition of *Verticillium suchlasporium* and other nematophagous fungi by bacteria colonizing *Heterodera avenae* females. *Nematologia Mediterranea, 18*, 233–237.
Lopez-Llorca, L. V., & Boag, B. (1993). Biological properties of a red pigment produced by the nematophagous fungus *Verticillium suchlasporium. Nematologia Mediterranea, 21*, 143–149.
Lopez-Llorca, L. V., & Claugher, D. (1990). Appressoria of the nematophagous fungus *Verticillium suchlasporium. Micron and Microscopica Acta, 21*, 125–130.
Lopez-Llorca, L. V., & Duncan, G. H. (1988). A study of the cereal cyst nematode (*Heterodera avenae*) by scanning electron microscopy. *Canadian Journal of Microbiology, 34*, 613–619.
Lopez-Llorca, L. V., & Duncan, G. H. (1991). Effects of fungal parasites on cereal cyst nematode (*Heterodera avenae* Woll.) from naturally infested soil – A scanning electron microscopy study. *Canadian Journal of Microbiology, 37*, 218–225.
Lopez-Llorca, L. V., & Jansson, H. B. (2007). Fungal parasites of invertebrates: Multimodal biocontrol agents? In G. D. Robson, P. van West, & G. M. Gadd (Eds.), *Exploitation of fungi* (pp. 310–335). Cambridge: Cambridge University Press.
Lopez-Llorca, L. V., & Robertson, W. M. (1992). Ultrastructure of infection of cyst nematode eggs by the nematophagous fungus *Verticillium suchlasporium. Nematologica, 39*, 65–74.
Lopez-Llorca, L., Olivares-Bernabeu, C., Salinas, J., et al. (2002a). Pre-penetration events in fungal parasitism of nematode eggs. *Mycological Research, 106*, 499–506.
Lopez-Llorca, L. V., Bordallo, J. J., Monfort, E., et al. (2002b). Use of light and scanning electron microscopy to examine colonisation of barley rhizosphere by the nematophagous fungus *Verticillium chlamydosporium. Micron, 3*, 61–67.
Lopez-Llorca, L. V., Maciá-Vicente, J. G., & Jansson, H. B. (2008). Mode of action and interactions of nematophagous fungi. In A. Ciancio & K. G. Mukerji (Eds.), *Integrated management and Biocontrol of vegetable and grains crops nematodes* (pp. 51–76). Heidelberg: Springer.
Luambano-Nyoni, N., Manzanilla-Lopez, R. H., Kimenju, J. W., et al. (2015). Effect of temperature, pH, carbon and nitrogen ratios on the parasitic activity of *Pochonia chlamydosporia* on *Meloidogyne incognita. Biological Control, 80*, 23–29.
Lumsden, R. D, Frias, T. G., & Gracia, E. R. et al. (1982). Biocontrol of *Pythium aphanidermatum* on cucumber by microbial isolates from Mexican soils. *Phytopathology, 72*, 1010 [Abstract].
Maciá-Vicente, J. G., Jansson, H. B., Talbot, N. J., et al. (2009a). Real-time PCR quantification and live-cell imaging of endophytic colonization of barley (*Hordeum vulgare*) roots by *Fusarium equiseti* and *Pochonia chlamydosporia. The New Phytologist, 182*, 213–228.

Maciá-Vicente, J. G., Rosso, L. C., Ciancio, A., et al. (2009b). Colonisation of barley roots by endophytic *Fusarium equiseti* and *Pochonia chlamydosporia*: Effects on plant growth and disease. *The Annals of Applied Biology, 155*, 391–401.

MacKintosh, G. M. C. D. (1960). The morphology of the Brassica eelworm *Heterodera cruciferae*, Franklin 1945. *Nematologica, 5*, 58–65.

Manzanilla-López, R. H., & Hunt, D. J. (2008). Tropical plant and soil nematodes: Diversity and interactions. In K. Del Claro, P. S. Oliveira, V. Rico-Gray, et al. (Eds.), *International Commission on Tropical Biology and Natural Resources, Encyclopaedia of Life Support Systems (EOLSS)*. Oxford: Eolss Publishers. [http://www.eolss.net].

Manzanilla-López, R. H., Rowe, J., Gravato-Nobre, M., et al. (1998). External secretions from the false root-knot nematode, *Nacobbus aberrans*. *Nematologica, 44*, 326–330.

Manzanilla-López, R. H., Atkins, S. D., Clark, I. M., et al. (2009). Measuring abundance, diversity and parasitic ability in two populations of the nematophagous fungus *Pochonia chlamydosporia* var. *chlamydosporia*. *Biocontrol Science and Technology, 19*, 391–406.

Manzanilla-López, R. H., Esteves, I., Powers, S. J., et al. (2011a). Effects of crop plants on abundance of *Pochonia chlamydosporia* and other fungal parasites of root-knot and potato cyst nematodes. *The Annals of Applied Biology, 159*, 118–129.

Manzanilla-López, R. H., Clark, I. M., Atkins, S. D., et al. (2011b). Exploring competitiveness and variation in the nematophagous fungus *Pochonia chlamydosporia* var. *chlamydosporia* and its significance for biological control. *Bulletin OILB/SROP, 63*, 37–40.

Manzanilla-López, R. H., Devonshire, J., Ward, E., et al. (2014). A combined cryo-scanning electron microscopy/cryoplaning approach to study the infection of *Meloidogyne incognita* eggs by *Pochonia chlamydosporia*. *Nematology, 16*, 1059–1067.

Mauchline, T. H., Kerry, B. R., & Hirsch, P. (2002). Quantification in soil and the rhizosphere of the nematophagous fungus *Verticillium chlamydosporium* by competitive PCR and comparison with selective plating. *Applied and Environmental Microbiology, 68*, 1846–1853.

Mauchline, T. H., Kerry, B. R., & Hirsch, P. (2004). The biocontrol fungus *Pochonia chlamydosporia* shows nematode host preference at the infraspecific level. *Mycological Research, 108*, 106–169.

Meyer, S. L., Huettel, R. N., Liu, X. Z., et al. (2004). Activity of fungal culture filtrates against soybean cyst nematode and root-knot nematode egg hatch and juvenile motility. *Nematology, 6*, 23–32.

Mi, Q., Yang, J., Ye, F., et al. (2010). Cloning and overexpression of *Pochonia chlamydosporia* chitinase gene pcchi44, a potential virulence factor in infection against nematodes. *Process Biochemistry, 45*, 810–814.

Mo, M., Xu, C., & Zhang, K. (2005). Effects of carbon and nitrogen sources, carbon-to-nitrogen ratio, and initial pH on the growth of nematophagous fungus *Pochonia chlamydosporia* in liquid culture. *Mycopathologia, 159*, 381–387.

Monfort, E., Lopez-Llorca, L. V., Jansson, H. B., et al. (2005). Colonisation of seminal roots of wheat and barley by egg-parasitic nematophagous fungi and their effects on *Gaemannomyces graminis* var. *tritici* and development of root-rot. *Soil Biology and Biochemistry, 37*, 1229–1235.

Morgan-Jones, G., White, J. F., & Rodriguez-Kabana, R. (1983). Phytonematode pathology: Ultrastructural studies. I. Parasitism of *Meloidogyne arenaria* eggs by *Verticillium chlamydosporium*. *Nematropica, 13*, 245–260.

Morton, C. O., Hirsch, P. R., Peberdy, J. P., & Kerry, B. R. (2003a). Cloning of and genetic variation in protease VCP1 from the nematophagous fungus *Pochonia chlamydosporia*. *Mycological Research, 107*, 38–46.

Morton, C. O., Mauchline, T. H., Kerry, B. R., & Hirsch, P. R. (2003b). PCR-based DNA fingerprinting indicates host-related genetic variation in the nematophagous fungus *Pochonia chlamydosporia*. *Mycological Research, 107*, 198–205.

Mukhtar, T., & Pervaz, I. (2003). *In vitro* evaluation of ovicidal and larvicidal effects of culture filtrate of *Verticillium chlamydosporium* against *Meloidogyne javanica*. *International Journal of Agricultural and Biology, 5*, 576–579.

Niblack, T. L., & Karr, A. L. (1994). Source of antimicrobial activity in the gelatinous matrix of *Heterodera glycines*. *Journal of Nematology, 26*, 561.
O'Flaherty, S. M., Hirsch, P. R., & Kerry, B. R. (2003). The influence of the root-knot nematode *Meloidogyne incognita*, the nematicide aldicarb and the nematophagous fungus *Pochonia chlamydosporia* on heterotrophic bacteria in soil and the rhizosphere. *European Journal of Soil Science, 54*, 759–766.
Olivares-Bernabeu, C. M., & Lopez-Llorca, L. V. (2002). Fungal egg-parasites of plant-parasitic nematodes from Spanish soils. *Revista Iberoamericana de Micología, 19*, 104–110.
Orion, D., & Kritzman, G. (1991). Antimicrobial activity of *Meloidogyne javanica* gelatinous matrix. *Revue de Nématologie, 14*, 481–483.
Papert, A., & Kok, C. J. (2000). Population size and community metabolic profile of the bacterial population of *Meloidogyne hapla* egg masses. *Nematology, 2*, 581–584.
Perry, J. N. (1978). A population model for the effect of parasitic fungi on numbers of the cereal cyst-nematode, *Heterodera avenae*. *Journal of Applied Ecology, 15*, 781–787.
Perry, R. N. (2002). Cuticle, moulting and exsheathment. In L. D. Lee (Ed.), *The biology of nematodes* (p. 153). Boca Raton: CRC Press Taylor & Francis Group.
Puertas, A., de la Noval, B. M., Martínez, B., et al. (2006). Interacción de *Pochonia chlamydosporia* var. *catenulata* con *Rhizobium* sp., *Trichoderma harzianum* y *Glomus clarum* en el control de *Meloidogyne incognita*. *Revista de Protección Vegetal, 21*, 80–89.
Renker, C., Otto, P., Schneider, K., et al. (2005). Oribatid mites as potential vectors for soil microfungi: Study of mite-associated fungal species. *Microbial Ecology, 50*, 518–528.
Rodríguez-Kábana, R., & Morgan-Jones, G. (1988). Potential for nematode control by mycofloras endemic in the tropics. *Journal of Nematology, 20*, 191–203.
Rosso, L. C., Finetti-Sialer, M. M., Hirsch, P. R., et al. (2011). Transcriptome analysis shows differential gene expression in the saprotrophic to parasitic transition of *Pochonia chlamydosporia*. *Applied Microbiology and Biotechnology, 90*, 1981–1994.
Sánchez-Moreno, S., Nicola, N. L., Ferris, H., et al. (2009). Effects of agricultural management on nematode–mite assemblages: Soil food web indices as predictors of mite community composition. *Applied Soil Ecology, 41*, 107–117.
Santana Nunes, J., Rocha de Brito, M., Cunha Zied, D., et al. (2017). Evaluation of the infection process by *Lecanicillium fungicola* in *Agaricus bisporus* by scanning microscopy. *Revista Iberoamericana de Micología, 34*, 36–42.
Sasser, J. N., & Freckman, D. W. (1987). A world perspective on nematology: The role of the society. In J. Veech & D. W. Dickson (Eds.), *Vistas on nematology* (pp. 7–14). De Leon Spring: Society of Nematologists.
Schomaker, C. H., & Been, T. (2013). Quantitative nematology and management. In R. N. Perry & M. Moens (Eds.), *Plant nematology* (2nd ed., pp. 301–330). Wallingford: CABI.
Segers, R., Butt, T. M., Keen, J. N., et al. (1995). The subtilisins of the invertebrate mycopathogens *Verticillium chlamydosporium* and *Metarhizium anisopliae* are serologically and functionally related. *FEMS Microbiology Letters, 126*, 227–232.
Segers, R., Butt, T. M., Carder, J. H., et al. (1999). The subtilisins of the fungal pathogens of insects, nematodes and plants: Distribution, and variation. *Mycological Research, 103*, 395–402.
Sellitto, V. M., Curto, G., DallaValle, E., et al. (2016). Effect of Pochonia chlamydosporia-based formulations and the regulation of root-knot nematodes and plant growth response. *Frontiers in Life Science, 9*(3), 177–181. doi:10.1080/21553769.2016.11193827.
Shamim, Q. A., Vicar, S., Ara, J., et al. (2012). Nematicidal potential of culture filtrates of soil fungi associated with rhizosphere and rhizoplane of cultivated and wild plants. *Pakistan Journal of Botany, 44*, 1041–1046.
Sharma, P., Jha, A. B., Dubey, R. S., et al. (2012). Reactive oxygen species, oxidative damage, and antioxidative defense mechanisms under stressful conditions. *Journal of Botany, 2012*, 1–26.
Sharon, E., & Spiegel, Y. (1993). Glycoprotein characterization of the gelatinous matrix in root-knot nematode *Meloidogyne javanica*. *Journal of Nematology, 25*, 585–589.

Sharon, E., Orion, D., & Spiegel, Y. (1993). Binding of soil micro-organisms and red blood cells by the gelatinous matrix and eggs of *Meloidogyne javanica* and *Rotylenchus reniformis*. *Fundamental and Applied Nematology, 16*, 5–9.
Siddiqui, I. A., Atkins, S. D., & Kerry, B. R. (2009). Relationship between saprotrophic growth in soil of different biotypes of *Pochonia chlamydosporia* and the infection of nematode eggs. *The Annals of Applied Biology, 155*, 131–141.
Song, Z., Shen, L., Zhong, Q., et al. (2016). Liquid culture production of microsclerotia of *Purpureocillium lilacinum* for use as a bionematicide. *Nematology, 18*, 719–726.
Spiegel, Y., & McClure, M. A. (1995). The surface-coat of plant-parasitic nematodes – Chemical composition, origin, and biological role – A review. *Journal of Nematology, 27*, 127–134.
Stadler, M., Hans-Volker, T., Katsiou, E., et al. (2003). Chemotaxonomy of *Pochonia* and other conidial fungi with *Verticillium*-like anamorphs. *Mycological Progress, 2*, 95–122.
Stirling, G. R. (1991). *Biological control of nematodes*. Wallingford: CAB International.
Stirling, G. R. (2014). *Biological control of plant-parasitic nematodes: Soil ecosystem management in sustainable agriculture* (2nd ed.). Wallingford: CABI.
Stirling, G. R., & Smith, L. (1998). Field tests of formulated products containing either *Verticillium chlamydosporium* or *Arthrobotrys dactyloides* for biological control of root-knot nematodes. *Biological Control, 11*, 231–239.
Sutherland, E. D., & Papavizas, G. C. (1991). Evaluation of oospore hyperparasites for the control of *Phytophthora* crown rot on pepper. *Journal of Phytopathology, 131*, 33–39.
Sykes, D. (1994). *The growth and sporulation of* Verticillium chlamydosporium. MSc Thesis, University of Manchester.
Tahseen, Q., Clark, I. M., Atkins, S. D., et al. (2005). Impact of the nematophagous fungus *Pochonia chlamydosporia* on nematode and microbial populations. *Communications in Agricultural and Applied Biological Sciences, 70*, 81–87.
Tobin, J. D., Haydock, P. P. J., Hare, M. C., et al. (2008). Effect of the fungus *Pochonia chlamydosporia* and fosthiazate on the multiplication rate of potato cyst nematodes (*Globodera pallida* and *G. rostochiensis*) in potato crops grown under UK field conditions. *Biological Control, 46*, 194–201.
Van Damme, V., Hoedekie, A., & Viaene, N. (2005). Long-term efficacy of *Pochonia chlamydosporia* for management of *Meloidogyne javanica* in glasshouse crops. *Nematology, 7*, 727–736.
Verdejo-Lucas, S., Sorribas, F. J., Ornat, C., et al. (2003). Evaluating *Pochonia chlamydosporia* in a double-cropping system of lettuce and tomato in plastic houses infested with *Meloidogyne javanica*. *Plant Pathology, 52*, 521–528.
Viaene, N. M., & Abawi, G. S. (2000). *Hirsutella rhossiliensis* and *Verticillium chlamydosporium* as biocontrol agents of the root-knot nematode *Meloidogyne hapla* on lettuce. *Journal of Nematology, 32*, 85–100.
Vilchis-Martínez, K., Manzanilla-López, R. H., Powers, S., et al. (2013). Effect of the addition of crude plant extracts on the parasitism of *Pochonia chlamydosporia* var. *chlamydosporia* on *Meloidogyne incognita*. *Nematropica, 43*, 254–260.
Wang, K., Riggs, R. D., & Crippen, D. (2005). Isolation, selection, and efficacy of *Pochonia chlamydosporia* for control of *Rotylenchulus reniformis* on cotton. *Phytopathology, 95*, 890–893.
Ward, E., Kerry, B., Manzanilla-López, R., et al. (2012). The *Pochonia chlamydosporia* serine protease Gene *vcp1* is subject to regulation by carbon, nitrogen and pH: Implications for nematode biocontrol. *PloS One, 7*(4), e35657.
Zare, R., & Gams, W. (2007). *Pochonia globispora* sp. nov. *Nova Hedwigia, 84*, 421–428.
Zare, R., Gams, W., & Evans, H. C. (2001). A revision of *Verticillium* section Prostrata V. The genus *Pochonia*, with notes on *Rotiferophthora*. *Nova Hedwigia, 73*, 51–86.
Zou, C. S., Mo, M. H., Gu, Y. Q., et al. (2007). Possible contributions of volatile-producing bacteria to soil fungistasis. *Soil Biology and Biochemistry, 39*, 2371–2379.

Chapter 4
Interactions Between *Pochonia chlamydosporia* and Nematodes

Thalita S. Avelar Monteiro, Everaldo A. Lopes, Harry C. Evans, and Leandro Grassi de Freitas

Abstract *Pochonia* parasitizes the eggs and females of plant-parasitic nematodes (PPN). The fungus-nematode interaction involves a complex series of events which can be affected by the environment. Thus, an understanding of the bionomics of this interaction is essential in order to improve the efficiency of the biological management of PPN. In this chapter, we provide an overview of the process of infection of nematodes by *Pochonia* and the role of the environment on the fungus-nematode interaction. Firstly, we focus on the events and the mechanisms underlying adhesion, penetration and colonisation of nematodes by *Pochonia*. We discuss how the infection process is driven by both mechanical forces, induced by the appressoria, and enzymatic activity, specifically by the serine proteases (P32, VCP1, SCP1) and chitinases. Environmental factors have a profound influence on the *Pochonia*-nematode interaction and these are discussed in detail. Temperature, pH, soil type, soil microbiota and roots can enhance or reduce the parasitism of the nematode by the fungal antagonist. Finally, we discuss how the method of application of *Pochonia* and its timing can impact on the establishment of the fungus in the soil and, consequently, on the control of nematodes.

T.S. Avelar Monteiro • L. Grassi de Freitas (✉)
Departamento de Fitopatología, Universidade Federal de Viçosa, Minas Gerais, Brazil
e-mail: thalita.monteiro@ufv.br; leandro@ufv.br

E.A. Lopes
Instituto de Ciências Agrárias, Campus de Rio Paranaíba, Universidade Federal de Viçosa, Rio Paranaíba, Minas Gerais, Brazil
e-mail: everaldolopes@ufv.br

H.C. Evans
CAB International, Egham, Surrey TW20 9TY, UK
e-mail: h.evans@cabi.org

R.H. Manzanilla-López, L.V. Lopez-Llorca (eds.), *Perspectives in Sustainable Nematode Management Through* Pochonia chlamydosporia *Applications for Root and Rhizosphere Health*, Sustainability in Plant and Crop Protection,
DOI 10.1007/978-3-319-59224-4_4

4.1 Introduction

The genus *Pochonia* was created by the Brazilian mycologists A.C. Batista and O.M. Fonseca, over 50 years ago (Batista and Fonseca 1965), based on the type species *P. humicola* isolated from soil in North-east Brazil. However, it was not until decades later that it was realised that this taxon had been described much earlier and assigned to the genus *Verticillium*, under the name *V. chlamydosporium* (Zare et al. 2000). Based on both morphological and molecular characteristics, *V. chlamydosporium* was transferred to *Pochonia* (Zare et al. 2001). Currently, the genus *Pochonia* comprises a single species split into five varieties (see Chap. 2). Due to its potential as a biocontrol agent of PPN and its worldwide occurrence, *P. chlamydosporia* var. *chlamydosporia* has been the most studied taxon in the genus.

Pochonia chlamydosporia is associated with the rhizosphere, a micro-habitat formed by roots and their associated microbiota. The rhizosphere is subject to modifications resulting from the interactions of the plant and its environment. Therefore, this biocontrol agent needs to establish specific interactions with its host and also requires phenotypic plasticity to deal with competition from other organisms and pressures due to the continual modifications to its habitat.

This fungus has a multitrophic lifestyle, behaving as a saprophyte, a parasite and an endophyte. Behavioural alterations are regulated by complex interactions. The change from the saprophytic to the parasitic phase, for example, can occur for nutritional reasons, which is evidenced by an increase in the density of *P. chlamydosporia* as a consequence of an increase in the population of nematodes in the roots. This stimulation can be affected by changes in the exudation profile of the root system (Kerry 2000; Manzanilla-López et al. 2011).

The interactions of *Pochonia* with plants are linked to diverse phenomena of both scientific and practical interest. Induction of resistance, priming for enhanced defence, growth promotion and absorption of nutrients, such as phosphorus, are some of the known effects of the endophytic colonisation by the fungus in plants (Monteiro 2013; dos Santos et al. 2014; Escudero et al. 2014; Dallemole-Giareta et al. 2015; Larriba et al. 2015; Medeiros et al. 2015; Zavala-Gonzalez et al. 2015).

This chapter deals with the bionomics of this interaction and considers the various events in the infection process, notably those involving the mechanisms of parasitism and the factors that influence this interaction.

4.2 Parasitic Events

Pochonia chlamydosporia parasitizes the eggs and females of PPN (Morgan-Jones et al. 1983; Lopez-Llorca et al. 2002). The fungus preferentially colonizes fertilised eggs and also those with developed juveniles (Morgan-Jones et al. 1983; Manzanilla-López et al. 2013). The events involved in parasitism by *Pochonia* comprise adhesion, penetration and colonization.

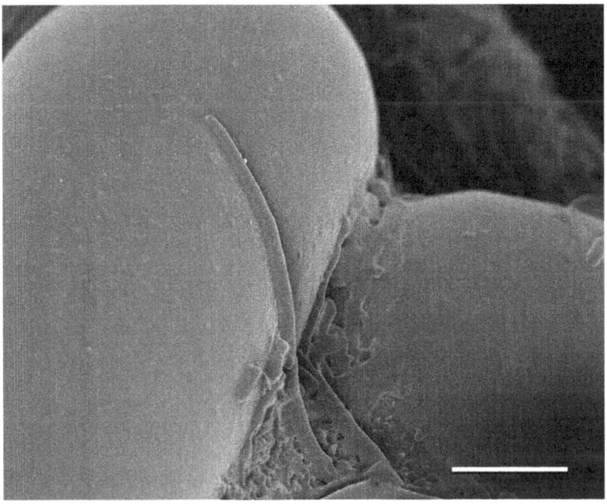

Fig. 4.1 Cryo-Scanning Electron Microscopy micrograph showing hypha of *Pochonia chlamydosporia* on the surface of *Meloidogyne incognita* eggs, 24 h after inoculation. Scale bar 10 µm (After Manzanilla-López et al. 2014. Copyright Rothamsted Research Ltd.)

4.2.1 Adhesion

Adhesion is a critical step in the infection process allowing the fungus to gain a foothold for successful parasitism and, ultimately, for its survival. *Pochonia* produces glycoproteins that are responsible for the adhesion of conidia and hyphae to the eggs (Fig. 4.1) (Lopez-Llorca et al. 2002). These metabolites aid the fixation of the spore germ tube or the hypha to the substrate whilst also offering protection against desiccation. In general, the adhesion process begins in a moist environment, suggesting that the adhesive substances may be insoluble in water (Nicholson and Moraes 1980).

The development of the fungal germ tube and its differentiation – especially the formation of the appressoria – can occur in response to surface signals; including hardness, hydrophobicity and topography (Tucker and Talbot 2001). A study of the effect of hydrophobicity on the formation of appressoria by *P. rubescens* (= *Metapochonia*, Kepler et al. 2014; see Chapter 2) and *P. chlamydosporia* indicated that hydrophobic surfaces induced more appressoria differentiation than hydrophilic ones (Lopez-Llorca et al. 2002), despite the capacity of these fungi to produce appressoria on all surfaces.

4.2.2 Penetration

Pochonia produces appressoria and secretes extracellular enzymes for host penetration, i.e., the same strategy as used by entomopathogenic and plant-pathogenic fungi. This similarity in pathogenesis mechanisms may explain why the genetic homology between *P. chlamydosporia* and the entomopathogenic fungus

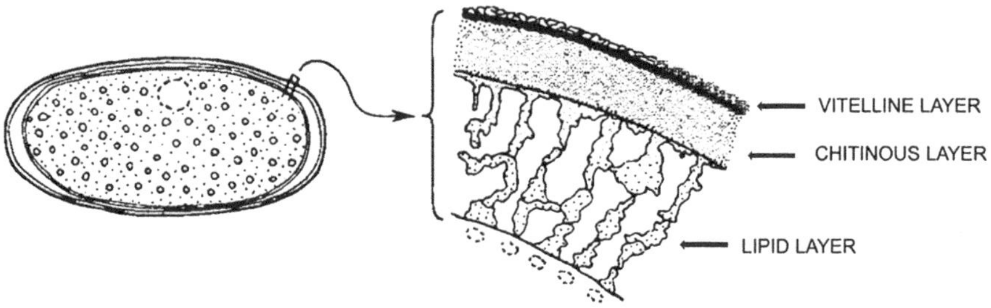

Fig. 4.2 Diagrammatic reconstruction of a cross section of the egg shell of plant-parasitic nematodes (Adapted from Bird and Bird 1991)

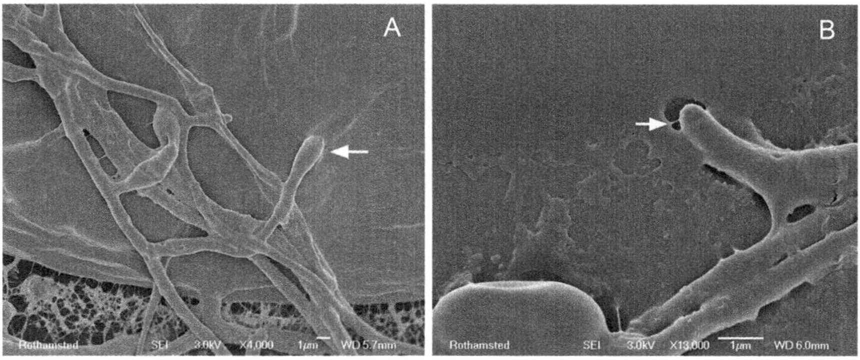

Fig. 4.3 (**A, B**) Cryo-Scanning Electron Microscopy micrograph showing appressoria (*arrows*) of *Pochonia chlamydosporia* on the surface of *Meloidogyne incognita* eggs 24 h after inoculation (Copyright Rothamsted Research Ltd.)

Metarhizium is higher than that between *P. chlamydosporia* and the nematode-trapping fungus *Arthrobotrys oligospora*, which produces adhesive trapping nets to capture motile nematode stages (Larriba et al. 2014).

The processes regulating the mechanical force used by the appressoria of *Pochonia* to penetrate the egg shell have not yet been determined, but the secreted enzymes play an important role in overcoming the vitelline membrane and exposing the chitin layer (Fig. 4.2). The physical and chemical processes work together for the penetration of the germ tube into the egg (Lopez-Llorca et al. 2002; Manzanilla-López et al. 2011).

Appressoria can develop at the tip of the germ tubes, as well as on the lateral and terminal hyphae of *Pochonia* (Fig. 4.3). This was first reported by Lopez-Llorca and Claugher in 1990 (Lopez-Llorca and Robertson 1992a; Segers et al. 1996). The morphology and number of appressoria formed depends on the nematode host. For example, on the egg surface of *Meloidogyne incognita* appressoria are more numerous and larger than on the eggs of *Globodera rostochiensis* (Segers et al. 1996).

Appressoria formed by *Pochonia* are rudimentary, similar to those produced by necrotrophic plant pathogens (Gayad 1961) such as *Bipolaris sorokiniana* (Kumar et al. 2002). These fungi do not need to produce complex appressoria – expending more energy or resources – since they also release enzymes that destroy the plant tissues, facilitating host penetration and colonisation. Similarly, *Pochonia* also produces extracellular enzymes that facilitate parasitism of nematode eggs and does not depend solely on the formation of appressoria.

4.2.3 Colonization

Following penetration, the fungus colonizes the host tissues in order to obtain nutrients and thereby increase its reproductive capacity. *Pochonia chlamydosporia* uses available sugars in the egg as a carbon source. Trehalose, for example, is a disaccharide commonly found in the egg shell and in the juveniles, that plays an important role in protecting nematodes against desiccation and which is associated with anhydrobiosis and the maintenance of second-stage juvenile (J2) dormancy within the egg shell (Yen et al. 1996; Wharton 2011).

Pochonia chlamydosporia colonizes both the eggs and embryos, as well as J2 within the eggs (Morgan-Jones et al. 1983). The fungus also parasitizes females and colonizes the specialized feeding cells of sedentary endoparasitic nematodes that accumulate plant nutrients (Fig. 4.4) (Nasu 2013).

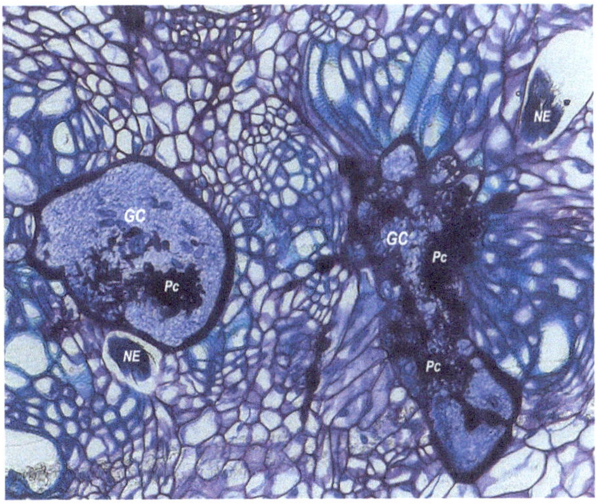

Fig. 4.4 Micrograph of a soybean root inoculated with the nematode *Meloidogyne incognita* and the fungus *Pochonia chlamydosporia*, showing a female nematode body (NE) and giant cells (GC), with fungal hyphae (Pc) inside the giant cells. Micrograph material embedded in methacrylate and stained with toluidine blue (Source: Nasu 2013)

4.3 Enzymes

Extracellular enzymes secreted by nematophagous fungi are important in all stages of the parasitism of eggs (Huang et al. 2004; Morton et al. 2004; Yang et al. 2007; Lopez-Llorca et al. 2010) and also during endophytism (Lopez-Llorca et al. 2010; Larriba et al. 2014). The nematode egg shell is composed of an outer proteinaceous vitelline layer; a middle chitinous layer, composed of a protein matrix embedded with chitin microfibrils; and an inner lipid layer (Fig. 4.2) (Bird and McClure 1976). Proteins and chitin account for about 80% of the dry weight of the egg shell of *M. incognita* (Bird and McClure 1976). Thus, the simultaneous production of proteases and chitinases is fundamental in the process of penetration and colonization of eggs by *Pochonia* (Clarke et al. 1967; Olivares-Bernabeu and Lopez-Llorca 2002; Esteves et al. 2009).

4.3.1 Serine Proteases

The primary amino acid serine is a non-essential component of proteins that serves as the nucleophilic amino acid at the active site of serine proteases. These enzymes cut the peptide bonds in proteins and are divided into various families which are grouped according to their structural similarity and biological function (Di Cera 2009). Serine proteases are the largest category of proteases encoded by the *P. chlamydosporia* genome (Larriba et al. 2014). The serine proteases P32 (Lopez-Llorca 1990), VCP1 (Segers et al. 1994) and SCP1 (Lopez-Llorca et al. 2010) have already been identified in *Pochonia*.

4.3.1.1 P32 Serine Protease

This enzyme was first purified in 1990 and its molecular mass was empirically estimated at 32 kDa (Lopez-Llorca 1990), although this value was predicted as 28.7 kDa after cloning and analysis of the P32 gene and its encoded protein (Larriba et al. 2012). The role of P32 in the pathogenicity of *P.* [= *Metapochonia*] *suchlasporia* to eggs of *Heterodera schachtii* was reported 2 years later (Lopez-Llorca and Robertson 1992a). P32 belongs to the S8 family (Lopez-Llorca 1990), which is also known as the subtilisin family. Nematophagous and entomopathogenic fungi produce subtilisin-like serine proteases to disrupt the outer layers of the hosts during the infection process (Li et al. 2010). P32 shows homology with the subtilisin-like serine proteases VCP1 (76%) of *P. chlamydosporia*, and with Pr1A (86%) of the entomopathogenic fungus *Metarhizium anisopliae* (Larriba et al. 2012). VCP1 removes the vitelline membrane of the egg shell (Segers et al. 1994, 1996; Ward et al. 2012) and exposes the chitin layer (Fig. 4.2). Thus, P32 may also have a similar function in the parasitism of *P.* [= *Metapochonia*] *rubescens*.

4.3.1.2 VCP1 Serine Protease

This 33 kDa serine protease was first purified in 1994 (Segers et al. 1994) and also belongs to the S8 family. It is produced by *P. chlamydosporia* during the initial stages of egg infection, causing the disruption of the vitelline layer (Fig. 4.2), facilitating hyphal penetration and subsequent infection of the nematode (Segers et al. 1994, 1996; Ward et al. 2012). Thus, this enzyme can act as a pathogenicity factor. Eggs of *M. incognita* pre-treated with purified VCP1 are colonized more rapidly than non-treated ones (Segers et al. 1996). VCP1 is also important in the endophytic phase of the fungus (Lopez-Llorca et al. 2010).

Carbon and nitrogen sources, as well as pH, regulate the production of many of the extracellular enzymes produced by fungi (Aro et al. 2005). The presence of ammonium chloride initially reduces VCP1 expression but then rapidly increases its production (Ward et al. 2012). In a glucose-rich environment, the fungus reduces the production of VCP1. *Pochonia chlamydosporia* prefers to use the readily available carbon source rather than to colonize nematode eggs (Ward et al. 2012). Thus, readily-metabolisable carbon sources and low pH in the environment may compromise nematode parasitism by *P. chlamydosporia* (Ward et al. 2012).

Genetic variation in VCP1 between isolates from different nematode hosts may explain host preference in *P. chlamydosporia*. Amino acid polymorphisms were found between VCP1 in isolates of the fungus obtained from root-knot nematodes (*Meloidogyne* spp.) and from cyst nematodes (*Heterodera schachtii* and *Globodera rostochiensis*). The presence of a glycine in isolates from root-knot nematodes, instead of an alanine in isolates from cyst nematodes, changes the amino acid affinities at the S3 substrate-binding region (Morton et al. 2003). Differences in amino acid sequences affect enzyme properties and substrate use (Segers et al. 1995; Liu et al. 2007). Thus, amino acid polymorphisms in VCP1 may have an influence on host preference between biotypes of *P. chlamydosporia*.

4.3.1.3 Serine Protease SCP1

The serine protease carboxypeptidase (SCP1) belongs to the S10 family (serine carboxypeptidases) and it was first purified in 2010 (Lopez-Llorca et al. 2010). The secreted form of this enzyme has a molecular mass of 60.5 kDa (Larriba et al. 2012). It belongs to a different protease subfamily (Lopez-Llorca et al. 2010) and it has homology of 88.1% and 86.5% with serine carboxypeptidases of *Metarhizium anisopliae* and *M. acridum*, respectively (Larriba et al. 2012). SCP1 is produced during the endophytic colonization of roots by *P. chlamydosporia* (Lopez-Llorca et al. 2010) and during the parasitism of root-knot nematode eggs (Escudero et al. 2016). Furthermore, activity of this protease is increased in the presence of chitosan. The serine protease SCP1 is involved in the endophytic and parasitic phases of the fungus (Lopez-Llorca et al. 2010; Escudero et al. 2016).

4.3.2 Chitinases

Through the action of VCP1 and P32 on the egg shell, the chitinous layer is exposed (Fig. 4.2). Chitin is a polymer of N-acetylglucosamine present in the intermediate layer of the shell of nematode eggs (Yang et al. 2007; Perry and Wharton 2011). Thus, in order to breach this barrier, the fungus needs to produce endochitinases, exochitinases and N-acetyl-β-D-glucosaminidases (Segers et al. 1994; Tikhonov et al. 2002; Khan et al. 2004; Mi et al. 2010; Palma-Guerrero et al. 2010). Endochitinases randomly cut the chitin chain, then exochitinases cleave chitobiose, and finally glucosamine monomers are excised from chitobiose by the N-acetyl-β-D-glucosaminidases (Morton et al. 2004).

The first chitinase purified from *Pochonia* was a 43 kDa endochitinase – CHI43 (Tikhonov et al. 2002). This enzyme was shown to have a synergistic interaction with the serine protease P32 in damaging eggs of *Globodera pallida* (Tikhonov et al. 2002). *Pochonia chlamydosporia* and *P.* [*Metapochonia*] *rubescens* both produce chitinases, although the activity of N-acetyl-β-D-glucosaminidases from *M. rubescens* on *Globodera* egg shells is higher than that of *P. chlamydosporia* (Tikhonov et al. 2002). This difference in the expression of chitinases between these two fungal species may explain the apparent specialization of *M. rubescens* and *P. chlamydosporia* for infection of *Globodera* and *Meloidogyne*, respectively.

The 44 kDa endochitinase PCCHI44 from *P. chlamydosporia* was cloned by Mi et al. (2010) who noted that this enzyme may be a potential virulence factor. This chitinase causes furrowing and flaking of the shell of *M. incognita* eggs followed by the appearance of large vacuoles which can interfere with the normal development of nematodes (Mi et al. 2010).

Entomopathogenic fungi can transform chitin from the cuticle of their hosts into chitosan using chitin deacetylases. The chitosan formed is then degraded by fungal chitosanases during host penetration (Nahar et al. 2004). Possibly, this process also occurs during parasitism of nematodes because the composition of the nematode egg shell is similar to the insect cuticle and *P. chlamydosporia* has the capacity to produce chitosanases in vitro (Palma-Guerrero et al. 2008, 2010). *Pochonia chlamydosporia* produces chitosanases only in culture media containing chitosan, whereas chitinases are produced in media containing either chitosan or chitin (Palma-Guerrero et al. 2010). Interestingly, the chitinase concentration was greater, approximately twice as much, in the medium with chitosan, suggesting that this medium induces the production of chitinase (Palma-Guerrero et al. 2010). Chitosan also induces the production of enzymes besides chitinase, such as VCP1 and SPC1, which is important for egg parasitism (Palma-Guerrero et al. 2010; Escudero et al. 2016).

Information on the pathway involved is lacking, however, and this warrants further research. Chitosan has been used in the control of plant pathogens and could possibly be integrated with *Pochonia* for the management of plant disease complexes caused by fungi and nematodes. Transgenic plants expressing *Pochonia* enzymes may be another alternative for the management of nematodes. For instance, tomato plants expressing the PjCHI-1 chitinase gene, isolated from the non-nematophagous fungus *Paecilomyces javanicus* (= *Isaria javanica*), suppressed the reproduction and the embryonic development of *M. incognita* (Chan et al. 2010).

4.3.3 Other Enzymes

Pochonia secretes other enzymes during parasitism and endophytism. However, little is known about these metabolites and their role in the fungus-nematode and fungus-plant interactions. It is likely that the degradation of the inner lipid layer of the egg shell (Fig. 4.2) involves the activity of lipases (Perry and Trett 1986). Proline represents 32.8% of the amino acid content of tylenchid egg shell (Bird and McClure 1976) and prolyl aminopeptidases (S33) are the most abundant serine protease family in the *P. chlamydosporia* genome (Larriba et al. 2014). Several families of expressed proteases in *Pochonia* have relation to the endophytic behaviour of fungi, such as rhomboid proteases (S54) and metalloproteases (M28, M35, M36 and M43) (Larriba et al. 2014).

4.4 Secondary Metabolites

Secondary metabolites are structurally heterogeneous molecules, usually of low molecular weight, that are not directly required for the growth of the producer organism. Fungal and bacterial inhabitants of the soil produce a range of secondary metabolites, principally to compete with other organisms for space and nutrients (Brakhage 2013).

Pochonia is known to produce secondary metabolites (Khambay et al. 2000; Hellwig et al. 2003; Niu et al. 2010). Some of these have been investigated for their medicinal uses, for example, for the treatment of herpes and cancer (Hellwing et al. 2003; Moulin et al. 2005; Huang et al. 2008), whilst others, such as phomalactone and aurovertin, have been tested as nematicides (Khambay et al. 2000; Wang et al. 2015). Phomalactone has nematicidal activity *in vitro*, and can kill 84% J2 of *M. incognita* within 96 h. Aurovertins A3 and A4 have shown toxicity to the free-living nematode *Panagrellus redivivus*, although none of the aurovertin-type metabolites studied thus far has been found to inhibit eclosion of *M. incognita* juveniles (Khambay et al. 2000; Niu et al. 2010). Aurovertin D has lethal activity against *M. incognita* juveniles (Wang et al. 2015). Nevertheless, more research is necessary in order to elucidate the role of secondary metabolites in nematode control and the ability of these toxins to function under natural conditions (see Chap. 7, Secondary metabolites).

4.5 Factors Affecting the Parasitic Activity of *Pochonia chlamydosporia*

Different conditions favour the establishment and guaranteed efficacy of biological control against soil pathogens. There is a considerable amount of information on the conditions favouring the development of *Pochonia* and of nematodes. However, our knowledge of the factors that affect *Pochonia*-nematode interactions is inadequate and further research is necessary.

Soil conditions (pH, temperature and soil type), fungistatic activity of the soil, endophytic activity and induced resistance, interactions of *Pochonia* with nematode hosts and other organisms, timing and methods of application of the fungus to the soil, are all increasingly important factors which need to be investigated for the appropriate management of PPN.

4.5.1 Soil pH

The pH is a critical variable for the establishment of compatible interactions. Indirectly, pH regulates enzymatic activity since each enzyme has an optimum pH at which this activity is maximized. VCP1, an enzyme which has been well studied and characterized, is produced by *P. chlamydosporia* during egg parasitism and has its greatest activity at high pH values (Ward et al. 2012). However, the percentage of *M. incognita* eggs infected by *P. chlamydosporia* was about four times higher at pH 6.6 than at pH 5.8 or 7.1 (Luambano et al. 2015).

Pochonia grows between pH 3.0–7.0 in artificial conditions and most strains prefer an alkaline pH (Lopez-Llorca et al. 1993; Olivares-Bernabeu and Lopez-Llorca 2002). Current data show that the results are complex when the pH is variable. It is also important to take into consideration the fact that different fungal isolates can have different optimum conditions for growth and development (Kerry et al. 1986). Thus, more research is necessary in order to understand better this factor and thereby improve nematode management.

4.5.2 Temperature

Another important factor affecting enzymatic activity is temperature. Nagesh and co-workers (2007), working with isolates from fields infested with root-knot and reniform nematodes in tropical regions of India, found that *P. chlamydosporia* grows well at temperatures between 25-35 °C, with an optimum at 30 °C. However, Kerry et al. (1986), working with isolates from *Heterodera avenae* in England, found differences for optimum temperature ranging from 18–25 °C. Unfortunately, the relationship between the optimum temperature for growth and that for parasitism has not been made. For one of the studied isolates, optimum growth occurred at

25 °C, whilst for parasitism the optimum was 12 °C (Kerry et al. 1986; Irving and Kerry 1986). Isolates of *P. chlamydosporia* differ in their climatic requirements according to the environmental conditions where the nematode host and fungus co-evolved. Nevertheless, fungi are generally less efficient when soil temperatures are either too low, because of the reduced rate of fungus development, or when temperatures are over 30 °C, because it accelerates embryonic development of the nematode, thereby allowing the target to escape fungal colonization after hatching (Kerry 2000; Verdejo-Lucas et al. 2003).

4.5.3 Soil Type

Soil is the medium that supports and supplies the plants with water and nutrients, as well as the microorganisms that live there. The importance of soil in the survival of microorganisms is so critical that soil type and its properties determine the inhabitants of this complex environment. An example of this is the preference of nematodes for sandy soils, which is why they usually cause most damage to crops cultivated in this soil type. This phenomenon is considered to be due to the relative ease of locomotion in these soils and the low organic matter content which decreases the number, and therefore impact, of natural enemies of the nematodes whilst stressing the plant due to poor nutrition. Just like nematodes and other soil organisms, *P. chlamydosporia* also shows preferences for soil type and it develops best in sandy soil, although it is also capable of growing in clay and highly saline soils (Nasu 2013; Ceiro et al. 2014). *Pochonia chlamydosporia* penetrates into the deeper layers in both sandy and clay soils, although it develops best in the former (Nasu 2013).

Various factors could be involved in this preference for sandy soils, including better porosity and thus good aeration, lower levels of toxic elements, such as aluminium, iron and manganese, and better drainage, which helps the fungus to move freely within the soil profile and reducing the chances of anoxia or lack of oxygen, since the fungus is an aerobic organism. The nematode also develops best in sandy soils. Thus, the highest parasitism of nematodes by *P. chlamydosporia* also occurs in these soils, as the nematode presence increases the cfu (colony forming units) of the fungus in the soil (Nasu 2013). *Pochonia chlamydosporia* can also grow in Luvisol – a soil type with a high level of NaCl – which makes this biological control agent a suitable candidate for the management of plant-parasitic nematodes in soils with salinity problems (Ceiro et al. 2014).

4.5.4 Fungistatic Activity of the Soil

The fungistatic activity of the soil dictates the establishment of biocontrol agents introduced into the new environment for management of nematodes. This activity refers to the biotic and/or abiotic properties of the soils that inhibit germination of the fungal propagules. In sterile soil, *P. chlamydosporia* develops better than in natural soils and, as a consequence, nematode control is more efficient (Monfort et al. 2006; Zou et al. 2007; Podestá et al. 2009). Native soil microorganisms may produce toxic metabolites that inhibit or harm other organisms. For example, *Verticillium suchlasporium* (= *Metapochonia suchlasporia*) has its growth suppressed in the presence of *Pseudomonas* spp. (Lopez-Llorca and Boag 1990).

4.5.5 Endophytic Activity and Plant Induced Resistance

The host plant is very important for the establishment of the fungus in the soil, and, as a consequence, it can facilitate the control of nematodes. *Pochonia chlamydosporia* grows endophytically within the roots and plant species can either be good or bad hosts, as indicated by the cfu of the fungus within the root. When the roots support more than 200 cfu/cm^2 of root, it is considered a good host, but if lower than this it is categorized as a bad host (Kerry 2000). Beans, cabbage, crotalaria, kale, pigeon pea, potato, pumpkin, and tomato are good hosts for *P. chlamydosporia*. Aubergine, okra, soybean, sorghum, and wheat are considered bad hosts, whilst chilli, sweet potato, cowpea, rye, tobacco, and cotton are moderate hosts (Bourne et al. 1996; Kerry 2000).

The fungus protects the plant against nematode action in the rhizosphere and has the ability to colonise the roots and thus to affect control. In addition, the fungus promotes plant growth by increasing the uptake of nutrients. For example, *P. chlamydosporia* increases by up to 24.5% the absorption of phosphorus by the plant (Monteiro 2013). This could be explained, at least in part, by its ability to solubilise this nutrient. This can lead to a cascade of benefits for the plant, for example, improved growth and reduced flowering time, so that the plant becomes productive earlier (Zavala-Gonzalez et al. 2015).

Endophytic colonization by *P. chlamydosporia* induces the formation of plant structures resistant to plant pathogens, such as papillae (Fig. 4.5). Interestingly, the presence of callose is not related to the colonization of roots by the fungus, indicating that *Pochonia* has an indirect action in the induction of host defences to pathogens (Bordallo et al. 2002; Lopez-Llorca et al. 2002b; Monfort et al. 2005; Escudero and Lopez-Llorca 2012). This phenomenon was observed by Medeiros et al. (2015) when *P. chlamydosporia* induced localized resistance to *M. javanica* and activated priming in tomato plants, as witnessed by an increase of polyphenoloxidase and peroxidase enzyme activities. Larriba et al. (2015) also reported a moderate induced resistance in barley by *P. chlamydosporia* through gene up-regulation of the genes involved in the jasmonic acid metabolism, amongst other factors.

Another intriguing phenomenon observed in the rhizosphere is that the population density of *P. chlamydosporia* increases with the nematode population in the roots. As a result of nematode infection and production of the egg mass, root exudation increases and enhances the colonization of the root surface by the fungus (Wang and Bergeson 1974; Kerry and Bourne 1996; Bourne and Kerry 1999).

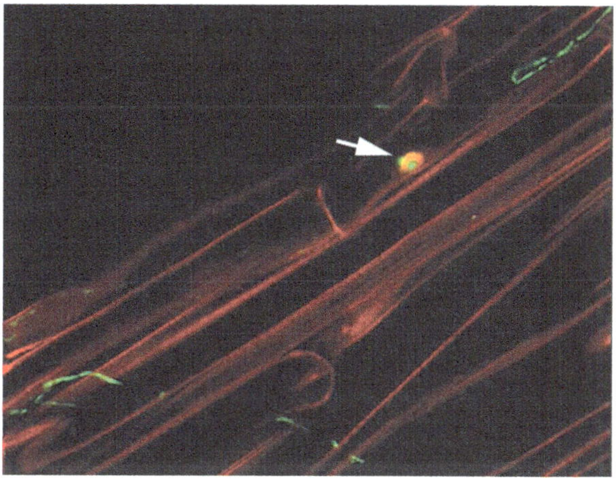

Fig. 4.5 Papilla (*arrow*) associated with root colonization by *Pochonia chlamydosporia* (transformed with GFP) (Courtesy N. Escudero)

4.5.6 Interactions with Other Soil Organisms

The use of combinations or 'cocktails' of biocontrol agents can be an interesting option, because the employment of different organisms – often having different mechanisms of control – can result in synergistic interactions that promote the management of plant pathogens. However, some combinations of antagonists may be incompatible. Therefore, prior to implementing a management strategy it is important to understand the interactions between biocontrol agents. Thus, studies of compatibility between antagonists will be essential in the development of novel methods, for the management of plant-parasitic nematodes.

The combined use of *P. chlamydosporia* and *Duddingtonia flagrans* – a nematophagous fungus studied for the control of important helminths of domestic animals (see Chap. 17) – increased the control of the root-knot nematodes compared to when these fungi were tested separately (Monteiro 2013). This synergism can be explained by the fact that these two fungi possess distinct mechanisms of action, *P. chlamydosporia* parasitizing eggs and females whereas *D. flagrans* is a predator of juveniles, capturing them with hyphae on sticky three-dimensional nets. The ideal conditions for the development of each organism are variable and, in this way, the use of more than one organism can help to stabilize biocontrol given that, when field conditions are unfavourable for one organism, they might be favourable for another.

The mycoparasite *Trichoderma* may reduce the mycelial growth of *P. chlamydosporia* by the production of volatile metabolites *in vitro* (Ferreira et al. 2008). However, when they are applied to the soil together for the control of root-knot nematodes, *P. chlamydosporia* and *Trichoderma* cause a greater reduction in the number of eggs than when they are applied separately (Alves 2016), indicating that the simultaneous use of these fungi could be considered commercially (see also Chap. 3).

In all probability, most cases of natural control are the result of mixtures of antagonists and not by large populations of a single agent, complexes of microor-

Table 4.1 Thickness (μm) of some plant-parasitic nematode egg shells

Species	Approx. total thickness (μm)	References
Meloidogyne javanica	600	Bird and McClure (1976)
Globodera pallida	500	Lopez-Llorca and Robertson (1992b)
Heterodera glycines	400	Perry and Trett (1986)
Tylenchulus semipenetrans	400	Bird and McClure (1976)
Rotylenchulus reniformis	350	Bird and McClure (1976)
Heterodera schachtii	310	Perry and Trett (1986)
Heterodera avenae	250	Lopez-Llorca and Robertson (1992b)
Pratylenchus neglectus	210	Bird and McClure (1976)

ganisms being more stable and tending to have a better ecological balance compared to single organisms (Cook and Baker 1983). Thus, when environmental conditions are unfavourable for one antagonist, but not for another, natural control can still operate and vice versa (Cook and Baker 1983).

4.5.7 *Interactions of* Pochonia *with Nematode Hosts*

Different species of nematodes are parasitized by *P. chlamydosporia*, although there is variation in the *Pochonia*-nematode host interaction (Arora et al. 1996; Morton et al. 2003; Manzanilla-López et al. 2009a, b). The fungus infects eggs and for this it needs to breach the various layers or barriers in the egg shell (Fig. 4.2). The thickness of the egg shell varies among different plant-parasitic nematodes (Table 4.1). In some genera of nematodes, the egg shell is thicker and can hinder parasitism as, for example, in *Meloidogyne*.

The cross-linking of the egg shell proteins and polymorphism in serine proteases may have an influence on the susceptibility of nematodes to infection by *Pochonia* (Lopez-Llorca and Fry 1988; Segers et al. 1994, 1996; Morton et al. 2003). VCP1 removes the outer vitelline membrane of *M. incognita* eggs (Segers et al. 1994), but does not cause the same effect on *G. pallida* eggs (Segers et al. 1996). It is likely, therefore, that host preference is driven by differences in the composition of the nematode egg shell (Morton et al. 2004).

4.5.8 *Timing and Methods of Application of* Pochonia chlamydosporia *to the Soil*

The factors that influence the performance of *P. chlamydosporia* are not only genotypic or environmental: management practices are also fundamental. The method of application and its timing are essential for good establishment of the fungus in the soil and, ultimately, for successful biological control of PPN targets.

A study has shown that when the fungus is applied to sterile soil up to 20 days before the tomato crop, there is improved control of gall nematodes and, as a consequence, less damage when compared to application at planting (Podestá 2010). The explanation behind this is that the fungus has more time to colonize the soil and parasitise the nematode in the absence of the plant host and so, when the crop is sown, the nematode population is likely lower.

4.6 Concluding Remarks

Pochonia chlamydosporia has been effective in reducing root-knot and cyst nematode populations and increasing soybean and carrot yields in field tests in Brazil. In order to improve our understanding of the fungal mode of action in nature under different cropping systems, new techniques such as DNA probes, Real Time PCR, and handheld devices that can read strands of DNA could help to detect and to quantify strains of the fungus, both in the soil and in the roots of different plant species. The resulting information would help scientists, agronomists and farmers to plan management schemes that would enhance *Pochonia* populations and thus nematophagous activity for a more sustainable agriculture. We expect to see some growers replacing chemical nematicides with *P. chlamydosporia* in their production systems in the future. The simple implementation of changing to biological products to reduce the use of chemicals, and the amount of toxic residues in nature, would be a good start for a more environmentally-friendly farming system. A study of combining *P. chlamydosporia* with other biocontrol microorganisms, such as *Trichoderma*, *Metarhizium*, *Beauveria*, *Bacillus*, and other, as yet undiscovered taxa, has the potential to reduce nematodes, as well as other pathogens and pests, in an integrated pest management strategy. New technologies in the biopesticide industry are needed in order to improve mass production and formulation, thus helping the fungus to express its full potential in the field for better control of nematodes. *Pochonia*, due to its suspected high diversity (see Chapter 2), and potential for biocontrol, presents a challenge to researchers in terms of what remains to be discovered, especially about its interactions with nematodes in different agroecosystems.

References

Alves PS. (2016). Compatibilidade entre *Pochonia chlamydosporia* e *Trichoderma* spp. no controle de *Meloidogyne javanica* em tomateiro. MSc dissertation, Universidade Federal de Viçosa, MG, Brazil.

Aro, N., Pakula, T., & Penttila, M. (2005). Transcriptional regulation of plant cell wall degradation by filamentous fungi. *FEMS Microbiology Reviews, 29*, 719–739.

Arora, D. K., Hirsch, P. R., & Kerry, B. R. (1996). PCR-based molecular discrimination of *Verticillium chlamydosporium* isolates. *Mycological Research, 100*, 801–809.

Batista, A. C., & Fonseca, O. M. (1965). *Pochonia humicola* n. gen. e n. sp., uma curiosa entidade fúngica dos solos do Nordeste do Brasil. *Public Institute of Micological Recife, 462*, 1–11.
Bird, A. F., & Bird, J. (1991). *The structure of nematodes*. San Diego: Academic Press.
Bird, A. F., & McClure, M. A. (1976). The tylenchid (Nematoda) egg shell: structure, composition and permeability. *Parasitology, 72*, 19–28.
Bordallo, J. J., Lopez-Llorca, L. V., Jansson, H.-B., et al. (2002). Colonization of plant roots by egg-parasitic and nematode-trapping fungi. *The New Phytologist, 154*, 491–499.
Bourne, J. M., & Kerry, B. R. (1999). Effect of the host plant on the efficacy of *Verticillium chlamydosporium* as a biological control agent of root-knot nematodes at different nematode densities and fungal application rates. *Soil Biology and Biochemistry, 31*, 75–84.
Bourne, J. M., Kerry, B. R., & De Leij, F. A. A. M. (1996). The importance of the host plant on the interaction between root-knot nematodes (*Meloidogyne* spp.) and the nematophagous fungus *Verticillium chlamydosporium* Goddard. *Biocontrol Science and Technology, 6*, 539–548.
Brakhage, A. A. (2013). Regulation of fungal secondary metabolism. *Nature Reviews Microbiology, 11*, 21–32.
Ceiro, W. G., Arévalo, J., Puertas, A. N., et al. (2014). Efecto de concentraciones de NaCl sobre el crecimiento micelial y la esporulación de *Pochonia chlamydosporia* (Goddard) Zare y Gams en medio PDA y suelo. *Revista de Protección Vegetal, 29*, 122–127.
Chan, Y. L., Cai, D., Taylor, P. W. J., et al. (2010). Adverse effect of the chitinolytic enzyme PjCHI-1 in transgenic tomato on egg mass production and embryonic development of *Meloidogyne incognita*. *Plant Pathology, 59*, 922–930.
Clarke, A. J., Cox, P. M., & Shepherd, A. M. (1967). The chemical composition of the egg shells of the potato cyst-nematode, *Heterodera rostochiensis* Woll. *The Biochemical Journal, 104*, 1056–1060.
Cook, R. J., & Baker, K. F. (1983). Why biological control? In R. J. Cook & K. F. Baker (Eds.), *The nature and practice of biological control of plant pathogens* (pp. 1–29). St. Paul: American Phytopathological Society.
Dallemole-Giaretta, R., Freitas, L. G., Lopes, E. A., et al. (2015). *Pochonia chlamydosporia* promotes the growth of tomato and lettuce plants. *Acta Scientiarum Agronomy, 37*, 417–423.
Di Cera, E. (2009). Serine proteases. *International Union Biochemical Molecular Biological Life*, (5), 510–515.
Dos Santos, V. C., Curtis, R. H. C., & Abrantes, I. (2014). The combined use of *Pochonia chlamydosporia* and plant defense activators – a potential sustainable control strategy for *Meloidogyne chitwoodi*. *Phytopathologia Mediterranea, 1*, 66–74.
Escudero, N., & Lopez-Llorca, N. V. (2012). Effects on plant growth and root-knot nematode infection of an endophytic GFP transformant of the nematophagous fungus *Pochonia chlamydosporia*. *Symbiosis, 57*, 33–42.
Escudero, N., Marhuenda-Egea, F. C., Ibanco-Cañete, R., et al. (2014). A metabolomic approach to study the rhizodeposition in the tritrophic interaction: Tomato, *Pochonia chlamydosporia* and *Meloidogyne javanica*. *Metabolomics, 10*, 788–804.
Escudero, N., Ferreira, S. R., & Lopez-Moya, F. (2016). Chitosan enhances parasitism of *Meloidogyne javanica* eggs by the nematophagous fungus *Pochonia chlamydosporia*. *Fungal Biology, 120*, 572–585.
Esteves, I., Peteira, B., Atkins, S. D., et al. (2009). Production of extracellular enzymes by different isolates of *Pochonia chlamydosporia*. *Mycological Research, 113*, 867–876.
Ferreira, P. A., Ferraz, S., Lopes, E. A., et al. (2008). Parasitismo de ovos de *Meloidogyne exigua* por fungos nematófagos e estudo da compatibilidade entre os isolados fúngicos. *Revista Trópica: Ciencias Agrárias e Biológicas, 3*, 15–21.
Gayad, S. K. (1961). Production of symptoms of barley leaf spot disease by cultural filtrates of *Helminthosporium sativum*. *Nature, 191*, 725–726.
Hellwig, V., Mayer-Bartschmid, A., Muller, H., et al. (2003). Pochonins A-F, new antiviral and anti-parasitic resorcylic acid lactones from *Pochonia chlamydosporia* var. *catenulata*. *Journal of Natural Products, 66*, 829–837.

Huang, X., Zhao, N., & Zhang, K. (2004). Extracellular enzymes serving as virulence factors in nematophagous fungi involved in infection of the host. *Research in Microbiology, 155*, 811–816.

Huang, T.-C., Chang, H.-Y., Hsu, C.-H., et al. (2008). Targeting therapy for breast carcinoma by ATP synthase inhibitor Aurovertin B. *Journal of Proteome Research, 7*, 1433–1444.

Irving, F., & Kerry, B. R. (1986). Variation between strains of the nematophagous fungus *Verticillium chlamydosporium* Goddard. Factors affecting parasitism of cyst nematode eggs. *Nematologica, 32*, 474–485.

Kepler, R. M., Humber, R. A., Bischoff, J. F., et al. (2014). Clarification of generic and species boundaries for *Metarhizium* and related fungi through multigene phylogenetics. *Mycologia, 106*, 811–829.

Kerry, B. R. (2000). Rhizosphere interactions and the exploitation of microbial agents for the biological control of plant-parasitic nematodes. *Annual Review of Phytopathology, 38*, 423–441.

Kerry, B. R., & Bourne, J. M. (1996). Importance of rhizosphere interactions in the biological control of plant parasitic nematodes: a case study using *Verticillium chlamydosporium. Pesticide Science, 47*, 69–75.

Kerry, B. R., Irving, F., & Hornsey, J. C. (1986). Variation between strains of the nematophagous fungus *Verticillium chlamydosporium* Goddard. Factors affecting growth *in vitro*. *Nematologica, 32*, 461–473.

Khambay, B. P. S., Bourne, J. M., Cameron, S., et al. (2000). A nematicidal metabolite from *Verticillium chlamydosporium. Pest Management Science, 56*, 1098–1099.

Khan, A., Williams, K. L., & Nevalainen, H. K. M. (2004). Effects of *Paecilomyces lilacinus* protease and chitinase on the eggshell structures and hatching of *Meloidogyne javanica* juveniles. *Biological Control, 31*, 346–352.

Kumar, J., Schafer, P., Huckelhoven, R., et al. (2002). *Bipolaris sorokiniana*, a cereal pathogen of global concern: cytological and molecular approaches towards better control. *Molecular Plant Pathology, 3*, 185–195.

Larriba, E., Martín-Nieto, J., & Lopez-Llorca, L. V. (2012). Gene cloning, molecular modeling, and phylogenetics of serine protease P32 and serine carboxypeptidase SCP1 from nematophagous fungi *Pochonia rubescens* and *Pochonia chlamydosporia. Canadian Journal of Microbiology, 58*, 815–827.

Larriba, E., Jaime, M. D. L. A., Carbonell-Caballero, J., et al. (2014). Sequencing and functional analysis of the genome of a nematode egg-parasitic fungus, *Pochonia chlamydosporia. Fungal Genetics and Biology, 65*, 69–80.

Larriba, E., Jaime, M. D. L. A., Nislow, C., et al. (2015). Endophytic colonization of barley (*Hordeum vulgare*) roots by the nematophagous fungus *Pochonia chlamydosporia* reveals plant growth promotion and a general defense and stress transcriptomic response. *Journal of Plant Research, 128*, 665–678.

Li, J., Yu, L., Yang, J., et al. (2010). New insights into the evolution of subtilisin-like serine protease genes in Pezizomycotina. *BMC Evolutionary Biology, 10*, 1.

Liu, S. Q., Meng, Z. H., Yang, J. K., et al. (2007). Characterizing structural features of cuticle-degrading proteases from fungi by molecular modeling. *BMC Structural Biology, 7*, 1.

Lopez-Llorca, L. V. (1990). Purification and properties of extracellular proteases produced by the nematophagous fungus *Verticillium suchlasporium. Canadian Journal of Microbiology, 36*, 530–537.

Lopez-Llorca, L. V., & Boag, B. (1990). Inhibition of *Verticillium suchlasporium* and other nematophagous fungi by bacteria colonizing *Heterodera avenae* females. *Nematologia Mediterranea, 18*, 233–237.

Lopez-Llorca, L. V., & Claugher, D. (1990). Appressoria of the nematophagous fungus *Verticillium suchlasporium. Micron and Microscopica Acta, 21*, 125–130.

Lopez-Llorca, L. V., & Fry, S. C. (1988). Dityrosine, trityrosine and tetratyrosine, potential cross-links in structural proteins of plant-parasitic nematodes. *Nematologica, 35*, 165–179.

Lopez-Llorca, L. V., & Robertson, W. M. (1992a). Immumocytochemical localization of a 32-kDa protease from the nematophagous fungus *Verticillium suchlasporium* in infected nematode eggs. *Experimental Mycology, 16*, 261–267.

Lopez-Llorca, L. V., & Robertson, W. M. (1992b). Ultrastructure of infection of cyst nematode eggs by the nematophagous fungus *Verticillium suchlasporium*. *Nematologica, 39*, 65–74.
Lopez-Llorca, L. V., Moya, M., & Llinares, A. (1993). Effect of pH on growth and pigment production of nematophagous and entomogenous fungi. *Micologia e Vegetazione Mediterranea, 8*, 107–112.
Lopez-Llorca, L. V., Olivares-Bernabeu, C., Salinas, J., et al. (2002). Pre-penetration events in fungal parasitism of nematode eggs. *Mycological Research, 106*, 499–506.
Lopez-Llorca, L., Gómez-Vidal, S., Monfort, E., et al. (2010). Expression of serine proteases in egg-parasitic nematophagous fungi during barley root colonization. *Fungal Genetics and Biology, 47*, 342–351.
Luambano, N. D., Manzanilla-López, R. H., Kimenju, J. W., et al. (2015). Effect of temperature, pH, carbon and nitrogen ratios on the parasitic activity of *Pochonia chlamydosporia* on *Meloidogyne incognita*. *Biological Control, 80*, 23–29.
Manzanilla-López, R. H., Atkins, S. D., Clark, I. M., et al. (2009a). Measuring abundance, diversity and parasitic ability in two populations of the nematophagous fungus *Pochonia chlamydosporia* var. *chlamydosporia*. *Biocontrol Science and Technology, 19*, 391–406.
Manzanilla-López, R. H., Clark, I. M., Atkins, S. D., et al. (2009b). Rapid and reliable DNA extraction and PCR fingerprinting methods to discriminate multiple biotypes of the nematophagous fungus *Pochonia chlamydosporia* isolated from plant rhizospheres. *Letters in Applied Microbiology, 48*, 71–76.
Manzanilla-López, R. H., Esteves, I., Powers, S. J., et al. (2011). Effects of crop plants on abundance of *Pochonia chlamydosporia* and other fungal parasites of root-knot and potato cyst nematodes. *Annals of Applied Biology, 159*, 118–129.
Manzanilla-López, R. H., Esteves, I., Finetti-Sialer, M. M., et al. (2013). *Pochonia chlamydosporia*: Advances and challenges to improve its performance as a biological control agent of sedentary endo-parasitic nematodes. *Journal of Nematology, 45*, 1–7.
Manzanilla-López, R. H., Devonshire, J., Ward, E., et al. (2014). A combined cryo-scanning electron microscopy/cryoplaning approach to study the infection of *Meloidogyne incognita* eggs by *Pochonia chlamydosporia*. *Nematology, 16*, 1059–1067.
Medeiros, H. A., Resende, R. S., Ferreira, F. C., et al. (2015). Induction of resistance in tomato against *Meloidogyne javanica* by *Pochonia chlamydosporia*. *Nematoda, 2*, 10015–10022.
Mi, Q., Yang, J., Ye, F., et al. (2010). Cloning and over expression of *Pochonia chlamydosporia* chitinase gene pcchi44, a potential virulence factor in infecting against nematodes. *Process Biochemistry, 45*, 810–814.
Monfort, E., Lopez-Llorca, L. V., Jansson, H. B., et al. (2005). Colonisation of seminal roots of wheat and barley by egg-parasitic nematophagous fungi and their effects on *Gaemannomyces graminis* var. *tritici* & development of root-rot. *Soil Biology and Biochemistry, 37*, 1229–1235.
Monfort, E., Lopez-Llorca, L. V., Jansson, H. B., et al. (2006). *In vitro* soil receptivity assays to egg parasitic nematophagous fungi. *Mycological Progress, 5*, 18–23.
Monteiro TSA. (2013). Controle biológico do nematoide das galhas, *Meloidogyne javanica*, e promoção de crescimento vegetal com os fungos *Pochonia chlamydosporia* e *Duddingtonia flagrans*. MSc Dissertation, Universidade Federal de Viçosa, MG, Brazil.
Morgan-Jones, G., White, J. F., & Rodriguez-Kabana, R. (1983). Phytonematode pathology: ultrastructural studies. I. Parasitism of *Meloidogyne arenaria* eggs by *Verticillium chlamydosporium*. *Nematropica, 13*, 245–260.
Morton, C. O., Hirsch, P. R., Peberdy, J. P., et al. (2003). Cloning of and genetic variation in protease VCP1 from the nematophagous fungus *Pochonia chlamydosporia*. *Mycological Research, 107*, 38–46.
Morton, C. O., Hirsch, P. R., & Kerry, B. R. (2004). Infection of plant-parasitic nematodes by nematophagous fungi – review of the application of molecular biology to understand infection process and to improve biological control. *Nematology, 6*, 161–170.
Moulin, E., Barluenga, S., & Winssinger, N. (2005). Concise synthesis of Pochonin A, an HSP90 inhibitor. *Organic Letters, 25*, 5637–5639.

Nagesh, M., Hussaini, S. S., Ramanujam, B., et al. (2007). Molecular identification, characterization, variability and infectivity of Indian isolates of the nematophagous fungus *Pochonia chlamydosporia*. *Nematologia Mediterranea, 35*, 47–56.

Nahar, P., Ghormade, V., & Deshpande, M. V. (2004). The extracellular constitutive production of chitin deacetylase in *Metarhizium anisopliae*: possible edge to entomopathogenic fungi in the biological control of insect pests. *Journal of Invertebrate Pathology, 85*, 80–88.

Nasu EGC. (2013). Tratamento de sementes de soja e algodão com *Pochonia chlamydosporia* no controle de *Meloidogyne incognita* e histopatologia da inteiração tritrófica. Doctoral thesis, Universidade Federal de Viçosa, MG, Brazil.

Nicholson, R. L., & Moraes, W. B. C. (1980). Survival of *Colletotrichum graminicola*: importance of the spore matrix. *Phytopathology, 70*, 255–261.

Niu, X., Wang, Y., Chu, Y., et al. (2010). Nematode toxic aurovertin-type metabolites from a root-knot nematode parasitic fungus *Pochonia chlamydosporia*. *Journal of Food Agricultural Chemistry, 58*, 828–834.

Olivares-Bernabeu, C. M., & Lopez-Llorca, L. V. (2002). Fungal egg-parasites of plant- parasitic nematodes from Spanish soils. *Revista Iberoamericana de Micología, 19*, 104–110.

Palma-Guerrero, J., Jansson, H-B., Salinas, J., et al. (2008). Effect of chitosan on hyphal growth and spore germination of plant pathogenic and biocontrol fungi. *Journal of Applied Microbiology 104*, 541–553.

Palma-Guerrero, J., Gómez-Vidal, S., Tikhonov V. E., et al. (2010). Comparative analysis of extracellular proteins from *Pochonia chlamydosporia* grown with chitosan or chitin as main carbon and nitrogen sources. *Enzyme and Microbial Technology, 46*, 568–574.

Perry, R. N., & Trett, M. W. (1986). Ultrastructure of the egg shell of *Heterodera schachtii* and *H. glycines* (Nematoda: Tylenchida). *Revue de Nématologie, 9*, 399–403.

Perry, R. N., & Wharton, D. A. (2011). Survival of parasitic nematodes outside the host. In R. N. Perry & M. Moens (Eds.), *Molecular and physiological basis of nematode survival* (pp. 1–22). Wallingford: CAB International.

Podestá GS. (2010). Aplicação de *Pochonia chlamydosporia* em pré-plantio para potencializar o controle de *Meloidogyne javanica* em tomate e alface. MSc dissertation, Universidade Federal de Viçosa, MG, Brazil.

Podestá, G. S., Dallemole-Giaretta, R., Freitas, L. G., et al. (2009). Atividade nematófaga de *Pochonia chlamydosporia* em solo natural ou autoclavado sobre *Meloidogyne javanica*. *Nematologia Brasileira, 32*, 191–193.

Segers, R., Butt, T., Kerry, B. R., et al. (1994). The nematophagous fungus *Verticillium chlamydosporium* produces a chymoeslastase-like protease which hydrolyses host nematode proteins in situ. *Microbiology, 140*, 2715–2723.

Segers, R., Butt, T. M., Kerry, B. R., et al. (1995). The subtilisins of the invertebrate mycopathogens *Verticillium chlamydosporium* and *Metarhizium anisopliae* are functionally and serologically related. *FEMS Microbiology Letters, 126*, 227–231.

Segers, R., Butt, T. M., Kerry, B. R., et al. (1996). The role of the proteinase VCPI produced by the nematophagous fungus *Verticillium chlamydosporium* in the infection process of nematode eggs. *Mycological Research, 100*, 421–428.

Tikhonov, V. E., Lopez-Llorca, L. V., Salinas, J., et al. (2002). Purification and characterization of chitinases from the nematophagous fungi *Verticillium chlamydosporium* and *V. suchlasporium*. *Fungal Genetics and Biology, 35*, 67–78.

Tucker, S. L., & Talbot, N. J. (2001). Surface attachment and pre-penetration stage development by plant pathogenic fungi. *Annual Review of Phytopathology, 39*, 385–417.

Verdejo-Lucas, S., Sorribas, F. J., Ornat, C., et al. (2003). Evaluating *Pochonia chlamydosporia* in a double-cropping system of lettuce and tomato in plastic houses infested with *Meloidogyne javanica*. *Plant Pathology, 52*, 521–528.

Wang, E. L. H., & Bergeson, G. B. (1974). Biochemical changes in root exudate and xylem sap of tomato plants infected with *Meloidogyne incognita*. *Journal of Nematology, 6*, 194–202.

Wang, Y. L., Li, L. F., Li, D. X., et al. (2015). Yellow pigment aurovertins mediate interactions between the pathogenic fungus *Pochonia chlamydosporia* and its nematode host. *Journal of Agricultural and Food Chemistry, 63*, 6577–6587.

Ward, E., Kerry, B. R., Manzanilla-López, R. H., et al. (2012). The *Pochonia chlamydosporia* serine protease gene *vcp1* is subject to regulation by carbon, nitrogen and pH: implications for nematode biocontrol. *PloS One, 7*, e35657.

Wharton, D. A. (2011). Cold tolerance. In R. N. Perry & D. A. Wharton (Eds.), *Molecular and physiological basis of nematode survival* (pp. 182–204). Wallingford: CAB International.

Yang, J., Tian, B., Liang, L., et al. (2007). Extracellular enzymes and the pathogenesis of nematophagous fungi. *Applied Microbiology and Biotechnology, 75*, 21–31.

Yen, J.-H., Niblack, T. L., Karr, A. L., et al. (1996). Seasonal biochemical changes in eggs of *Heterodera glycines* in Missouri. *Journal of Nematology, 28*, 442–450.

Zare, R., Gams, W., & Culham, A. (2000). A revision of *Verticillium* sect. *Prostrata* I. Phylogenetic studies using ITS sequences. *Nova Hedwigia, 71*, 465–480.

Zare, R., Gams, W., & Evans, H. C. (2001). A revision of *Verticillium* section *Prostrata*. V. The genus *Pochonia*, with notes on *Rotiferophthora*. *Nova Hedwigia, 73*, 51–86.

Zavala-Gonzalez, E. A., Escudero, N., Lopez-Moya, F., et al. (2015). Some isolates of the nematophagous fungus *Pochonia chlamydosporia* promote root growth and reduce flowering time of tomato. *The Annals of Applied Biology, 166*, 472–483.

Zou, C.-S., Mo, M.-H., Y-Q, G., et al. (2007). Possible contributions of volatile-producing bacteria to soil fungistasis. *Soil Biology and Biochemistry, 39*, 2371–2379.

Part III
-*Omics* and Other Approaches

Chapter 5
Regulatory Factors in *Pochonia chlamydosporia*-Induced Gene Expression

Aurelio Ciancio, Isabella Pentimone, Mariantonietta Colagiero, and Laura Rosso

Abstract Soil microorganisms respond to, and release signals in, the rhizosphere, affecting root growth and their interactions with parasites. Understanding the gene expression patterns that are active in roots is fundamental for successful exploitation of beneficial associations. *Pochonia chlamydosporia* is a facultative parasite of nematode eggs and a growth-promoting endophyte. Transcriptomic studies from roots colonized by *P. chlamydosporia* showed that the fungus differentially regulates several genes. These included transcription factors and microRNAs differentially expressed during root endophytism. Both are transcriptional regulators and form integral parts of signalling webs, modulating many biological processes. The transcription factors involved in defence, resistance and plant growth include WRKYs, a family of genes expressed either in *P. chlamydosporia*-colonized or control roots. *In vitro* studies on endophytism also showed differential expression of 26 miRNAs, with 154 potential target genes involved in apoptosis, metabolism and binding, including transcription factors. The differential gene expression induced by *P. chlamydosporia* in the presence of nematodes or other pathogens may disclose novel pest and disease management strategies. The fungus transcriptomic analyses also support the production of industrial and commercial bioformulations for plant protection, through the induction of endogenous plant defence mechanisms.

A. Ciancio (✉) • I. Pentimone • M. Colagiero • L. Rosso
Istituto per la Protezione Sostenibile delle Piante, CNR,
Via G. Amendola 122/D, 70126 Bari, Italy
e-mail: aurelio.ciancio@ipsp.cnr.it; isabella.pentimone@ipsp.cnr.it; mariantonietta.colagiero@ipsp.cnr.it; laura.rosso@ipsp.cnr.it

R.H. Manzanilla-López, L.V. Lopez-Llorca (eds.), *Perspectives in Sustainable Nematode Management Through* Pochonia chlamydosporia *Applications for Root and Rhizosphere Health*, Sustainability in Plant and Crop Protection,
DOI 10.1007/978-3-319-59224-4_5

5.1 Introduction

Biological pest management is a promising field in crop protection, providing alternative tools that appear to be more sustainable and safer than the majority of pesticides currently in use. Although the use of beneficial fungi in nematode management is a relatively 'old' idea – discussed by L.C. Duddington as early as 1956 – many aspects concerning their biology and interactions with roots and nematodes have been only recently explored. Data are required at the biochemical and molecular levels on the several fungi interacting with plants and other soil organisms, in order to warrant reproducible and consistent exploitation of most useful associations.

Pochonia chlamydosporia is a rhizosphere colonizer of many plants where it occurs as a facultative parasite of nematode eggs. The fungus is also a widespread root endophyte with growth-promoting effects observed on different plant species (Maciá-Vicente et al. 2009; Escudero and Lopez-Llorca 2012). Its capacity for root colonization may vary depending on isolate and host plants. Rotations have been proposed as a means to increase its density in soil, treating winter crops with chlamydospores (Kerry and Bourne 1996; Kerry 2000). A detailed insight into the fungus biology is, however, needed to determine its potential and validate its application.

Recent studies have shown that the genome of *P. chlamydosporia* is highly enriched in genes encoding hydrolytic enzymes, expressed during root colonization or egg degradation (Larriba et al. 2014). In many isolates of *P. chlamydosporia*, a high endophytic capacity is coupled to high rates of root colonization and root-knot nematode (RKN) parasitism. These traits suggest a potential of *P. chlamydosporia* in the biocontrol of nematode pests and as a rhizosphere microbial component underpinning both plant health and nutrition (Manzanilla-Lopez et al. 2009; 2013). In this chapter, some aspects of the fungus biology are reviewed, with particular attention to regulating factors affecting gene expression in colonized roots.

5.2 Transcription Factors

Significant changes are induced by *P. chlamydosporia* when colonizing roots, as shown by a number of studies carried out by means of transcriptomic analyses (Rosso et al. 2011; Ciancio et al. 2013; Larriba et al. 2014, 2015).

Microarray data on barley roots endophytically colonized by *P. chlamydosporia* showed that 1.2% of the 22,792 array probesets were differentially expressed. These included 13 upregulated probesets annotated as Gene Onthology (GO) 'molecular functions' related to Transcription Factors (TF) gene families. Furthermore, a functional gene classification based on MapMan onthology (Ramšak et al. 2014) showed 55 differentially regulated probesets involved in the response to biotic and abiotic stress. These were mainly upregulated heat shock proteins (HSP) and other probesets related to protein degradation, with 19 differentially regulated expressed sequence tags involved in the response to abiotic stress, including heat shock factors (HSF), GATA, MADS, C2H2-ZNF and WRKY[11] TFs (Larriba et al. 2015).

Next generation sequencing (NGS) data from dixenic *in vitro* assays with tomato roots showed that the fungus induces a differential expression of many genes and a deep metabolic re-programming. A number of metabolic pathways specific for transport, secondary metabolism, stress and defence response were highly expressed in the presence of endophytic *P. chlamydosporia*, while photosynthesis-related genes were down-regulated (Ciancio et al. 2013). The study showed that, within the 87 WRKY annotated in the *Solanum lycopersicum* genome, 20 were differentially expressed in the presence of *P. chlamydosporia*. WRKYs play a fundamental role in the activation of the plant immune system and also affect many processes needed for growth and development (Eulgem et al. 2000; Bhattarai et al. 2010). Mechanisms of signalling and transcriptional regulation uncovered WRKY protein functions via interactions with a diverse array of partners, including MAP kinases 14-3-3 proteins, calmodulin, histone deacetylases, resistance proteins and others. In *Arabidopsis*, more than 70% of the 100 genes forming the WRKY family respond to infections by pathogens and to treatments with salicylic acid (Chen et al. 2010). Actual knowledge on WRKY systems supports further studies on the genetic bases of their activation at the cellular level.

In tomato, common mechanisms regulate the transcriptional activation of WRKYs associated with defence response, including a group of DNA-binding proteins containing WRKY domains. However, the role of these interacting proteins in the *P. chlamydosporia* regulation of WRKYs and of their action is still unclear. It is possible that these TFs act in association with specific target genes during biological processes such as endophytism or parasitism. A particularly important aspect concerns the integration of these dynamic and complex protein-protein and protein-DNA interactions in the WRKY-mediated transcription of important target genes. Research on this aspect is in progress, to develop a comprehensive understanding of the WRKYs signalling and transcriptional regulatory networks, through chromatin immunoprecipitation and other molecular or genomic tools. A framework for WRKYs signalling pathways in the root-fungus interaction must also be constructed based on data on regulation by signalling proteins, activating or repressing target genes in the cell nuclei affected by the endophyte.

Since many TFs regulate a large number of genes, a further issue concerns the identification of new TFs-target gene associations under different stress conditions, as well as the elucidation of their regulation patterns (Table 5.1). To date, WRKYs have been associated with many biotic and abiotic stress factors in plants (Li et al. 2012). Among the genes involved in nematode resistance observed in tomato by microarray analysis, WRKY72 acts as an *Mi-1* gene activating factor, and utilizes gene mechanisms independent from the salicylic acid defensive pathways (Bhattarai et al. 2010). *Mi-1* is also expressed in healthy tomato roots colonized endophytically by *P. chlamydosporia* (see next paragraph). This complex of resistance genes and associated

[1]WRKY: a superfamily of TFs unique for plants, involved in regulation of plant development and response to biotic or abiotic stress.

Table 5.1 WRKY transcription factors found as differentially expressed in tomato roots with endophytic *Pochonia chlamydosporia* (Pc), with corresponding classification and functional conditions as reported in the literature

TFs	Group	Associated stress conditions reported	References
Up regulated			
WRKY3	I	n.a.	
WRKY5	I	Drought	Huang et al. (2012)
WRKY23	II-c	Salt, *Clavibacter* infection or fungal stress	Huang et al. (2012)
WRKY29	II-e	n.a.	
WRKY34	I	Pollen-specific, male gametogenesis	Guan et al. (2014)
WRKY37	II-e	n.a.	
WRKY73	II-b	Tomato spotted wilt virus (TSWV), in roots only	Huang et al. (2012)
WRKY75	II-c	TSWV	Huang et al. (2012)
Down regulated			
WRKY2	I	TSWV, in roots only	Huang et al. (2012)
WRKY12	II-c	n.a.	
WRKY13	II-c	n.a.	
WRKY16	II-b	TSWV	Huang et al. 2012
WRKY17	II-b	Fungal stress (at 4 days only); negative regulator of resistance in *Arabidopsis*	Huang et al. (2012), Journot-Catalino et al. (2006)
Expressed only in roots with Pc			
WRKY31	I	Drought	Huang et al. (2012)
WRKY32	I	Drought	Huang et al. (2012)
WRKY72	II-b	RKN	Bhattarai et al. (2010)

TFs suggests a possible means of exploitation in managing RKN in tomato crops. However, *Mi* genes become inactivated at temperatures higher than 28 °C and in many field situations they cannot be fully exploited for RKN control (Branch et al. 2004).

Recent data concerning the genome-wide mapping of protein-DNA interactions enable the identification of many transcriptional regulation mechanisms. A precise map of binding sites for TFs and other DNA binding proteins is vital for deciphering the gene regulatory networks that underlie various biological processes, including the plant defence response to pathogens (Zang et al. 2009; Palii et al. 2010). To date, there are few data available on the transcriptome of *P. chlamydosporia* in variable conditions, although the mechanisms of WRKY activation in the presence of different stress factors have been identified (Li et al. 2012). These pathways possibly underpin the growth promotion effects observed for *P. chlamydosporia*. The fungus was found to promote root and shoot weight in wheat, irrespective of infection by the root pathogenic fungus *Gaeumannomyces graminis* var. *tritici* (Monfort et al. 2005). However, details about the molecular pathways active in the response of parasitized roots in the presence of endophytic *P. chlamydosporia* are as yet unknown.

Next generation sequencing (NGS) data from tomato roots suggest that several WRKYs are involved in the tri-trophic plant-pathogen-endophyte interaction. However, the beneficial role of *P. chlamydosporia* in eliciting an early reaction to a root invading pathogen is just being probed. This aspect is particularly intriguing, since the fungus activity as a defence promoter is possibly the result of an evolutionary adaptation to its specific telluric microhabitat. Some hypotheses may be formulated in this regard. The first one considers that this behaviour might have arisen as the adaptation of a former pathogen that evolved to become a safe endosymbiont (eliciting a minimal defence reaction). Alternatively it may be the result of a complex adaptation of a soil biocontrol agent that evolved to become a root 'alerting' endophyte. The fungus metabolism as an egg parasite indeed contrasts with its activity as an early inducer of RKN resistance (see next paragraph). An early plant defensive response would in fact reduce the number of nematode eggs produced (limiting by this way the fungus's own food source). This complex behaviour may hence hide a more complex evolutionary scenario, a trace of which may be found in the cascade of genes and TFs activated in roots. Similarities may exist with the ecology of *Metarhizium* spp., a genus phylogenetically close to *Pochonia* (Larriba et al. 2014). These fungi act as hyperparasites of insects in soil. However, they return to the roots (through an endophytic phase) the nutrients that the insects took away, with a probable positive benefit (Behie et al. 2012). When this scenario can be shown for *P. chlamydosporia*, it will shed light on the evolutionary adaptations underpinning the observed interactions.

5.3 Resistance Genes

A global analysis of transcription profiles, applied to roots of tomato inoculated with *P. chlamydosporia*, show that the fungus induces the expression of several transcripts related to genes involved in a plant defence response. Up-regulated transcripts include members of the disease resistance gene family encoding nucleotide-binding site leucine-rich repeats (NBS-LRR), proteins like coiled-coil (CC) motifs CC-NBS-LRR, TIR-NBS-LRR (Toll/Interleukin-1 Receptor homology) and bed finger-NBS-LRR. Plant NBS-LRR proteins act through a network of signalling pathways and induce a series of hypersensitive defence responses, such as the activation of an oxidative burst, calcium and other ion fluxes, and mitogen-associated protein kinase cascade, as well as the induction of pathogenesis-related genes (Hammond-Kosack and Parker 2003; Belkhadir et al. 2004; Pedley and Martin 2005). In particular, transcripts differentially expressed in tomato roots inoculated with *P. chlamydosporia* encode proteins of the CC-NBS-LRR (CNLs) subfamily, whose products contain CC motifs in the amino-terminal domain and are involved in pathogens recognition (Rosso et al. 2013).

Genes involved in compatible or incompatible nematode responses are expressed in *P. chlamydosporia*-inoculated tomato roots, including *Mi*, a RKN and aphid resistance gene partially regulated by a salicylic acid (SA)-dependent defence pathway (Branch et al. 2004; Bhattarai et al. 2008). It confers resistance to RKN through PRF and Mi proteins, and holds several structural motifs linked to stress response, including a nucleotide binding site (NBS) and a carboxy-terminal LRR region. Endophytism by *P. chlamydosporia* also induces expression of several genes involved in salicylic acid (SA)-dependent signalling. The SA signal transduction pathway is crucial for resistance to foliar pathogens, through SA-responsive PR (Pathogenesis-related) genes, including PR-1 (unknown function), PR-2 (a β-1,3-glucanase) and PR-5 (a thaumatin-like protein) (Cao et al. 1998; Glazebrook 2005).

Many PR-5 proteins possess antifungal activity as shown by *in vitro* experiments. A PR-5 protein from *Arabidopsis thaliana* ATLP3, expressed in recipient bacterial cells, showed considerable antifungal activity against pathogenic fungi such as *Verticillium alboatrum*, *V. dahliae* and *Fusarium oxysporum*, as well as against non-pathogenic fungi such as *Trichoderma reesei* and *Candida albicans* (Xu et al. 1997). The antifungal activity was associated with an increased permeability of fungal membranes by pore-formation mechanisms (Anžlovar et al. 1998; Selitrennikoff 2001; Anžlovar and Dermastia 2003).

In tomato roots inoculated with *P. chlamydosporia*, transcripts for thaumatin-like proteins (TLP-proteins) are differentially repressed or induced at different levels. The TLP-protein (PR5) family includes the closely related proteins permatin, zeamatin, and osmotin, the latter showing up- or down-regulated transcripts. Zeamatin has been shown to cause fungal hyphae to leak and rupture by forming pores in the plasma membrane (Roberts and Selitrennikoff 1990). Osmotin causes membrane leakage and dissipation of pH gradient across the membrane (Abad et al. 1996).

Gene expression studies done on *P. chlamydosporia* in a saprotrophic-to-parasitic transition show differences among several genes involved in cellular signals, transport or DNA repair, with a distinct cluster of genes commonly expressed during parasitic or poor nutritional conditions (Rosso et al. 2011). In this case, the fungus was tested on different media, simulating starvation or presence of nematode eggs, and not in its endophytic phase. The factors inducing the switch of *P. chlamydosporia* from a saprophytic or endophytic trophism towards parasitism are still unknown, but soil nutrients and root exudates were observed to affect its growth and behaviour. As with other parasitic fungi, molecules rich in carbon inhibit the production of enzymes involved in nematode infection, whereas high levels of organic matter in soil appear to stimulate growth of *P. chlamydosporia*, but not its biocontrol capacity (Kerry and Bourne 1996).

The chemical composition of the host surface is a key factor in the induction of the specific hydrolytic enzymes involved in infection (Tunlid and Jansson 1991). In the *P. chlamydosporia* and nematode interaction, most of the knowledge about the fungus metabolism concerns VCP1, a specific enzyme degrading outer proteins of the egg shell. Fast metabolised C sources like glucose or, for a few hours, ammonium chloride and unfavourable acidic pH in the rhizosphere or in the egg-mass appeared to compromise the levels of nematode parasitism by affecting the production of VCP1 (Ward et al. 2012). The release of VCP1 is a single step in a complex biochemical cascade of gene induction and activation processes that is largely unknown (Lopez-Llorca et al. 2010; Ward et al. 2012). Assays on healthy and RKN-parasitized plants showed the early induction of several specialized genes, including a phytase activated 4 and 8 h after incubation. Phytic acid was observed to induce a mycelial growth stimulation at 5–10 ppm, suggesting that the fungus can rapidly rely on this molecule as a suitable P source (Rosso et al. 2014).

5.4 Small RNAs

Small RNAs (sRNAs) regulate critical effector genes in many infection processes (Ruiz-Ferrer and Voinnet 2009; Peláez and Sanchez 2013; Yang and Huang 2014; Weiberg et al. 2014). In particular, the microRNAs (miRNAs), are involved in a variety of developmental processes in plants and animals (Axtell et al. 2011). They are endogenous 21–24 nucleotide non-coding, single-stranded RNA species that act as gene regulators. Usually, miRNAs derive from precursor transcripts by the action of two RNase III-type proteins, Drosha and Dicer. Drosha cleaves primary transcripts (pri-miRNA) to yield a stem-loop structure intermediate, known as the precursor miRNA (pre-miRNA). The pre-miRNA is further cleaved by Dicer to release the miRNA/miRNA* duplex (Ketting et al. 2001; Lee et al. 2003). One strand of the RNA duplex, the miRNA, is incorporated stably into the RNA-induced silencing complex (RISC), while the other strand, the miRNA*, is degraded. The RISC loaded with miRNAs targets mRNAs for translational repression or mRNA cleavage at the post transcriptional level. In this process mRNAs are recognized in either near perfect or lacking full base complementarity by their miRNAs (Schwarz et al. 2003).

The regulatory effects of miRNAs and other non-coding (nc)RNAs in the interactions between plant and endophytic *P. chlamydosporia* were studied by NGS of tomato roots inoculated with the fungus and controls. Endophytism affects the pattern of miRNAs expression in tomato, with 26 miRNAs differentially expressed between the two treatments. Among 154 potential tomato target genes, there are those involved in apoptosis, primary metabolism and binding functions, i.e. Squamosa TF proteins. Comparative analysis with previous RNAseq data show that 48 out of 5055 *P. chlamydosporia* down-regulated tomato genes are miRNA targets. Furthermore, five miRNAs (sly-miR9473-5p, sly-miR169c, sly-miR169a, sly-miR9476-5p and sly-miR1918) are only present in the fungus colonized roots (Pentimone et al. 2015a).

Other sRNAs, including transfer RNA (tRNA)-derived sRNAs, are also differentially expressed between control and fungus colonized tomato roots. Data suggest that miRNAs and other sRNAs play an active role in the complex interactions between roots and endophyte (Pentimone et al. 2015b).

5.5 Conclusions and Future Perspectives

A practical aspect of the *P. chlamydosporia* genomic studies is represented by exploitation of the fungus as a biopesticide or as a plant growth promoter. *Pochonia chlamydosporia* has already reached the stage of industrial production in the EU

(Italy, Spain, Portugal), Cuba, Brazil and other countries, and there are efforts aimed at optimizing its production, formulation and application (see Chap. 15). 'Green products' based on *P. chlamydosporia* aim at IPM (integrated pest management) practices that are competitive with the soil fumigants and pesticides which today are almost banned in the EU. Given the lack of suitable nematode management tools, there is a commercial interest in developing safer, biological practices that may fill the market space left by the banned or restricted chemicals. A better understanding of *P. chlamydosporia* behaviour in the rhizosphere is hence needed at the applied level so as to improve its formulation and/or use in the field. Besides, gene expression data may represent a screening tool, for comparison of isolate suitability, based on known fungus-root-nematode interactions.

Next generation sequencing data are expected to yield key information on plant cellular processes and gene expression, as well as on root physiology and metabolism. At the cellular level, there are still questions concerning the functioning and activation of WRKYs and of their activated gene targets, an obligate step for the elucidation of the plant response to biotic stress.

A practical aspect concerns the exploitation of the molecular data produced. Many pathogens are severe limiting factors for many crops worldwide and the study of useful soil microorganisms represents a research challenge. Actual IPM efforts are underpinned by the quest for safe microorganisms, suitable for sustainable management of pests and diseases, as these are needed worldwide. Any new data produced will therefore useful in sustainable IPM in the field, supporting industrial production by providing species or isolates that act as efficient and effective biopesticides or as plant growth promoters.

The microbial complexity of rhizosphere and root endophytes is a natural resource that has been poorly exploited thus far, since microorganisms represent an advanced solution to specific crop problems. Experimental evidence has revealed beneficial effects of treatments with fungi like *P. chlamydosporia*, including isolates available for industrial production. These effects have been shown on horticultural crops in particular, as productivity increases in the range of 30% were observed in field assays on glasshouse zucchini in Southern Italy (Pietrantonio et al. 2013). Root endophytes and egg parasites will therefore provide the basis for commercial products with rhizosphere amendment properties, reconstructing a balance in soils that are often compromised by previous misuse or excessive pesticide use. *Pochonia chlamydosporia* is not pathogenic to vertebrates and offers the double advantage of being involved in plant nutrition and protection, yet without risks for operators and the environment, issues that are highly attractive to the general public. These products therefore meet the demand from growers and consumers concerning 'green' technical factors in sustainable agriculture and to increase the productivity of nurseries and horticultural crops.

The molecular studies on *P. chlamydosporia* and related species are justified by possible benefits that may be derived from their application. MiRNAs or other products may be also proposed as putative signal molecules, useful in management of pests and diseases and/or promoting plant growth. Given their specifically targeted effects, artificial manipulation of miRNAs through differential gene expression may emerge as a novel alternative approach in crop management or pest control. The development of research activities in this field will undoubtedly contribute to increase benefits from microbial activity and diversity, allowing a safer use of soil and natural resources through knowledge-based approaches.

References

Abad, L. R., D'Urzo, M. P., Liu, D., et al. (1996). Antifungal activity of tobacco osmotin has specificity and involves plasma membrane permeabilization. *Plant Science, 118*, 11–23.

Anžlovar, S., & Dermastia, M. (2003). The comparative analysis of osmotins and osmotin-like PR-5 proteins. *Plant Biology, 5*, 116–124.

Anžlovar, S., Serra, M. D., Dermastia, M., et al. (1998). Membrane permeabilizing activity of pathogenesis-related protein linusitin from flax grain. *Molecular Plant-Microbe Interactions, 11*, 610–617.

Axtell, M. J., Westholm, J. O., & Lai, E. C. (2011). Vive la difference: Biogenesis and evolution of microRNAs in plants and animals. *Genome Biology, 12*, 221. doi:10.1186/gb-2011-12-4-221.

Behie, S. W., Zelisko, P. M., & Bidochka, M. J. (2012). Endophytic insect-parasitic fungi translocate nitrogen directly from insects to plants. *Science, 336*, 1576–1577.

Belkhadir, Y., Subramaniam, R., & Dangl, J. L. (2004). Plant disease resistance protein signaling: NBS-LRR proteins and their partners. *Current Opinion in Plant Biology, 7*, 391–399.

Bhattarai, K. K., Xie, Q. G., Mantelin, S., et al. (2008). Tomato susceptibility to root-knot nematodes requires an intact jasmonic acid signaling pathway. *Molecular Plant-Microbe Interactions, 21*, 1205–1214.

Bhattarai, K. K., Atamian, H. S., Kaloshian, I., et al. (2010). WRKY72-type transcription factors contribute to basal immunity in tomato and *Arabidopsis* as well as gene-for-gene resistance mediated by the tomato R gene Mi-1. *The Plant Journal, 63*, 229–240.

Branch, C., Hwang, C. F., Navarre, D. A., & Williamson, V. M. (2004). Salicylic acid is part of the mi-1-mediated defense response to root-knot nematode in tomato. *Molecular Plant-Microbe Interactions, 17*, 351–356.

Cao, H., Li, X., & Dong, X. (1998). Generation of broad-spectrum disease resistance by overexpression of an essential regulatory gene in systemic acquired resistance. *Proceedings of the National Academic Sciences USA, 95*, 6531–6536.

Chen, H., Lai, Z., Shi, J., et al. (2010). Roles of *Arabidopsis* WRKY18, WRKY40 and WRKY60 transcription factors in plant responses to abscisic acid and abiotic stress. *BMC Plant Biology, 10*, 281.

Ciancio, A., Colagiero, M., Ferrara, M., et al. (2013, July 21–25). Transcriptome changes in tomato roots during colonization by the endophytic fungus *Pochonia chlamydosporia*. 5th congress of European Microbiologists (FEMS), Leipzig, Germany.

Escudero, N., & Lopez-Llorca, L. V. (2012). Effects on plant growth and root-knot nematode infection of an endophytic GFP transformant of the nematophagous fungus *Pochonia chlamydosporia*. *Symbiosys, 57*, 33–42.

Eulgem, T., Rushton, P. J., Robatzek, S., et al. (2000). The WRKY superfamily of plant transcription factors. *Trends in Plant Science, 5*, 199–206.

Glazebrook, J. (2005). Contrasting mechanisms of defense against biotrophic and necrotrophic pathogens. *Annual Review of Phytopathology, 43*, 205–227.

Guan, Y., Meng, X., Khanna, R., et al. (2014). Phosphorylation of a WRKY transcription factor by MAPKs is required for pollen development and function in *Arabidopsis*. *PLoS Genetics, 10*, e1004384.

Hammond-Kosack, K. E., & Parker, J. E. (2003). Deciphering plant-pathogen communication: Fresh perspectives for molecular resistance breeding. *Current Opinion in Biotechnology, 14*, 177–193.

Huang, S., Gao, Y., Liu, J., et al. (2012). Genome-wide analysis of WRKY transcription factors in *Solanum lycopersicum. Molecular Genetics and Genomics, 287*, 495–513.

Journot-Catalino, N., Somssich, I. E., Roby, D., et al. (2006). The transcription factors WRKY11 and WRKY17 act as negative regulators of basal resistance in *Arabidopsis thaliana. Plant Cell, 18*, 3289–3302.

Kerry, B. R. (2000). Rhizosphere interactions and the exploitation of microbial agents for the biological control of plant-parasitic nematodes. *Annual Review of Phytopathology, 38*, 423–424.

Kerry, B. R., & Bourne, J. M. (1996). The importance of rhizosphere interactions in the biological control of plant parasitic nematodes: A case study using *Verticillium chlamydosporium. Pesticide Science, 47*, 69–75.

Ketting, R. F., Fischer, S. E. J., Bernstein, E., et al. (2001). Dicer functions in RNA interference and in synthesis of small RNA involved in developmental timing in *C. elegans. Genes & Development, 15*, 2654e2659.

Larriba, E., Jaime, M. D., Carbonell-Caballero, J., et al. (2014). Sequencing and functional analysis of the genome of a nematode egg-parasitic fungus, *Pochonia chlamydosporia. Fungal Genetics and Biology, 65*, 69–80.

Larriba, E., Jaime, M. D. L. A., Nislow, C., et al. (2015). Endophytic colonization of barley (*Hordeum vulgare*) roots by the nematophagous fungus *Pochonia chlamydosporia* reveals plant growth promotion and a general defense and stress transcriptomic response. *Journal of Plant Research, 128*, 665–678.

Lee, Y., Ahn, C., Han, J. J., et al. (2003). The nuclear RNase III Drosha initiates microRNA processing. *Nature, 425*, 415e419.

Li, J., Luan, Y., & Jin, H. (2012). The tomato SlWRKY gene plays an important role in the regulation of defense responses in tobacco. *Biochemical and Biophysical Research Communications, 427*, 671–676.

Lopez-Llorca, L. V., Gómez-Vidal, S., Monfort, E., et al. (2010). Expression of serine proteases in egg-parasitic nematophagous fungi during barley root colonization. *Fungal Genetics and Biology, 47*, 342–351.

Maciá-Vicente, J., Rosso, L. C., Ciancio, A., et al. (2009). Colonisation of barley roots by endophytic *Fusarium equiseti* and *Pochonia chlamydosporia*: Effects on plant growth and disease. *The Annals of Applied Biology, 155*, 391–401.

Manzanilla-López, R. H., Atkins, S. D., Clark, I., et al. (2009). Measuring abundance, diversity and parasitic ability in two populations of the nematophagous fungus *Pochonia chlamydosporia* var. *chlamydosporia. Biocontrol Science and Technology, 19*, 391–406.

Manzanilla-López, R. H., Esteves, I., Finetti-Sialer, M. M., et al. (2013). *Pochonia chlamydosporia*: Advances and challenges to improve its performance as a biological control agent of sedentary endo-parasitic nematodes. *Journal of Nematology, 45*, 1–7.

Monfort, E., Lopez-Llorca, L. V., Jansson, H. B., et al. (2005). Colonisation of seminal roots of wheat and barley by egg-parasitic nematophagous fungi and their effects on *Gaemannomyces graminis* var. *tritici* & development of root-rot. *Soil Biology and Biochemistry, 37*, 1229–1235.

Palii, C. G., Perez-Iratxeta, C., Yao, Z., et al. (2010). Differential genomic targeting of the transcription factor TAL1 in alternate haematopoietic lineages. *The EMBO Journal, 30*, 494–509.

Pedley, K. F., & Martin, G. B. (2005). Role of mitogen-activated protein kinases in plant immunity. *Current Opinion in Plant Biology, 8*, 541–547.

Peláez, P., & Sanchez, F. (2013). SmallRNAs in plant defense responses during viral and bacterial interactions: Similarities and differences. *Frontiers in Plant Science, 4*, 343. doi:10.3389/fpls.2013.00343.

Pentimone, I., Lebrón, R., Hackeberg, M., et al. (2015a). Expression profile of non-coding small RNAs in tomato roots during *Pochonia chlamydosporia* endophytism. *Nematropica, 45*, 13–14.

Pentimone, I., Rosso, L., Nigro, F. et al (2015b, June 7–11) Role and activity of miRNAs in the interaction of tomato roots with endophytic *Pochonia chlamydosporia*. *FEMS*, 6th congress of European microbiologists, Maastricht.

Pietrantonio, L., Di Cillo, P., Pignoli, G., et al (2013, October 22–23) Green technologies for sustainable management of field crops: Use and potentialities of the fungus *Pochonia chlamydosporia*. Italian Forum on Industrial Biotechnology and Bioeconomy (IFIB), Naples, Italy.

Ramšak, Ž., Baebler, Š., Rotter, A., et al. (2014). GoMapMan: Integration, consolidation and visualization of plant gene annotations within the MapMan ontology. *Nucleic Acids Research, 42*(D1), D1167–D1175.

Roberts, W. K., & Selitrennikoff, C. P. (1990). Zeamatin, an antifungal protein from maize with membrane-permeabilizing activity. *Journal of General Microbiology, 136*, 1771–1778.

Rosso, L. C., Finetti-Sialer, M. M., Hirsch, P. R., et al. (2011). Transcriptome analysis shows differential gene expression in the saprotrophic to parasitic transition of *Pochonia chlamydosporia*. *Applied Microbiology and Biotechnology, 90*, 1981–1994.

Rosso, L. C., Pentimone, I., Colagiero, M., et al. (2013). Expression of *Meloidogyne incognita* resistance genes induced by endophytic *Pochonia chlamydosporia* in tomato. *Nematropica, 43*, 322.

Rosso, L. C., Colagiero, M., Salatino, N., et al. (2014). Effect of trophic conditions on gene expression of *Pochonia chlamydosporia*. *The Annals of Applied Biology, 164*, 232–243.

Ruiz-Ferrer, V., & Voinnet, O. (2009). Roles of plant small RNAs in biotic stress responses. *Annual Review of Plant Biology, 60*, 485–510.

Schwarz, D. S., Hutvagner, G., Du, T., et al. (2003). Asymmetry in the assembly of the RNAi enzyme complex. *Cell, 115*, 199e208.

Selitrennikoff, C. P. (2001). Antifungal proteins. *Applied and Environmental Microbiology, 67*, 2883–2894.

Tunlid, A., & Jansson, S. (1991). Proteases and their involvement in the infection and immobilization of nematodes by the nematophagous fungus *Arthrobotrys oligospora*. *Applied and Environmental Microbiology, 57*, 2868–2872.

Ward, E., Kerry, B. R., Manzanilla-López, R. H., et al. (2012). The *Pochonia chlamydosporia* serine protease gene *vcp1* is subject to regulation by carbon, nitrogen and pH: Implications for nematode biocontrol. *PloS One, 7*, e35657.

Weiberg, A., Wang, M., Bellinger, M., et al. (2014). Small RNAs: A new paradigm in plant-microbe interactions. *Annual Review of Phytopathology, 52*, 495–516.

Yang, L., & Huang, H. (2014). Roles of small RNAs in plant disease resistance. *Journal of Integrative Plant Biology, 56*, 962–970.

Zang, C., Schones, D. E., Zeng, C., et al. (2009). A clustering approach for identification of enriched domains from histone modification ChIP-seq data. *Bioinformatics, 25*, 1952–1958.

Chapter 6
Proteomics

Nuria Escudero, Sonia Gómez-Vidal, and Luis V. Lopez-Llorca

Abstract Fungal proteomics research is growing as a result of the large number of fungal sequenced genomes of well annotated proteins available today. The proteome of any organism is dynamic as proteins differ depending on environmental conditions, unlike genomes which are practically constant for all the cells of an organism. In this chapter we have reviewed the 'state-of-the-art' of fungal proteomics, including sample preparation, protein separation and identification. We have given examples of proteomics of entomopathogenic and nematophagous fungi. We have also focused our attention on the proteomic study of *Pochonia chlamydosporia* carried out to date. In this study, the fungus was grown in chitin or chitosan as the main carbon and nitrogen nutrient sources, and the secretome of the fungus in both conditions analyzed. Proteins were concentrated using TCA/acetone. Two-dimensional, sodium dodecyl sulphate polyacrylamide gel electrophoresis and differential gel electrophoresis separated proteins for size and isolectric point. Some of the proteins overexpressed with chitosan that were identified using MALDI/TOF-TOF and LC-MS, were related with carbohydrate or protein degradation. The

N. Escudero (✉)
Laboratory of Plant Pathology, Multidisciplinary Institute for Environmental Studies (MIES) Ramón Margalef, Department of Marine Sciences and Applied Biology, University of Alicante, Alicante, Spain

Department of Agri-Food Engineering and Biotechnology, Universitat Politècnica de Catalunya, Catalunya, Spain
e-mail: nuria.escudero@upc.edu

S. Gómez-Vidal
Laboratory of Plant Pathology, Multidisciplinary Institute for Environmental Studies (MIES) Ramón Margalef, Department of Marine Sciences and Applied Biology, University of Alicante, Alicante, Spain

Research Technical Services, University of Alicante, Alicante, Spain
e-mail: Sonia.gomez@ua.es

L.V. Lopez-Llorca
Laboratory of Plant Pathology, Multidisciplinary Institute for Environmental Studies (MIES) Ramón Margalef, Department of Marine Sciences and Applied Biology, University of Alicante, Alicante, Spain
e-mail: lv.lopez@ua.es

R.H. Manzanilla-López, L.V. Lopez-Llorca (eds.), *Perspectives in Sustainable Nematode Management Through* Pochonia chlamydosporia *Applications for Root and Rhizosphere Health*, Sustainability in Plant and Crop Protection, DOI 10.1007/978-3-319-59224-4_6

recently available complete *Pochonia chlamydosporia* genome sequence could help with protein identification of fungal secretomes under various conditions.

6.1 Introduction

The term "proteomics", was first coined in 1997 by Peter H. James to make an analogy with genomics, the study of the genome for the complete set of DNA of an organism, including all of its genes. Proteomics also refers to "the large-scale comprehensive study of a specific proteome, including information on protein abundances, their variations and modifications, along with their interacting partners and networks, in order to understand cellular processes" (James 1997). Proteomics is more complicated than genomics because an organism's genome is more or less constant, whereas the proteome differs between cells and with time and environmental conditions (Bini et al. 2014).

The systematic analysis of the proteome allows qualitative and quantitative measurements of large numbers of proteins that directly influence cellular biochemistry, and thus provide an accurate analysis of cellular state or system changes during growth, development, and response to environmental factors. We can say that the proteome is dynamic, reflecting the conditions to which a cell is exposed or, for example, a specific disease process. There is, therefore, a potentially huge number of proteomes for each cell type (Bhadauria et al. 2007).

6.1.1 Fungal Proteomics

Proteomics, by virtue of its capacity to yield definitive information on protein identity, localization, post-translational modifications and the accuracy of *in silico* gene model prediction in fungi, has become an integral component of all large-scale '*-omics*' and systems approaches for understanding the rich complexity of fungal biochemistry, plant-fungal interactions, pathogenesis and fungal colonization (Bhadauria et al. 2007; Doyle 2011).

Fungal proteomics research, especially from filamentous fungi, has increased over the past years. This has been due to the recent availability of multiple fungal genome sequences (http://jgi.doe.gov/our-science/science-programs/fungal-genomics/1000-fungal-genomes/), the advent of next-generation nucleic acid sequencing and the availability of powerful proteomics technologies, especially tandem mass spectrometry (MS), liquid chromatography-mass spectrometry (LC-MS) MS (Doyle 2011).

The earliest proteomic studies of filamentous fungi dealt with *Trichoderma reesei* cell envelope proteins (Lim et al. 2001) and *Aspergillus fumigatus* glycosylphosphatidylinositol-anchored proteins (Bruneau et al. 2001). Kim et al. (2007) reviewed the proteomic studies made on filamentous fungi. For instance, *Aspergillus* spp., *Botrytis cinerea*, *Neurospora crassa*, *Penicillium expansum*,

Phanerochaete chrysosporium, *Pleurotus sapidus*, *Sclerotinia sclerotiorum*, and *Trichoderma* spp. were all cited in these studies. The number of publications on filamentous fungi proteomics has increased ever since.

6.1.2 Sample Preparation

Protein extraction and sample preparation are the most critical steps for any proteomics study because they can reduce protein yield, affecting biological activities and structural integrity of specific target proteins. For instance, proteins for two-dimensional electrophoresis analysis should be denatured, reduced and solubilised or rehydrated in order to obtain a complete disruption of their intra- and intermolecular secondary bonds and, hence, to provide assurance that each spot in the gel represents an individual polypeptide (Gómez-Vidal et al. 2008; Bianco and Perrotta 2015).

For total protein extraction, an ideal protocol would reproducibly capture all protein species in a proteome with low contamination from other molecules (Bhadauria et al. 2007). For this purpose, several procedures for protein extraction, precipitation or solubilisation have been reported in the scientific literature in order to overcome the limitations mentioned above and maximise protein yield (Bianco and Perrotta 2015).

Protein extraction is achieved by different techniques dependent on cell type. The literature shows the use of different systems to break the cell, such as the traditional mechanical systems (e.g., French press, grinding with glass beads and liquid nitrogen, sonication), and chemical (detergent lysis) or enzymatic extraction (Westermeier and Naven 2002; Simpson 2003). In mycological research, the robust cell walls of most fungi need to be disrupted. For instance, enzymatic digestion or mechanical lysis in a mortar with liquid nitrogen with or without glass beads has been used for yeast and filamentous fungi (Kim et al. 2007; Salazar 2008; Martínez-Gomariz et al. 2009; Barros et al. 2010; Qiu et al. 2012; Su et al. 2013; Rodrigues et al. 2015).

When a proteomic study is focused on proteins within specific organelles like the plasma membrane, outer membrane, mitochondria, etc., authors use the terms "subproteomics" and "subproteome" (Kim et al. 2007; Bhadauria et al. 2007). The procedure for protein extraction from fungal organelles has been described in several works (Hernández-Macedo et al. 2002; Grinyer et al. 2004; De Oliveira and De Graaff 2011).

In 2000 Tjalsma et al. coined the term "secretome" in their study of the eubacterium *Bacillus subtilis*. They defined the secretome as the totality of the proteins secreted and the secretory machinery of the bacterium. Depending on the authors, this secretome protein population either includes any protein strictly external to the plasma membrane or only soluble proteins outside the cell wall. Fungi are exodigesters so they must produce a plethora of extracellular enzymes, an important part of their secretome, which allows efficient degradation of biopolymers encountered during the interaction with their hosts or in their environments. Deciphering fungal secretomes has become an important research goal ever since secreted proteins were identified as the main effectors responsible for interactions (pathogenic or symbiotic) between plants and fungi (Girard et al. 2013). These effectors are small proteins that modulate the immune system of plants allowing the establishment of microorganisms therein. Effectors often interfere with signal pathways, either those required for host invasion or those that trigger host resistance (Giraldo and Valent 2013). Effectors are also present in animal-fungi interactions. Engelmann and Pujo (2010)) summarized the available information about the pathways and effectors involved in the immune system of *Caenorhabditis elegans*, focusing on the immune response to nematophagous fungi.

Extensive advances in technology, emergence of genomics, proteomics and protein-tagging technology, have allowed the determination of the complex composition of the secretome (Adav and Sze 2013). Using proteomic approaches, the main technical difficulties in the study of a microorganism interacting with a host are two: (i) the ratio of microbe over host in terms of respective cell concentration; and (ii) the quality of the genomic annotation for the two partners. This is often overcome by *in vitro* assays where the microorganism is cultured in the presence of purified fractions of its host (Girard et al. 2013).

Concerning protein sample preparation, both cell and secreted proteins are usually precipitated with trichloroacetic acid (TCA) in acetone with 2-mercaptoethanol or dithiothreitol (DTT) in order to concentrate the sample and remove interfering compounds such as salts, phenolics, organic and fatty acids, pigments and polysaccharides. In addition, proteolysis and other protein modifications are reduced. TCA/Acetone precipitation method, as previously described by Damerval et al. (1986), is a very useful method for minimizing protein degradation and removing contaminants. However, protein losses due to incomplete precipitation and/or resolubilization of proteins can occur. Chaotropes (urea, thiourea) and zwitterionic detergents (CHAPS) are usually used to circumvent this problem during sample preparation by increasing protein solubilisation.

6.1.3 Protein Separation and Identification

In proteomic research, proteins are usually identified by the mass-to-charge ratio of their peptides and fragments. Consequently, sample separation prior to mass analysis is generally required to reduce sample biological complexity in order to lower the risk of unambiguous identifications. The major separation technologies commonly used in proteomics can be divided into gel-based and gel-free approaches (Bianco and Perrotta 2015).

6.1.3.1 Gel-Based Proteomics

Gel-based separation techniques rely on one-dimensional (1-D) or two-dimensional (2D), sodium dodecyl sulphate polyacrylamide gel electrophoresis (SDS-PAGE), both being very popular methods because of their robustness and simplicity (Bianco and Perrotta 2015).

Two-dimensional electrophoresis is the current dominant technique in gel-based proteomic studies and separates proteins according to two independent parameters, i.e., isoelectric point (pI) in the first dimension and molecular mass (M_r) in the second dimension by coupling isoelectric focusing (IEF) and SDS-PAGE (Görg and Weiss 2004). Since these two parameters are unrelated, it is possible to obtain an almost uniform distribution of protein spots across a two-dimensional gel. The resulting map of protein spots can be considered as the protein fingerprint of that sample (Baggerman et al. 2005). The general work flow, after gel-separation, involves protein visualization by staining, gel image analysis, protein spot excision from a 2D electrophoresis gel, digestion into fragments by specific proteases such as trypsin, analysis of the peptide mass by Mass Spectrometry (MS), database search, and further validation. Two-dimensional electrophoresis remains a well-established technique in proteomic analysis of complex samples (Adav and Sze 2013).

Grinyer et al. (2004) published the first proteomics study of a filamentous fungus, *Trichoderma harzianum* carried out with 2D electrophoresis technology (reviewed by Bianco and Perrotta 2015). Further studies mapping intracellular and extracellular fungal proteomes were published using this established method and were promoted by numerous fungal genome projects, which increased the number of protein identifications. Two-dimensional gel electrophoresis has been used to generate protein maps and also in comparative studies where differences between diverse experimental conditions, strains or mutants have been examined.

The introduction of differential gel electrophoresis (DIGE technology), a variant of 2D electrophoresis, addresses some of the drawbacks found with the original technique, mainly the inherent variation between gels. For example, two identical protein samples run on separated 2D electrophoresis gels produce similar but not identical protein maps. DIGE provides a platform for controlling variation due to sample preparation, protein separation and difference detection by comparing up to three fluorescently staining protein samples running simultaneously on the same 2D electrophoresis gel. This technique has the advantage of running an internal standard, a mix of all protein samples of the experiment co-resolved together with two protein samples on the gels. This means that the internal standard sample comprises equal amounts of each sample to be compared. The standard is then mixed with two of the protein samples that have to be compared and applied to 2D electrophoresis. This leads to an enormous increase in accuracy and multiplexing of samples using different fluorescent cyanine dyes (i.e., Cy2, Cy3, and Cy5). Differential gel electrophoresis (DIGE) technology has reduced the systematic variation and increased the sensitivity and resolution of the gel-based proteomic approaches (Marouga et al. 2005; Minden et al. 2009; Bianco and Perrotta 2015). As 2D electrophoresis, DIGE technology work flow comprises gel image analysis and MS identification.

Differential gel electrophoresis technology was applied by Palma-Guerrero et al. (2010) to investigate the extracellular proteins of the filamentous fungus *Pochonia chlamydosporia* growing on semi-liquid medium with chitosan as the main carbon and nitrogen source versus a medium with chitin. Results will be discussed in detail in Sect. 6.3 of this chapter.

As already mentioned, after gel-separation, proteins should be visualized by staining with Coomassie blue dye, silver stains, fluorescent dyes, immunochemical detection, or radiolabelling (Görg and Weiss 2004). For quantitative analysis, imaging software is required to align gel spots and measure their intensities. To do this, gels are captured using a scanner recording light transmitted through or reflected from the stained gel or fluorescent scanner. For gel image analysis several software packages can be used, e.g., PDQuest™ 2D (Bio-Rad Laboratories), Phoretix™, SameSpots™ (NonLinear Dynamics) or ImageMaster and DeCyder 2D (GE Healthcare). These software packages allow image alignment and spot matching across the gels, normalization, background adjustment and noise removal, spot detection and quantification by calculation of the spot volumes, and statistical analysis to highlight differentially present proteins. Protein spots of interest are excised manually or using an automated spot cutter from 2D electrophoresis gels and then the spots are in-gel digested, usually using trypsin. Tryptic peptides are analysed by mass spectrometry and further examples, using *P. chlamydosporia* and similar fungi, can be found in: Gómez-Vidal et al. 2009; Martínez-Gomariz et al. 2009; Palma-Guerrero et al. 2010; Abdallah et al. 2012.

Mass spectrometry is an analytical technique that measures the molecular weight of molecules based upon the motion of a charged particle in an electric or magnetic field. The molecules are converted into, and separated according to, their mass:charge ratio (*m/z*). This technique is performed by a mass spectrometer (Westermeier and Naven 2002).

A typical mass spectrometer consists of an ion source, a mass analyser and a detector. Two "soft ionization" methods, namely matrix-assisted laser desorption ionisation (MALDI) and electro spray ionization (ESI), volatilize and ionize large biomolecules such as peptides and proteins. Different mass analyzers are used in proteomics research, i.e., time of fly (TOF), ion trap (IT), quadrupole (Q), orbitrap, etc. MALDI is usually coupled to TOF and TOF/TOF analyzers and ESI have mostly been coupled to ion traps and triple quadrupole or hybrid TOF-MS. Nevertheless, compared to MALDI, ESI has a significant advantage in its ease of coupling to separation techniques such as liquid chromatography (LC) and high-performance liquid chromatography (HPLC). Typically, a mixture of proteins is first separated by LC followed by tandem MS (MS/MS). In this procedure, a mixture of charged peptides is separated in the first MS according to their *m/z* ratios to create a list of the most intense peptide peaks. In the second MS analysis, the instrument is adjusted so that only a specific *m/z* species is directed into a collision cell to generate "daughter" ions derived from the "parent" species. Database search for protein identification is usually done via the MASCOT (http://www.matrixscience.com) search engine (Bhadauria et al. 2007).

6.1.3.2 Gel-Free Proteomics

Gel-based proteomics have technical limitations in the separation and visualization of certain classes of proteins. Therefore, in recent years, most developmental efforts have been focused on alternative approaches, such as gel-free proteomics. A typical 2D electrophoresis gel cannot visualize the entire proteome, especially so for proteins present in extremely low concentrations, or those that cannot be separated on a 2D electrophoresis gel due to their physicochemical properties (e.g., pI, hydrophobicity, molecular weight) preventing detection. Among the different possible approaches to study proteins, MS-based proteomics, an entirely new toolbox, has now become available for quantitative analysis. Complex peptide fractions, generated after protein proteolytic digestion, can be resolved using different fractionation strategies, i.e., ion-exchange chromatography (strong cation-exchange chromatography [SCX]), reversed-phase chromatography (RP), two-dimensional liquid chromatography (2D–LC), multidimensional protein identification technology (MudPit), and offgel electrophoresis (OGE). Although these approaches were initially targeted as replacements for gel-based methods, they should probably be regarded as complements to, rather than as mere replacements of, 2D electrophoresis. In fact, there is no single method that can provide exhaustive information for all the protein components in a complex mixture and different methodologies usually offer complementary information which can then be integrated into more detailed analyses. (Baggerman et al. 2005; Abdallah et al. 2012; Bianco and Perrotta 2015).

Quantitative proteomic approaches can be classified as either gel-based or gel-free methods, the latter being broadly categorized into "label-free" or "label-based" methods, of which the second can be further subdivided into the various types of labelling approaches, such as chemical and metabolic labelling (Abdallah et al. 2012). Accurate protein quantitation is currently one of the most challenging and rapidly changing areas of proteomics. The choice of methods for quantitative proteomics depends on multiple factors, including the source and the number of the samples, the number of treatments to be compared, the type of equipment available, time requirement and, most importantly, cost. Several quantitative methods, including isotope labelling approaches like Isotope-Coded Affinity Tag (ICAT), Stable Isotope Labelling by Amino Acids in Cell Culture (SILAC), $^{15}N/^{14}N$ metabolic labelling, $^{18}O/^{16}O$ enzymatic labelling, Isotope Coded Protein Labelling (ICPL), Tandem Mass Tags (TMT), Isobaric Tags for Relative and Absolute Quantification (iTRAQ), and other chemical labelling have been adopted in quantitative proteomics (Adav and Sze 2013).

The aforementioned techniques have, with the exception of iTRAQ, rarely been used in fungal proteomic analysis. This label-based quantitative proteomics method has been used to investigate the proteome of filamentous fungi such as *Fusarium graminearum* or *Aspergillus fumigatus* (reviewed by Bianco and Perrotta 2015).

6.2 Proteomics Applied to Fungal Pathogens of Invertebrates

6.2.1 Entomopathogenic Fungi

At least 90 genera and more than 700 species of fungi are pathogens of insects and they are distributed in practically every important fungal taxonomic group except for the higher basidiomycetes (St. Leger et al. 2011). The entomopathogenic fungi, also called entomophagous or entomogenous fungi, is the fungal group most exploited commercially for biological control purposes (Maciá-Vicente et al. 2011).

Entomopathogenic fungi vary considerably in their mode of action and virulence. The adherence and penetration ability of a fungus to the host integument determines the success of the infection process. During the degradation of the insect integument or cuticle a variety of extracellular enzymes are produced. The life cycle of entomopathogenic fungi on the insect cuticle begins with spore germination and penetration, followed by a rapid proliferation of fungal cells which ultimately results in the death of the host. Host death may be followed by the production of infective spores that can infect immediately to repeat the cycle, or by the production of resting spores or other resistant structures that require a period of dormancy (Shahid et al. 2012).

Various unexpected roles have been reported for entomopathogenic fungi, including endophytism, pathogenesis of protozoans, antagonism of plant pathogens, rhizosphere associations and plant growth promotion (Vega et al. 2009; St. Leger et al. 2011). Species of *Acremonium*, *Beauveria*, *Cladosporium*, *Clonostachys* and *Isaria* have been isolated as endophytes, despite being traditionally regarded as entomopathogens (Vega et al. 2009). We have provided evidence of the endophytic behaviour of the entomopathogenic fungi *B. bassiana*, *Lecanicillium dimorphum* and *L. psalliotae* in leaf petioles of *Phoenix dactylifera* L., in both laboratory and field assays, without causing evident harm to the host-plant. These fungi survived up to 30 days after inoculation within the host tissues and were able to move away from the inoculation site (Gómez-Vidal et al. 2006). Proteomic analysis of endophytically colonised date palms indicated that entomopathogenic fungi modulated plant defence responses, energy metabolism and the expression of cell division-related proteins (Gómez-Vidal et al. 2009).

To date, relatively few studies on the proteomes of entomopathogenic fungi have been carried out, but these include a revision of fungal proteomics by Doyle (2011) to identify protein functions. Entomopathogens were mainly represented by *Metarhizium* spp. proteomics, possibly because of the importance of this group as biological control agents.

Murad et al. (2008) used 2D electrophoresis and mass spectrometry to identify *M. anisopliae* enzymes, including reductases, proteases and acetyltransferases, that were secreted in the presence of the insect pest, *Callosobruchus maculatus*. Differential immune-proteomics using 2D electrophoresis followed by MS/MS enables identification of proteins secreted by *M. anisopliae* during the infection of arthropods (Santi et al. 2010). Barros et al. (2010) compared the proteome profiles of *M. acridum* conidia and mycelia and identified proteins expressed in these two developmental stages by 2D electrophoresis and MALDI-TOF spectrometry. Su et al. (2013) carried out a very similar study with *M. anisopliae* conidia and mycelia. Both changes in protein expression in the interactions between *M. anisopliae* and the greyback canegrub (*Dermolipida albohirtum*), and secretome of the fungus in the presence of insect, were evaluated by 2D electrophoresis and MALDI-TOF-MS and/or MALDI-TOF/TOF-MS (Manalil et al. 2009, 2010).

By using a similar proteomic strategy, as in Barros et al. (2010), proteins differentially expressed in *Aschersonia placenta* conidia and mycelia were studied by Qiu et al. (2012). Quin et al. (2009) identified proteins from conidia of the entomopathogen *Nomuraea rileyi* isolated from infected silkworm larvae using 2D electrophoresis and MALDI-TOF-MS.

Proteomics is, therefore, a powerful tool to enhance our knowledge of fungal biology and in particular of fungus-invertebrate interactions.

6.2.2 Nematophagous Fungi

Nematophagous fungi are the most diverse nematode antagonists. They are found in many different taxonomic groups within the fungal kingdom and use a variety of mechanisms to capture and kill nematodes (Stirling 2014). Depending on their mode of attacking nematodes, the nematophagous fungi can be divided into four groups: i) nematode-trapping (sometimes called predacious or predatory fungi); ii) endoparasitic; iii) egg- and female-parasitic; and iv) toxin-producing fungi (Jansson and Lopez-Llorca 2001). In all four groups, nematode parasitism results in a completely digested nematode which supplies the fungus with nutrients and energy for continued growth (Lopez-Llorca et al. 2008).

The nematode-trapping fungi have developed sophisticated hyphal structures, such as hyphal nets, knobs, branches or rings, with which nematodes are captured either by adhesion or mechanically (Nordbring-Hertz et al. 2006). Traditionally, the nematode-trapping fungi were assigned to three genera (*Arthrobotrys*, *Dactylella*, and *Monacrosporium*) on the basis of the morphology of their conidia and conidiophores (Stirling 2014). Extracellular serine proteases have been purified and characterized from several nematode-trapping fungi, such as *A. oligospora* (Tunlid et al. 1994; Minglian et al. 2004), *A. conoides* (Yang et al. 2007a), *Dactylella shizishanna* (Wang et al. 2006a) and *D. varietas* (Yang et al. 2007b), *Monacrosporium cystosporium* (Yang et al. 2008) and *M. microscaphoides* (Wang et al. 2006b). In the case of *A. oligospora*, different lectins were also purified and characterized (Borrebaeck et al. 1984; Rosen et al. 1992). Khan et al. (2008) used 2D electrophoresis and peptide mass fingerprinting to identify the *M. lysipagum* proteome of mycelium containing sticky knobs for nematode capturing. Yang et al. (2011) sequenced the genome of the nematode-trapping fungus *A. oligospora*. They also identified some proteins involved in trap formation using 2D electrophoresis and MALDI-TOF. These authors also found proteins involved in translocation, post-translational modification, amino acid metabolism, carbohydrate metabolism, energy conversion, cell wall and membrane biogenesis. Recently, a LC/MS study revealed an extended set of virulence related proteins, such as adhesins and serine proteases, on the cell wall related to trap formation (Liang et al. 2013).

Endoparasitic fungi use their spores (conidia or zoospores) to infect nematodes. Their spores either adhere to the nematode cuticle, the spore contents are injected into the nematode, or the spores are swallowed by the host and then germinate. Most of these fungi are obligate parasites of nematodes and live their entire vegetative lives inside infected nematodes (Lopez-Llorca et al. 2008). Among the endoparasites, *Drechmeria coniospora* and *Hirsutella rhossiliensis* are the most important species that infect nematodes with their spores and spend their vegetative lives inside infected nematodes. Irrespective of the infection method, the result is always the same: the death of the nematode (Nordbring-Hertz et al. 2006). In *D. coniospora* surface proteins involved in the infection process, such as chymotrypsin-like proteases were detected by gelatin-containing gel electrophoresis (Jansson and Friman 1999). An alkaline (Wang et al. 2009) and a neutral (Wang et al. 2007) serine proteases from *H. rhossiliensis* were characterized and their encoding genes cloned.

The egg- and female-parasitic fungi infect nematode females and the eggs they contain using appressoria (*Pochonia* spp., *Paecilomyces* spp.) or zoospores (*Nematophthora gynophila*) (Lopez-Llorca et al. 2008). Proteomic studies from *Pochonia* spp. will be discussed in detail in the next section of this chapter. A basic serine protease was purified from *Purpureocillium lilacinum* (= *Paecilomyces lilacinus*) with biological activity against *Meloidogyne hapla* eggs (Bonants et al. 1995). The identification and characterization of *P. lilacinum* chitinases was performed on non-denaturing one-dimensional (1D) and two-dimensional (2D) gels using a sandwich assay with glycol chitin as a substrate (Khan et al. 2003). Nematode eggs treated with purified chitinases displayed large vacuoles in the chitin layer and their vitelline layer was split and lost its integrity (Khan et al. 2004).

Finally, the toxin-producing fungi immobilize the nematodes by a toxin prior to hyphal penetration through the nematode cuticle. *Pleurotus* spp. produces toxic droplets and *Coprinus* spp. a toxin, in "spiny structures" (Lopez-Llorca et al. 2008). In a proteomic study in *Pleurotus cornucopiae* a lectin similar to the one from *A. oligospora* was cloned (Iijima et al. 2002) and expressed in *Pichia pastoris* (Iijima et al. 2003).

6.3 *Pochonia chlamydosporia* Proteomics

Proteases and chitinases are considered putative pathogenicity factors of *Pochonia chlamydosporia* (Casas-Flores and Herrera-Estrella 2007), with subtilisins as key proteinases secreted by *P. chlamydosporia* (Segers et al. 1996) and *Pochonia rubescens* [= *Metapochonia*] (Lopez-Llorca and Robertson 1992); these enzymes have been widely studied by proteomics techniques.

6.3.1 *Enzyme Purification*

The first protein purified for *Pochonia* spp. was P32, a serine protease of *P. rubescens* [= *Metapochonia*] (Lopez-Llorca 1990) and then VCP1, a serine protease from *P. chlamydosporia* (Segers et al. 1994). In the case of P32 the fungus was grown in a semi-liquid medium with gelatin as substrate and P32 was purified from culture filtrates by gel filtration and ion exchange chromatography. For VCP1 purification *P. chlamydosporia* was grown in soy peptone medium, and VCP1 was purified from culture filtrates by isoelectric focusing. Both, VCP1 and P32 serine proteases were cloned (Morton et al. 2003; Larriba et al. 2012). A new, ca 60 kDa, serine carboxypeptidase from *Pochonia* spp. (SCP1) was identified from barley roots colonized by the fungus (Lopez-Llorca et al. 2010).

CHI43, a main 43 kDa chitinase from *Pochonia* spp., was purified from culture filtrates of these fungi growing with chitin as main C and N source (Tikhonov et al. 2002). Purified CHI43 was found to degrade the shell of *Globodera pallida* eggs. The purified chitinase plus the serine protease caused more damage than each enzyme acting separately. A chitinase of *P. chlamydosporia*, PCCHI44, was cloned using degenerate primers and a genome walking approach, and overexpressed in *Escherichia coli*. The treatment of *M. incognita* eggs using this purified chitinase resulted in degradation of *M. incognita* eggs, as shown by the large vacuoles that were formed after 3 h to PCCHI44 exposure (Mi et al. 2010).

6.3.2 *Role of* Pochonia chlamydosporia *Enzymes in Pathogenicity*

Pochonia spp. proteases are able to degrade egg shell proteins from *G. pallida* (Lopez-Llorca 1990) and *M. incognita* (Segers et al. 1996). VPC1 was also immunolocalized in *M. javanica* eggs infected by *P. chlamydosporia* (Segers et al. 1995). The P32 serine protease immunolocation in appressoria of *Pochonia* spp. infecting nematode eggs supports its role as a determinant of pathogenicity (Lopez-Llorca and Robertson 1992). As previously mentioned, *Pochonia* spp. chitinases are thought to act cooperatively with proteases to degrade the nematode egg shell (Tikhonov et al. 2002). SCP1, a serine carboxypeptidase from *P. chlamydosporia*, was recently immunolocalized in *Meloidogyne javanica* eggs parasitized by the fungus (Escudero et al. 2016).

6.3.3 *Role of* Pochonia chlamydosporia *Enzymes in Endophytism*

Pochonia chlamydosporia can also endophytically colonize plant roots. Using Western blotting and immunolocalization *P. chlamydosporia* was found to secrete VCP1 and SCP1 proteases when the fungus was found endophytically colonizing barley roots (Lopez-Llorca et al. 2010).

6.3.4 Pochonia chlamydosporia *Secretome Analysed by Difference Gel Electrophoresis*

Qualitative and quantitative differences in protein accumulation between both culture filtrates of *P. chlamydosporia* grown with either chitin or chitosan as the main carbon and nitrogen sources were found using proteomic techniques (Palma-Guerrero et al. 2010). This process is summarized in Fig. 6.1.

Filtrates from 19-day-old semi-liquid medium cultures were used for protein extraction by precipitation with TCA/Acetone. Fungal secretomes are obtained as very dilute protein solutions and filtrates must therefore be concentrated. This procedure is also useful to separate proteins from the large amounts of metabolites secreted by fungi that potentially alter the secretome extract (Girard et al. 2013). As mentioned above, TCA/Acetone precipitation is a very useful method for minimizing protein degradation and removing contaminants (Damerval et al. 1986).

Nonetheless, other gel-based separation techniques, 2D electrophoresis and DIGE (Difference Gel Electrophoresis) technology, were used to compare proteins from culture filtrates. First, protein separation was performed in 2D electrophoresis using IPG strips 3–10 linear pH gradient and gels were stained with Sypro Ruby, a highly sensitive fluorescent stain. A maximum of 323 protein spots were detected with both qualitative (i.e., spot presence or absence) and quantitative differences (i.e., higher or lower spot intensity) between treatments. Seven new spots and 15 protein spots with significantly higher expression were found in chitosan versus chitin media. One protein, which was new in chitosan media, was identified by MS-MS (MALDI-TOF-TOF) as homologous to a hypothetical protein, a putative glycosyl hydrolase of the family 18, type II chitinases that hydrolyze chitin. In a parallel analysis of chitinolytic and chitosanolytic enzyme assays, the authors found a higher chitinolytic activity in chitosan versus chitin media. This newly expressed protein could contribute to this higher activity. Four highly expressed proteins in chitosan medium were successfully identified by MS-MS (MALDI-TOF-TOF). Two proteins, with different molecular weight and similar pI, appeared as homologous to a mitochondrial ATP synthase subunit 4. Their presence in the extracellular media could be explained by cell autolysis, and the higher levels in chitosan versus chitin media could be related to the higher level of proteins requiring ATP to regulate protein signalling. Two more proteins were related to protein signalling regulation, a homologue to ribosomal RNA large subunit methyltransferase and a

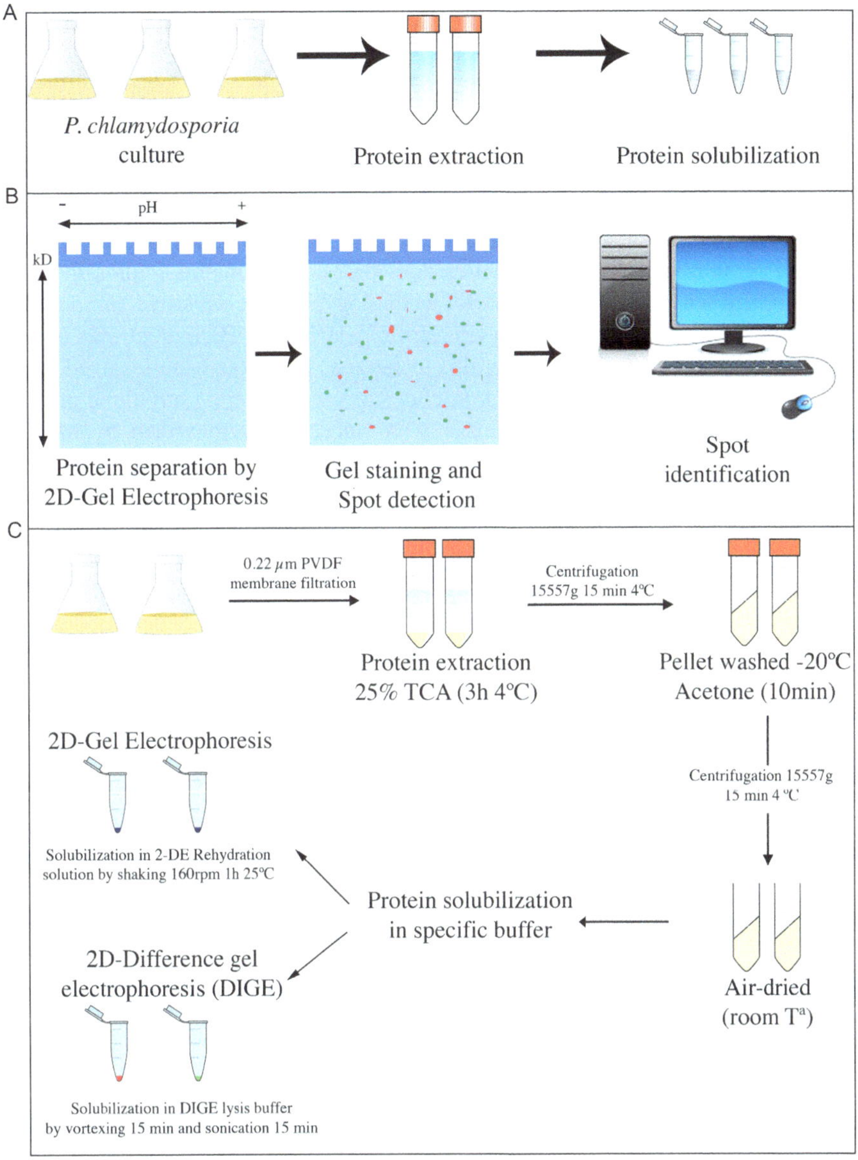

Fig. 6.1 Flow-chart of main steps followed in Proteomics Experimentation with *Pochonia chlamydosporia*. (**A**) The fungus was grown in liquid culture (either with chitin or chitosan). Proteins were extracted from culture filtrates by precipitation with 25% trichloroacetic acid (TCA). They were resuspended in buffer for 2D Gel Electrophoresis; (**B**) 2D Electrophoresis was carried out to separate proteins spots by their pI and molecular weight. Spots were detected with fluorochromes (Sypro Ruby stain or DIGE technology) to identify spot differences between samples. Spots were excised and subjected to Mass Spectrometry analysis, and database search for protein identification; (**C**) *P. chlamydosporia* protein extraction protocol, cultures were filtered through a 0.22 μm Polyvinylidene Difluoride (PVDF) Membrane. Proteins in culture filtrates were then precipitated using 25% TCA for 3 h at 4 °C. Precipitates were centrifuged at 15557 *g* for 15 min at 4 °C. Resulting protein pellets were washed three times with cold acetone (–20 °C) for 10 min and a final centrifugation (15557 *g* for 15 min at 4 °C). Proteins in-pellets were air-dried at room temperature and then solubilised in either: 2D Gel Electrophoresis rehydration solution 160 rpm shaking 1 h at 25 °C, or 2D DIGE lysis buffer 15 min vortexing and 15 min sonication (See Palma-Guerrero et al. (2010) for further details)

homologue to a Tyrosine-protein kinase transforming protein FPS. The significance of their higher expression on chitosan media could be related to changes in protein regulation by chitosan, and its presence in the extracellular media could also be explained by cell autolysis.

Two-dimensional electrophoresis gel in the pI range 3–10 did not resolve in detail pI 7–10 proteins, and this region had special interest because of the high presence of protein spots with clear differences between treatments (chitosan versus chitin). Difference gel Electrophoresis technology, a more sensitive and accurate tool than 2D electrophoresis, using IPG strips 7–10 linear pH gradient, was used to resolve this shorter pI range. Using this technique, 164 spots were resolved, 30 of these showing increased expression in chitosan versus chitin. Only three proteins with increased expression in chitosan were successfully identified by MS-MS (MALDI-TOF-TOF). One of the spots was identified as a hypothetical protein of unknown function and the other two could be involved in chromosome condensation (Probable Fe (2+)-trafficking protein and hypothetical Rad21 Rec8). Its increased expression in chitosan media could be related to changes in hyphal growth and morphogenesis induced by chitosan. In this respect, it is known that chitosan affects hyphal growth and conidia germination of fungi, including *P. chlamydosporia* (Palma-Guerrero et al. 2008). The presence of these two proteins in the extracellular media could be associated with cell autolysis.

Since a low level of identification was obtained with MALDI, LC/MS-MS analysis was applied for some spots using nanoLCnanoESI-QTOF, six proteins with increased expression in chitosan in the pI 7–10 range were identified. One of the proteins was identified as similar to a putative glycosyl hydrolase of the family of alpha-1,6-mannases, suggesting that chitosan induced mannanase activity and another one as a homologous to Protein kinase DC1, which is involved in protein signalling regulation. Two proteins were identified as the same product, a hypothetical protein with a WSC domain involved in carbohydrate binding and present in fungal exoglucanases, suggesting that either this protein could be involved in chitosan degradation or that it is induced by chitosan but is not able to degrade it. Another two proteins were both homologous to VCP1, the main alkaline serine protease of the nematophagous fungus *P. chlamydosporia.* One of these could be a proteolytic fragment of the other. The higher expression of VCP1 in chitosan medium could be related to chitosan induced chitinolytic activity since chitinases and proteases are co-expressed in the presence of chitin (Tikhonov et al. 2002; Esteves et al. 2009).

In this first approach to study the extracellular proteins expressed by *P. chlamydosporia* in a medium containing chitosan using proteomics, pathogenesis proteins, included a chitinase, a mannanase, an exoglucanase and an alkaline serine protease, and proteins related with signalling regulation, were identified as proteins that were overexpressed or newly induced.

The fact that no previous genomic or proteomic data were available for *P. chlamydosporia* made the identification of proteins more difficult and only 14 of 52 (~27%) spots analysed could be identified using Swiss-Prot and NCBInr databases and, in all cases, by homology with other species. The recent *P. chlamydosporia* genome sequencing (Larriba et al. 2014) opens new possibilities for protein identification in the proteomes of this fungus.

6.4 Concluding Remarks

Proteomics is envisaged as a powerful tool to understand the multitrophic biology of *P. chlamydosporia*. It can complement previous work performed on 1D Electrophoresis with extracellular enzymes involved in infection such as proteases or chitinases.

Proteomics is also useful to understand *P. chlamydosporia* endophytism. Finally, proteomics has a potential role in analysing secretomes of *P. chlamydosporia* when the fungus is used as a cell factory. New, fast sequencing techniques will help to understand protein polymorphism in the fungus.

Acknowledgements This research was funded by the Spanish Ministry of Economy and Competitiveness Grant AGL 2015-66833.

References

Abdallah, C., Dumas-Gaudot, E., Renaut, J. et al. (2012). Gel-based and gel-free quantitative proteomics approaches at a glance. International Journal of Plant Genomics. ID 494572 doi:10.1155/2012/494572.

Adav, S. S., & Sze, S. K. (2013). Fungal secretome for biorefinery: Recent advances in proteomic technology. *Mass Spectrometry Letters, 4*, 1–9.

Baggerman, G., Vierstraete, E., De Loof, A., et al. (2005). Gel based versus gel-free proteomics: A review. *Combinatorial Chemistry & High Throughput Screening, 8*, 669–677.

Barros, B. H., da Silva, S. H., dos ReisMarques, E. R., et al. (2010). A proteomic approach to identifying proteins differentially expressed in conidia and mycelium of the entomopathogenic fungus *Metarhizium acridum*. *Fungal Biology, 114*, 572–579.

Bhadauria, V., Zhao, W. S., Wang, L. X., et al. (2007). Advances in fungal proteomics. *Microbiological Research, 162*, 193–200.

Bianco, L., & Perrotta, G. (2015). Methodologies and perspectives of proteomics applied to filamentous fungi: From sample preparation to secretome analysis. *International Journal of Molecular Sciences, 16*, 5803–5829.

Bini, L., Calvete, J. J., Hochstrasser, D., et al. (2014). The magic of words. *Journal of Proteomics, 107*, 1–4.

Bonants, P. J. M., Fitters, P. F. L., Thijs, H., et al. (1995). A basic serine protease from *Paecilomyces lilacinus* with biological activity against *Meloidogyne hapla* eggs. *Microbiology, 141*, 775–784.

Borrebaeck, C. A. K., Mattiason, B., & Nordbring-Hertz, B. (1984). Isolation and partial characterization of a carbohydrate-binding protein from a nematode-trapping fungus. *Journal of Bacteriology, 159*, 53–56.

Bruneau, J.-M., Magnin, T., Tagat, E., Legrand, R., Bernard, M., Diaquin, M., Fudali, C., & Latgé, J.-P. (2001). Proteome analysis of Aspergillus fumigatus identifies glycosylphosphatidylinositol-anchored proteins associated to the cell wall biosynthesis. *Electrophoresis, 22*, 2812–2823.

Casas-Flores, S., & Herrera-Estrella, A. (2007). Antagonism of plant parasitic nematodes by fungi. In C. P. Kubicek & I. S. Druzhinina (Eds.), *Environmental and microbial relationships, 2nd edn, The Mycota IV* (pp. 147–157). Heidelberg: Springer.

Damerval, C., De Vienne, D., Zivy, M., et al. (1986). Technical improvements in two-dimensional electrophoresis increase the level of genetic variation detected in wheat-seedling proteins. *Electrophoresis, 7*, 52–54.

De Oliveira, J. M. P. F., & de Graaff, L. H. (2011). Proteomics of industrial fungi: Trends and insights for biotechnology. *Applied Microbiology and Biotechnology, 89*, 225–237.

Doyle, S. (2011). Fungal proteomics: From identification to function. *FEMS Microbiology Letters, 321*, 1–9.

Engelmann, I., & Pujo, N. (2010). Innate immunity in *C. elegans*. In K. Söderhäll (Ed.), *Invertebrate Immunity* (pp. 105–121). Heidelberg: Landes Bioscience and Springer Science+Business Media.

Escudero, N., Ferreira, S. R., Lopez-Moya, F., et al. (2016). Chitosan enhances parasitism of *Meloidogyne javanica* eggs by the nematophagous fungus *Pochonia chlamydosporia*. *Fungal Biology*. doi:10.1016/j.funbio.2015.12.005.

Esteves, I., Peteira, B., Powers, S., et al. (2009). Effects of osmotic and matric potential on radial growth and accumulation of endogenous reserves in three isolates of *Pochonia chlamydosporia*. *Biocontrol Science and Technology, 19*, 185–199.

Giraldo, M. C., & Valent, B. (2013). Filamentous plant pathogen effectors in action. *Nature Reviews. Microbiology, 11*, 800–814.

Girard, V., Dieryckx, C., Job, C., et al. (2013). Secretomes: The fungal strike force. *Proteomics, 13*, 597–608.

Gómez-Vidal, S., Lopez-Llorca, L. V., Jansson, H.-B., et al. (2006). Endophytic colonization of date palm (*Phoenix dactylifera* L.) leaves by entomopathogenic fungi. *Micron, 37*, 624–632.

Gómez-Vidal, S., Tena, M., Lopez-Llorca, L. V., et al. (2008). Protein extraction from *Phoenix dactylifera* L. leaves, a recalcitrant material, for two-dimensional electrophoresis. *Electrophoresis, 29*, 448–456.

Gómez-Vidal, S., Salinas, J., Tena, M., et al. (2009). Proteomic analysis of date palm (*Phoenix dactylifera* L.) responses to endophytic colonization by entomopathogenic fungi. *Electrophoresis, 30*, 2996–3005.

Görg, A., & Weiss, W. (2004). Protein profile comparisons of microorganism, cells and tissues using 2D gels. In D. W. Speicher (Ed.), *Proteome analysis. Interpreting the genome* (pp. 20–74). Amsterdam: Elsevier BV.

Grinyer, J., McKay, M., Herbert, B., et al. (2004). Fungal proteomics: Mapping the mitochondrial proteins of a *Trichoderma harzianum* strain applied for biological control. *Current Genetics, 45*, 170–175.

Hernández-Macedo, M. L., Ferraz, A., Rodriguez, J., et al. (2002). Iron-regulated proteins in *Phanerochaete chrysosporium* and *Lentinula edodes*: Differential analysis by sodium dodecyl sulfate polyacrylamide gel electrophoresis and two-dimensional polyacrylamide gel electrophoresis profiles. *Electrophoresis, 23*, 655–661.

Iijima, N., Yoshino, H., Ten, L. C., et al. (2002). Two genes encoding fruit body lectins of *Pleurotus cornucopiae*: Sequence similarity with the lectin of a nematode-trapping fungus. *Bioscience, Biotechnology, and Biochemistry, 66*, 2083–2089.

Iijima, N., Amano, K., Ando, A., et al. (2003). Production of fruiting-body lectins of *Pleurotus cornucopiae* in methylotrophic yeast *Pichia pastoris*. *Journal of Bioscience and Bioengineering, 95*, 416–418.

James, P. (1997). Protein identification in the post-genome era: The rapid rise of proteomics. *Quarterly Reviews of Biophysics, 30*, 279–331.

Jansson, H. B., & Friman, E. (1999). Infection-related surface proteins on conidia of the nematophagous fungus *Drechmeria coniospora*. *Mycological Research, 103*, 249–256.

Jansson, H., & Lopez-Llorca, L.V. (2001). Biology of nematophagous fungi. Trichomycetes and other fungal groups: Robert W. Lichtwardt commemoration, pp. 144–173.

Khan, A., Williams, K., Molloy, M. P., et al. (2003). Purification and characterization of a serine protease and chitinases from *Paecilomyces lilacinus* and detection of chitinase activity on 2D gels. *Protein Expression and Purification, 32*, 210–220.

Khan, A., Williams, K. L., & Nevalainen, H. K. M. (2004). Effects of *Paecilomyces lilacinus* protease and chitinase on the eggshell structures and hatching of *Meloidogyne javanica* juveniles. *Biological Control, 3*, 346–352.

Khan, A., Williams, K., & Soon, J. (2008). Proteomic analysis of the knob-producing nematode-trapping fungus *Monacrosporium lysipagum*. *Mycological Research, 112*, 1447–1452.

Kim, Y., Nandakumar, M. P., & Marten, M. R. (2007). Proteomics of filamentous fungi. *Trends in Biotechnology, 25*, 395–400.

Larriba, E., Martín-Nieto, J., & Lopez-Llorca, L. V. (2012). Gene cloning, molecular modeling, and phylogenetics of serine protease P32 and serine carboxypeptidase SCP1 from nematophagous fungi *Pochonia rubescens* and *Pochonia chlamydosporia*. *Canadian Journal of Microbiology, 58*, 815–827.

Larriba, E., Jaime, M. D. L. A., Carbonell-Caballero, J., et al. (2014). Sequencing and functional analysis of the genome of a nematode egg-parasitic fungus, *Pochonia chlamydosporia*. *Fungal Genetics and Biology, 65*(C), 69–80.

Liang, L., Wu, H., Liu, Z., et al. (2013). Proteomic and transcriptional analyses of *Arthrobotrys oligospora* cell wall related proteins reveal complexity of fungal virulence against nematodes. *Applied Microbiology and Biotechnology, 97*, 8683–8692.

Lim, D., Hains, P., Walsh, B., et al. (2001). Proteins associated with the cell envelope of *Trichoderma reesei*: A proteomic approach. *Proteomics, 1*, 899–910.

Lopez-Llorca, L. (1990). Purification and properties of extracellular proteases produced by the nematophagous fungus *Verticillium suchlasporium*. *Canadian Journal of Microbiology, 36*, 530–537.

Lopez-Llorca, L., & Robertson, W. (1992). Immunocytochemical localization of a 32-kDa protease from the nematophagous fungus *Verticillium suchlasporium* in infected nematode eggs. *Experimental Mycology, 16*, 261–267.

Lopez-Llorca, L. V., Maciá-Vicente, J. G., Jansson, H., et al. (2008). Mode of action and interactions of nematophagous fungi. In A. Ciancio & K. G. Mukerji (Eds.), *Integrated Management of Plant Pests and Diseases, vol. 2. Of the series integrated management and biocontrol of vegetable and grain crops nematodes* (pp. 51–76). Netherlands: Springer.

Lopez-Llorca, L. V., Gómez-Vidal, S., Monfort, E., et al. (2010). Expression of serine proteases in egg-parasitic nematophagous fungi during barley root colonization. *Fungal Genetics and Biology, 47*, 342–351.

Maciá-Vicente, J. G., Palma-Guerrero, J., Gómez-Vidal, S., et al. (2011). New insights on the mode of action of fungal pathogens of invertebrates for improving their biocontrol performance. In K. G. Davies & Y. Spiegel (Eds.), *Biological control of plant-parasitic nematodes: building coherence between microbial ecology and molecular mechanisms, progress in biological control* (pp. 203–225). Netherlands: Springer.

Manalil, N. S., Te'o, V. S. J., Braithwaite, K., et al. (2009). A proteomic view into infection of greyback canegrubs (*Dermolepida albohirtum*) by *Metarhizium anisopliae*. *Current Genetics, 55*, 571–581.

Manalil, N. S., Te'o, V. S. J., Braithwaite, K., et al. (2010). Comparative analysis of the *Metarhizium anisopliae* secretome in response to exposure to the greyback cane grub and grub cuticles. *Fungal Biology, 114*, 637–645.

Marouga, R., David, S., & Hawkins, E. (2005). The development of the DIGE system: 2D fluorescence difference gel analysis technology. *Analytical and Bioanalytical Chemistry, 382*, 669–678.

Martínez-Gomariz, M., Perumal, P., Mekala, S., et al. (2009). Proteomics analysis of cytoplasmic and surface proteins from yeast cells, hyphae, and biofilms of *Candida albicans*. *Proteomics, 9*, 2230–2252.

Mi, Q., Yang, J., Ye, F., et al. (2010). Cloning and overexpression of *Pochonia chlamydosporia* chitinase gene pcchi44, a potential virulence factor in infection against nematodes. *Process Biochemistry, 45*, 810–814.

Minden, J. S., Dowd, S. R., Meyer, H. E., et al. (2009). Difference gel electrophoresis. *Electophoresis, 30*, S156–S161.

Minglian, Z., Minghe, M., & Keqin, Z. (2004). Characterization of a neutral serine protease and its full-length cDNa from the nematode-trapping fungus *Arthrobotrys oligospora*. *Mycologia, 96*, 16–22.

Morton, C., Hirsch, P., Peberdy, J., et al. (2003). Cloning of and genetic variation in protease VCP1 from the nematophagous fungus *Pochonia chlamydosporia*. *Mycological Research, 107*, 38–46.

Murad, A. M., Noronha, E. F., Miller, R. N. G., et al. (2008). Proteomic analysis of *Metarhizium anisopliae* secretion in the presence of the insect pest *Callosobruchus maculatus*. *Microbiology, 154*, 3766–3774.

Nordbring-Hertz, B., Jansson, H. B., & Tunlid, A. (2006). Nematophagous fungi. In: eLS. Wiley, Chichester. doi:10.1002/9780470015902.a0000374.pub3.

Palma-Guerrero, J., Jansson, H. B., Salinas, J., et al. (2008). Effect of chitosan on hyphal growth and spore germination of plant pathogenic and biocontrol fungi. *Journal of Applied Microbiology, 104*, 541–553.

Palma-Guerrero, J., Gómez-Vidal, S., Tikhonov, V. E., et al. (2010). Comparative analysis of extracellular proteins from *Pochonia chlamydosporia* grown with chitosan or chitin as main carbon and nitrogen sources. *Enzyme and Microbial Technology, 46*, 568–574.

Qiu, J., Su, Y., Gelbic, I., et al. (2012). Proteomic analysis of proteins differentially expressed in conidia and mycelium of the entomopathogenic fungus. *Aschersonia placenta*. *Canadian Journal of Microbiology, 58*, 1327–1334.

Quin, L., Liu, X., Li, J., et al. (2009). Protein profile of *Nomuraea rileyi* spore isolation from infected silkworm. *Current Microbiology, 58*, 578–585.

Rodrigues, A. M., Kubitschek-Barreira, P. H., Fernandes, G. F., et al. (2015). Two-dimensional gel electrophoresis data for proteomic profiling of *Sporothrix* yeast cells. *Data in Brief, 2*, 32–38.

Rosen, S., Ek, B., Rask, L., et al. (1992). Purification and characterization of a surface lectin from the nematode-trapping fungus *Arthrobotrys oligospora*. *Journal of General Microbiology, 138*(12), 2663–2672.

Salazar, O. (2008). Bacteria and yeast cell disruption using lytic enzymes. In A. Posh (Ed.), *Methods in Molecular Biology* (pp. 23–34). Clifton NJ: Humana Press.

Santi, L., Silva, W. O. B., Pinto, A. F. M., et al. (2010). *Metarhizium anisopliae* host-pathogen interaction: Differential immunoproteomics reveals proteins involved in the infection process of arthropods. *Fungal Biology, 114*, 312–319.

Segers, R., Butt, T., Kerry, B., et al. (1994). The nematophagous fungus *Verticillium chlamydosporium* produces a chymoelastase-like protease which hydrolyses host nematode proteins in situ. *Microbiology, 140*, 2715–2723.

Segers, R., Butt, T., Keen, J., et al. (1995). The subtilisins of the invertebrate mycopathogens *Verticillium chlamydosporium* and *Metarhizium anisopliae* are serologically and functionally related. *FEMS Microbiology Letters, 126*, 227–231.

Segers, R., Butt, T. M., Kerry, B. R., et al. (1996). The role of the proteinase VCP1 produced by the nematophagous *Verticillium chlamydosporium* in the infection process of nematode eggs. *Mycological Research, 100*, 421–428.

Shahid, A. A., Rao, Q. A., & Bakhsh, A. (2012). Entomopathogenic fungi as biological controllers: New insights into their virulence and pathogenicity. *Archives of Biological Sciences, 64*, 21–42.

Simpson, R. J. (2003). Preparation of cellular and subcellular extracts. In R. J. Simpson (Ed.), *Proteins and proteomics. A laboratory manual* (pp. 91–142). New York: Cold Spring Harbor Laboratory Press.

St. Leger, R. J., Wang, C., & Fang, W. (2011). New perspectives on insect pathogens. *Fungal Biology Reviews, 25*, 84–88.

Stirling, G. R. (2014). *Biological control of plant-parasitic nematodes: Soil ecosystem management in sustainable agriculture* (2nd ed.). London: CABI.

Su, Y., Guo, Q., Tu, J., et al. (2013). Proteins differentially expressed in conidia and mycelia of the entomopathogenic fungus *Metarhizium anisopliae sensu stricto*. *Canadian Journal of Microbiology, 59*, 443–448.

Tikhonov, V., Lopez-Llorca, L., Salinas, J., et al. (2002). Purification and characterization of chitinases from the nematophagous fungi *Verticillium chlamydosporium* and *V. suchlasporium*. *Fungal Genetics and Biology, 35*, 67–78.

Tjalsma, H., Bolhuis, A., Jongbloed, J. D. H., et al. (2000). Signal peptide-dependent protein transport in *Bacillus subtilis*, a genome-based survey of the secretome. *Microbiology and Molecular Biology Reviews, 64*, 515–547.

Tunlid, A., Rosen, S., Ek, B., et al. (1994). Purification and characterization of an extracellular serine protease from the nematode-trapping fungus *Arthrobotrys oligospora*. *Microbiology, 140*, 1687–1695.

Vega, F. E., Goettel, M. S., Blackwell, M., et al. (2009). Fungal entomopathogens: New insights on their ecology. *Fungal Ecology, 2*, 149–159.

Wang, M., Yang, J., & Zhang, K. Q. (2006a). Characterization of an extracellular protease and its cDNA from the nematode-trapping fungus *Monacrosporium microscaphoides*. *Canadian Journal of Microbiology, 52*, 130–139.

Wang, R. B., Yang, J. K., Lin, C., et al. (2006b). Purification and characterization of an extracellular serine protease from the nematode-trapping fungus *Dactylella shizishanna*. *Letters in Applied Microbiology, 42*, 589–594.

Wang, B., Wu, W., & Liu, X. (2007). Purification and characterization of a neutral serine protease with nematicidal activity from *Hirsutella rhossiliensis*. *Mycopathologia, 1163*, 169–176.

Wang, B., Liu, X., Wu, W., et al. (2009). Purification, characterization, and gene cloning of an alkaline serine protease from a highly virulent strain of the nematode-endoparasitic fungus *Hirsutella rhossiliensis*. *Microbiological Research, 164*, 665–673.

Westermeier, R., & Naven, T. (2002). Expression proteomics. In R. Westermeier & T. Naven (Eds.), *Proteomics in practice: A laboratory manual of proteome analysis* (pp. 11–160). Weinheim: Wiley-VCH Verlag-GmbH.

Yang, J., Li, J., Liang, L., et al. (2007a). Cloning and characterization of an extracellular serine protease from the nematode-trapping fungus *Arthrobotrys conoides*. *Archives of Microbiology, 188*, 167–174.

Yang, J., Liang, L., Zhang, Y., et al. (2007b). Purification and cloning of a novel serine protease from the nematode-trapping fungus *Dactylellina varietas* and its potential roles in infection against nematodes. *Applied Microbiology and Biotechnology, 75*, 557–565.

Yang, J. K., Ye, F. P., Mi, Q. L., et al. (2008). Purification and cloning of an extracellular serine protease from the nematode-trapping fungus *Monacrosporium cystosporium*. *Journal of Microbiology and Biotechnology, 18*, 852–858.

Yang, J., Wang, L., Ji, X., et al. (2011). Genomic and proteomic analyses of the fungus *Arthrobotrys oligospora* provide insights into nematode-trap formation. *PLoS Pathogens, 7*(9), e1002179.

Chapter 7
Secondary Metabolites from *Pochonia chlamydosporia* and Other Species of *Pochonia*

Xue-Mei Niu

Abstract This chapter aims to present classification, occurrence, biological activities and functions of 34 secondary metabolites from *Pochonia chlamydosporia* (= *Verticillium chlamydosporium*) and 105 secondary metabolites from the other species of *Pochonia* reported up to the end of 2015. The secondary metabolites from *P. chlamydosporia* mainly include resorcylic acid lactone, pyranones, alkaloid and phenolics, while those from the other species of *Pochonia* belong to polycyclic aromatic compounds, nonaromatic polyketides, phenol-terpenoid hybrids, β-carotene-type neurosporaxanthin, pentanorlanostane triterpenoids, dahiane type diterpenoids, cyclodepsipeptides, verticillin-type diketopiperazines, linear lipopeptide and polyhydroxylated pyrrolizidine. Many of these natural products have attracted much attention for their fascinating molecular architectures and attractive biological activities such as antibacterial, antifungal, antioxidative activities, anti-malarial, antinematicidal, antivirus, antitumour, and other activities.

7.1 Classification and Occurrence of Secondary Metabolites from *Pochonia chlamydosporia* (= *Verticillium chlamydosporium*)

The secondary metabolites from *P. chlamydosporia* can be grouped into several categories which include resorcylic acid lactone (RAL), pyranones, and spirocyclic alkaloid (Table 7.1). Many of these compounds have been reported to be produced by other fungal species. For example, radicicol (**1**), the most purified and famous metabolite (Winssinger and Barluenga 2007) from *P. chlamydosporia* strains, was first purified as a metabolite of *Monosporium bonorden*, and named initially as

X.-M. Niu (✉)
State Key Laboratory for Conservation and Utilization of Bio-resources and Key Laboratory for Microbial Resources of the Ministry of Education, Yunnan University, Kunming 650091, People's Republic of China
e-mail: xmniu@ynu.edu.cn

R.H. Manzanilla-López, L.V. Lopez-Llorca (eds.), *Perspectives in Sustainable Nematode Management Through* Pochonia chlamydosporia *Applications for Root and Rhizosphere Health*, Sustainability in Plant and Crop Protection,
DOI 10.1007/978-3-319-59224-4_7

Table 7.1 Secondary metabolites purified from *Pochonia chlamydosporia* (= *Verticillium chlamydosporium*)

Metabolite class	Metabolite name	References
Resorcylic acid lactones	Radicicol/Monorden (**1**)	Khambay et al. (2000), Hellwig et al. (2003), and Shinonaga et al. (2009a, b)
	Tetrahydromonorden (**2**)	Hellwig et al. (2003)
	Hexahydromonorden (**3**)	Hellwig et al. (2003)
	6,13-Dichloro-5,9,14,16-tetrahydroxy-3-methyl-3,4,5,6,9,10-hexahydro-1H-2-benzoxacyclotetradecine-1,11(12H)-dione (**4**)	Hellwig et al. (2003)
	13-Chloro-5,6,9,14,16-pentahydroxy-3-methyl-3,4,5,6,9,10-hexahydro-1H-2-benzoxacyclotetradecine-1,11(12H)-dione (**5**)	Hellwig et al. (2003)
	Monorden analogue-1 (**6**)	Shinonaga et al. (2009b)
	Monorden E (**7**)	Shinonaga et al. (2009b)
	Pochonin A (**8**)	Hellwig et al. (2003)
	Pochonin B (**9**)	Hellwig et al. (2003)
	Pochonin C (**10**)	Hellwig et al. (2003)
	Pochonin D (**11**)	Hellwig et al. (2003)
	Pochonin E (**12**)	Hellwig et al. (2003) and Shinonaga et al. (2009b)
	Pochonin F (**13**)	Hellwig et al. (2003) and Shinonaga et al. (2009b)
	Pochonin G (**14**)	Shinonaga et al. (2009a)
	Pochonin H (**15**)	Shinonaga et al. (2009a)
	Pochonin I (**16**)	Shinonaga et al. (2009a)
	Pochonin J (**17**)	Shinonaga et al. (2009a)
	Pochonin K (**18**)	Shinonaga et al. (2009b)
	Pochonin L (**19**)	Shinonaga et al. (2009b)
	Pochonin M (**20**)	Shinonaga et al. (2009b)
	Pochonin N (**21**)	Shinonaga et al. (2009b)
	Pochonin O (**22**)	Shinonaga et al. (2009b)
	Pochonin P (**23**)	Shinonaga et al. (2009b)
	Monocillin I (**24**)	Shinonaga et al. (2009b)
	Monocillin II (**25**)	Hellwig et al. (2003) and Shinonaga et al. (2009b)
	Monocillin III (**26**)	Hellwig et al. (2003) and Shinonaga et al. (2009b)
	Monocillin IV (**27**)	Shinonaga et al. (2009b)
	Monocillin II glycoside (**28**)	Hellwig et al. (2003) and Shinonaga et al. (2009b)

(continued)

Table 7.1 (continued)

Metabolite class	Metabolite name	References
Pyranones	Phomalactone (**29**)	Khambay et al. (2000)
	Aurovertin D (**30**)	Niu et al. (2010) and Wang et al. (2015)
	Aurovertin F (**31**)	Niu et al. (2010) and Wang et al. (2015)
	Aurovertin E (**32**)	Niu et al. (2010) and Wang et al. (2015)
	Aurovertin I (**33**)	Niu et al. (2010) and Wang et al. (2015)
Alkaloid	Pseurotin A (**34**)	Bloch and Tamm (1976)

monorden (Delmotte and Delmotte-Plaquee 1953). It was independently purified from *Nectria radicicola* Wollenw. [= *Ilyonectria radicicola*]; *Cylindrocarpon destructans* (*Zinssm.*) *Scholten* [= *Ilyonectria destructans*] and from mycorrhizal tuberous roots of the saprophytic orchid *Dipodium punctatum* and given the name of radicicol (Evans and White 1966; Mirrington et al. 1964). As the original structure of monorden was incorrect, the name radicicol has prevailed. Radicicol (**1**) and some of its derivatives were also found to be synthesized by fungi, including *Penicillium luteoaurantium* G. Sm. [= *Penicillium resedanum*] (Nozawa and Nakajima 1979), a destructive mycoparasite, *Monocillium nordinii* (Bourch) W. Gams, of pine stem rusts, a mycoparasite *Humicola fuscoatra* Traaen NRRL 22980 isolated from an *Aspergillus flavus* Link sclerotium buried for 3 years in a Georgia field planted to corn (Wicklow et al. 1998), two *Chaetomium chiversii* (J.C. Cooke) A. Carter and *Paraphaeosphaeria quadriseptata* M.E. Barr associated with a Sonoran desert plant (Turbyville et al. 2006). It has been assumed that the occurrence of radicicol (monorden, **1**) and its derivatives as major constituents of the fermentation extracts of two mycoparasite fungi *H. fuscoatra* and *M. nordinii* indicated that monorden may play a key role in the mycoparasitic action (Wicklow et al. 1998) since radicicol (**1**) was inhibitory to the growth of numerous fungal strains, including *Aspergillus niger* Tiegh., *A. flavus*, and *Eurotium repens* De Bary 1870 (Wicklow et al. 1998). Fungi in unrelated fungal taxa could benefit from the same classes of secondary metabolites with similar ecological functions, and one might anticipate finding evolutionary convergence in pathways leading to their biosynthesis. For this purpose, occurrences of the purified secondary metabolites from *P. chlamydosporia* strains are also listed (Figs. 7.1 and 7.2).

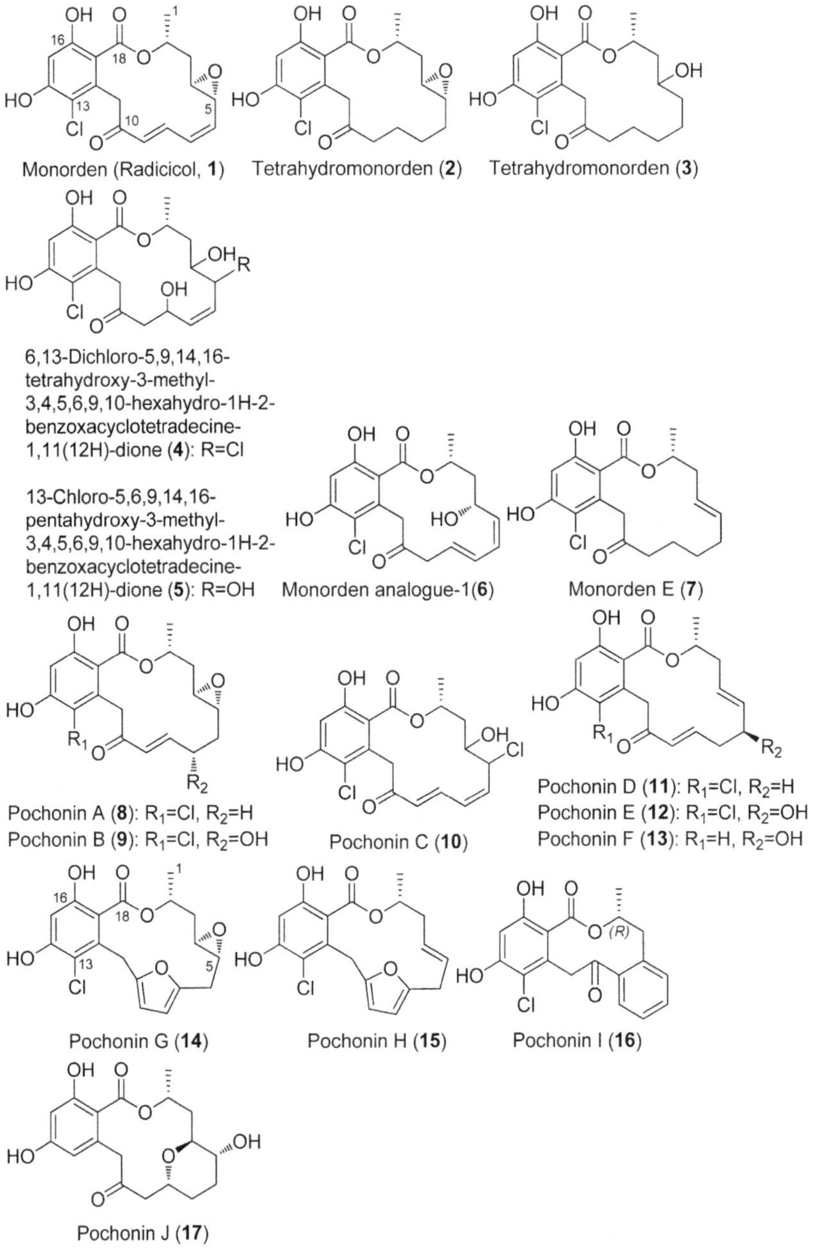

Fig. 7.1 The chemical structure of secondary metabolites (**1**–**34**) from *Pochonia chlamydosporia*

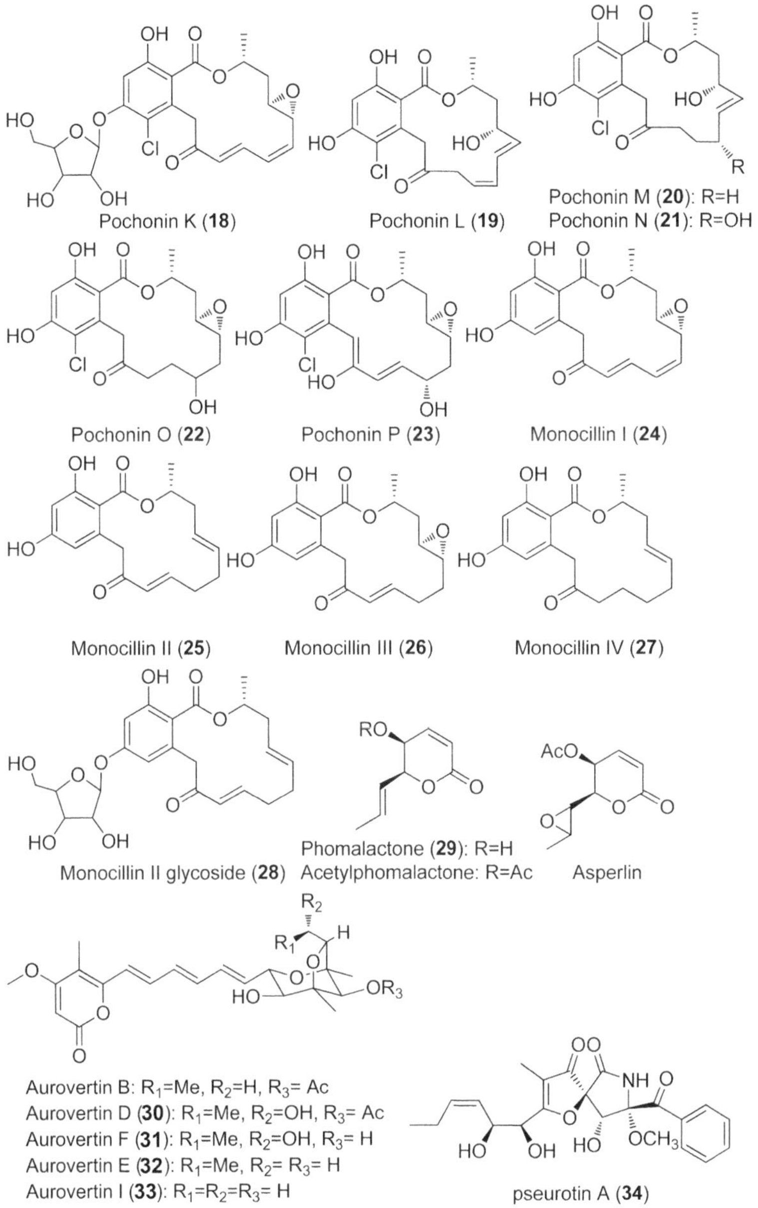

Fig. 7.1 (continued)

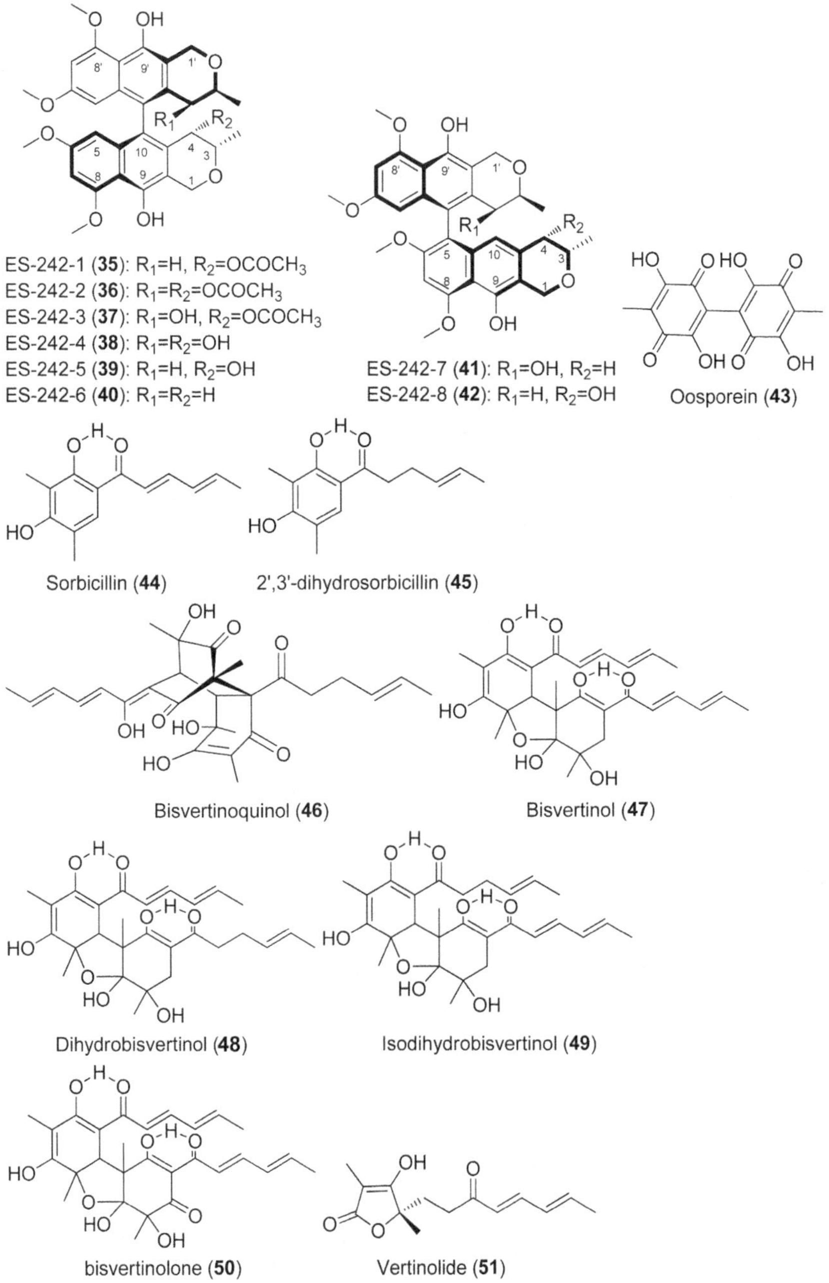

Fig. 7.2 The chemical structure of secondary metabolites (**35**–**105**) from *Pochonia* spp. other than *P. chlamydosporia*

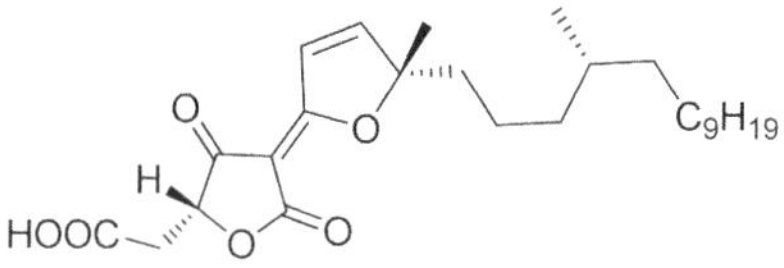
Lowdenic acid (**52**)

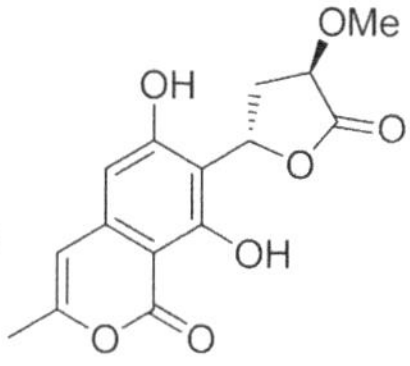
Canescin A (**53**)

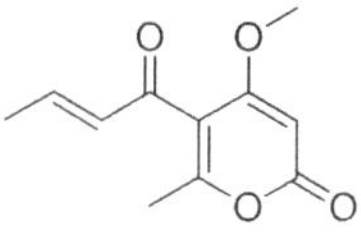
Pyrenocine A (**54**)

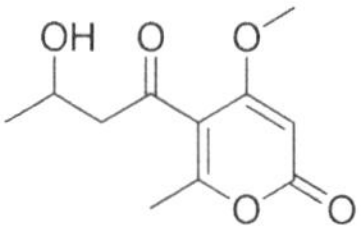
Pyrenocine B (**55**)

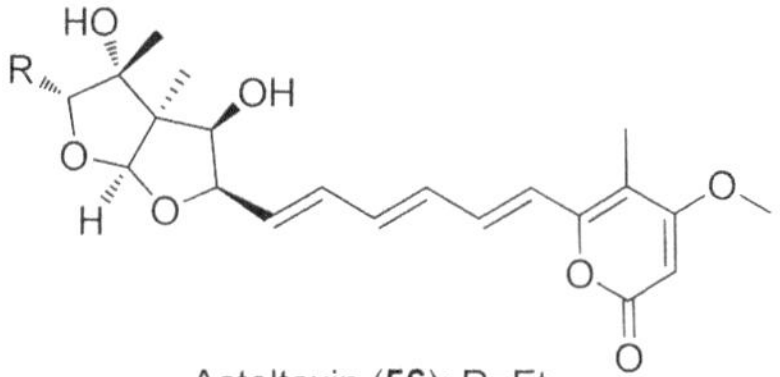
Asteltoxin (**56**): R=Et
Asteltoxin C (**57**): R=Me

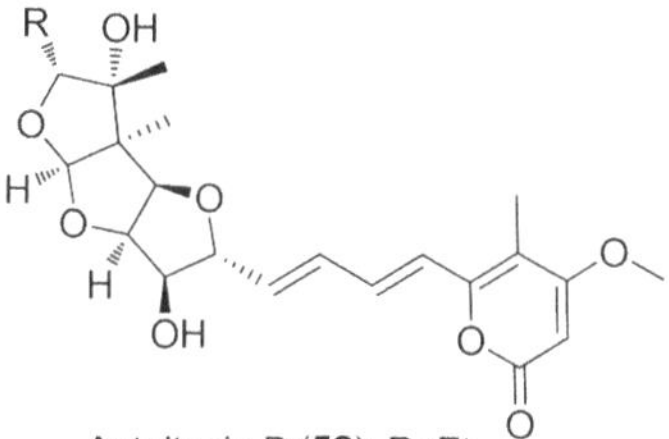
Asteltoxin B (**58**): R=Et
Asteltoxin D (**59**): R=Me

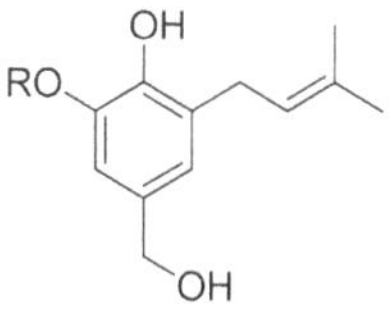
Bigutol (**60**): R=H
Methylbigutol (**61**); R= Me

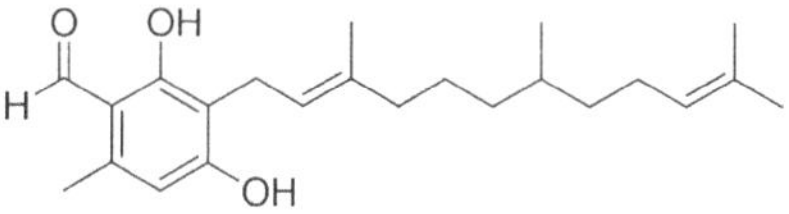
LL-Z1272β (**62**)

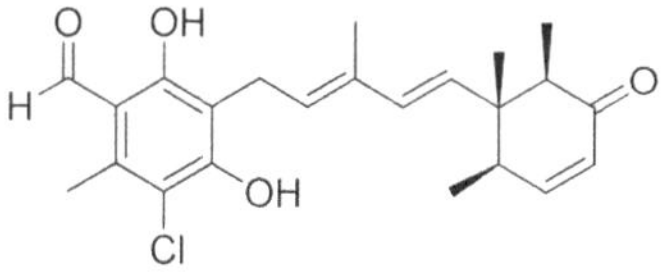
8',9'-dehydroascochlorin (**63**)

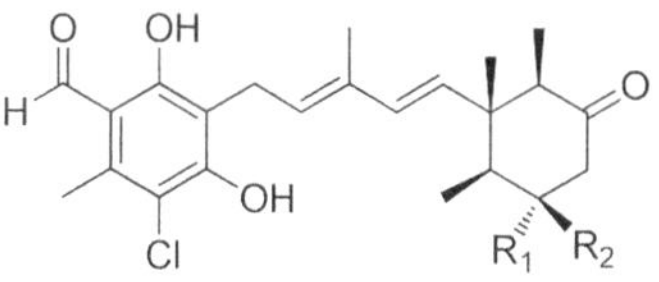
Ascochlorin/LL-Z1272γ (**64**): $R_1=R_2=H$
8'-Acetoxyascochlorin/LL-Z1272ζ (**65**): $R_1=OAc$, $R_2=H$
8'-Hydroxyascochlorin (**66**): $R_1=OH$, $R_2=H$

Fig. 7.2 (continued)

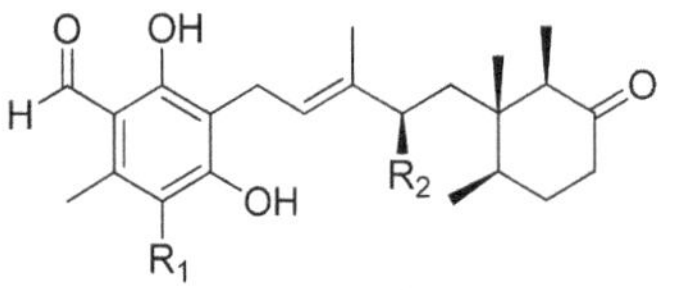

LL-Z1272ε (**67**): $R_1=R_2=H$
4',5'-dihydroascochlorin/LL-Z1272δ (**68**): $R_1=Cl$, $R_2=H$
4',5'-dihydro-4'-hydroxyascochlorin (**69**): $R_1=Cl$, $R_2=OH$

Vertihemipterin A (**70**): $R_1=Cl$, $R_2=$ 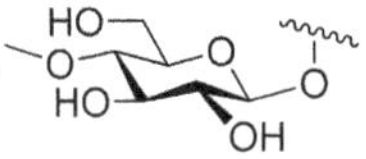

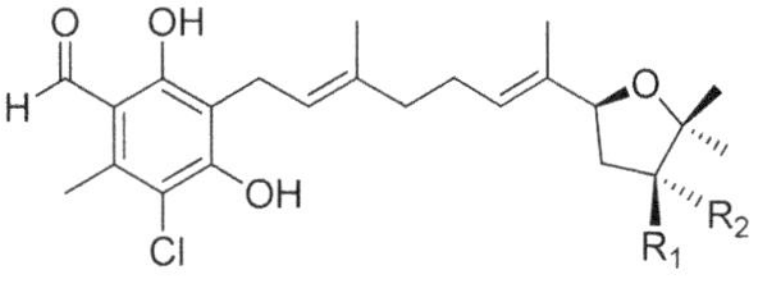

Ascofuranone (**71**): $R_1=R_2=O$
Ascofuranol (**72**): $R_1=OH$, $R_2=H$

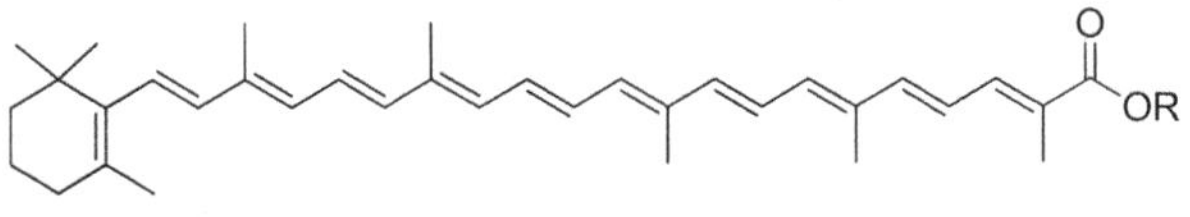

β-Apo-4'-carotenoic acid (**73**): R=H
β-Apo-4'-carotenoic acid methyl ester (**74**): R=Me

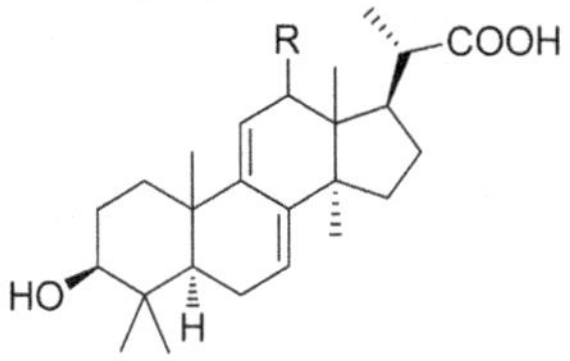

3β-Hydroxyl-23,24,25,26,27-Pentanorlanost-7,9(11)-diene acid (**75**): R=H
3β,12-Dihydroxyl-23,24,25,26,27-pentanorlanost-7,9(11)-diene acid (**76**): R=OH

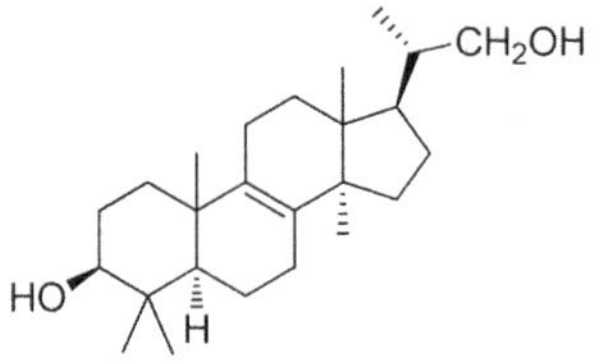

23,24,25,26,27-Pentanorlanost-8-en-3β,22-diol (**77**)

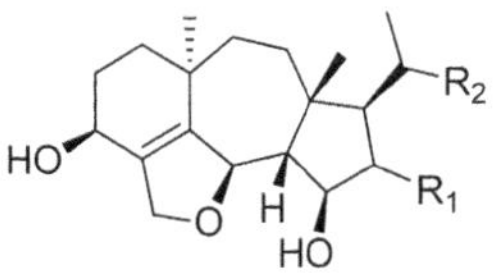

Dahliane A (**78**): R_1=β–OH, $R_2=H$
Dahliane B (**79**): R_1=β–OH, $R_2=OH$
Dahliane C (**80**): R_1=α–OH, $R_2=H$
Dahliane D (**81**): R_1=α–OH, $R_2=OH$

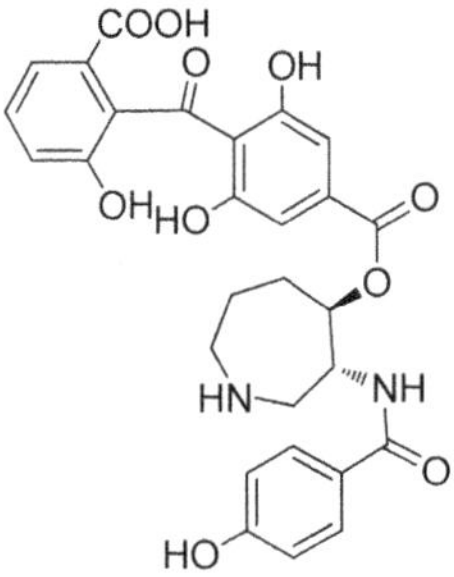

Balanol (**82**)

Fig. 7.2 (continued)

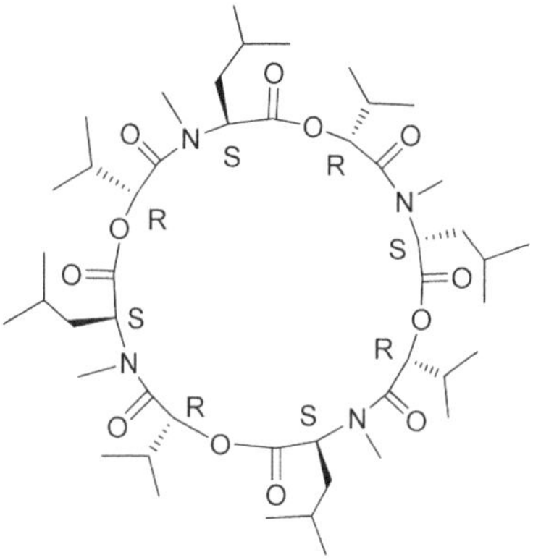

Bassianolide (**83**)

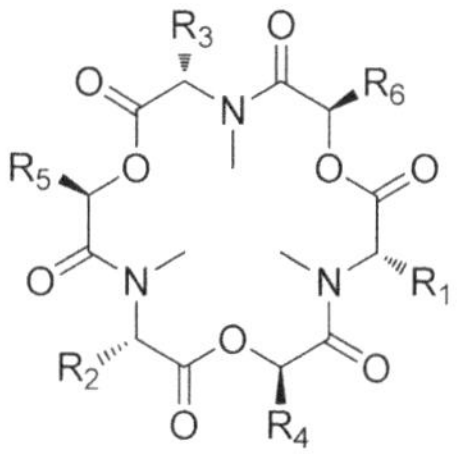

enniatin B (**84**): $R_1=R_2=R_3=R_4=R_5=R_6=$iPr
enniatin B_4 (D, **85**): $R_1=R_2=R_4=R_5=R_6=$iPr, $R_3=$iBu
enniatin C (**86**): $R_1=R_2=R_3=$iBu,$R_4=R_5=R_6=$iPr
enniatin G (**87**): $R_1=R_2=$iBu,$R_3=R_4=R_5=R_6=$iPr
enniatin MKI688 (**88**): $R_1=R_2=R_3=$iPr, $R_4=R_5=R_6=$sBu
enniatin H (**89**): $R_1=R_2=R_3=R_5=R_6=$iPr, $R_4=$sBu
enniatin I (**90**): $R_1=R_2=R_3=R_6=$iPr, $R_4=R_5=$sBu
enniatin O_1 (**91**): $R_1=$iBu,$R_2=R_3=R_5=R_6=$iPr, $R_4=$sBu
enniatin O_2 (**92**): $R_1=$iBu,$R_2=R_3=R_4=R_6=$iPr, $R_5=$sBu
enniatin O_3 (**93**): $R_1=$iBu,$R_2=R_3=R_4=R_5=$iPr, $R_6=$sBu

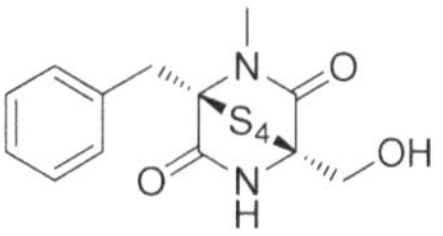

1-Demethylhyalodendrin tetrasulfide (**94**)

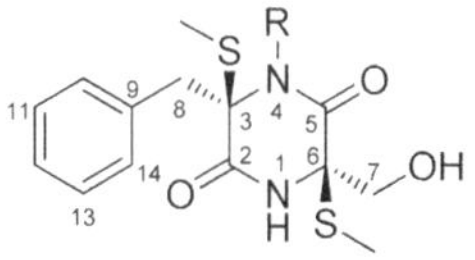

Bisdethiodi (methylthio)-1-demethylhyalodendrin (**95**): R=Me
Bisdethiodi (methylthio)hyalodendrin (**96**): R=H

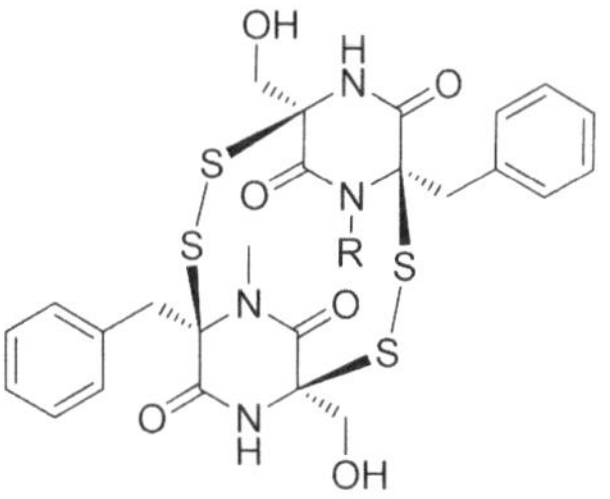

Vertihemiptellide A (**97**): R= Me
Vertihemiptellide B (**98**): R= H

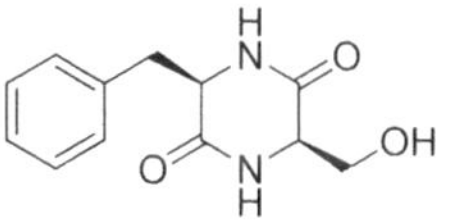

Demethylhyalodendrin (**99**)

Fig. 7.2 (continued)

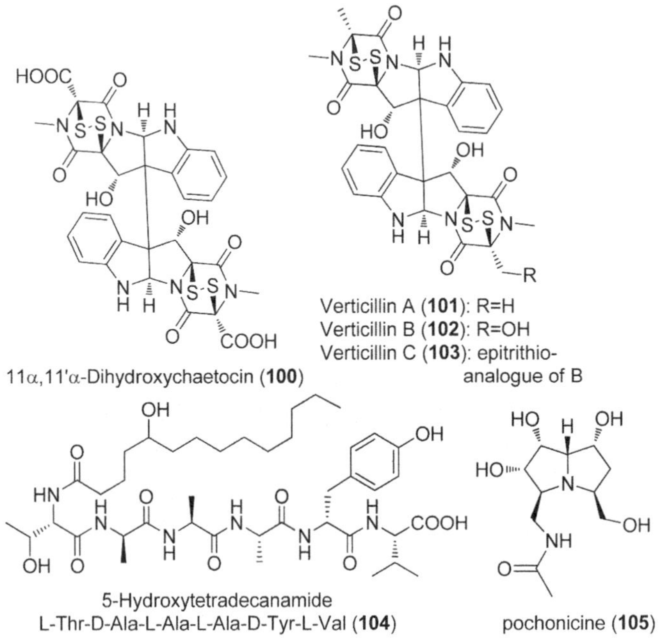

Fig. 7.2 (continued)

7.1.1 Resorcylic Acid Lactones

Resorcylic acid lactones (RALs) are a family of compounds possessing a resorcylate moiety fused within a 12- or 14-membered macrolactone ring. They have been known for decades since the first purification of monorden (radicicol, **1**) (Delmotte and Delmotte-Plaquee 1953), followed by zearalenone (Stob et al. 1962), LL-Z1640–2 (Ellestad et al. 1978), and hypothemycin (Nair and Carey 1980). In recent years, more and more resorcylic acid lactone metabolites with bioactivities have been purified from fungi and structurally characterized. In 2007, the chemistry and biology of resorcylic acid lactones from fungi were both reviewed (Winssinger and Barluenga 2007). In 2014, several reviews on bioactivities, biosynthesis, and some representative chemical syntheses of this class of the resorcylic acid lactones were reported (Xu et al. 2014a, b). This chapter mainly describes the characterization and occurrence of resorcylic acid lactones (RALs) in *P. chlamydosporia* according to their categories, including radicicol (monordens), pochonins and monocillins.

7.1.1.1 Radicicol (Monorden) and Derivatives

In 2000, Khambay et al. (2000) first reported the purification of radicicol from *Verticillium chlamydosporium* (= *P. chlamydosporia*) when searching for the nematicidal components from nematophagous fungi. Radicicol (**1**) was found to be the major component in 14 day old potato dextrose cultures of the fungus *V. chlamydosporium*. However, radicicol (**1**) did not exhibit nematicidal activity in a bioassay using the root-knot nematode *Meloidogyne incognita* (Kofoid & White) Chitwood.

A bioassay was carried out in 2003 in search for antiviral agents against Herpes Simplex Virus 1 (HSV) and the fraction analyses of the culture broths (7–8 days) of *P. chlamydosporia* var. *catenulata* strain P 0297 led to isolation and identification of radicicol (**1**) as the main metabolite, together with other RALs including one radicicol derivative, naturally occurring tetrahydromonorden (**2**) (Hellwig et al. 2003). The paper also reported that hydrogenation of radicicol (**1**) with H_2/Pd-C under atmospheric pressure also yielded tetrahydromonorden (**2**) with an intact oxirane ring, as well as hexahydromonorden (**3**) with a secondary alcohol function from an opened oxirane ring. Two artificial derivatives (**4** and **5**) with a microscale epoxide ring opening came from radicicol (**1**) with HCl in methanol. Radicicol (**1**) showed potent inhibitory activity in the nM range in a cellular replication assay against HSV1, and was also cytotoxic at the concentration necessary for HSV1 inhibition. Tetrahydromonorden (**2**) with the epoxy moiety as in radicicol (**1**) also exhibited inhibitory bioactivities with an IC_{50} value of 1.5 μM against HSV1 and selectivity towards the parasite *Eimeria tenella* (taxonomically related to *Plasmodium*) at 10 ppm (Hellwig et al. 2003). However, hexahydromonorden (**3**), with a secondary alcohol function instead of an intact oxirane ring as in radicicol (**1**), was found to be devoid of activity in the assays.

A secretory glycoprotein of the WNT family (wingless-type mouse mammary tumour virus integration site family), WNT-5A (member 5A), was highly expressed in the dermal papillae of depilated skin. WNT-5A expression inhibitor promotes the proliferation of dermal papilla cells. In 2009, in the search for a hair-growth stimulant using the inhibitory activity against WNT-5A expression as a bioassay guide, radicicol (**1**) was purified and identified as a potently active compound from a 6 day culture broth of the fungus *P. chlamydosporia* var. *chlamydosporia* strain TF-0480 (Shinonaga et al. 2009a, b). An analogue of radicicol (**1**), monorden analogue-1 (**6**), together with monorden E (**7**), was also purified from the culture broth of the fungus strain TF-0480 (Shinonaga et al. 2009b). The relative stereochemistry of monorden analogue-1 (**6**) was first elucidated to be a-methyl (C-2) and a-hydroxyl (C-4) based on ^{1}H-^{1}H coupling constants and NOESY (Nuclear Overhauser Effect Spectroscopy) data. This metabolite with only a planar structure was originally reported as being obtained from a mycoparasite, *Humicola fuscoatra* NRRL 22980, purified from a sclerotium of *Aspergillus flavus* buried in a cornfield near Tifton, Ga (Wicklow et al. 1998). Monorden E (**7**), as well as radicicol (**1**), were originally isolated from the fermentation broth of *Humicola* sp. FO-2942 that produced amidepsines, inhibitors of diacylglycerol acyltransferase. These two compounds were found to show antifungal activity only against *Aspergillus niger* among 16 microorganisms tested, with IC_{50} values (i.e., concentration needed to inhibit a biological or biochemical function by half) of 12 and 70 μM, respectively (Arai et al. 2003; Yamamoto et al. 2003). However, monorden analogue-1 (**6**) and monorden E (**7**) did not show any obvious active biological activity during the inhibitory bioassays against WNT-5A expression. With few exceptions, the production of radicicol (**1**) and some of its derivatives appears to be a rather constant feature in cultures of worldwide strains of *P. chlamydosporia* (Khambay et al. 2000).

7.1.1.2 Pochonins

As the name implies, pochonins (**8–23**) were first purified and identified from *P. chlamydosporia* strains (Hellwig et al. 2003; Shinonaga et al. 2009a, b). All the pochonins, except for pochonins F (**13**) and J (**17**), were chlorine-containing RALs, and pochonin K (**18**) was a 14-aldofuranose derivative of radicicol (**1**). Among these secondary metabolites, pochonins G (**14**) and H (**15**) were the first two examples in the radicicol family to have a furan ring, and pochonins L-N (**19–21**) are the first three examples of radicicol analogues in which the double bond has an E-configuration at C5-C6. Pochonin I (**16**) has a single benzene moiety in the macrolide ring.

In 2003, pochonins A-F (**8–13**) were also extracted from culture broths of *P. chlamydosporia* var. *catenulata* strain P 0297 (Hellwig et al. 2003). Pochonins A-F (**8–13**), except for pochonin D (**11**), inhibited HSV1 with IC_{50} values 1.5–10 μM, less than that of radicicol (**1**) (0.2–0.8 μM). Although it can be speculated that the HSV1 inhibitory activity may originate from Hsp90 (heat shock protein 90) inhibition, pochonin D (**11**) that was later reported to be an Hsp90 inhibitor had no activity in the assay (Moulin et al. 2005). On the other hand, pochonin C (**10**), which was a poor Hsp90 inhibitor, retained activity in the HSV1 replication assay while being the least cytotoxic compound (90 mM). As with many of the radicicol type RALs, the pochonins were also found to be devoid of the oestrogenic activity which accompanied the zearalenone type RALs. A library based on the pochonin D (**11**) scaffold led to the identification of moderate kinase inhibitors (low mM) for Src (Sarcoma-family kinases), VEGFR (vascular endothelial growth factor), auroras A and B. Although pochonin D (**11**) is an Hsp90 inhibitor, the analogues that showed kinase activity were poor Hsp90 inhibitors and vice versa (Moulin et al. 2006). In addition, pochonin A (**8**) inhibited *Eimeria tenella* at 10 ppm.

In 2009, pochonins G-P (**14–23**), together with pochonins B (**9**) and D-F (**11–13**), were purified and characterized from *P. chlamydosporia* strain TF-0480 using the inhibitory activity against WNT-5A (a cysteine-rich secreted glycoprotein growth factor) expression as a bioassay guide (Shinonaga et al. 2009a, b). Pochonins B (**9**), D (**11**), E (**12**), F (**13**), K (**18**), L (**19**) and O (**22**) showed a strong WNT-5A expression inhibitory activity with IC_{50} values ranging from 8 to 18 μM, weaker than that of radicicol (**1**), with IC_{50} value at 0.19 μM. The data suggested that the 4,5-epoxide or 4,5-E-olefin moieties present in Pochonins K (**18**) and O (**22**), or E (**12**) and F (**13**), respectively, may be necessary for radicicol-type compounds designed to inhibit WNT-5A expression. The chlorine atom at C-13 may decrease the toxicity against dermal papilla cells.

Pochonins occur exclusively in *Pochonia* spp. (Hellwig et al. 2003). With few exceptions, the production of some pochonins, as well as radicicol (**1**), appears to be a rather constant feature in the cultures of *P. chlamydosporia* worldwide. These chemotaxonomical investigations supported the generic segregation of fungal species with verticillium-like anamorphs, all of which were included in *Verticillium* sect. *Prostrata*. This was also deduced from minisatellite PCR fingerprinting studies on *Pochonia* and other conidial fungi with *Verticillium*-like anamorphs (Hellwig et al. 2003).

7.1.1.3 Monocillins

The class of monocillin metabolites, including monocillin I (**24**), monocillin II (**25**), monocillin III (**26**), and monocillin IV (**27**), were non chlorine-containing RALs originally purified from *Monocillium nordinii* (Bourch.) W. Gams, together with radicicol (**1**) (Ayer et al. 1980, Ayer and Peña-Rodriguez 1987). The monocillin I (**24**), monocillin III (**26**) and radicicol (**1**), were found to be the three major secondary metabolites in potato dextrose broth with 0.2% yeast extract cultures of the fungus. Radicicol (**1**) and monocillin I (**24**) showed pronounced activity against a wide variety of fungi, including *Ceratocystis ulmi* (= *Ophiostoma ulmi*), the causal agent of Dutch elm disease, and could also cause complete inhibition of spore germination of the rust, *Endocronartium harknessii* (J.P. Moore) Y. Hirats, the host of *M. nordinii*, at 10 and 50 μg/ml, respectively. This indicated that RALs might play a key role in the mycoparasitic action of *M. nordinii* on rust sori in nature (Ayer et al. 1980; Ayer and Peña-Rodriguez 1987).

In 2003, monocillins II-III (**25–26**) were first purified as major secondary metabolites besides radicicol (**1**), as well as a novel monocillin II glycoside (**28**) and the non chlorine-containing pochonin F (**13**) as minor secondary metabolites, from the culture broth of *P. chlamydosporia* var. *catenulata* strain P 0297 fermented in bromide-containing culture media, which led to a shift in secondary metabolite production and yielded much production of nonchlorine-containing RALs (Hellwig et al. 2003). Monocillin III (**26**) was the dechloro analogue of pochonin A (**8**) and monocillin II (**25**) that of pochonin D (**11**). Monocillin II glycoside (**28**) was identified as an aldofuranose moiety linked to position 14 of monocillin II (**25**), according to the HMBC signal from the anomeric proton H-1′ to the quaternary carbon C-14. Exact determination of the sugar was not performed due to the limited amount of sample available.

Monocillin III (**26**) was found to show potent inhibitory activity in the HSV1 replication assay with IC_{50} value of 0.4 μM, as strong as radicicol (**1**), while monocillin II (**25**) did not show any inhibitory activity. This provided further evidence that the chlorine substituent is not essential for the antiviral activity. Monocillin II glycoside (**28**), like pochonin D (**11**) that contains a double bond instead of the epoxide ring, showed only cytostatic effects.

In 2009, monocillins I-IV (**24–27**) were also purified from *P. chlamydosporia* strain TF-0480 and evaluated for their inhibitory activities against WNT-5A expression and cytotoxicities against dermal papilla cells (Shinonaga et al. 2009b).

Monocillins I-III (**24–26**) showed a strong WNT-5A expression inhibitory activity with IC_{50} values 1.93, 7.36 and 9.43 μM, respectively. These three secondary metabolites also displayed cytotoxicities against dermal papilla cells with TC_{50} values 2.90, 17.62, and 28.87 μM, respectively. However, monocillin IV (**27**) did not display any biological activity.

In addition, monocillin I (**24**), as well as radicicol (**1**), showed pronounced activity against a wide variety of fungi, such as *Phycomyces blaksleeanus* (*Zygomycotina*), *Pythium debaryanum* (= *Globisporangium debaryanum*, *Mastigomycotina*), *Ceratocystis ulmii* (= *Ophiostoma ulmi*, *Ascomycotina*) as the cause of Dutch elm disease, and *Phellinus pini* (= *Porodaedalea pini*, *Basidiomycota*), indicating that their antifungal activities are nonspecific. The inhibitory effect of radicicol (**1**) is only slightly stronger than that of monocillin I (**24**) toward most of the test fungi, suggesting that the chlorine on the aromatic ring of radicicol (**1**) did not play an important role in the bioactivity (Ayer et al. 1980; Ayer and Peña-Rodriguez 1987). The antifungal activity of monocillins II-IV (**25–27**) was negligible. This suggested that alterations in the macrocyclic ring of radicicol (**1**) and monocillin I (**24**) led to a significant loss of activity of monocillins II-V (**25–28**).

7.1.2 Pyranones

Pyranone secondary metabolites are a family of 2-pyrone (α-pyrone or pyran-2-one) compounds, widely distributed in bacteria and fungi. They are capable of binding to specific protein domains to exert a wide range of biological effects. Two compounds, 4-phenylethynyl-4-tetrahydropyranyl propargyl ether- and 4-ethynyl-6-methyl-2-pyrones, have excellent potential as a new class of anticancer agents (Fairlamb and McGlacken 2005; Fang et al. 2015). Most recently, α-pyrones were identified as a new class of signalling molecules in bacterial communication, similar to quorum sensing (Brachmann et al. 2013). However, the natural functions of most of α-pyrone secondary metabolites in fungi still remain unknown. The characterization and occurrence of two pyranones in *P. chlamydosporia* is described below.

7.1.2.1 Phomalactones

Phomalactone (**29**) also known as 6-(l-propenyl)-5,6-dihydro-5-hydroxypyran-2-one, was first purified from the plant-pathogenic fungus, *Nigrospora* sp. (Evans et al. 1969), and subsequently also purified from *Phoma minispora* in 1970 (Yamamoto et al. 1970; Yamano et al. 1971). Phomalactone (**29**) was also found in fungi, e.g., two entomopathogenic fungi, *Hirsutella thompsonii* var. *synnematosa* (Krasnoff and Gupta 1994) and *Paecilomyces cateniobliquus* (= *Isaria cateniobliqua*) YMF1.01799 (Wu et al. 2012), and *Nigrospora sphaerica* (= *Khuskia oryzae*) (Kim et al. 2001). Since these fungal genera are unrelated, it is likely that the production of phomalactone (**29**) evolved independently.

The first study on the isolation of phomalactone from the fungus *V. chlamydosporium* (= *P. chlamydosporia*) was reported in 2000. A bioassay-directed fractionation of the 14 day old potato dextrose cultures of the fungus, combined with the bioassay against the root-knot nematode *Meloidogyne incognita*, led to isolation of phomalactone (**29**) as the nematicidal metabolite. Another metabolite, radicicol (**1**), was also reported to be obtained as the major metabolite in this experiment. Even though the content of phomalactone (**29**) in the culture broth was less than half of radicicol (**1**), only phomalactone (**29**) showed the mortality of *M. incognita* achieving up to 84% in 96 h at a concentration of 500 mg/l. Phomalactone (**29**) also showed dose-dependent insecticidal activity against apple maggot flies, *Rhagoletis pomonella*, with IC_{50} of 2 mg/ml for 72 h, and was mildly toxic to tephritid fruit flies. It also showed inhibitory activity against spores of *B. bassiana* with IC_{50} of 468 μg/ml and *Metarhizium anisopliae* with IC_{100} at 2 mg/ml. However, under the same conditions, phomalactone (**29**) failed to inhibit germination of spores of two other entomopathogenic fungi of interest, *V. lecanii* (= *Lecanicillium lecanii*) and *Paecilomyces fumosoroseus* (= *Isaria fumosorosea*) (Krasnoff and Gupta 1994).

Acetylphomalactone and Asperline, two compounds structurally closely related to phomalactone, were purified from *Aspergillus caespitosus*. Asperline (= asperlin), an acetyl derivative of phomalactone (**29**) with an epoxide at the olefinic double bond in the propenyl side chain, was originally obtained from *A. nidulans* (Argoudelis and Zieserl 1966) and has subsequently been purified from several other *Aspergillus* spp.: *A. elegans* and *A. carneus* (Yamamoto et al. 1970), and *A. caespitosus* (Mizuba et al. 1975). The same absolute configurations (5S, 6S) at carbons 5 and 6 in naturally occurring phomalactone (**29**), acetylphomalactone, and asperline, suggest a common biosynthetic route. Phomalactone (**29**) inhibited growth of a wide range of microorganisms, including fungi, bacteria, and a protozoan. Acetylphomalactone and asperline also show a similar spectrum of antimicrobial activity. Asperlin was also found to have antitumour activity (Owen and Bhuyan 1965).

7.1.2.2 Aurovertins

The chemical structures of aurovertin-type metabolites are partly similar to phomalactone, which contains a carbon skeleton of 5-methyl-6-((E)-prop-1-enyl)pyran-2-one. The first four aurovertin secondary metabolites were initially identified from a polyphagous fungicolous fungus, *Calcarisporium arbuscula* NRRL 3705 (Baldwin et al. 1964). The structure of aurovertin B was proposed on the basis of its chemical properties and both ^{13}C and ^{1}H NMR spectroscopic data (Mulheirn et al. 1974), and characterized as an unusual 6-ethyl-1,5-dimethyl-4,7-dioxabicyclo[3.2.1] octane ring system with a conjugated R-pyrone moiety. The absolute configuration of aurovertin B was finally determined by total synthesis (Nishiyama et al. 1988). All the structures of aurovertins were determined on the basis of spectroscopic comparison with aurovertin B (Wang et al. 2005). Aurovertin D (**30**), together with three new aurovertins F (**31**), G and H, was obtained from the entomopathogenic fungus *Metarhizium anisopliae* (Azumi et al. 2008). The absolute configuration of aurovertin F (**31**) was determined by analysis of the CD spectrum. Another aurovertins, including aurovertins E (**32**) and I-S, together with aurovertins B-C, were identified from the basidiomycete *Albatrellus confluens* (Wang et al. 2005; Guo et al. 2013). It has been assumed that these aurovertin-type compounds reported in the literature probably share the same absolute configuration on biosynthetic grounds. The biosynthetic origin of aurovertins was proposed, and subsequently verified from labelling studies, to derive from a polyketide pathway. Aurovertin E (**32**) represents the structurally simplest member of the family and is the biosynthetic precursor to other derivatives. The bicyclic ether moiety in **32** is fused to a methylated α-pyrone via a triene linker and can be differentially substituted to yield various aurovertin congeners. The most recent study has revealed that only four enzymes are required to produce aurovertin E (**32**). The core polyketide synthase produces a polyene α-pyrone. Following pyrone O-methylation by a methyltransferase, a flavin-dependent mono-oxygenase and an epoxide hydrolase can iteratively transform the terminal triene portion of the precursor into the dioxabicyclo[3.2.1] octane scaffold (Mao et al. 2015).

During the past three decades, this class of yellow pigment secondary metabolites has attracted much interest because aurovertin B binds to the β subunit of the F_1-ATPase found on the inner membrane of mitochondria, inhibiting the ATPase and thereby uncoupling oxidative phosphorylation in *in vitro* bioassays (Ebel and Lardy 1975; Van Raaij et al. 1996; Johnson et al. 2009). Furthermore, aurovertin B strongly inhibits the proliferation of breast cancer cell lines and arrests cell cycles at the G0/G1 phase (Huang et al. 2008). Among the most recent identified aurovertins (J-S), only aurovertin P, together with aurovertin B, was reported to exhibit moderate cytotoxicities compared to positive control cisplatin.

The first isolation of aurovertin-type secondary metabolites (D, E, E and I, **30**–**33**) was reported from a root-knot nematode parasitic fungus, *P. chlamydosporia* strain YMF 1.00613, collected from root-knots of tobacco infected by *M. incognita* (Niu et al. 2010). Aurovertin I (**33**) was the first natural product with an aurovertin skeleton with one carbon atom less in the unusual 6-ethyl-1,5-dimethyl-4,7-dioxabicyclo[3.2.1] octane ring system. Aurovertins D (**30**) and F (**31**) were found to be the major aurovertin constituents produced by the nematode-parasitic fungus *P. chlamydosporia*, and show toxicity towards the freeliving nematode *Panagrellus redivivus* with IC_{50} values 41.7 and 88.6 μg/ml, respectively. In the nematicidal bioassay, the internal structures of nematodes exposed to aurovertin D (**30**) or F (**31**) were observed to disintegrate and many vacuoles formed within their body, leading to only empty cuticles of the dead nematodes remaining after 48 h. However, all four aurovertins D, F, E and I (**30**–**33**) falied to show any obvious inhibitory effects on egg hatching of the root-knot nematode *M. incognita*. Compared with phomalactone (**29**), aurovertin secondary metabolites displayed stronger nematode toxicity activities, which suggested that the long chain attached to the moiety of 5-methyl-6-((E)-prop-1-enyl)pyran-2-one in aurovertin-type secondary metabolites would increase their nematicidal activity.

During the course of the aurovertin metabolic profiling of *P. chlamydosporia* (Niu et al. 2010), it has been noted that aurovertin D (**30**) was the richest aurovertin-type component in the chloroform fraction of the extract of mycelia of *P. chlamydosporia*, while aurovertin F (**31**) was the major metabolite in the corresponding part of the fermentation broth. Because aurovertin F (**31**) was the O-deacetyl derivative of aurovertin D (**30**), it seemed that the fungus changed the composition of aurovertin-type metabolites before they released these secondary metabolites to the culture broth. It can be expected that acetyltransferase was possibly involved in the modifying process. It is not clear why the composition of secondary metabolites aurovertin of the mycelia of *P. chlamydosporia* is different from that of the culture broth (Niu et al. 2010).

Wang et al. (2015) reported that most *P. chlamydosporia* isolates did not produce aurovertin secondary metabolites. Among the 17 *P. chlamydosporia* isolates investigated, which were obtained from three different ecological niches including root-knot nematode females, root-knot nematode eggs, or soils, only three fungal strains yielded distinctive luminous yellow fermentation broths. Interestingly, only the three *P. chlamydosporia* strains obtained from nematode females instead of eggs showed strong inhibitory activities toward juvenile nematodes. Chemical investigation of luminous yellow fermentation broths of these three *P. chlamydospori*a strains demonstrated that the yellow pigments consisted of aurovertin secondary metabolites. Among them, four aurovertins D, F, E, and I (**30**–**33**), were identified from all three nematicidal *P. chlamydosporia* strains. Further HPLC-DAD (High-performance liquid chromatography with diode-array detection) analysis of the aurovertin metabolite profiles of the *P. chlamydosporia* strains indicated that these three nematicidal *P. chlamydosporia* strains obtained from the nematode worms tended to yield a total yellow pigment aurovertin production exceeding the inhibitory concentration shown in nematicidal bioassays, and aurovertin D (**30**) became the main component among the yellow pigment secondary metabolites of the aurovertin-producing *P. chlamydosporia* strains grown for ≥5 days. Further bioassay revealed that aurovertin D (**30**) caused significant death of *M. incognita* second-stage juveniles (J2) (LC_{50} value 16.45 μg/ml) and also inhibited *Caenorhabditis elegans* (LC_{50} 33.50 μg/ml), indicating a stronger inhibitory activity toward root-knot than free-living nematodes. In addition, aurovertin D (**30**) exerted profound and detrimental effects on the viability of *C. elegans*, even at a subinhibitory concentration. Aurovertin D (**30**) could also trigger DAF-16/FOXO transcription factor in nematodes (Wang et al. 2015).

7.1.3 Spirocyclic Alkaloid-Pseurotin A

Pseurotin A (**34**) of mixed polyketide synthase-nonribosomal peptide synthetase (PKS-NRPS) origin was originally obtained from the culture filtrates of *Pseudeurotium ovalis* (*Ascomycetes*) (Bloch and Tamm 1976). The structure of pseurotin A (**34**) was identified by spectral data and chemical transformations, and by X-ray analysis of its dibromo derivative, to be a novel, highly substituted and functionalized spirocyclic system containing oxygen and nitrogen atoms. The spirocyclic alkaloid (**34**) was subsequently purified from the submerged cultures of an osmophilic *Aspergillus fumigatus* strain HA 57–88 as an inhibitor of chitin synthase during a search for selective antifungal compounds with a screening using *Coprinus cinereus* (Schaeff.) Gray (= *Coprinopsis cinerea*) solubilized chitin synthase EC 2.4.1:16 (Wenke et al. 1993). Recently, pseurotin A (**34**) was reported to be extracted from a marine-derived *A. fumigatus* (Boot et al. 2007), and also obtained from an endophytic fungus, *Penicillium janczewskii* K.M. Zalessky, from the phloem of the Chilean gymnosperm, *Prumnopitys andina* (Schmeda-Hirschmann et al. 2008). Several studies reported that pseurotin A (**34**) showed no obvious antibacterial or antifungal activity except for its moderate effect against the phytopathogenic bacteria *Erwinia carotovora* and *Pseudomonas syringae*, with IC_{50} values of 220 and 112 μg/ml, respectively.

In 2003, Hellwig reported that pseurotin A (**34**) was a main metabolite in most *P. chlamydosporia* isolates propagated in Q6-medium (D-glucose 0.2%, glycerol 1%, cotton seed meal 0.5%; tap water 1 l, pH 7.2). Compound **34** was also subjected to all the biological investigations performed with the pochonins in a cellular replication assay against HSV1, and found to be inactive (Hellwig et al. 2003).

A new derived compound, 11-O-methylpseurotin A, along with pseurotin A (**34**), was also identified from the marine-derived *Aspergillus fumigatus*, using a bioassay-guided fractionation employing wild-type and *hof1*Δ strains of *Saccharomyces cerevisiae* in a yeast halo assay (Boot et al. 2007). Hof1 is required for cytokinesis and regulates actomyosin ring dynamics and septin localization and interacts with the formins, Bni1p and Bnr1p, and with Cyk3p, Vrp1p, and Bni5p. It is synthetically lethal with other cytokinesis-related genes, including CYK3 and BNI5, and the type II myosin-encoding gene MYO1, which is also required for cytokinesis and cell separation. Compounds with selective toxicity to the *hof1*Δ strain are therefore likely to target pathways parallel to Hof1 and may inhibit cytokinesis. Pseurotin A (**34**) was inactive against both strains, while 11-O-methylpseurotin A inhibited *hof1*Δ yeast and was twofold less toxic to wild-type yeast. The identification of the cellular target(s) of 11-O-methylpseurotin A are currently under way (Boot et al. 2007).

7.2 Other Secondary Metabolites from *Pochonia* (*Verticillium*) Species

The types of secondary metabolites from the genus *Pochonia* (*Verticillium*) strains consisted mainly of polycyclic aromatic compounds, nonaromatic polyketides (including bisvertinols, bisvertinoqinol, sorbicillins, pyrenocines, vertinolide and lowdenic acid), *β*-carotene-type neurosporaxanthin, pentanorlanostane triterpenoids, diterpenoids, and cyclopeptides (including bassianolide and epidithiodioxopiperazine-type verticillins). Many of these chemical compounds have been found to have a myriad of biological activities, including antibacterial, antifungal, antioxidative activities, antivirus and antitumour. However, studies with regard to the nematicidal activities of the natural compounds from this fungus have rarely been reported and discussed. The unique types of metabolites from *P. chlamydosporia* and the chemical components of other former *Pochonia* (*Verticillium*) species are summarized in Table 7.2.

Table 7.2 Secondary metabolites from former species included under *Pochonia* (*Verticillium* spp.)

Secondary metabolite class	Metabolite name	Occurrence	References
Aromatic compounds	ES-242-1 (**35**)	*Verticillium* sp. SPC-15898	Toki et al. (1992a, b)
	ES-242-2 (**36**)	*Verticillium* sp. SPC-15898	Toki et al. (1992b)
	ES-242-3 (**37**)	*Verticillium* sp. SPC-15898	Toki et al. (1992b)
	ES-242-4 (**38**)	*Verticillium* sp. SPC-15898	Toki et al. (1992b)
	ES-242-5 (**39**)	*Verticillium* sp. SPC-15898	Toki et al. (1992b)
	ES-242-6 (**40**)	*Verticillium* sp. SPC-15898	Toki et al. (1992b)
	ES-242-7 (**41**)	*Verticillium* sp. SPC-15898	Toki et al. (1992b)
	ES-242-8 (**42**)	*Verticillium* sp. SPC-15898	Toki et al. (1992b)
	Oosporein (**43**)	*Verticillium psalliotae*	Wainwright and Betts (1986)
Vertinoids	Sorbicillin (**44**)	*V. intertextum* ATCC 46284	Trifonov et al. (1983)
	2′,3′-dihydrosorbicillin (**45**)	*V. intertextum* ATCC 46284	Trifonov et al. (1983)
	Bisvertinoquinol (**46**)	*V. intertextum* ATCC 46284	Trifonov et al. (1983)
	Bisvertinol (**47**)	*V. intertextum* ATCC 46284	Trifonov et al. (1986)
	Dihydrobisvertinol (**48**)	*V. intertextum* ATCC 46284	Trifonov et al. (1986)
	Isodihydrobisvertinol (**49**)	*V. intertextum* ATCC 46284	Trifonov et al. (1986)
	Bisvertinolone (**50**)	*V. intertextum* ATCC 46284	Trifonov et al. (1986)
Furanone and Pyranone	Vertinolide (**51**)	*V. intertextum* ATCC 46284	Trifonov et al. (1982)
	Lowdenic acid (**52**)	*Verticillium* sp. (MYC-406= NRRL 29280 = CBS 102427)	Angawi et al. (2003)
	Canescin A (**53**)	*Verticillium* sp. (MYC-406= NRRL 29280 = CBS 102427)	Angawi et al. (2003)
	Pyrenocine A (**54**)	*V. hemipterigenum* (teleomorph: *T. hemipterigena*) BCC 1449	Nilanonta et al. (2003b)
	Pyrenocine B (**55**)	*V. hemipterigenum* (teleomorph: *T. hemipterigena*) BCC 1449	Nilanonta et al. (2003b)
	Asterltoxin (**56**)	*P. bulbillosa* 8-H-28	Adachi et al. (2015)
	Asterltoxin C (**57**)	*P. bulbillosa* 8-H-28	Adachi et al. (2015)
	Asterltoxin B (**58**)	*P. bulbillosa* 8-H-28	Adachi et al. (2015)
	Asterltoxin D (**59**)	*P. bulbillosa* 8-H-28	Adachi et al. (2015)

(continued)

Table 7.2 (continued)

Secondary metabolite class	Metabolite name	Occurrence	References
Phenol-Terpenoid hybrids	Bigutol (**60**)	*V. biguttatum*	Morris et al. (1995)
	Methylbigutol (**61**)	*V. biguttatum*	Morris et al. (1995)
	LL-Z1272β (**62**)	*Verticillium* sp. FO-2787	Takamatsu et al. (1994)
	8′,9′-dehydroascochlorin (**63**)	*Verticillium* sp. FO-2787	Takamatsu et al. (1994)
	Ascochlorin/LL-Z1272γ (**64**)	*Verticillium* sp. FO-2787	Takamatsu et al. (1994)
		V. hemipterigenum BCC 2370	Seephonkai et al. (2004)
	8′-Acetoxyascochlorin/ LL-Z1272ζ (**65**)	*Verticillium* sp. FO-2787	Takamatsu et al. (1994)
		V. hemipterigenum BCC 2370	Seephonkai et al. (2004)
	8′-Hydroxyascochlorin (**66**)	*V. hemipterigenum* BCC 2370	Seephonkai et al. (2004)
	LL-Z1272ε (**67**)	*Verticillium* sp. FO-2787	Takamatsu et al. (1994)
	4′,5′-dihydroascochlorin/ LL-Z1272δ (**68**)	*Verticillium* sp. FO-2787	Takamatsu et al. (1994)
	4′,5′-dihydro-4′-hydroxyascochlorin (**69**)	*V. hemipterigenum* BCC 2370	Seephonkai et al. (2004)
	Vertihemipterin A (**70**)	*V. hemipterigenum* BCC 2370	Seephonkai et al. (2004)
	Ascofuranone (**71**)	*V. hemipterigenum* BCC 2370	Seephonkai et al. (2004)
	Ascofuranol (**72**)	*V. hemipterigenum* BCC 2370	Seephonkai et al. (2004)

Terpenoids	β-apo-4′-carotenoic acid (**73**)	*V. agaricinum*	Valadon and Mummery (1977)
	β-apo-4′-carotenoic acid methyl ester (**74**)	*V. agaricinum*	Valadon and Mummery (1977)
	3β-hydroxyl-23,24,25,26,27-pentanorlanost-7,9(11)-diene acid (**75**)	*V. lecanii*	Grove (1984)
	3β,12-Dihydroxyl-23,24,25,26,27-pentanorlanost-7,9(11)-diene acid (**76**)	*V. lecanii*	Grove (1984)
	23,24,25,26,27-Pentanorlanost-8-en-3β,22-diol (**77**)	*V. lecanii*	Grove (1984)
	Dahliane A (**78**)	*V. dahliae*	Wu et al. (2015)
	Dahliane B (**79**)	*V. dahliae*	Wu et al. (2015)
	Dahliane C (**80**)	*V. dahliae*	Wu et al. (2015)
	Dahliane D (**81**)	*V. dahliae*	Wu et al. (2015)
Nitrogen-containing phenolic compound	Balanol (**82**)	*V. balanoides*	Kulanthaivel et al. (1993)
Cyclodepsipeptides	Bassianolide (**83**)	*V. lecanii* (*Lecanicillium* sp. now)	Suzuki et al. (1977)
	Enniatin B (**84**)	*V. hemipterigenum* BCC 1449	Nilanonta et al. (2003b)
	Enniatin B4 (D, **85**)	*V. hemipterigenum* BCC 1449	Nilanonta et al. (2003b)
	Enniatin C (**86**)	*V. hemipterigenum* BCC 1449	Nilanonta et al. (2003b)
	Enniatin G (**87**)	*V. hemipterigenum* BCC 1449	Nilanonta et al. (2003b)
	Enniatin MK1688 (**88**)	*V. hemipterigenum* BCC 1449	Nilanonta et al. (2003b)
	Enniatin H (**89**)	*V. hemipterigenum* BCC 1449	Nilanonta et al. (2003b)
	Enniatin I (**90**)	*V. hemipterigenum* BCC 1449	Nilanonta et al. (2003b)
	Enniatin O_1 (**91**)	*V. hemipterigenum* BCC 1449	Supothina et al. (2004)
	Enniatin O_2 (**92**)	*V. hemipterigenum* BCC 1449	Supothina et al. (2004)
	Enniatin O_3 (**93**)	*V. hemipterigenum* BCC 1449	Supothina et al. (2004)

(continued)

Table 7.2 (continued)

Secondary metabolite class	Metabolite name	Occurrence	References
Diketopiperazines	1-Demethylhyalodendrin tetrasulfide (**94**)	*V. hemipterigenum* BCC 1449	Nilanonta et al. (2003b)
	Bisdenthiodi (methylthio)-1-demethylhyalodendrin (**95**)	*V. hemipterigenum* BCC 1449	Nilanonta et al. (2003b)
	Bisdethiodi (methylthio) hyalodendrin (**96**)	*V. hemipterigenum* BCC 1449	Nilanonta et al. (2003b)
	Vertihemiptellide A (**97**)	*V. hemipterigenum* BCC 1449	Nilanonta et al. (2003b)
	Vertihemiptellide B (**98**)	*V. hemipterigenum* BCC 1449	Nilanonta et al. (2003b)
	Demethylhyalodendrin (**99**)	*V. hemipterigenum* BCC 1449	Nilanonta et al. (2003b)
	$11\alpha,11'\alpha$-Dihydroxychaetocin (**100**)	*V. tenerum*	Hauser et al. (1972)
	Verticillin A (**101**)	*Verticillium* sp. TM-759	Minato et al. (1973)
	Verticillin B (**102**)	*Verticillium* sp. TM-759	Minato et al. (1973)
	Verticillin C (**103**)	*Verticillium* sp. TM-759	Minato et al. (1973)
Linear lipopeptide	5-Hydroxytetradecanamide L-Thr-D-Ala-Ala-Ala-D-Tyr-L-Val (**104**)	*P. bulbillosa* 38G272	Koehn et al. (2008)
Polyhydroxylated pyrrolizidine	Pochonicine (**105**)	*P. suchlasporia* var. *suchlasporia* TAMA 87	Usuki et al. (2009)

7.2.1 Aromatic Compounds

A series of dimeric naphthopyran bioxanthracenes with the trivial names ES-242-1-8 (**35–42**) were purified and identified from an undescribed *Verticillium* sp. SPC-15898 (Toki et al. 1992a, b). These bioxanthracene analogues and their monomers were also purifieded from the insect pathogenic fungus *Cordyceps pseudomilitaris* BCC1620 (Isaka et al. 2001). Amongst these secondary metabolites, ES-242-1 at 1.0 μM showed neuroprotective activity against glutamate induced death of cultured nerve cells by inhibiting ligand binding to the N-methyl-D-aspartate receptor (Toki et al. 1992b).

Another type of aromatic compound is oosporein (**43**), a red pigment originally described in 1944 as a dye produced by the fungus *Oospora colorans*. It was identified from several fungi as a mycotoxin to plants and poultry from cultures of *Beauveria* (Vining et al. 1962) and *Chaetomium* sp. (Cole et al. 1974). In 1986, oosporein (**43**) was characterized from *V. psalliotae* (= *Lecanicillium psalliotae*) that was selected as an antagonist against *Phytophthora infestans*, a pathogen which causes late blight of tomato, *Lycopersicon esculentum* (= *Solanum lycopersicum*) (Wainwright and Betts 1986). Precursor-feeding studies using intact cells of *Chaetomium aureum* suggested a polyketide origin for oosporein, with the common fungal metabolite orsellinic acid as a likely intermediate that undergoes multiple oxidations and dimerization by oxidative coupling to yield the final dibenzoquinone product (Steiner et al. 1974). Oosporein exhibited a strong inhibitory activity especially against *P. infestans* with MIC at 16 μm. Oosporein (**43**) also inhibited the Ca-dependent ATPases of erythrocyte membranes at relatively high concentrations (200 μg/ml) possibly by disturbing membrane structure.

7.2.2 Vertinoids

In 1983, three secondary metabolites, the hexaketides yellow sorbicillin (**44**) and its derivative 2′,3′-dihydrosorbicillin (**45**), and the dimeric hexaketide yellow bisvertinoquinol (**46**) were reported from the culture medium of *V. intertextum* ATCC 46284 (Trifonov et al. 1983). Sorbicillin (**44**) had been observed previously as a minor metabolite of *Penicillium chrysogenum* (formerly known as *P. notatum*). The structure of complex bisvertinoquinol (**46**) was obtained from an X-ray structure analysis. In 1986, another four new dimeric hexaketide secondary metabolites, bisvertinol (**47**), dihydrobisvertinol (**48**), isodihydrobisvertinol (**49**) and bisvertinolone (**50**), were found in the culture medium of the same *V. intertextum* strain (Trifonov et al. 1986). These bisvertinols (**46–49**) are closely related to the metabolite bisvertinolone (**50**). The entire group of compounds **44–50** from *V. Intertextum* is referred to as the vertinoids, which are understood to be hexaketide-derived secondary metabolites with two additional methyl groups, one at C(2) and the other at C(4) (from the carboxylic end) of the C12-chain: **44–45** are monomeric (C14) and **46–50** are dimeric (C28) vertinoids.

7.2.3 *Furanone and Pyranone*

1n 1982, a new tetronic acid derivative, vertinolide (**51**), was purified as the main constituent of the chloroform extract from the culture of *V. intertextum* ATCC 46284 (Trifonov et al. 1982). The structure of vertinolide (**51**) was established by X-ray diffraction analysis and contained a 4-hydroxy-3,5-dimethyl-2(5H)-furanon-5-yl and an (*E*,*E*)-2,4-hexadienon-1-yl substructures with a dimethylene bridge in between.

The vertinolide (**51**) producing organism *V. intertextum* was originally isolated as a laboratory contaminant; its identity with *V. intertextum* was established by W. Gams (Centraalburean voor Schimmelcultures, Utrech, Holland). The fungus was grown in static liquid culture on 2.4% potato-dextrose Difco for 21 days in a constant temperature room at 27 °C under fluorescent lamps (Trifonov et al. 1981).

Another metabolite belonging to furanone type was an antifungal metabolite with a unique skeleton, namely lowdenic acid (**52**). It has been obtained from non-sporulating cultures of a undescribed fungicolous *Verticillium* sp. (MYC-406 = NRRL 29280 = CBS 102427) (Angawi et al. 2003). The gross structure of **52**, from a mixed biogenetic origin, was proposed by analysis of NMR data and confirmed by X-ray diffraction analysis, which enabled assignment of relative stereochemistry. Lowdenic acid (**52**) has an unusual bicyclic structure containing a furylidene ring linked via a carbon-carbon double bond to a tetrahydrofurandione ring and occurs as an equilibrium *E*/*Z* mixture (Angawi et al. 2003). Lowdenic acid (**52**) exhibited inhibitory activity against *Aspergillus flavus* (NRRL 6541) with an MIC value of approximately 6 μg/ml. It also caused zones of inhibition in standard disk assays at 200 μg/disk against *Candida albicans* ATCC 90029, *Staphylococcus aureus* ATCC 29213 and *Bacillus subtilis* ATCC 6051.

Along with the unique lowdenic acid (**52**), another antifungal metabolite, canescin A (**53**), possessing a pyrone ring fused with a phenol, was also purified from the same fungicolous *Verticillium* sp. under bioassay-guided fractionation of the extract. Canescin A (**53**) was originally purified from *Penicillium canescens* (Birch et al. 1969). A biogenetic pathway has also been proposed for canescin A (**53**) involving aldol condensation of a polyketide with a tricarboxylic acid unit through a multistep process (Lewis et al. 1988).

The pyrones pyrenocines A and B (**54–55**) were first described in 1979 and are better known as the broadly phytotoxic metabolites of *Pyrenochaeta terrestris* (= *Setophoma terrestris*), the causative agent of pink root disease in onions (Sato et al. 1979). The structures of both compounds were finally established with X-ray crystallography (Sato et al. 1981). Pyrenocine A (**54**) was found to be identical to citreopyrone, 5-crotonoyl-4-methoxy-6-methyl-2-pyrone, which was reported by Niwa et al. (1980) as a metabolite of *Penicillium citreovirens*. In 2003, pyrenocines A and B (**54–55**), identified as cytotoxins, were isolated from a culture broth of the entomopathogenic fungus *V. hemipterigenum* (teleomorph: *Torrubiella hemipterigena*) BCC 1449 collected from Thailand (Nilanonta et al. 2003a). Pyrenocines were also purified from plant associated *Alternaria* species (Tal and Robeson 1986),

Curvularia inaequalis isolated from Zoysia grass (Kim et al. 2000), the endophytic fungus *Penicillium paxilli* PSU-A71 (Rukachaisirikul et al. 2007), *Penicillium waksmanii* Zaleski OUPS-N133 isolated from the brown alga *Sargassum ringgoldianum* (Amagata et al. 1998), *Paecilomyces* sp. FKI-3573 isolated from a soil sample (Hashida et al. 2010), and *Colletotrichum* sp. HCCB03289 (Yang et al. 2014).

Pyrenocines show several biological activities, reportedly being phytotoxic, antifungal, antibacterial, cytotoxic, antimalarial, and antitrypanosomal (Sparace et al. 1987; Krohn et al. 2008). Pyrenocine A (**54**) showed the most potent antitrypanosomal activity (IC_{50} 0.12 mg/ml). Isotopic labelling feeding experiments suggested that pyrenocine carbon scaffolds originate from acetate via the polyketide pathway; however, the pyrone ring and the acyl side chain derive from separate polyketide chains with the pyrone originating from an unreduced triketide and the acyl chain is produced as a reduced diketide (Shizuri et al. 1984).

Asteltoxin type metabolites having a trienic α-pyrone structure are related to citreoviridin and aurovertins. However, unlike citreoviridins and aurovertins, asteltoxins have a unique, highly functionalized 2,8-dioxabicyclo[3.3.0]octane ring containing a quaternary carbon embedded in an array of six stereogenic centres. Asteltoxin (**56**) was first purified as a mycotoxin metabolite of the fungus *Aspergillus stellatus* Curzi (Kruger et al. 1979), and showed inhibitory activity against *Escherichia coli* BF1-ATPase (Satre 1981). In 2015, four asteltoxin type metabolites (**56–59**) including two new asteltoxins (**57** and **59**), were isolated from the culture broth of the fungus *Pochonia bulbillosa* (=*Metapochonia bulbilosa*) 8-H-28, which was obtained from the fruiting body of *Elaphocordyceps capitata* (= *Tolypocladium capitatum*) collected in Japan, during screening for cytotoxic substances (Adachi et al. 2015). Metabolites **58** and **59** have a novel tricyclic ring system connected to a dienyl α-pyrone structure. Compounds **56** and **57** have the same 2,8-dioxabicyclo[3.3.0]octane ring. Compound **57** showed potent antiproliferative activity against NIAS-SL64 cells derived from the fat body of *Spodoptera litura* larvae, while compounds **58** and **59** were inactive.

7.2.4 Phenol-Terpenoid Hybrids

Antifungal metabolites of mixed biosynthetical pathways, bigutol (**60**) and its derivative methylbigutol (**61**), were purified from the mycoparasitic biocontrol agent *V. biguttatum* (Morris et al. 1995). The structures of bigutol (**60**) and methylbigutol (**61**) contained a prenylated 4-(hydroxymethyl)benzene-1,2-diol. Both metabolites inhibited mycelial growth of *Rhizoctonia solani* (MIC 138 μg/ml) and that of other plant pathogenic fungi on potato dextrose agar. Production of these antifungals by *V. biguttatum* suggests that antibiosis may play a role during biocontrol by this mycoparasite, particularly in plant diseases caused by *R. solani*.

Ascochlorin type metabolites are a class of a 2,4-dihydroxy-5-chloro 6-methylbenzaldehyde (or 5-chloroorcylaldehyde) substituted at C-3 with a sesquiterpene side chain. Some ascochlorin metabolites lack the chlorine at C-5. In 1968, ascochlorin type compounds were first purified from the fungus *Ascochyta viciae* (= *Septoria viciae*) (Tamura et al. 1968), and have subsequently been purified from a variety of fungi including *Ascochyta viciae* (Sasaki et al. 1974), *Acremonium luzulae* (= *Gliomastix luzulae*) (Cagnoli-Bellavita et al. 1975), a sponge-derived fungus *Acremonium* sp. (Zhang et al. 2009), *Fusarium* sp. LL-Z1272 (Ellestad et al. 1969), *Cephalosporium diospyri* (= *Nalanthamala diospyri*) IFO 6118 (Kawagishi et al. 1984), *Colletotrichum nicotianae* (= *Colletotrichum tabacum*) (Kosuge et al. 1973), *Cylindrocarpon lucidum* (= *Thelonectria lucida*) (Singh et al. 1996), *Cylindrocladium ilicicola* (= *Calonectria pyrochroa*) MFC-870 (Hayakawa et al. 1971; Minato et al. 1972), *Nectria coccinea* (= *Neonectria coccinea*) (Aldridge et al. 1972), and a leafhopper pathogenic fungus, *Microcera* sp. BCC 17074 (Isaka et al. 2015), They are also structurally related to the moverastins, cancer cell migration inhibitors from *Aspergillus* sp. (Takemoto et al. 2005).

In 1994, a series of ascochiorin type compounds, including a new ascochiorin, 8′,9′-dehydroascochiorin (**63**), and five know ascochiorins, LL-Z 1272β (**62**), ascochiorin /12 LL-Z 1272γ (**64**), 8′-Acetoxyascochlorin/ LL-Z 1272ζ (**65**), LL-Z 1272ε (**67**), and LL-Z 1272δ (**68**), were discovered from the cultured mycelium of *Verticillium* sp. FO-2787 in the search for new non-steroidal inhibitors of T-5α-reductase (Takamatsu et al. 1994). In 2004, a new ascochlorin, 8′-hydroxyascochlorin (**66**), and a novel ascochlorin glycoside, vertihemipterin A (**70**), together with six known ascochlorins, including the aglycone of vertihemipterin A, namely 4′,5′-dihydro-4′-hydroxyascochlorin (**69**), 8′-acetoxyascochlorin/LL-Z 1272ζ (**65**), ascochlorin (**64**), 8′,9′-dehydroascochlorin (**63**), and the biogenetically related ascofuranone (**71**) and ascofuranol (**72**), were purified from the insect pathogenic fungus *V. hemipterigenum* (= *Torrubiella hemipterigena*) BCC 2370 (Seephonkai et al. 2004).

Members of ascochiorin-type compounds have been known to exhibit antifungal (Bal-Tembe et al. 1999), antiviral and antitumour activities (Takatsuki et al. 1969). They are also known to be Ras farnesyl-protein transferase (FPTase) inhibitors as well as testosterone 5α-reductase inhibitors (Takamatsu et al. 1994). Ascochlorin (**64**) is known to be active against Newcastle Disease Virus (Takatsuki et al. 1969).

7.2.5 Terpenoids

In 1969–1972, a type of carotenoids, including β-apo-4′-carotenoic acid and its methyl ester (**73–74**) were purified from *V. agaricinum* (Valadon and Mummery 1977). This is the first demonstration of neurosporaxanthin in mitosporic fungi.

In 1984, 23,24,25,26,27-pentanorlanost-7,9(11)-diene acids (**75–76**) were isolated from the mycelium of the entomopathogenic fungus *V. lecanii* (Zimm.) (= *Lecanicillium lecanii*). In the course of an unsuccessful attempt to increase the yields of these acids by incorporating lanosterol in the culture medium, a novel 23,24,25,26,27-pentanorlanost-8-ene-3β,22-diol (**77**), was obtained from the mycelium produced in both lanosterol-enriched cultive media of the same fungus (Grove 1984).

Guanacastane-type diterpenoids were first purified from an endophytic fungus isolated from a *Daphnopsis americana* tree (Brady et al. 2000), and were subsequently found mainly in the genus *Coprinus* (Liu et al. 2014). Some exhibited significant antibiotic and cytotoxic activities. In 2015, four new guanacastane-type diterpenoids, namely dahlianes A-D (**78**–**81**), were obtained from cultures of *V. dahliae*, a fungus isolated from the gut of *Coridius chinensis* (Hemiptera) (Wu et al. 2015). Their structures were elucidated on the basis of extensive spectroscopic data analysis. In cytotoxicity evaluation against human tumour cell lines, compounds **79** and **80** exhibited significant cytotoxicities against MCF-7 cell lines with IC_{50} values of 3.35 and 4.72 mM, respectively.

7.2.6 Nitrogen-Containing Phenolic Compounds

An unusal metabolite, balanol (**82**), was discovered from the fungus *V. balanoides* (= *Tolypocladium balanoides*) collected from *Pinus palustris* needles in the search for novel and potent inhibitors for Protein kinase C (PKC), a family of serine/threonine-specific kinases (Kulanthaivel et al. 1993). Balanol (**82**) consisted of an unusal (3R,4R)-3-aminoazepan-4-ol condensed with a 4-(2-carboxy-6-hydroxybenzoyl)-3,5-dihydroxybenzoic acid and a 4-hydroxybenzoic acid. Balanol (**82**) was also obtained from three closely related *Fusarium* species, *F. merismoides* Corda (= *Fusicolla merismoides*) NR6356, *F. aquaeductuum* Lagh. (= *Fusiculla aquaeductuum*) IMI 103658 and *Fusarium* sp. NR7222 (Ohshima et al. 1994). First purified from *Cordyceps ophioglossoides* (= *Tolypocladium ophioglossoides*) as ophiocordin, the compound was later found to be identical to balanol (**82**) (Boros et al. 1994).

The biosynthesis of balanol (**82**) was proposed to involve the convergence of two different biosynthetic pathways: the benzophenone generated from polyketide biosynthesis and the 3-amino-hexahydroazepine-4-ol with the N-terminal p-hydroxybenzoate protecting group from non-ribosomal peptide biosynthesis. The hexahydroazepine originates from intramolecular cyclization of 3-hydroxylysine by amide formation between the terminal amino group of the side chain and the main chain carboxylate, followed by reduction. Coupling of the two parallel products is then achieved by esterification of the hexahydroazepine-4-ol intermediate (Molnár et al. 2010).

Balanol (**82**) is one of the most potent ATP-competitive inhibitors of protein kinase C (PKC) and protein kinase A (PKA) at nanomolar concentrations. The human PKC enzyme comprises a number of isozymes and inappropriate activation of PKC has been linked to a variety of disorders (Kulanthaivel et al. 1993).

7.2.7 Depsipeptides

A depsipeptide is generally a molecule that has both peptide and ester linkages in proximity in amino acid-containing structure. The array of depsipeptide natural products discovered to date is extensive, and many have been tested as entry points into particular areas of therapeutic discovery. In 1977, bassianolide (**83**), a octadepsipeptide composed of four moles each of L-N-methylleucine and D-α-hydroxyisovaleric acid, was obtained from the insect pathogenic fungi *Beauveria bassiana* and *Lecanicillium* sp. (= *V. lecanii*) in the course of screening for insecticidal metabolites of fungi (Suzuki et al. 1977). Bassianolide (**83**) inhibited acetylcholine-induced smooth muscle contraction in a manner that did not involve ionophoric interactions (Nakajyo et al. 1983). The NRPS of bassianolide (**83**) was recently characterized from the wood-decaying fungus *Xylaria* sp. (Jirakkakul et al. 2008), and from *B. bassiana* (Xu et al. 2009).

Enniatins, the cyclooligomer hexadepsipeptides, are another group of cyclic trimeric esters of a dipeptidol monomer consisting of an N-methyl amino acid (Leu, Val, Ile) and a D-2-hydroxycarboxylic acid derived from these branched chain amino acids (Firáková et al. 2007). Enniatins were named, and for the first time obtained, in 1947 from *Fusarium orthoceras* var. *enniatinum*. Enniatins are secondary metabolites produced predominantly by various *Fusarium* strains isolated from different sources, e.g. marine algae *Codium fragile*, walnut tree leaves, and foliage of the balsam fir, *Abies balsamea* (Tomoda et al. 1992; Visconti et al. 1992). These secondary metabolites are also produced by *Halosarpheia* strains (*Halosphaeriaceae*, *Ascomycota*) (Firáková et al. 2007).

The insect pathogenic fungus *V. hemipterigenum* is the asexual state of *Torrubiella hemipterigena* that only infects leafhoppers in the Indian Ocean region. In 2004, two new analogues, enniatins H and I (**89–90**), bearing respectively one and two Hmp residues, instead of Hiv, together with the known enniatins B and B4 (**84–85**), were identified during the early investigation of the antimalarial constituents of *V. hemipterigenum* BCC 1449 (Nilanonta et al. 2003a). Studies on precursor-directed biosynthesis with *V. hemipterigenum* strain BCC 1449 led to the production and identification of three analogs, enniatins C (**86**), G (**87**) and MK1688 (**88**). Enniatin C (**86**) is a cyclic trimer of the same dipeptidol unit (Nilanonta et al. 2003a).

Optimal culture conditions for enniatin production using *V. hemipterigenum* BCC 1449 have been investigated. Among various liquid media tested, highest enniatin production was obtained using yeast extract sucrose. Application of this condition to large-scale fermentation resulted in the isolation of three new analogs, O_1, O_2 and O_3 (**91–93**), which are closely related isomers that were characterized as an inseparable mixture, along with seven known enniatins (**84–90**) (Supothina et al. 2004). Enniatin B (**84**) was the major secondary metabolite in the extract of the fungus *V. hemipterigenum* BCC 1449 and was most active against the malaria parasite and it is most likely that this compound was responsible for the antimalarial activity detected.

This class of enniatin compounds, which has been found to exhibit antibiotic, insecticidal, and phytotoxic activities, can also inhibit acyl-CoA: cholesterol acyltransferase (ACAT). Enniatin B (**84**), along with another Enniatin B1, has been shown to protect human lymphoblastoid cells in HIV-1-induced cell death, and enniatins MK1688 and I (**88** and **90**) exhibited strong inhibitory activities against HIV-1 integrase (Shin et al. 2009).

7.2.8 Diketopiperazines

Diketopiperazines are frequent microbial metabolites formed by the intramolecular cyclization of dipeptides. During the investigation of *V. hemipterigenum* BCC 1449, two new diketopiperazines, 1-demethylhyalodendrin tetrasulfide and bisdenthiodi (methylthio)-1-demethylhyalodendrin (**94–95**), were obtained together with some enniatins from ethyl acetate extracts of the culture filtrate grown in PDB (Nilanonta et al. 2003b). Compound **94** bears a S4-bridge and **95** has two methylthio groups attached to the same side of the six-membered ring. They were 1-demethyl analogs of the known hyalodendrin tetrasulfide and bisdethiodi (methylthio) hyalodendrin, respectively, which had previously been obtained from *Hyalodendron* sp. (Strunz et al. 1974) and *Penicillium turbatum* (Michel et al. 1974). Upon changing the liquid medium to yeast extract sucrose, two additional new compounds, vertihemiptellides A and B (**97–98**), and two known compounds, bisdethiodi(methylthio)hyalodendrin (**96**) and demethylhyalodendrin (**99**), were also obtained from the same strain of *V. hemipterigenum* BCC 1449. Vertihemiptellides (**97–98**) are symmetrical dimeric compounds and its *N*-demethyl analogue. Their structures possess a unique skeleton where two diketopiperazines are linked via dithio bridges (Isaka et al. 2005). The unusual structure of **97** was confirmed by X-ray crystallographic analysis, which also established the absolute configuration of this compound. Dimers **97** and **98** exhibited moderate antituberculous activity (MIC 12.5 μg/ml) and moderate cytotoxicity.

In 1972, a new antibacterial and antimitotic dimeric epipolythiopiperazinedione metabolite, $11\alpha,11'\alpha$-dihydroxychaetocin (**100**), was purified from the fungus *V. tenerum* (= *Acrostalaqmus luteoalbus*) (Hauser et al. 1972). In 1973, three new dimeric epidithiodioxopiperaine metabolites, verticillins A-C (**101–103**) were purified from a species of *Verticillium* strain TM-759, an imperfect fungus isolated from a basidiocarp of *Coltricia cinnamomea* (Minato et al. 1973). Verticillins A and B (**101–102**), are derivatives of bi-(3,11a–epidithio-l,4-dioxopyrazino[l',2':1,5]pyrrolo-[2,3-b]indol-l0b-yl). Verticillin A (**101**) is structurally very similar to Chaetocin from *Chaetomium minutum*, differing only in the position of two hydroxyl group, and verticillin C (**103**) is thought to be an epitrithio-analogue of verticillin B (**102**).

A series of epidithiodioxopiperaine metabolites were also reported from *Chaetomium* spp. (Saito et al. 1988), *Bionectria byssicola* F120 (Zheng et al. 2007) and *Gliocladium* spp. (Joshi et al. 1999). Recently, the question as to whether

Verticillium spp. could produce verticillin A (**101**) has been raised because they seem to lack the necessary genes involved in epipolythiodioxopiperazine biosynthesis (Schenke et al. 2011). Verticillin A (**101**) has been shown to be produced by *Clonostachys rosea*, formerly known as *Gliocladium roseum*, and this fungus is not only morphologically similar to, but also a mycoparasite of, *Verticillium*. It has thus been assumed that the original verticillin A (**101**)-producing *Verticillium* was either confused with another fungus or contaminated with *Clonostachys rosea*. Verticillins A and B (**101–102**) showed the strong cytotoxicity effects against HeLa cells. Recently, verticillin A (**101**) was found to be an inducer of hepatoma cell apoptosis *in vitro* and an inhibitor of tumour xenograft growth *in vivo*. In mice, verticillin A (**101**) alone was adequate to kill cancer cells, but the required dose made the mice sick, a common problem with many cancer therapies. However, when a lower dose was paired with Tumour Necrosis Factor (TNF)-related apoptosis-inducing ligand (TRAIL), which induces cancer cells to self destruct, it became a powerful, more tolerable combination that killed previously resistant cells. Metabolite **101** sensitizes cancer cells to TRAIL, and appears to keep cancer cells from developing resistance to TRAIL. Verticillin A (**101**) appears to work by upregulating DNA Methylation-Dependent BN1P3, a gene that promotes cell death, and makes the cancer cells sensitive to TRAIL (Liu et al. 2011).

7.2.9 Lipopeptides

Lipopeptides constitute a structurally diverse group of metabolites produced by various bacterial and fungal genera. A key structural feature of lipopeptide antibiotics is the long-chain fatty acid attached to a peptide portion. In the past decades, research on lipopeptides has been fueled by their antimicrobial, antitumour, immunosuppressant and surfactant activities (Raaijmakers et al. 2010). Bioassay-directed fractionation of a liquid culture of *P. bulbillosa* (= *Metapochonia bulbilosa*) 38G272 led to the isolation of a series of structurally novel, prospective cell wall-active lipopeptides (Koehn et al. 2008). The main component of this suite is a linear hexapeptide with a δ-hydroxymyristic acid amide substituted N-terminus (**104**). The structure of 5-hydroxytetradecanmide L-Thr-D-Ala-L-Ala-L-Ala-D-Tyr-L-Val (**104**) was deduced using high-field microsample NMR, Fourier transform mass spectrometry, and microscale chemical degradation. The compound was expected to be a potent inhibitor for cell wall synthesis.

7.2.10 Polyhydroxylated Pyrrolizidine Alkaloids

The distribution of most polyhydroxylated pyrrolizidine alkaloids, including australines, alexines, and hyacinthacines, has been restricted to plants (Kato et al. 2007). In 2009, a new polyhydroxylated pyrrolizidine alkaloid, designated as pochonicine

(**105**), was obtained from a solid culture of the fungal strain *P. suchlasporia* var. *suchlasporia* (= *Metapochonia suchlasporia*) TAMA 87 (Usuki et al. 2009). The structure of pochonicine (**105**) was determined using NMR and MS techniques as (1R*, 3S*, 5S*, 6S*, 7R*, 7aS*)-5-acetamidomethyl-3-hydroxymethyl-1,6,7-trihydroxypyrrolizidine. Pochonicine (**105**) showed potent inhibition against β-N-acetylglucosaminidases (GlcNAcases) of various organisms including insects, fungi, mammals, and a plant, but showed no inhibition against β-glucosidase of almond, α-glucosidase of yeast, or chitinase of *Bacillus* sp. The GlcNAcase inhibitory activity of pochonicine (**105**) was comparable to nagstatin, a potent GlcNAcase inhibitor of fungal origin. Pochonicine (**105**) is the first example of such a compound purified from a fungus.

7.3 Conclusions and Future Perspectives

In this chapter the secondary metabolites from *P. chlamydosporia* and other *Pochonia* spp. have been described. Most of these compounds displayed an impressive array of biological activities including antitumoral, antifungal, and enzyme inhibitory capabilities. The structure-activity relationship has been established for only a few classes of *Pochonia* spp. metabolites. In fact, the many other *Pochonia* spp. still remain unexplored. In most cases, both the structural diversity of the secondary metabolites and their biological activities from *Pochonia* fungi have been studied only superficially. Though, the rest of metabolites in the other known species of *Pochonia* need to be investigated in detail, it is clear that the structure types of natural products from *P. chlamydosporia* share little with secondary metabolites from the other species of *Pochonia* (*Verticillium*). The dominant type of secondary metabolites from *P. chlamydosporia* is the family of resorcylic acid lactones (RAL), which have also been reported to be produced by a variety of fungi, such as *Monocillium*, *Hypomyces* or *Penicillium*. However, there is as yet no evidence of their occurrence in other species of *Pochonia* (*Verticillium*). The challenging task for natural product chemists and biologists is to clarify the biosynthetic origin of these RALs.

The bioactivity of RALs makes them promising candidates for clinical drug development. For example, the well-known radicicol (**1**) and monocillin I (**24**) exhibited potent inhibitory activity against Hsp90. Heat shock protein 90 is believed to be an efficient target for the treatment of cancer. In consequence pharmaceutical industries and research institutions have invested great effort in finding novel inhibitors of Hsp90 in recent years. *Pochonia* spp. have recently entered the omics-era (Chaps. 5, 6, 8, 9, 10). Application of genome mining strategies to *Pochonia* and their cultural, genetic, or epigenetic manipulations will surely lead to the identification of novel compounds for evaluation of pharmaceutically relevant bioactivities. Given the ongoing efforts and interest exhibited by chemists and pharmacologists, it is highly probable that further natural or synthetic multi-membered RALs will enter into clinical trials and consequently their biosynthetic pathways will be understood more clearly in the near future.

However to date, no purified RALs have been reported to cause significant mortality or microscopic pathological alterations upon injection into insects or nematodes. In fact, of the *Pochonia* spp. metabolites, only pyranones from some *P. chlamydosporia* strains, phomalactone and aurovertins, have been reported to show nematicidal acitivity. Further investigation revealed that aurovertin secondary metabolites could be produced by some *P. chlamydosporia* isolates (Wang et al. 2015). Of the 17 tested *P. chlamydosporia* isolates, only three of the strains, obtained from root-knot nematode females rather than root-knot nematode eggs and soil, yielded the distinctive nematicidal luminous yellow pigments. *Pochonia chlamydosporia* can survive as a saprophyte in the soil in the absence of both plant and nematode hosts. Therefore, many strains of the fungus, which is present worldwide, may thus survive in the soil despite the lack of nematotoxic metabolites. This might be attributed to adaptation driving forces from diverse ecological niches, suggesting that the secondary metabolite production of *P. chlamydosporia* strains (and those from other *Pochonia* spp.) may vary across geographical and smaller spatial scales. This is in good agreement with the fact that metabolite variations occurred not only in response to long-term selective processes, but also in response to short-term environmental variation. It therefore seems important to explore further the role of *P. chlamydosporia* environmental variation in secondary metabolites production. However, there are relatively few studies investigating the nature and significance of the secondary metabolites with this variation from the structure and function of ecosystems.

Research efforts have greatly advanced the field of biological control, as witnessed by the availability of a number of commercial products, yet efficacy issues continue to be one of the most important obstacles to broad acceptance of fungal biological control agents. Since the rate of killing with pathogenetic fungi is slower than that with chemical pesticides, these biocontrol agents are better suited for preventing large infestations rather than for treating existing problems. Better understanding of secondary metabolites and their natural functions in *P. chlamydosporia*, together with knowlege of their biosynthesis, would aid in achieving highly effective biological control agents. This would target secondary metabolites as virulence factors to plant and animal-parasitic nematodes that are known to be parasitized by the fungus. Engineering fungi for increased virulence (reduced time to kill in lower doses) by expressing transgene-encoded toxin regulators, or by elevated production of native metabolite virulence factors, are expected to provide more attractive biocontrol agents.

References

Adachi, H., Doi, H., Kasahara, Y., et al. (2015). Asteltoxins from the entomopathogenic fungus *Pochonia bulbillosa* 8-H-28. *Journal of Natural Products, 78*, 1730–1734.

Aldridge, D. C., Borrow, A., Foster, R. G., et al. (1972). Metabolites of *Nectria coccinea*. *Journal of the Chemical Society, Perkin Transactions, 1*, 2136–2141.

Amagata, T., Minoura, K., & Numata, A. (1998). Cytotoxic metabolites produced by a fungal strain from a *Sargassum* alga. *The Journal of Antibiotics, 51*, 432–434.

Angawi, R. F., Swenson, D. C., Gloer, J. B., et al. (2003). Lowdenic acid: A new antifungal polyketide-derived metabolite from a new fungicolous *Verticillium* sp. *Journal of Natural Products, 66*, 1259–1262.

Arai, M., Yamamoto, K., Namatame, I., et al. (2003). New monordens produced by amidepsine-producing fungus *Humicola* sp. FO-2942. *The Journal of Antibiotics, 56*, 526–532.

Argoudelis, A. D., & Zieserl, J. F. (1966). The structure of U-13,933, a new antibiotic. *Tetrahedron Letters, 18*, 1969–1973.

Ayer, W. A., & Peña-Rodriguez, L. (1987). Minor metabolites of *Monocillium nordinii. Phytochemistry, 26*, 1353–1355.

Ayer, W. A., Lee, S. P., Tsuneda, A., et al. (1980). The isolation, identification, and bioassay of the antifungal metabolites produced by *Monocillium nordinii. Canadian Journal of Microbiology, 26*, 766–773.

Azumi, M., Ishidoh, K., Kinoshita, H., et al. (2008). Aurovertins F-H from the entomopathogenic fungus *Metarhizium anisopliae. Journal of Natural Products, 71*, 278–280.

Baldwin, C., Weaver, L., Brooker, R., et al. (1964). Biological and chemical properties of aurovertin, a metabolic product of *Calcarisporium abuscula. Lloydia, 27*, 88–95.

Bal-Tembe, S., Kundu, S., Roy, K., et al. (1999). Activity of the ilicicolins against plant pathogenic fungi. *Pesticide Science, 55*, 645–647.

Birch, A. J., Hirkinshaw, J. H., Chaplen, P., et al. (1969). The structures of canescins-A and -B. *Australian Journal of Chemistry, 22*, 1933–1941.

Bloch, P., & Tamm, C. (1976). Pseurotin, a new metabolite of *Pseudeurotium ovalis* Stolk having an unusual hetero-spirocyclic system. *Helvetica Chimica Acta, 59*, 133–137.

Boot, C. M., Gassner, N. C., Compton, J. E., et al. (2007). Pinpointing pseurotins from a marine-derived *Aspergillus* as tools for chemical genetics using a synthetic lethality yeast screen. *Journal of Natural Products, 70*, 1672–1675.

Boros, C., Hamilton, S. M., Katz, B., et al. (1994). Comparison of balanol from *Verticillium balanoides* and ophiocordin from *Cordyceps ophioglossoides. The Journal of Antibiotics, 47*, 1010–1016.

Brachmann, A. O., Brameyer, S., Kresovic, D., et al. (2013). Pyrones as bacterial signaling molecules. *Nature Chemical Biology, 9*, 573–578.

Brady, S. F., Singh, M. P., Janso, J. E., et al. (2000). Guanacastepene, a fungal-derived diterpene antibiotic with a new carbon skeleton. *Journal of the American Chemical Society, 122*, 2116–2117.

Cagnoli-Bellavita, N., Ceccherelli, P., Fringuelli, R., et al. (1975). Ascochlorin: a terpenoid metabolite from *Acremonium luzulae. Phytochemistry, 14*, 807.

Cole, R. J., Kirksey, J. W., Cutler, H. G., et al. (1974). Toxic effects of oosporein from *Chaetomium trilatelare. Journal of Agricultural and Food Chemistry, 22*, 517–522.

Delmotte, P., & Delmotte-Plaquee, J. (1953). A new antifungal substance of fungal origin. *Nature, 171*, 344.

Ebel, R., & Lardy, H. (1975). Influence of aurovertin on mitochondrial ATPase activity. *The Journal of Biological Chemistry, 250*, 4992–4995.

Ellestad, G. A., RHJr, E., & Kunstmann, M. P. (1969). Some new terpenoid metabolites from an unidentified *Fusarium* species. *Tetrahedron, 25*, 1323–1334.

Ellestad, G. A., Lovell, F. M., Perkinson, N. A., et al. (1978). New zearalenone related macrolides and isocoumarins from an unidentified fungus. *The Journal of Organic Chemistry, 43*, 2339–2343.

Evans, G., & White, N. H. (1966). Radicicolin and radicicol, two new antibiotics produced by *Cylindrocarpon radicicola. Transactions of the British Mycological Society, 49*, 563–576.

Evans, R. H., Ellested, G. A., & Kunstmann, M. P. (1969). Two new metabolites from an unidentified *Nigrospora* species. *Tetrahedron Letters, 10*, 1791–1794.

Fairlamb, I. J., & McGlacken, G. P. (2005). 2-pyrones natural products and mimetics: Isolation, characterisation and biological activity. *Natural Product Reports, 22*, 369–385.

Fang, S., Chen, L., Yu, M., et al. (2015). Synthesis, antitumor activity, and mechanism of action of 6-acrylic phenethyl ester-2-pyranone derivatives. *Organic & Biomolecular Chemistry, 13*, 4714–4726.

Firáková, S., Proksa, B., & Šturdíková, M. (2007). Biosynthesis and biological activity of enniatins. *Pharmazie, 62*, 563–568.

Grove, J. F. (1984). 23, 24, 25, 26, 27-Pentanorlanost-8-en-3β,22-diol from *Verticillium lecanii*. *Phytochemistry, 23*, 1721–1723.

Guo, H., Feng, T., Li, Z. H., et al. (2013). Ten new aurovertins from cultures of the basidiomycete *Albatrellus confluens*. *Natural Products and Bioprospecting, 3*, 8–13.

Hashida, J., Niitsuma, M., Iwatsuki, M., et al. (2010). Pyrenocine I, a new pyrenocine analogue produced by *Paecilomyces* sp. FKI-3573. *The Journal of Antibiotics, 63*, 559–561.

Hauser, D., Loosli, H. R., & Niklaus, P. (1972). Isolierung von 11α, 11′α-Dihydroxychaetocin aus *Verticillium tenerum*. *Helvetica Chimica Acta, 55*, 2182–2187.

Hayakawa, S., Minato, H., & Katagiri, K. (1971). The ilicicolins, antibiotics from *Cylindrocladium ilicicola*. *The Journal of Antibiotics, 24*, 653–654.

Hellwig, V., Mayer-Bartschmid, A., Müller, H., et al. (2003). Pochonins A-F, new antiviral and antiparasitic resorcylic acid lactones from *Pochonia chlamydosporia* var. *catenulata*. *Journal of Natural Products, 66*, 829–837.

Huang, T. C., Chang, H. Y., Hsu, C. H., et al. (2008). Targeting therapy for breast carcinoma by ATP synthase inhibitor aurovertin B. *Journal of Proteome Research, 7*, 1433–1444.

Isaka, M., Kongsaeree, P., & Thebtaranonth, Y. (2001). Bioxanthracenes from the insect pathogenic fungus *Cordyceps pseudomilitaris* BCC 1620. II. Structure elucidation. *The Journal of Antibiotics, 54*, 36–43.

Isaka, M., Palasarn, S., Rachtawee, P., et al. (2005). Unique diketopiperazine dimers from the insect pathogenic fungus *Verticillium hemipterigenum* BCC 1449. *Organic Letters, 7*, 2257–2260.

Isaka, M., Yangchum, A., Supothina, S., et al. (2015). Ascochlorin derivatives from the leafhopper pathogenic fungus *Microcera* sp. BCC 17074. *The Journal of Antibiotics, 68*, 47–51.

Jirakkakul, J., Punya, J., Pongpattanakitshote, S., et al. (2008). Identification of the nonribosomal peptide synthetase gene responsible for bassianolide synthesis in wood-decaying fungus *Xylaria* sp. BCC1067. *Microbiology, 154*, 995–1006.

Johnson, K. M., Swenson, L., Opipari, A. W., Jr., et al. (2009). Mechanistic basis for differential inhibition of the F1FO-ATPase by aurovertin. *Biopolymers, 91*, 830–840.

Joshi, B. K., Gloer, J. B., & Wicklow, D. T. (1999). New verticillin and glisoprenin analogues from *Gliocladium catenulatum*, a mycoparasite of *Aspergillus flavus* sclerotia. *Journal of Natural Products, 62*, 730–733.

Kato, A., Kato, N., Adachi, I., et al. (2007). Isolation of glycosidase-inhibiting hyacinthacines and related alkaloids from *Scilla socialis*. *Journal of Natural Products, 70*, 993–997.

Kawagishi, H., Sato, H., Sakamura, S., et al. (1984). Isolation and structure of a new diprenyl phenol, colletorin B produced by *Cephalosporium diospyri*. *Agricultural and Biological Chemistry, 48*, 1903–1904.

Khambay, B., Bourne, J., Cameron, S., et al. (2000). A nematicidal metabolite from *Verticillium chlamydosporium*. *Pest Management Science, 56*, 1098–1099.

Kim, J. C., Choi, G. J., Kim, H. T., et al. (2000). Pathogenicity and pyrenocine production of *Curvularia inaequalis* isolated from zoysia grass. *Plant Disease, 84*, 684–688.

Kim, J. C., Choi, G. J., Park, J. H., et al. (2001). Activity against plant pathogenic fungi of phomalactone isolated from *Nigrospora sphaerica*. *Pest Management Science, 57*, 554–559.

Koehn, F. E., Kirsch, D. R., Feng, X., et al. (2008). A cell wall-active lipopeptide from the fungus *Pochonia bulbillosa*. *Journal of Natural Products, 71*, 2045–2048.

Kosuge, Y., Suzuki, A., Hirota, S., & Tamura, S. (1973). Structure of colletochlorin from *Colletotrichum nicotianae*. *Agricultural and Biological Chemistry, 37*, 455–456.

Krasnoff, S., & Gupta, S. (1994). Identification of the antibiotic phomalactone from the entomopathogenic fungus *Hirsutella thompsonii* var. *synnematosa*. *Journal of Chemical Ecology, 20*, 293–302.

Krohn, K., Sohrab, M. H., Draeger, S., et al. (2008). New pyrenocines from an endophytic fungus. *Natural Product Communications, 3*, 1689–1692.
Kruger, G. J., Steyn, P. S., Vleggaar, R., et al. (1979). X-ray crystal structure of asteltoxin, a novel mycotoxin from *Aspergillus stellatus* Curzi. *Journal of the Chemical Society, Chemical Communications, 10*, 441–442.
Kulanthaivel, P., Hallock, Y. F., Boros, C., et al. (1993). Balanol: A novel and potent inhibitor of protein kinase C from the fungus *Verticillium balanoides*. *Journal of the American Chemical Society, 115*, 6452–6453.
Lewis, C., Staunton, J., & Sunter, D. (1988). Biosynthesis of canescin, a metabolite of *Aspergillus malignus*: Incorporation of methionine, acetate, succinate, and isocoumarin precursors, labelled with deuterium and carbon-13. *Journal of the Chemical Society, 4*, 747–754.
Liu, F., Liu, Q., Yang, D., et al. (2011). Verticillin A overcomes apoptosis resistance in human colon carcinoma through DNA methylation-dependent upregulation of BNIP3. *Cancer Research, 71*, 6807–6816.
Liu, Y. Z., CH, L., & Shen, M. (2014). Guanacastane-type diterpenoids from *Coprinus plicatilis*. *Phytochemistry Letters, 7*, 161–164.
Mao, X.-M., Zhan, Z.-J., Grayson, M. N., et al. (2015). Efficient biosynthesis of fungal polyketides containing the dioxabicyclo-octane ring system. *Journal of the American Chemical Society, 137*, 11904–11907.
Michel, K. H., Chaney, M. O., Jones, N. D., et al. (1974). Epipolythiopiperazinedione antibiotics from *Penicillium turbatum*. *The Journal of Antibiotics, 27*, 57–64.
Minato, H., Katayama, T., Hayakawa, S., et al. (1972). Identification of ilicicolins with ascochlorin and LL-Z 1272. *The Journal of Antibiotics, 25*, 315–316.
Minato, H., Matsumoto, M., & Katayama, T. (1973). Studies on the metabolites of *Verticillium* sp. structures of verticillins A, B, and C. *Journal of the Chemical Society. Perkin Transactions 1, 17*, 1819–1825.
Mirrington, R. N., Ritchie, E., Shoppee, C. W., et al. (1964). The constitution of radicicol. *Tetrahedron Letters, 7*, 365–370.
Mizuba, S., Lee, K., & Jiu, J. (1975). Three antimicrobial metabolites from *Aspergillus caespitosus*. *Canadian Journal of Microbiology, 21*, 1781–1787.
Molnár, I., Gibson, D. M., & Krasnoff, S. B. (2010). Secondary metabolites from entomopathogenic Hypocrealean fungi. *Natural Product Reports, 27*, 1241–1275.
Morris, R. A. C., Ewing, D. D. F., Whipps, J. M., et al. (1995). Antifungal hydroxymethyl-phenols from the mycoparasite *Verticillium biguttatum*. *Phytochemistry, 39*, 1043–1048.
Moulin, E., Zoete, V., Barluenga, S., et al. (2005). Synthesis, and biological evaluation of HSP90 inhibitors based on conformational analysis of radicicol and its analogues. *Journal of the American Chemical Society, 127*, 6999–7004.
Moulin, E., Barluenga, S., Totzke, F., et al. (2006). Diversity-oriented synthesis of pochonins and biological evaluation against a panel of kinases. *Chemistry – A European Journal, 12*, 8819–8834. doi:10.1002/chem.200600553.
Mulheirn, L., Beechey, R., & Leworthy, D. (1974). Aurovertin B, a metabolite of *Calcarisporium arbuscula*. *Journal of the Chemical Society, Chemical Communications, 21*, 874–876.
Nair, M. S. R., & Carey, S. T. (1980). Metabolites of pyrenomycetes XIII: Structure of (+)-hypothemycin, an antibiotic macrolide from *Hypomyces trichothecoides*. *Tetrahedron Letters, 21*, 2011–2012.
Nakajyo, S., Shimizu, K., Kometani, A., et al. (1983). On the inhibitory mechanism of bassianolide, a cyclodepsipeptide, in acetylcholine-induced contraction in guinea-pig *taenia coli*. *Japanese Journal of Pharmacology, 33*, 573–582.
Nilanonta, C., Isaka, M., Chanphen, R., et al. (2003a). Unusual enniatins produced by the insect pathogenic fungus *Verticillium hemipterigenum*: Isolation and studies on precursor-directed biosynthesis. *Tetrahedron, 59*, 1015–1020.
Nilanonta, C., Isaka, M., Kittakoop, P., et al. (2003b). New Diketopiperazines from the entomopathogenic Fungus *Verticillium hemipterigenum* BCC 1449. *The Journal of Antibiotics, 56*, 647–651.

Nishiyama, S., Toshima, H., Kanai, H., et al. (1988). Total synthesis and the absolute configuration of aurovertin B, an acute neurotoxic metabolite. *Tetrahedron, 44*, 6315–6324.

Niu, X. M., Wang, Y. L., Chu, Y. S., et al. (2010). Nematodetoxic aurovertin-type metabolites from a root-knot nematode parasitic fungus *Pochonia chlamydosporia. Journal of Agricultural and Food Chemistry, 58*, 828–834.

Niwa, M., Ogiso, S., Endo, T., et al. (1980). Isolation and structure of citreopyrone, a metabolite of *Penicillium citreo-viride* Biourge. *Tetrahedron Letters, 21*, 4481–4482.

Nozawa, K., & Nakajima, S. (1979). Isolation of radicicol from *Penicillium luteo-aurantium* and meleagrin, a new metabolite, from *Penicillium meleagrinum. Journal of Natural Products, 42*, 374–377.

Ohshima, S., Yanagisawa, M., Katoh, A., et al. (1994). *Fusarium merismoides* CORDA NR 6356, the source of the protein kinase C inhibitor, azepinostatin taxonomy, yield improvement, fermentation and biological activity. *The Journal of Antibiotics, 47*, 639–647.

Owen, S. P., & Bhuyan, B. K. (1965). Biological properties of a new antibiotic, U-13,933. *Antimicrobial Agents and Chemotherapy, 5*, 804–807.

Raaijmakers, J. M., De Bruijn, I., Nybroe, O., et al. (2010). Natural functions of lipopeptides from *Bacillus* and *Pseudomonas*: More than surfactants and antibiotics. *FEMS Microbiology Reviews, 34*, 1037–1062.

Rukachaisirikul, V., Kaeobamrung, J., Panwiriyarat, W., et al. (2007). A new pyrone derivative from the endophytic fungus *Penicillium paxilli* PSU-A71. *Chemical & Pharmaceutical Bulletin, 55*, 1383–1384.

Saito, T., Suzuki, Y., Koyama, K., et al. (1988). Chetracin A and chaetocins B and C, three new epipolythiodioxopiperazines from *Chaetomium* spp. *Chemical & Pharmaceutical Bulletin, 36*, 1942–1956.

Sasaki, H., Hosokawa, T., Sawada, M., et al. (1974). Isolation and structure of ascochlorin and its analogs. *Agricultural and Biological Chemistry, 38*, 1463–1466.

Sato, H., Konoma, K., & Sakamura, S. (1979). Phytotoxins produced by onion pink root fungus *Pyrenochaeta terrestris. Agricultural and Biological Chemistry, 43*, 2409–2411.

Sato, H., Konoma, K., Sakamura, S., et al. (1981). X-ray crystal structure of pyrenocine A, a phytotoxin from *Pyrenochaeta terrestris. Agricultural and Biological Chemistry, 45*, 795–797.

Satre, M. (1981). The effect of asteltoxin and citreomontanine, two polyenic a-pyrone mycotoxins, on *Escherichia coli* adenosine triphosphatase. *Biochemical and Biophysical Research Communications, 100*, 267–274.

Schenke, D., Bottcher, C., Lee, J., & Scheel, D. (2011). Verticillin A is likely not produced by *Verticillium* sp. *The Journal of Antibiotics, 64*, 523–524.

Schmeda-Hirschmann, G., Hormazabal, E., Rodriguez, J. A., et al. (2008). Cycloaspeptide A and pseurotin a from the endophytic fungus *Penicillium janczewskii. Zeitschrift für Naturforschung. Section C, 63*, 383–388.

Seephonkai, P., Isaka, M., Kittakoop, P., et al. (2004). A novel ascochlorin glycoside from the insect pathogenic fungus *Verticillium hemipterigenum* BCC 2370. *The Journal of Antibiotics, 57*, 10–16.

Shin, C.-G., An, D.-G., Song, H.-H., et al. (2009). Beauvericin and enniatins H, I and MK1688 are new potent inhibitors of human immunodeficiency virus type-1 integrase. *The Journal of Antibiotics, 62*, 687–690.

Shinonaga, H., Kawamura, Y., Ikeda, A., et al. (2009a). The search for a hair-growth stimulant: New radicicol analogues as WNT-5A expression inhibitors from *Pochonia chlamydosporia* var. *chlamydosporia. Tetrahedron Letters, 50*, 108–110.

Shinonaga, H., Kawamura, Y., Ikeda, A., et al. (2009b). Pochonins K-P: New radicicol analogues from *Pochonia chlamydosporia* var. *chlamydosporia* and their WNT-5A expression inhibitory activities. *Tetrahedron, 65*, 3446–3453.

Shizuri, Y., Kosemura, S., Yamamura, S., et al. (1984). Biosynthesis of citreothiolactone, citreopyrone and pyrenocine B. *Tetrahedron Letters, 25*, 1583–1584.

Singh, S. B., Ball, R. G., Bills, G. F., et al. (1996). Chemistry and biology of cylindrols: Novel inhibitors of Ras farnesyl-protein transferase from *Cylindrocarpon lucidum. The Journal of Organic Chemistry, 61*, 7727–7737.

Sparace, S. A., Reeder, R. D., & Khanizadeh, S. (1987). Antibiotic activity of the pyrenocines. *Canadian Journal of Microbiology, 33*, 327–330.

Steiner, E., Kalamar, J., Charollais, E., et al. (1974). Recherches sur la biochimie des champignons inférieurs IX. Synthèse de précurseurs marqués et biosynthèse de la phoenicine et de l'oosporéine. *Helvetica Chimica Acta, 57*, 2377–2387.

Stob, M., Baldwin, R. S., Tuite, J., et al. (1962). Isolation of an anabolic, uterotrophic compound from corn infected with *Gibberella zeae*. *Nature, 196*, 1318.

Strunz, G. M., Heissner, C. J., Kakushima, M., et al. (1974). Metabolites of *Hyalodendron* sp.: Bisdethiodi (methylthio) hyalodendrin. *Canadian Journal of Chemistry, 52*, 325–326.

Supothina, S., Isaka, M., Kirtikara, K., et al. (2004). Enniatin production by the Entomopathogenic fungus *Verticillium hemipterigenum* BCC 1449. *The Journal of Antibiotics, 57*, 732–738.

Suzuki, A., Kanaoka, M., Isogai, A., et al. (1977). Bassianolide, a new insecticidal cyclodepsipeptide from *Beauveria bassiana* and *Verticillium lecanii*. *Tetrahedron Letters, 18*, 2167–2170.

Takamatsu, S., Rho, M. C., Masuma, R., et al. (1994). A novel testosterone 5α-reductase inhibitor, 8′,9′-dehydroascochiorin produced by *Verticillium* sp. FO-2787. *Chemical & Pharmaceutical Bulletin, 42*, 953–956.

Takatsuki, A., Tamura, G., & Arima, K. (1969). Antiviral and antitumor antibiotics. XIV. Effects of ascochlorin and other respiration inhibitors on multiplication of Newcastle disease virus in cultured cells. *Applied Microbiology, 17*, 825–829.

Takemoto, Y., Watanabe, H., Uchida, K., et al. (2005). Chemistry and biology of moverastins, inhibitors of cancer cell migration, produced by *Aspergillus*. *Chemistry & Biology, 12*, 1337–1347.

Tal, B., & Robeson, D. J. (1986). The production of pyrenocines A and B by a novel *Alternaria* species. Z Naturforsch C. *Biosciences, 41*, 1032–1036.

Tamura, G., Suzuki, S., Takatsuki, A., et al. (1968). Ascochlorin, a new antibiotic, found by paper-disc agar-diffusion method. I. Isolation, biological and chemical properties of ascochlorin. *Journal of Antibiotics, 21*, 539–544.

Toki, S., Ando, K., Yoshida, M., et al. (1992a). ES-242-1, a novel compound from *Verticillium* sp., binds to a site on N-methyl-D-aspartate receptor that is coupled to the channel domain. *The Journal of Antibiotics, 45*, 88–93.

Toki, S., Ando, K., Kawamoto, I., et al. (1992b). ES-242-2, -3, -4, -5, -6, -7, and -8, novel bioxanpretreatment produced by *Verticillium* sp., which act on the N-methyl-D-aspartate receptor. *The Journal of Antibiotics, 45*, 1047–1054.

Tomoda, H., Nishida, H., Huang, X., et al. (1992). New cyclodepsipeptides, enniatins D, E and F produced by *Fusarium* sp. FO-1305. *The Journal of Antibiotics, 45*, 1207–1215.

Trifonov, L. S., Dreiding, A. S., Hoesch, L., et al. (1981). Isolation of four hexaketides from *Verticillium intertexturn*. *Helvetica Chimica Acta, 64*, 1843–1846.

Trifonov, L. S., Bieri, J. H., Prewo, R., et al. (1982). The constitution of vertinolide, a new derivative of tetronic acid, produced by *Verticillium intertextum*. *Tetrahedron, 38*, 397–403.

Trifonov, L. S., Bieri, J. H., Prewo, R., et al. (1983). Isolation and structure elucidation of three metabolites from *Verticillium intertextum*, sorbicillin dihydrosorbicillin and bisvertinoquinol. *Tetrahedron, 39*, 4243–4256.

Trifonov, L. S., Hilpert, H., Floersheim, P., et al. (1986). Bisvertinols: A new group of dimeric vertinoids from *Verticillium intertextum*. *Tetrahedron, 42*, 3157–3179.

Turbyville, T. J., Wijeratne, E. M., Liu, M. X., et al. (2006). Search for Hsp90 inhibitors with potential anticancer activity: Isolation and SAR studies of radicicol and monocillin I from two plant-associated fungi of the Sonoran desert. *Journal of Natural Products, 69*, 178–184.

Usuki, H., Toyo-oka, M., Kanzaki, H., et al. (2009). Pochonicine, a polyhydroxylated pyrrolizidine alkaloid from fungus *Pochonia suchlasporia* var. *suchlasporia* TAMA 87 as a potent b-N-acetylglucosaminidase inhibitor. *Bioorganic & Medicinal Chemistry, 17*, 7248–7253.

Valadon, L. R. G., & Mummery, R. S. (1977). Natural β apo 4′-carotenoic acid methyl ester in the fungus *Verticillium agaricinum*. *Phytochemistry, 16*, 613–614.

Van Raaij, M. J., Abrahams, J. P., Leslie, A. G. W., et al. (1996). The structure of bovine F1-ATPase complexed with the antibiotic inhibitor aurovertin B. *Proceedings of the National Academy of Sciences of the United States of America, 93*, 6913–6917.

Vining, L. C., Kelleher, W. J., & Schwarting, A. E. (1962). Oosporein production by a strain of *Beauveria bassiana* originally identified as *Amanita muscaria*. *Canadian Journal of Microbiology, 8*, 931–933.

Visconti, A., Blais, L. A., Apsimon, J. W., et al. (1992). Production of enniatins by *Fusarium acuminatum* and *Fusarium compactum* in liquid culture: Isolation and characterization of three new enniatins, B2, B3, and B4. *Journal of Agricultural and Food Chemistry, 40*, 1076–1082.

Wainwright, M., & Betts, R. P. (1986). Antibiotic activity of oosporein from *Verticillium psalliotae*. *Transactions of the British Mycological Society, 86*, 168–170.

Wang, F., Luo, D., & Liu, J. (2005). Aurovertin E, a new polyene pyrone from the basidiomycete *Albatrellus confluens*. *The Journal of Antibiotics, 58*, 412–415.

Wang, Y. L., Li, L. F., Li, D. X., et al. (2015). Yellow pigment aurovertins mediate interactions between the pathogenic fungus *Pochonia chlamydosporia* and its nematode host. *Journal of Agricultural and Food Chemistry, 63*, 6577–6587.

Wenke, J., Anke, H., & Sterner, O. (1993). Pseurotin A and 8-0-demethylpseurotin A from *Aspergillus fumigatus* and their inhibitory activities on chitin synthase. *Bioscience, Biotechnology, and Biochemistry, 57*, 961–964.

Wicklow, D. T., Joshi, B. K., Gamble, W. R., et al. (1998). Antifungal metabolites (monorden, monocillin IV, and cerebrosides) from *Humicola fuscoatra* Traaen NRRL 22980, a mycoparasite of *Aspergillus flavus* Sclerotia. *Applied and Environmental Microbiology, 64*, 4482–4484.

Winssinger, N., & Barluenga, S. (2007). Chemistry and biology of resorcylic acid lactones. *Chemical Communications, 7*, 22–26.

Wu, H. Y., Wang, Y. L., Tan, J. L., et al. (2012). Regulation of the growth of cotton bollworms by metabolites from an entomopathogenic fungus *Paecilomyces cateniobliquus*. *Journal of Agricultural and Food Chemistry, 60*, 5604–5608.

Wu, F. B., Li, T. X., Yang, M. H., et al. (2015). Guanacastane-type diterpenoids from the insect-associated fungus *Verticillium dahlia*. *Journal of Asian Natural Products Research, 18*, 117–124. doi:10.1080/10286020.2015.1061511.

Xu, Y., Orozco, R., Wijeratne, K. E. M., et al. (2009). Biosynthesis of the cyclooligomer depsipeptide bassianolide, an insecticidal virulence factor of *Beauveria bassiana*. *Fungal Genetics and Biology, 46*, 353–364.

Xu, Y., Zhou, T., Espinosa-Artiles, P., et al. (2014a). Insights into the biosynthesis of 12-membered resorcylic acid lactones from heterologous production in *Saccharomyces cerevisiae*. *ACS Chemical Biology, 9*, 1119–1127.

Xu, Y., Zhou, T., Zhang, S., et al. (2014b). Diversity-oriented combinatorial biosynthesis of benzenediol lactone scaffolds by subunit shuffling of fungal polyketide synthases. *Proceedings of the National Academy of Sciences of the United States of America, 34*, 12354–12359.

Yamamoto, I., Suide, H., Henmi, T., et al. (1970). Antimicrobial α/β-unsaturated δ-lactones from fungi. *Takeda Kenkyusho Ho, 29*, 1–10.

Yamamoto, K., Hatano, H., Arai, M., et al. (2003). Structure elucidation of new monordens produced by *Humicola* sp. FO-2942. *The Journal of Antibiotics, 56*, 533–538.

Yamano, T., Hemmi, S., Yamamoto, I. et al. (1971). *Fermentative production of antibiotic phomalactone*. Patent report, Japan. 71 32,800 (Ct. C 12d, A 61k, C 07g). Takeda Chemical Industries, Ltd.

Yang, Z., Bao, L., Yin, Y., et al. (2014). Pyrenocines N-O: Two novel pyrones from *Colletotrichum* sp. HCCB03289. *The Journal of Antibiotics, 67*, 791–793.

Zhang, P., Bao, B., Dang, H. T., et al. (2009). Anti-inflammatory sesquiterpenoids from a sponge-derived fungus *Acremonium* sp. *Journal of Natural Products, 72*, 270–275.

Zheng, C. J., Park, S. H., Koshino, H., et al. (2007). Verticillin G, a new antibacterial compound from *Bionectra byssicola*. *The Journal of Antibiotics, 60*, 61–64.

Chapter 8
Metabolomics

Nuria Escudero, Frutos Marhuenda-Egea, and Luis V. Lopez-Llorca

Abstract Metabolomics is the study of metabolites, small biomolecules (carbohydrates, lipids, amino acids and organic acids) present in a biological sample. Metabolomics tools include chromatography for separating metabolites and spectroscopy techniques for their identification. Metabolomics tools have allowed to analyze the composition of tomato root exudates in the tritrophic system: *Pochonia chlamydosporia*, *Meloidogyne javanica* and tomato (*Solanum lycopersicum*) and changes in root exudates that were due to the presence of the fungus, the nematode or both. Large amounts of fluorescent compounds were detected in tomato root exudates from plants with *M. javanica* root galls and egg masses. Profiles of root exudates in ^{1}H NMR included organic acids, sugars and amino acids. Acetate signal increased in root exudates with *M. javanica*. Using HPLC-MS metabolomic

N. Escudero (✉)
Laboratory of Plant Pathology, Department of Marine Sciences and Applied Biology, University of Alicante, Alicante, Spain

Department of Marine Sciences and Applied Biology, Multidisciplinary Institute for Environmental Studies (MIES) Ramón Margalef, University of Alicante, Alicante, Spain

Department of Agri-Food Engineering and Biotechnology, Universitat Politècnica de Catalunya, Catalunya, Spain
e-mail: nuria.escudero@upc.edu

F. Marhuenda-Egea
Department of Agrochemistry and Biochemistry, University of Alicante, Alicante, Spain

Laboratory of Plant Pathology, Department of Marine Sciences and Applied Biology, University of Alicante, Alicante, Spain

Department of Marine Sciences and Applied Biology, Multidisciplinary Institute for Environmental Studies (MIES) Ramón Margalef, University of Alicante, Alicante, Spain
e-mail: frutos@ua.es

L.V. Lopez-Llorca
Laboratory of Plant Pathology, Department of Marine Sciences and Applied Biology, University of Alicante, Alicante, Spain

Department of Marine Sciences and Applied Biology, Multidisciplinary Institute for Environmental Studies (MIES) Ramón Margalef, University of Alicante, Alicante, Spain
e-mail: lv.lopez@ua.es

R.H. Manzanilla-López, L.V. Lopez-Llorca (eds.), *Perspectives in Sustainable Nematode Management Through* Pochonia chlamydosporia *Applications for Root and Rhizosphere Health*, Sustainability in Plant and Crop Protection,
DOI 10.1007/978-3-319-59224-4_8

fingerprints of tomato root exudates were generated. Several *m/z* signals have been found and related to the presence of *M. javanica* and only one with the presence of *P. chlamydosporia*. Metabolomics data integrated with transcriptomics will help to understand rhizosphere signalling in multitrophic systems.

8.1 Introduction

After completion of the Human Genome Sequencing Project, we are now in the post-gen*omics* biology era. New '*-omics*' scientific disciplines have emerged, which allow us to understand the role of genes, proteins or metabolites, in biological systems, both in human and environmental samples (Weckwerth 2007; Griffiths 2008; Lindon et al. 2011; Lutz et al. 2013). Pathology is a suitable field for *-omics* development since it involves organismal interactions (Tan et al. 2009). As we will show in this chapter, *Pochonia chlamydosporia*-plant parasitic nematode-root multitrophic interactions is an ideal subject for metabolomics.

Metabolomics is the study of metabolites, small molecules, that occur in biological samples such as cells, biological fluids or tissues (Dixon and Strack 2003). They are the products of metabolism and include, for example, sugars (or carbohydrates), fats (or lipids), amino acids and their derivatives. The collection of all metabolites within a cell is called the metabolome. Scientists have begun to characterize the metabolome of pathosystems in order better to understand and diagnose disease and its associated metabolism (Fiehn 2002; Lindon et al. 2011).

Metabolomics makes use of bioinformatics and specialised statistical (including multivariate analysis) techniques known as Chemometrics (Madsen et al. 2010). Chemometrics helps the understanding and managing of biological information included in metabolite profiles in order to evaluate a given metabolic situation. A direct goal of Applied Metabolomics is, for instance, to find patterns that could be characteristic of a particular disease or pathological disorder. Metabolomics is a multidisciplinary topic involving biologists, computer scientists and organic/analytical chemists. Metabolomics tools separate, identify and quantify metabolites. These include chromatography, such as High Performance Liquid Chromatography (HPLC), for separating metabolites and spectroscopy techniques, mainly nuclear magnetic resonance (NMR) and mass spectrometry (MS), for their identification.

The main advantage of metabolomics for disease diagnosis, either in human or plant pathology, is that this approach scores the phenotype of two interacting organisms: host and pathogen. Phenotype is the expression of the genome modulated by the environment. When an organism gets sick or stressed, thereby triggering specific molecular changes, its phenotype is altered. This change may then, in principle, be measured using metabolomics and chemometric tools (Lindon et al. 2011).

8.1.1 Metabolomics

The metabolomic study of fungal-plant interactions is very complex due to the huge quantity of metabolites involved in this process. However, the study of organismal interactions using metabolomics reveals information that transcriptomics or proteomics cannot provide (Allwood et al. 2007). There are two main approaches in metabolomics studies: (i) the identification of a metabolite profile; and (ii) the detection of a metabolite fingerprint as a characteristic of a specific condition (Fiehn 2002).

8.1.2 Fungal Metabolomics

Metabolomics approaches are being included in transcriptomic studies for a better understanding of fungal-plant interactions (Rudd et al. 2015) or to detect and identify the metabolites involved in rhizosphere communications (van Dam and Bouwmeester 2016).

8.1.3 Sample Preparation

Samples for metabolomics studies usually require little adjustments. For instance, *P. chlamydosporia* culture filtrates or root exudates of plants colonized by the fungus should only be free of particulate material. They are usually centrifuged, filtered or both. Aqueous root exudates can be filtered using 0.22 µm pore size filters. These filtrates are ready for HPLC-MS analysis (see below and Escudero et al. 2014). For NMR, samples are usually concentrated by freeze-drying due to the low sensitivity of this technique compared with that of HPLC-MS. Once freeze-dried samples are resuspended in deuterated water (i.e., heavy water, D_2O). A further issue in sample preparation for metabolomics is the extraction in solvents of various polarities (e.g., methanol/ethyl acetate). By doing so, the sample to be analysed is enriched in compounds of a polarity similar to that of the solvent used as extractant. This is a very standardized protocol that involves aqueous samples mixing with organic solvents of analytical or HPLC grades. Dissociation of phases (water/solvent) is usually performed in a separating funnel. Solvent phase is collected and evaporated in a rotary evaporator under vacuum and at a moderate temperature (40–45 C). Care should be taken to avoid boiling the sample and exposure to intense light.

8.1.4 Metabolomic Separation and Identification

As discussed above, metabolomics techniques can achieve the detection of metabolite fingerprints or metabolite identification and quantification for profiling or other purposes. We will now describe the techniques we have used for the metabolomics analysis of the tritrophic interaction: tomato plants, nematodes and *P. chlamydosporia*.

8.1.4.1 Emission-Excitation Matrix (EEM) Fluorescence

We set emission and excitation wavelengths from 220–460 nm and 220–350 nm respectively. This range allows the detection of phenolic substances similar to humic and fulvic acids or the signature of the aromatic amino acids tryptophan and tyrosine (Martin-Mata et al. 2015). Chemometric tools such as Parallel Factor Analysis (PARAFAC) are used for Excitation-Emission Matrix (EEM) fluorescence spectrum analysis (Murphy et al. 2014).

8.1.4.2 Nuclear Magnetic Resonance

Through NMR we could identify some of the compounds present in the root exudates, such as sugars and amino acids. However, given the complexity of the samples, most of the metabolomic profiles showed overlap of metabolite signatures and this made individual identifications hard to make.

8.1.4.3 Metabolite Profiling by HPLC-MS

Filtered (0.22 μm) samples are injected into the HPLC column. A key issue in the HPLC-MS analysis, along with the mobile phases, is the column to be used. The availability of UPLC, the new materials for column fillings, and new devices to detect mass/charge (*m/z*) relationship represent an enormous advantage for the use of metabolomics in complex systems such as plant-fungal interactions. The designing of a protocol suitable for eluting the samples usually has the same polarity considerations as that of sample extraction (see Sect. 8.1.3). Solvents that are applied in a gradient help to resolve the complex mixture of metabolites present, for instance, in the tomato root exudates. There is a huge number of column types available commercially. In our case metabolites to be analyzed were rather polar and this limited the column choice. Further parameters to take in account include optimizing the intensity of the ionization source and knowing if the detection includes cations, anions or both.

8.2 Metabolomics and Fungal Pathogens of Invertebrates

8.2.1 Entomopathogenic Fungi

Two of the most studied entomopathogenic fungi are *Beauveria* and *Metarhizium* spp. Specifically, several metabolomics studies have been carried out to elucidate the secondary metabolites employed by entomopathogenic fungi that parasitize insects. We will explain in detail some of them and their characterization. This is especially relevant in view of the close evolutionary relationship of *Metarhizium* spp. with the nematophagous fungus *P. chlamydosporia* (Larriba et al. 2014).

Beauvericin is a depsipeptide from *B. bassiana* with antifungal and antibiotic activities. Its structure was established using chromatography and NMR (Hamill et al. 1969). However, it is not an essential virulence factor of the fungus (Xu et al. 2008). Entomopathogenic fungi other than *B. bassiana*, such as *Paecilomyces fumosoroseus*, also produce beauvericins (Bernardini et al. 1975). Other secondary metabolites, more specifically cyclodepsipeptides known as beauverolides, were isolated and characterized by Infrared (IR) and NMR spectroscopy from *B. bassiana* (Elsworth and Grove 1974, 1977). In the same way, another cyclodepsipeptide named Bassianolide, was identified in *B. bassiana* and *Verticillium* (= *Lecanicillium*) *lecanii* (both entomopathogenic and nematophagous fungus) by means of proton nuclear magnetic resonance (^{1}H NMR) and ^{13}C nuclear spectra (Kanaoka et al. 1978). Bassianolide displays insecticidal properties (Suzuki et al. 1977) and plays a significant role in fungal virulence to insects (Xu et al. 2009).

Beauveria bassiana can also behave as an endophyte (Gómez-Vidal et al. 2009; Vega et al. 2009). The fungus secretes a different metabolomic profile in infected insect pupae (*Euproctis pseudoconspersa* and *Bombyx mori*) than in exudates from asparagus and carrot roots colonised by the fungus. Sphingolipids and fungal fatty acid metabolites were higher in pupae extracts, however ornithine, arginine, and GABA (gamma-aminobutyric acid) were low in both substrates (Luo et al. 2015). Gas chromatography-mass spectrometry GC/MS and liquid chromatography-mass spectrometry LC/MS have been recently used to determine insect metabolic profiles after infection with *B. bassiana*. *Beauveria bassiana* infection in insects causes an increase of carbohydrates, amino acids, fatty acids, and lipids, and a decrease in eicosanoids and amines (Xu et al. 2015). No such studies are yet available for nematophagous fungi and their hosts.

From *M. anisopliae*, two cyclodepsipeptides, named destruxin A (DTX A) and destruxin B (DTX B), were purified with high toxic activity against *Bombyx mori* (Kodaira 1961). DTX A was characterized by NMR and IR spectroscopy (Suzuki et al. 1966) and DTX B was by IR spectroscopy (Tamura et al. 1964). Desmethyl-destruxin B, DTX C and DTX D were identified by IR Spectra and GC-MS (Suzuki et al. 1970). From that, several DTX and related secondary metabolites were characterized (Païs et al. 1981; Gupta et al. 1989; Wahlman and Davidson 1993). Reversed-Phase HPLC (Samuels et al. 1988) and capillary electrophoresis (CE) were applied to optimize the separation and detection of DTXs (Liu et al. 2004). In some isolates of *M. anisopliae*, DTX secretion was related to the fungal capacity to kill insects, indicating their possible role as pathogenic factors (Kershaw et al. 1999). However, a DTX gene knockout mutant of *M. robertsii* showed that DTX are dispensable for virulence in the insect host evaluated (Donzelli et al. 2012). *Metarhizium anisopliae* also produces aurovertins (Azumi et al. 2008) a type of polyketides, as in *P. chlamydosporia* (see below and Chap. 7 by Niu Xuemai). Aurovertin B has raised a lot of interest because inhibits proliferation of breast cancer cell lines (Huang et al. 2008). Using liquid chromatography coupled with quadrupole time-of-flight mass spectrometry, de Bekker et al. (2013) found that *B. bassiana* beauverolides are mainly secreted on living tissues. In contrast, *Metarhizium brunneum* destruxins are secreted on dead tissues. In conclusion, these closely related fungi have different strategies for using their secondary metabolites.

8.2.2 Nematophagous Fungi

Nematophagous fungi are well known producers of a wide array of secondary metabolites with diverse roles in their biology (for a review the reader is directed to Niu and Degenkolb, Chapter 7 and Vilcinskas 2016a, b).

Some of the secondary metabolites are, for instance, neurotoxins that paralyze nematodes so the fungus can parasitize them, as in the toxin droplet organs of *Pleurotus* (Barron and Thorn 1987). These fungi are wood (C, H rich poor in N) decomposers. Therefore, N rich nematode hosts confer a competitive advantage. Early research by Olthof and Estey (1963) on *Arthrobotrys oligospora*, perhaps the best known and widespread nematophagous fungus, also found evidence of nematotoxin production. Later, the nematoxin was identified as linoleic acid (Stadler et al. 1993). *Arthrobotrys oligospora* mechanically captures vermiform nematodes by means of adhesive hyphal networks (Nordbring-Hertz et al. 2006). Recently, secondary metabolites from the fungus have been found to have a developmental role. Arthrosporols, monocyclic sesquiterpenols, can have autoregulatory effects on the formation of conidiophores and the transition of hyphae to two-dimensional networks (Zhang et al. 2012). Other secondary metabolites, such as oligosporon and derivatives (farnesyl chains linked to cyclohexen rings), are wide spectrum antimicrobials (Niu and Zhang 2011). They are probably also a competitive advantage in the rhizosphere and soil where fierce microbial competition is always taking place. However, the role for most secondary metabolites of nematophagous fungi remains unknown. It is tempting to speculate that metabolomics can play a key role in understanding the signalling in the multitrophic lifestyle of many nematophagous fungi, such as *P. chlamydosporia*, which can play endophytic, parasitic, and saprotrophic roles in the environment.

8.3 *Pochonia chlamydosporia* Secondary Metabolites

Several secondary metabolites from *P. chlamydosporia* have been investigated. Chlamydocyl is a cyclic tetrapeptide and was the first secondary metabolite identified and characterized from *P. chlamydosporia* by IR and NMR spectroscopy (Closse and Huguenin 1974). This compound inhibits histone deacetylases (HDACs) and displays anti-tumoral properties for its high cytostatic activity (Stähelin and Trippmacher 1974). Diheteropeptin, a cyclictetrapeptide as chlamydocin was obtained by HPLC (Masuoka et al. 2000a), and its structure determined by NMR and MS (Masuoka et al. 2000b) and also displays cytostatic activity to Mv1Lu cells (Masuoka et al. 2000a). For a full review, we refer the reader to Chap. 7 by Niu).

Radicicol (monorden) showed WNT-5A (Wnt Family Member 5A Protein code gene) inhibitory activity. Search inhibitors of the glycoprotein WNT-5A could improve hair-growth, because WNT-5A promotes the proliferation of dermal papilla cells (Ikeda et al. 2003).

Pochonia chlamydosporia var. *catenulata* secretes secondary metabolites *in vitro* with antiviral and antiparasitic activities. These compounds, named pochonins, are resorcylic acid lactones derived from radicicol. Pochonins were identified and characterized using ^{1}H NMR spectra, reversed-phase (RP) HPLC and high-performance gel-permeation chromatography (HPGPC) (Hellwig et al. 2003). From *Pochonia chlamydosporia* var. *chlamydosporia* the new pochonins G-J (Shinonaga et al. 2009a) and K-P were determined (Shinonaga et al. 2009b). All showed high similarities to radicicol. The pochonins with 4,5-epoxide and 4,5-E-olefin moieties showed WNT-5A inhibitory activities, but were less cytotoxic than radicicol (Shinonaga et al. 2009a, b). 13-bromomonocillin I is a new metabolite from *P. chlamydosporia* var. *chlamydosporia* that showed WNT-5A inhibitory activity, its structure being determined using NMR, distortionless enhancement by polarization transfer (DEPT), correlated spectroscopy (COSY), heteronuclear multiple-quantum correlation (HMQC) and (heteronuclear multiple bond correlation (HMBC) spectra (Shinonaga et al. 2009c). Phomalactone has been isolated from *P. chlamydosporia* and found to be nematicidal (Khambay et al. 2000).

Pochonia [= *Metapochonia*] *suchlasporia* var. *suchlasporia* secrets a secondary metabolite named pochonicine. It is first the pyrrolizidine alkaloid found in nature with high glucosaminidase activity against insect, mammals, and plants. The chemical structure of pochonicine was determined using NMR and MS techniques (Usuki et al. 2009). The first proposed structure was revised by Kitamura et al. (2013). Methods for synthesis of pochonicine and eight stereoisomers were carried out with their NMR spectra (Zhu et al. 2013), Pochonicine 1D and its epimers 2D, 3D, and 4D showed capacity for hexosaminidase inhibition.

Aurovertins I, E, F and D were purified from *P. chlamydosporia* and their structures established by NMR and MS (Niu et al. 2010). Aurovertins F and D showed toxicity against the nematode *Panagrellus redivivus.* Wang et al. (2015) have recently determined by HPLC that aurovertin D is the most abundantly found aurovertin in *P. chlamydosporia* with strong toxicity against the root-knot nematode *M. incognita* and the model nematode *Caenorhabditis elegans*.

8.4 Metabolomics Applied to Tomato Plants Inoculated with *Pochonia chlamydosporia*

Several rhizosphere studies have been performed to improve the knowledge of plant-nematode interactions, using massive molecular analysis techniques such as transcriptomics (Jammes et al. 2005) or metabolomics (Hofmann et al. 2010). On the other hand, tomato root exudates have been tested on *Meloidogyne hispanica* juveniles (Duarte et al. 2015) or in the germination of fungal pathogens such as *Fusarium oxysporum* f. sp. *lycopersici* (Steinkellner et al. 2008). In Escudero et al. (2014) we analysed the composition of tomato root exudates in tritrophic systems (*P. chlamydosporia*-RKN-tomato (*Solanum lycopersicum*)) and identified changes due to the presence of the fungus, the nematode or both. In this respect, the nematode (RKN, *Meloidogyne javanica*) was the factor influencing most of the rhizodeposition profile.

We detected large amounts of fluorescent compounds in tomato root exudates from plants inoculated with *M. javanica* at the end of its life cycle (formation of root galls and egg masses). Some of these metabolic signatures found were the aromatic amino acids tryptophan and tyrosine. Amino acid biosynthesis has previously been associated with the presence of plant-parasitic nematodes (Hofmann et al. 2010). Fluorescence spectroscopy is a fast and easy technique since no sample processing is necessary. However it is limited to compounds with fluorochromes or able to emit fluorescence. For this reason, we also applied NMR and HPLC to root exudates. In ^{1}H NMR profiles of root exudates we identified several organic acids, sugars and amino acids. The acetate signal increased in the presence of *M. javanica*. Aromatic amino acids could also be part of plant peptide hormones (CLEs) stimulated by the presence of the nematodes (Gheysen and Mitchum 2011) or plant defences. Sugars and organic acids detected in the root exudates could be a source of nutrients for nematodes (Baldacci-Cresp et al. 2012).

Finally, in this work we generated a metabolomics fingerprint of tomato root exudates using HPLC-MS. We later focused our attention on those metabolites identified by their mass-to-charge ratio (*m/z*) signals with different intensity respect to root exudates from control plants. We found several *m/z* signals related to the presence of *M. javanica* and only one related to the presence of *P. chlamydosporia*. In future research the identification of these metabolites should be carried out – some of them could be plant defences blocked by the nematode. The best candidate *m/z* signals are those with a lower intensity in the root exudates from plants inoculated with *M. javanica*.

8.5 Future Perspectives: Metabolomics Data Integration with Other – Omics

It is envisaged that once metabolomics databases include enough entries they could be amenable to functional massive metabolomics interconnected with other *-omics* disciplines. Something similar to functional genomics/transcriptomics with GO (Gene Ontology) and KEGG (Kyoto Encyclopedia of Genes and Genomes) analyses. Their application to metabolomics includes several complications. Secondary metabolism is a complex topic for the variety of metabolites and their derivatives which are usually synthesized with low numbers of gene clusters. These encode multicomponent enzymatic systems, such as Polyketide Synthases which are very abundant in the *P. chlamydosporia* genome (Larriba et al. 2014). Various combinatorial multi-disciplinary approaches, involving bioinformatics, molecular biology, and metabolomics, could aid in understanding the biosynthesis of important secondary metabolites with biological activity from *P. chlamydosporia* molecules. The large amount of data available from transcriptomic and metabolomic studies can be integrated with the use of new bioinformatics and statistical tools. In this regard, Paintomics is a web server developed to facilitate the integration of transcriptomics and metabolomics data (http://www.paintomics.org/cgi-bin/main2.cgi), which is currently being tested (García-Alcalde et al. 2011).

Acknowledgements This research was funded by the Spanish Ministry of Economy and Competitiveness Grant AGL 2015-66833-R. Luis Vicente Lopez-Llorca was awarded a sabbatical grant (PR2015_00087) by the Spanish Ministry of Education, Culture and Sport.

References

Allwood, J. W., Ellis, D. I., & Goodacre, R. (2007). Metabolomic technologies and their application to the study of plants and plant–host interactions. *Physiologia Plantarum, 132*, 117–135.
Azumi, M., Ishidoh, K., Kinoshita, H., et al. (2008). Aurovertins F–H from the entomopathogenic fungus *Metarhizium anisopliae*. *Journal of Natural Products, 71*, 278–280.
Baldacci-Cresp, F., Chan, C., Maucourt, M., et al. (2012). (homo)glutathione deficiency impairs root-knot nematode development in *Medicago truncatula*. *PLoS Pathogens, 8*(1), e1002471.
Barron, G. L., & Thorn, R. G. (1987). Destruction of nematodes by species of *Pleurotus*. *Canadian Journal of Botany, 65*, 774–778.
Bernardini, M., Carilli, A., Pacioni, G., et al. (1975). Isolation of beauvericin from *Paecilomyces fumoso-roseus*. *Phytochemistry, 14*, 1865.
Closse, A., & Huguenin, R. (1974). Solierung und strukturaufklarung von chlamydocin (Isolation and elucidation of structure of chlamydocin). *Helvetica Chimica Acta, 57*, 533–545.
de Bekker, C., Smith, P. B., Patterson, A. D., et al. (2013). Metabolomics reveals the heterogeneous secretome of two entomopathogenic fungi to *ex vivo* cultured insect tissues. *PloS One, 8*(8), e70609.

Degenkolb, T., & Vilcinskas, A. (2016a). Metabolites from nematophagous fungi and nematicidal natural products from fungi as an alternative for biological control. Part I: Metabolites from nematophagous ascomycetes. *Applied Microbiology and Biotechnology, 100*, 3799–3812.

Degenkolb, T., & Vilcinskas, A. (2016b). Metabolites from nematophagous fungi and nematicidal natural products from fungi as alternatives for biological control. Part II: Metabolites from nematophagous basidiomycetes and non-nematophagous fungi. *Applied Microbiology and Biotechnology, 100*, 3813–3824.

Dixon, R. A., & Strack, D. (2003). Phytochemistry meets genome analysis, and beyond. *Phytochemistry, 62*, 815–816.

Donzelli, B. G. G., Krasnoff, S. B., Sun-Moon, Y., et al. (2012). Genetic basis of destruxin production in the entomopathogen *Metarhizium robertsii*. *Current Genetics, 58*, 105–116.

Duarte, A., Maleita, C., Abrantes, I., et al. (2015). Tomato root exudates induce transcriptional changes of *Meloidogyne hispanica* genes. *Phytopathologia Mediterranea, 54*, 104–108.

Eichinger, D. (1997). Encystation of entamoeba parasites. *BioEssays, 19*, 633–639.

Elsworth, J. F., & Grove, J. F. (1974). Search for biologically-active cyclodepsipeptides from *Beauveria bassiana*. *South African Journal of Science, 70*, 379.

Elsworth, J. F., & Grove, J. F. (1977). Cyclodepsipeptides from *Beauveria bassiana* Bals. Part 1. Beauverolides H and I. *Journal of the Chemical Society, Perkin Transactions, 1*, 270–273.

Escudero, N., Marhuenda-Egea, F. C., Ibanco-Cañete, R., et al. (2014). A metabolomic approach to study the rhizodeposition in the tritrophic interaction: Tomato, *Pochonia chlamydosporia* and *Meloidogyne javanica*. *Metabolomics, 10*, 788–804.

Fiehn, O. (2002). Metabolomics- the link between genotypes and phenotypes. *Plant Molecular Biology, 48*, 155–171.

García-Alcalde, F., García-López, F., Dopazo, J., et al. (2011). Paintomics: A web based tool for the joint visualization of transcriptomics and metabolomics data. *Bioinformatics, 27*(1.), btq594), 137–139.

Gheysen, G., & Mitchum, M. G. (2011). How nematodes manipulate plant development pathways for infection. *Current Opinion in Plant Biology, 14*, 415–421.

Gómez-Vidal, S., Salinas, J., Tena, M., et al. (2009). Proteomic analysis of date palm (*Phoenix dactylifera* L.) responses to endophytic colonization by entomopathogenic fungi. *Electrophoresis, 30*, 2996–3005.

Griffiths, W. J. (2008). *Metabolomics, metabonomics and metabolite profiling*. Cambridge: RSC Publishing.

Gupta, S., Roberts, D. W., & Renwick, J. A. A. (1989). Insecticidal cyclodepsipeptides from *Metarhizium anisopliae*. *Journal of the Chemical Society, Perkin Transactions 1, 12*, 2347–2358.

Hamill, R. L., Higgens, C. E., Boaz, H. E., et al. (1969). The structure of Beauvericin, a new depsipeptide antibiotic toxic to *Artemia salina*. *Tetrahedron Letters, 10*, 4255–4258.

Hellwig, V., Mayer-Bartschmid, A., Müller, H., et al. (2003). Pochonins A-F, new antiviral and antiparasitic resorcylic acid lactones from *Pochonia chlamydosporia* var. *catenulata*. *Journal of Natural Products, 66*, 829–837.

Hofmann, J., Ashry El, A. E. N., Anwar, S., et al. (2010). Metabolic profiling reveals local and systemic responses of host plants to nematode parasitism. *The Plant Journal, 62*, 1058–1071.

Huang, T. C., Chang, H. Y., Hsu, C. H., et al. (2008). Targeting therapy for breast carcinoma by ATP synthase inhibitor aurovertin B. *Journal of Proteome Research, 7*, 1433–1444.

Ikeda, A., Shinonaga, H., Fujimoto, N., Kasai, Y. (2003). PCT Gazette – Section I. Published international applications. WO 03/086334, 23 Oct 2003, p 62. http://www.wipo.int/edocs/pctdocs/en/2003/pct_2003_43-section1.pdf

Jammes, F., Lecomte, P., Almeida-Engler, J., et al. (2005). Genome-wide expression profiling of the host response to root-knot nematode infection in Arabidopsis. *The Plant Journal, 44*, 447–458.

Kanaoka, M., Isogai, A., & Murakoshi, S. (1978). Bassianolide, a new insecticidal cyclodepsipeptide from *Beauveria bassiana* and *Verticillium lecanii. Agricultural and Biological Chemistry, 42*, 629–635.

Khambay, B. P. S., Bourne, J. M., Cameron, S., et al. (2000). A nematicidal metabolite from *Verticillium chlamydosporium. Pest Management Science, 56*, 1098–1099.

Kershaw, M. J., Moorhouse, E. R., Bateman, R., et al. (1999). The role of destruxins in the pathogenicity of *Metarhizium anisopliae* for three species of insect. *Journal of Invertebrate Pathology, 74*, 213–223.

Kitamura, Y., Koshino, H., Nakamura, T., et al. (2013). Total synthesis of the proposed structure for pochonicine and determination of its absolute configuration. *Tetrahedron Letters, 54*, 1456.

Kodaira, Y. (1961). Biochemical studies on the muscardine fungi in the silkworms *Bombyx mori. Journal of the Faculty of Textile Science and Technology, Shinshu University Series A, Biology, 29*, 1–68.

Larriba, E., Jaime, M. D. L. A., Carbonell-Caballero, J., et al. (2014). Sequencing and functional analysis of the genome of a nematode egg-parasitic fungus, *Pochonia chlamydosporia. Fungal Genetics and Biology, 65*(C), 69–80.

Lindon, J. C., Nicholson, J. K., & Holmes, E. (2011). *The handbook of Metabonomics and metabolomics*. Amsterdam: Elsevier Science.

Liu, C. M., Huang, S. S., & Tzeng, Y. M. (2004). Analysis of destruxins produced from *Metarhizium anisopliae* by capillary electrophoresis. *Journal of Chromatographic Science, 42*, 140–144.

Luo, F., Wang, Q., Yin, C., et al. (2015). Differential metabolic responses of *Beauveria bassiana* cultured in pupae extracts, root exudates and its interactions with insect and plant. *Journal of Invertebrate Pathology, 130*, 1–11.

Lutz, N. W., Sweedler, J. V., & Wevers, R. A. (2013). *Methodologies for metabolomics: Experimental strategies and techniques*. Cambridge: Cambridge University Press.

Madsen, R., Lundstedt, T., & Trygg, J. (2010). Chemometrics in metabolomics – A review in human disease diagnosis. *Analytica Chimica Acta, 659*, 23–33.

Martin-Mata, J., Marhuenda-Egea, F. C., Moral, R., et al. (2015). Characterization of dissolved organic matter from sewage sludge using 3D-fluorescence spectroscopy and chemometric tools. *Communications in Soil Science and Plant Analysis, 46*, 188–196.

Masuoka, Y., Shin-Ya, K., Kim, J. B., et al. (2000a). Diheteropeptin, a new substance with TGF-ß-like activity, produced by a fungus, *Diheterospora chlamydosporia*. I. Production, Isolation and biological activities. *The Journal of Antibiotics, 53*, 788–792.

Masuoka, Y., Shin-Ya, K., Kim, J. B., et al. (2000b). Diheteropeptin, a new substance with TGF-ß-like activity, produced by a fungus, *Diheterospora chlamydosporia*. II. Physico-chemical properties and structure elucidation. *The Journal of Antibiotics, 53*, 793–798.

Murphy, K. R., Bro, R., & Stedmon, C. A. (2014). Chemometric analysis of organic matter fluorescence. In P. G. Coble, J. Lead, A. Baker, D. M. Reynolds, & R. G. M. Spencer (Eds.), *Aquatic organic matter fluorescence* (pp. 339–375). New York: Cambridge University Press.

Niu, X. M., Wang, Y. L., Chu, Y. S., et al. (2010). Nematodetoxic aurovertin-type metabolites from a root-knot nematode parasitic fungus *Pochonia chlamydosporia. Journal of Agricultural and Food Chemistry, 58*, 828–834.

Niu, X.-M., & Zhang, K.-Q. (2011). *Arthrobotrys oligospora*: A model organism for understanding the interaction between fungi and nematodes. *Mycology, 2*, 59–78.

Nordbring-Hertz, B., Jansson, H.B., Tunlid, A. (2006) Nematophagous fungi. eLS Citable reviews in the life sciences doi: 10.1002/9780470015902.a0000374.pub3.

Olthof, T. H. A., & Estey, R. H. A. (1963). A nematotoxin produced by the nematophagous fungus *Arthrobotrys oligospora* Fresenius. *Nature, 197*, 514–515.

Païs, M., Das, B. C., & Ferron, P. (1981). Depsipeptides from *Metarhizium anisopliae. Phytochemistry, 20*, 715–723.

Rudd, J. J., Kanyuka, K., Hassani-Pak, K., et al. (2015). Transcriptome and metabolite profiling of the infection cycle of *Zymoseptoria tritici* on wheat reveals a biphasic interaction with plant immunity involving differential pathogen chromosomal contributions and a variation on the hemibiotrophic lifestyle definition. *Plant Physiology, 167*, 1158–1185.

Samuels, R. I., Charnley, A. K., & Reynolds, S. E. (1988). Application of reversed-phase HPLC in separation and detection of the cyclodepsipeptide toxins produced by the entomopathogenic fungus *Metarhizium anisopliae*. *Journal of Chromatographic Science, 26*, 15–19.

Shinonaga, H., Kawamura, Y., Ikeda, A., et al. (2009a). The search for a hair-growth stimulant: New radicicol analogues as WNT-5A expression inhibitors from *Pochonia chlamydosporia* var. *chlamydosporia*. *Tetrahedron Letters, 50*, 108–110.

Shinonaga, H., Kawamura, Y., Ikeda, A., et al. (2009b). Pochonins K-P: New radicicol analogues from *Pochonia chlamydosporia* var. *chlamydosporia* and their WNT-5A expression inhibitory activities. *Tetrahedron, 65*, 3446–3453.

Shinonaga, H., Sakai, N., Nozawa, Y., et al. (2009c). 13-Bomomonocillin I: A New WNT-5A expression inhibitor produced by *Pochonia chlamydosporia* var. *chlamydosporia*. *Heterocycles, 78*(11). doi:10.3987/COM-09-11809.

Stadler, M., Anke, H., & Sterner, O. (1993). Linoleic acid – The nematicidal principle of several nematophagous fungi and its production in trap-forming submerged cultures. *Archives of Microbiology, 160*, 401–405.

Stähelin, H., & Trippmacher, A. (1974). Cytostatic activity of chlamydocin, a rapidly inactivated cyclic tetrapeptide. *European Journal of Cancer, 10*, 801–808.

Steinkellner, S., Mammerler, R., & Vierheilig, H. (2008). Germination of *Fusarium oxysporum* in root exudates from tomato plants challenged with different *Fusarium oxysporum* strains. *European Journal of Plant Pathology, 122*, 395–401.

Suzuki, A., Kuyama, S., Kodaira, Y., et al. (1966). Structural elucidation of destruxin A. *Agricultural and Biological Chemistry, 30*, 517–518.

Suzuki, A., Taguchi, H., & Tamura, S. (1970). Isolation and structure elucidation of three new insecticidal cyclodepsipeptides, destruxins C and D and desmethyldestruxin B, produced by *Metarrhizium anisopliae*. *Agricultural and Biological Chemistry, 34*, 813–816.

Suzuki, A., Kanaoka, M., Isogai, A., et al. (1977). Bassianolide, a new insecticidal cyclodepsipeptide from *Beauveria bassiana* and *Verticillium lecanii*. *Tetrahedron Letters, 18*, 2167–2170.

Tamura, S., Kuyama, S., Kodaira, Y., et al. (1964). Studies on destruxin B, an insecticidal depsipeptide produced by *Oospora destructor*. *Institute of Applied Microbiology (University of Tokyo) Symposium Microbiology, 6*, 127–140.

Tan, K.-C., Ipcho, S. V. S., Trengove, R. D., et al. (2009). Assessing the impact of transcriptomics, proteomics and metabolomics on fungal phytopathology. *Molecular Plant Pathology, 10*, 703–715.

Usuki, H., Toyo-oka, M., Kanzaki, H., et al. (2009). Pochonicine, a polyhydroxylated pyrrolizidine alkaloid from fungus *Pochonia suchlasporia* var. *suchlasporia* TAMA 87 as a potent ß-*N*-acetylglucosaminidase inhibitor. *Bioorganic & Medicinal Chemistry, 17*, 7248–7253.

van Dam, N. M., & Bouwmeester, H. J. (2016). Metabolomics in the Rhizosphere: Tapping into belowground chemical communication. *Trends in Plant Science, 21*, 256–265. dx.doi.org/10.1016/j.tplants.2016.01.008. Accessed 18 Oct 2016.

Vega, F. E., Goettel, M. S., Blackwell, M., et al. (2009). Fungal entomopathogens: New insights on their ecology. *Fungal Ecology, 2*, 149–159.

Wahlman, M., & Davidson, B. S. (1993). New destruxins from the entomopathogenic fungus *Metarhizium anisopliae*. *Journal of Natural Products, 56*, 643–647.

Wang, Y. L., Li, L. F., Li, D. X., et al. (2015). Yellow pigment aurovertins mediate interactions between the pathogenic fungus *Pochonia chlamydosporia* and its nematode host. *Journal of Agricultural and Food Chemistry, 63*, 6577–6587.

Weckwerth, W. (2007). *Metabolomics: Methods and protocols*. Totowa/NJ: Humana Press.

Xu, Y., Orozco, R., Wijeratne, E. M. K., et al. (2008). Biosynthesis of the cyclooligomer depsipeptide beauvericin, a virulence factor of the entomopathogenic fungus *Beauveria bassiana*. *Chemistry & Biology, 15*, 898–907.

Xu, Y., Orozco, R., Wijeratne, E. M. K., et al. (2009). Biosynthesis of the cyclooligomer depsipeptide bassianolide, an insecticidal virulence factor of *Beauveria bassiana*. *Fungal Genetics and Biology, 46*, 353–364.

Xu, Y. J., Luo, F., Gao, Q., et al. (2015). Metabolomics reveals insect metabolic responses associated with fungal infection. *Analytical and Bioanalytical Chemistry*. doi:10.1007/s00216-015-8648-8.

Zhang, H.-X., Tan, J.-L., Wei, L.-X., et al. (2012). Morphology regulatory metabolites from *Arthrobotrys oligospora*. *Journal of Natural Products, 75*, 1419–1423.

Zhu, J. S., Nakagawa, C. W., Adachi, I., et al. (2013). Synthesis of eight stereoisomers of Pochonicine: Nanomolar inhibition of β-N-Acetylhexosaminidases. *The Journal of Organic Chemistry, 78*, 10298–10309.

Chapter 9
Comparative Analysis of *Pochonia chlamydosporia* Mitogenome Reveals Dynamic Mitochondrial Evolution of the Nematophagous Fungi in *Hypocreales*

Runmao Lin, Xinyue Cheng, and Bingyan Xie

Abstract Mitochondria, the independent organelles, play diverse and crucial roles in eukaryotic cells. Their genetic materials can be used to provide hints concerning the evolutionary history of life. To understand the evolution of the nematophagous fungus *Pochonia chlamydosporia*, we analysed the organization of its mitogenome that encodes 14 typical protein-coding genes, two ribosomal RNAs, an intronic ribosomal protein (*rps3*) and a set of 23 transfer RNAs. Comparative analysis of mitogenomes between *P. chlamydosporia* and other four nematophagous fungi (*Lecanicillium saksenae*, *Acremonium implicatum* (= *Sarocladium implicatum*), *Hirsutella minnesotensis* and *Hirsutella rhossiliensis*) shows that they commonly encode 13 typical protein-coding genes and 12 types of tRNA and they naturally exhibit differences in gene numbers, introns and intergenic regions. The variable sizes of these five mitogenomes are mainly contributed by intergenic regions and coding regions. To identify any conflicts in phylogenetic relationships present in the invertebrate-pathogenic fungi and the plant pathogens between nuclear and mitochondrial phylogenies in *Hypocreales*, we performed a phylogenetic analysis based on the detected best model for amino acid sequences of 14 typical genes by manually

---

R. Lin
College of Life Sciences, Beijing Normal University, Beijing 100875, China

Institute of Vegetables and Flowers, Chinese Academy of Agricultural Sciences, Beijing 100081, China
e-mail: linrm2010@163.com

X. Cheng (✉)
College of Life Sciences, Beijing Normal University, Beijing 100875, China
e-mail: chengxy@bnu.edu.cn

B. Xie (✉)
Institute of Vegetables and Flowers, Chinese Academy of Agricultural Sciences, Beijing 100081, China
e-mail: xiebingyan@caas.cn

R.H. Manzanilla-López, L.V. Lopez-Llorca (eds.), *Perspectives in Sustainable Nematode Management Through* Pochonia chlamydosporia *Applications for Root and Rhizosphere Health*, Sustainability in Plant and Crop Protection,
DOI 10.1007/978-3-319-59224-4_9

improving the alignments. We found that the mitochondrial phylogeny is similar to the nuclear phylogeny, suggesting the same roles for both trees to investigate the phylogenetic relationships of species in *Hypocreales*. Moreover, the positively selected *rps3* genes with their special phylogenetic relationships that are different from those inferred from the other 14 protein-coding genes may contribute to explore further the evolutionary trajectory of the fungi in the *Hypocreales*.

9.1 Introduction

Mitochondria play diverse and crucial roles in eukaryotic cells, namely the three major functions of bioenergetics, biosynthesis and signalling (Chandel 2014). Their essential function in adenosine triphosphate (ATP) production maintains a high ATP/ADP ratio for driving biochemical reactions and functions involved in the biosynthesis of macromolecules (lipid, carbohydrates, proteins, nucleotides), as well as communicating biosynthetic and bioenergetics fitness to the rest of the cells. As an independent organelle in the cell, the mitochondrion contains its own genetic material. Furthermore, the inheritance of mitochondrial genes and genomes is different from nuclear genes and genomes, deviating from the Mendelian principles of segregation and independent assortment (Wilson and Xu 2012). In fungi, the diversity of mitochondrial DNA inheritance patterns has been reviewed (Wilson and Xu 2012), including the dominant pattern of uniparental inheritance, biparental inheritance, location-dependent inheritance, and progeny sex-dependent inheritance, which may suggest the dynamic evolution of fungal mitogenomes.

Comparative mitochondrial genomics approaches have been regarded as the effective way to investigate mitochondrial evolution (Gray 2012). In recent years, continuously increasing data on fungal genomics have been reported (http://www.ncbi.nlm.nih.gov/genome/browse/), making it possible to compare evolutionary analyses between nuclear and mitochondrial genomic resources and to compare mitochondrial characterisations between fungal species. In fungi, some efficient biological control agents (BCAs) within *Hypocreales* have been isolated and their genomes made available for comparative analysis (Table 9.1), such fungi including the nematophagous *Pochonia chlamydosporia* (Manzanilla-López et al. 2013; Larriba et al. 2014; Lin et al. 2015). Normally, the mitogenomes of these species contain protein-coding genes involved in oxidative phosphorylation (such as ubiquinone oxidoreductase (NADH), cytochrome c oxidoreductase (ubiquinol), O_2 oxidoreductase (cytochrome c) and ATP synthase, non-coding RNAs involving in translation (such as ribosomal RNAs, and transfer RNAs), introns and intergenic regions. Intergenic regions, introns and coding regions are the three main contributors to mitogenome size variation among fungi in *Sordariomycetes* (Lin et al. 2015). To understand better the structure of BCA mitogenomes and provide clues to their evolutionary history, we introduce the genomic structure of the *P. chlamydosporia* mitogenome and perform a phylogenetic-based approach to indicate the evolution

Table 9.1 List of biological control agents and plant pathogens in *Hypocreales* with published nuclear and mitochondrial genomes

Species	Host	Nuclear Genome Accession	Mitogenome Accession	References
Beauveria pseudobassiana	–	–	NC_022708	Oh et al. (2015)
Beauveria bassiana	*Dendrolimus punctatus*; aphids; planthoppers; spider mites	ADAH00000000	EU371503	Xu et al. (2009) and Xiao et al. (2012)
Cordyceps brongniartii	*Melolontha melolontha*	AZHA00000000	NC_011194	Ghikas et al. (2010) and Shang et al. (2016)
Cordyceps militaris	Lepidopteran pupae	ADNI00000000	NC_022834	Zheng et al. (2011) and Sung (2015)
Lecanicillium muscarium	Aphids	–	AF487277	Kouvelis et al. (2004)
Lecanicillium saksenae	Root-knot nematodes	–	KT585676	Xin et al. (2015)
Metarhizium anisopliae	Locusts; caterpillars; flies; crickets; beetles	ANDJ00000000	AY884128	Ghikas et al. (2006) and Gao et al. (2011)
Pochonia chlamydosporia	Root-knot nematodes; cyst nematodes	AOSW00000000	KF479445	Larriba et al. (2014) and Lin et al. (2015)
Hirsutella rhossiliensis	Soybean cyst nematode	–	KU203675	Wang et al. (2016a)
Hirsutella minnesotensis	Soybean cyst nematode	JPUM00000000	KR139916	Lai et al. (2014) and Zhang et al. (2015)
Acremonium implicatum	*Meloidogyne incognita*	–	KP164992	Yao et al. (2016)

of *P. chlamydosporia*, as well as investigate factors contributing to mitogenome size variation among the nematophagous fungi in *Hypocreales* by using comparative genomic methods.

9.2 The Organization and Structure of the *Pochonia chlamydosporia* Mitogenome

The complete circular mitogenome of *P. chlamydosporia* strain 170 is 25,615 bp in size, with GC content of 28.3% (Lin et al. 2015). A total of 40 genes are identified (Fig. 9.1A), including 15 protein-coding genes (seven ubiquinone oxidoreductases: *nad1*, *nad2*, *nad3*, *nad4*, *nad4L*, *nad5*, *nad6*; one cytochrome c oxidoreductase: *cob*; three O_2 oxidoreductases: *cox1*, *cox2*, *cox3*; three ATP synthases: *atp6*, *atp8*, *atp9*; one Ribosomal protein: *rps3*), the small and large ribosomal RNAs (*rns*, *rnl*), and 23 transfer RNAs (*trnA*, *C*, *D*, *E*, *F*, *G*, *H*, *I*, *K*, *L*, *M* (3), *N*, *P*, *Q*, *R* (2), *S*, *T*, *V*, *W*, *Y*). In this mitogenome, 14,385 bp (56.16%), 1,697 bp (6.63%), 6,232 bp (24.33%), 335 bp (1.31%) and 4,618 bp (18.03%) of coding regions, tRNAs, introns and intergenic regions are found, respectively (Table 9.2), with the *rps3* gene encoded in the intron region of *rnl*. For gene distribution, 23 tRNAs are clustered into three major groups of YDN, VISWP and TEMMLAFKQHM (Fig. 9.1A).

A comparison of mitochondrial genomic organizations between *P. chlamydosporia* and other nematophagous fungi shows that they commonly encode 13 typical protein-coding genes (*nad1–5*, *nad4L*, *cob*, *cox1–2*, *atp6*, *atp8–9* and *rps3*) and 12 types of tRNAs (*trnT*, *E*, *M*, *L*, *A*, *F*, *K*, *Q*, *H*, *R*, *C* and *Y*) and that they naturally exhibit differences in gene numbers, introns and intergenic regions (Fig. 9.1). A large number of new genes occur in other fungi, such as 14 hypothetical protein-coding genes in *H. minnesotensis* (Zhang et al. 2015a), and *trnQ* near *cox3* in *L. saksenae*. Meanwhile, gene loss events have occurred in other fungi, for example: *cox3* and *nad6* are lost in *A. implicatum* (= *Sarocladium implicatum*) (Yao et al. 2016). Diverse distributions of intergenic regions are visible, e.g. the remarkably large regions between *cox2* and *trnR* in *L. saksenae* and *A. implicatum*; an obviously large region between *nad6* and *rnl* in *P. chlamydosporia*; and a conspicuously large region between *rnl* and *nad2* in *H. rhossiliensis* (Fig. 9.1), findings that may suggest change and mobility of intergenic region sequences in mitogenomes during evolution. Furthermore, introns that even encode new hypothetical open reading frames (ORFs) are identified in *rnl*, *nad1*, *nad3-5*, *nad4L*, *cox1*, *cox3* and *cob* (Fig. 9.1; Table 9.2), indicating that various encoding-forms are shown with a preference for special genes in the mitogenomes and that there may be a mitochondrial gene response to functional requirements in the cell.

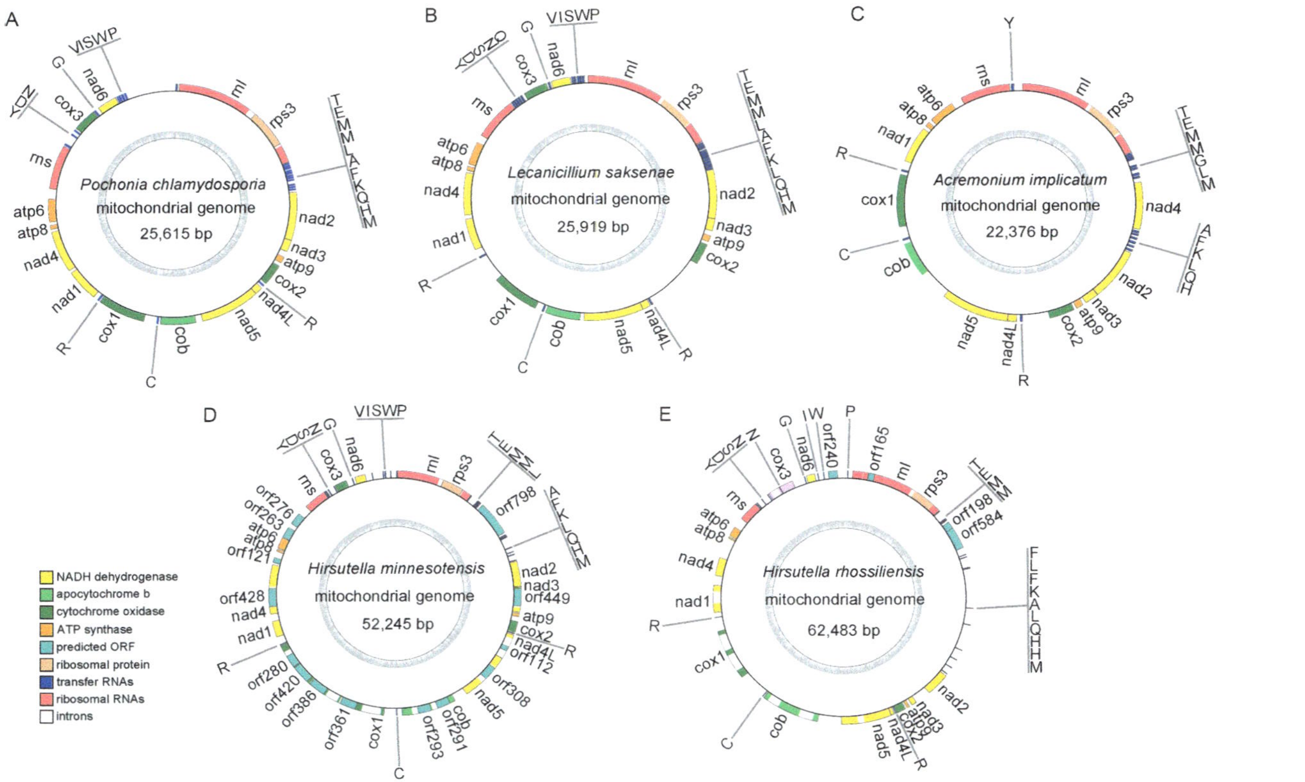

Fig. 9.1 Functional maps of the complete mitogenomes of five nematophagous fungi. **(A–E)**: display the maps of *Pochonia chlamydosporia*, *Lecanicillium saksenae*, *Acremonium implicatum*, *Hirsutella minnesotensis* and *Hirsutella rhossiliensis* mitogenome, respectively. The circles were originally generated using OrganellarGenomeDRAW (Lohse et al. 2013) and were further manually improved using Inkscape software (https://inkscape.org/)

Table 9.2 Gene and intergenic region sizes from the complete mitogenomes of five nematophagous fungi. Some genes share overlap sequences, and length of overlap sequences among coding regions is calculated once

Species	Length (bp)	CRs (bp)	tRNAs (bp)	rRNAs (bp)	Introns (bp)	Intergenic (bp)	Genes contain introns
Pochonia chlamydosporia	25,615	14,385 (56.16%)	1,697 (6.63%)	6,232 (24.33%)	335 (1.31%)	4,618 (18.03%)	*rnl*
Hirsutella minnesotensis	52,245	29,961 (57.35%)	1,863 (3.57%)	6,716 (12.85%)	5,272 (10.09%)	10,251 (19.62%)	*rnl*, *nad3*, *nad4L*, *nad5*, *cob*, *cox1*, *nad4*
Hirsutella rhossiliensis	62,483	24,106 (38.58%)	1,922 (3.08%)	9,024 (14.44%)	4,935 (7.90%)	25,119 (40.20%)	*rnl*, *nad5*, *cob*, *cox1*, *nad1*, *cox3*
Lecanicillium saksenae	25,919	14,334 (55.30%)	1,926 (7.43%)	6,034 (23.28%)	354 (1.37%)	4,941 (19.06%)	*rnl*
Acremonium implicatum	22,376	12,639 (56.48%)	1,255 (5.61%)	5,429 (24.26%)	561 (2.51%)	4,185 (18.70%)	*rnl*

bp base pairs, *CR*s coding regions

9.3 Mitogenome Size Variations for the Nematophagous Fungi in *Hypocreales*

Previous studies (Belcour et al. 1997; Hausner 2003; Formighieri et al. 2008) found that the lengths of introns and intergenic regions result in different sizes of mitogenomes. In *Sordariomycetes*, another factor of CRs (Coding Regions) contributing to mitogenome size variations was found (Lin et al. 2015). A comparison of mitogenome sizes between the nematophagous fungi in *Hypocreales* shows that intergenic regions and CRs are the two major contributors (Table 9.3). For example, the *H. rhossiliensis* mitogenome is 36,868 bp larger than the *P. chlamydosporia* mitogenome, and the size of the intergenic regions of the *H. rhossiliensis* mitogenome is 20,501 bp larger than that of *P. chlamydosporia* (Table 9.2), therefore the intergenic regions contribute 55.61% (20,501/36,868) of the total size variation. These results can be observed in the functional maps in Fig. 9.1.

9.4 Phylogenetic Relationships of Biological Control Agents and Plant Pathogens in *Hypocreales*

Several studies have reported research on the mitochondrial and nuclear genomes of nematophagous fungi, entomopathogenic fungi and plant pathogens in *Hypocreales* (Table 9.1). However, the phylogenetic placement of some species were shown to be different in the nuclear and mitochondrial data analyses, and also in mitochondrial data analyses using various methods, such as the opposite placement of *H. minnesotensis* in *Hypocreales* using nuclear and mitochondrial phylogenies (Lai et al. 2014; Zhang et al. 2015a), and the different positions of *P. chlamydosporia* and *Beauveria bassiana* in the maximum likelihood (ML) phylogeny and the neighbour-joining phylogeny (Lin et al. 2015; Wang et al. 2016a), thereby raising questions as to whether there are conflicts of phylogenetic relationships for fungi in *Hypocreales* between nuclear and mitochondrial phylogenies and whether there is currently a best model for constructing mitochondrial phylogeny.

To answer these questions, we chose 14 protein-coding genes of *nad1-6*, *nad4L*, *cox1-3*, *cob*, *atp6*, and *atp8-9* from 16 species containing nematophagous fungi, entomopathogenic fungi and plant pathogens and used them for phylogenetic analysis (Fig. 9.2). The amino acid sequences of each gene were aligned by performing MUSCLE v3.8.31 (Edgar 2004) analysis and they were concatenated using an in-house perl script. The alignments were manually confirmed to avoid mismatched positions and the gap regions were removed. Then the sequences (3,689 aa) were used for the ML phylogeny construction using Mega v6.06 with the cpREV + F + G model (Adachi et al. 2000; Tamura et al. 2013) which was identified as the best model from ProtTest v3.4 analysis (Darriba et al. 2011). The generated mitochondrial phylogeny was similar to the nuclear phylogeny (Wang et al. 2016b) based on 855 single-copy family genes with the Dayhoff model implemented in the TREE-

Table 9.3 The three primary contributors to size variations of mitogenomes of the nematophagous fungi *Acremonium implicatum* (= *Sarocladium implicatum*), *Pochonia chlamydosporia*, *Lecanicillium saksenae*, and *Hirsutella minnesotensis*

	Acremonium implicatum			*Pochonia chlamydosporia*			*Lecanicillium saksenae*			*H. minnesotensis*		
	1	2	3	1	2	3	1	2	3	1	2	3
H. rhossiliensis	IRs	CRs	introns	IRs	CRs	introns	IRs	CRs	introns	IRs	rRNAs	tRNAs
	52.20%	28.59%	10.91%	55.61%	26.37%	12.48%	55.19%	26.73%	12.53%	145.22%	22.54%	0.58%
H. minnesotensis	CRs	IRs	introns	CRs	IRs	introns	CRs	IRs	introns			
	57.99%	20.31%	15.77%	58.49%	21.15%	18.54%	59.36%	20.17%	18.68%			
L. saksenae	CRs	IRs	tRNAs	IRs	tRNAs	introns						
	47.84%	21.34%	18.94%	106.25%	75.33%	6.25%						
P. chlamydosporia	CRs	rRNAs	tRNAs									
	53.91%	24.79%	13.65%									

IRs intergenic regions, *CRs* coding regions

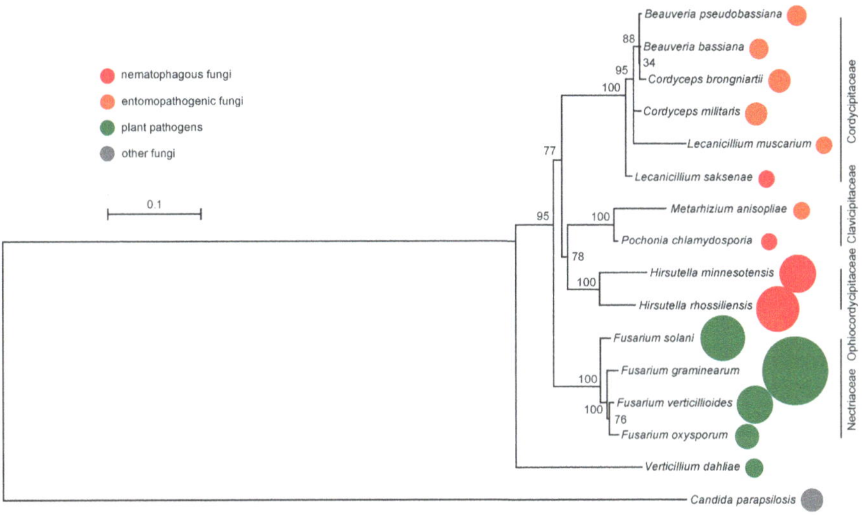

Fig. 9.2 Phylogenetic relationships between *Pochonia chlamydosporia* and other fungi in *Hypocreales*. The tree shows that *P. chlamydosporia* is most closely related to *Metarhizium anisopliae*, an entomopathogenic fungus. The mitogenome sizes are indicated by circles. The ML phylogeny was constructed by 14 typical proteins based on the cpREV + F + G model implemented in Mega with bootstrap value of 1000 (Adachi et al. 2000; Tamura et al. 2013). This model was shown to be the best model by performing a ProtTest analysis (Darriba et al. 2011). *Candida parapsilosis* was used as an outgroup

PUZZLE program (Schmidt et al. 2002), indicating the possibility of the same roles for mitochondrial and nuclear trees to investigate the phylogenetic relationships of fungi in *Hypocreales*. In the tree, four nematophagous fungi of two *Hirsutella* species, *P. chlamydosporia* and *L. saksenae* were clustered into three clades representing the three families *Ophiocordycipitaceae*, *Clavicipitaceae* and *Cordycipitaceae*, respectively (Fig. 9.2), without including analysis of *A. implicatum* (= *Sarocladium implicatum*) because of its loss of *cox3* and *nad6*. Both the nuclear and mitochondrial phylogenies confirm that *P. chlamydosporia* is more closely related to *Hirsutella* than to *L. saksenae* and *B. bassiana*, although the mitogenome size of *P. chlamydosporia* is similar to that of *L. saksenae*. In addition, the positions of species in *Cordycipitaceae* shown in the tree are slightly different from the phylogeny results based on 12 protein-coding genes (without *cox3* and *nad6*) (Xin et al. 2015), suggesting that the more genetic information that is used, the more correct the obtained tree is for demonstrating phylogenetic relationships. Moreover, the phylogeny confirms that the invertebrate-pathogenic fungi and the plant pathogens in *Hypocreales* form a monophyletic group sharing a most recent common ancestor, as found in a previous study (Lin et al. 2015).

During the construction of the phylogeny for fungi in the *Hypocreales*, another protein-coding gene, *rps3*, was not included due to the detection of positive selection signals in this gene (Lin et al. 2015). The *rps3* genes in *Hypocreales* fungi share high sequence similarity, but their phylogenetic relationships are different from those inferred from the other 14 protein-coding genes. For these *rps3* encoded in the group I introns within *rnl* genes, their nearly ubiquitous presence has been attributed to a vertical mode of inheritance rather than horizontal inheritance during evolution (Sethuraman et al. 2009). Another study (Ran et al. 2010) of mitochondrial *rps3* genes in gymnosperms also detected positive selection signals. The positively selected genes may contribute to further study to explore the evolutionary trajectory of the fungi in the *Hypocreales*.

9.5 Future Perspectives

Pochonia chlamydosporia has naturally occurring multitrophic lifestyles (saprophyte, parasite and endophyte) (Manzanilla-López et al. 2013; Larriba et al. 2014) and its reference nuclear and mitochondrial genomes have been reported (Larriba et al. 2014; Lin et al. 2015), thereby making it a valuable resource for investigating genetic variation and evolutionary processes. Mitochondria are of significant importance in basic cellular functions. The *P. chlamydosporia* mitogenome is small in size (25,615 bp), but its function and evolution remain poorly understood and many interesting questions are posed. For example: do any mitochondrial recombination events happen in *P. chlamydosporia*? At the population level, would any sequence variations related to specific phenotypes/biotypes be detected in mitogenomes and/or mitochondrial transcriptomes? In a fungal cell, one of the most obvious characters is that several copies of mitogenomes are present (Wilson and Xu 2012; Fritsch et al. 2015). The multiple copies of DNA and the biparental inheritance patterns may result in the phenomenon of recombination (Wilson and Xu 2012; Fritsch et al. 2015), a factor that could drive evolution. In *Saccharomyces cerevisiae* and *Fusarium fujikuroi* mitogenomes, evidence of recombination has been discovered (Dujon et al. 1974; Fourie et al. 2013; Fritsch et al. 2015), and should provide suggestions relevant to further studies on *Pochonia*. In previous genomic analyses (Larriba et al. 2014; Lin et al. 2015), the close phylogenetic relationships between *Pochonia* and *Metarhizium* has been determined. However, little is known about the origin and diversification of *Pochonia*. A study of genetic variation among *Pochonia* populations, using the PCR approach, by Manzanilla-López et al. (2009) discovered some relations between isolates and their hosts and/or geographical origins. The development of high throughput sequencing technologies and bioinformatics approaches provide the possibility of exploring functional roles of mitogenomes and their evolution at a population level. In the human genome, mitogenomes involved in regulation, transcription and posttranscriptional processes have been elucidated by analysing sequence variations of mitochondrial DNA and transcriptomes (Rackham et al. 2012; Hodgkinson et al. 2014); whilst in *Phytophthora infestans* populations, their genetic diversity and evolutionary history has been investigated using mitochondrial and nuclear genomic data (Yoshida et al. 2013). Future studies focusing on the *P. chlamydosporia* mitogenomes at the population level will doubtless shed light on their origin, diversification and functional roles.

Moreover, analysis of encoded genes (including protein-coding genes and tRNAs) in mitogenomes may represent diverse functions and complex evolutionary history (Gray 2012), such as research on functional features and evolutionary mode of fungal *rps3* genes (Sethuraman et al. 2009; Lin et al. 2015). Study of protein-protein interactions and phylogenetic findings will facilitate our understanding of mitochondrial proteins. The mitochondrial roles of cellular energy production through oxidative phosphorylation interplay with the interactions of proteins in both the nucleus and the mitochondria (Chandel 2014; Lotz et al. 2014; Fritsch et al. 2015). Several such interactions in human, mouse and *Drosophila* mitogenomes have been reported (Lotz et al. 2014), making possible the identification of mitochondrial proteome interactions in *P. chlamydosporia* and other nematophagous fungi using the interolog and domain-domain interaction methods (Matthews et al. 2001; Deng et al. 2002; Ng et al. 2003; Lei et al. 2014). If the mitochondrial interaction maps of nematophagous fungi are available, a comparison of the interactions between species may indicate the dynamics of these factors during evolution. Additionally, phylogenetic evidence of fungal mitochondrial horizontal gene transfers of introns in protein-coding genes (such as *cob*) have been reported (Wu et al. 2015; Zhang et al. 2015b), suggesting that mitogenomes are evolutionarily highly chimeric. Multiple introns in protein-coding genes were identified in both *Hirsutella* mitogenomes (Fig. 9.1; Table 9.2), implying the existence of some evolutionary patterns for these factors still awaiting exploration. For mitochondrial tRNAs in the tadpole shrimp (*Triops cancriformis*), their dynamics were revealed by analysis of tRNA-derived fragments (TRFs) in small RNA data (Hirose et al. 2015). For analysis of fungal small RNA data, abundant TRFs were also identified (Supplementary Table S1 in study by Lin et al. 2016), suggesting some clues as to future studies of functions and expression patterns, as well as the evolution of mitochondrial tRNAs in nematophagous fungi for adaptation to different environments.

References

Adachi, J., Waddell, P. J., Martin, W., et al. (2000). Plastid genome phylogeny and a model of amino acid substitution for proteins encoded by chloroplast DNA. *Journal of Molecular Evolution, 50*, 348–358.

Belcour, L., Rossignol, M., Koll, F., et al. (1997). Plasticity of the mitochondrial genome in *Podospora*. Polymorphism for 15 optional sequences: Group-I, group-II introns, intronic ORFs and an intergenic region. *Current Genetics, 31*, 308–317.

Chandel, N. S. (2014). Mitochondria as signaling organelles. *BMC Biology, 12*, 34.

Darriba, D., Taboada, G. L., Doallo, R., et al. (2011). ProtTest 3: fast selection of best-fit models of protein evolution. *Bioinformatics, 27*, 1164–1165.

Deng, M., Mehta, S., Sun, F., et al. (2002). Inferring domain-domain interactions from protein-protein interactions. *Genome Research, 12*, 1540–1548.

Dujon, B., Slonimski, P. P., & Weill, L. (1974). Mitochondrial genetics IX: A model for recombination and segregation of mitochondrial genomes in *Saccharomyces cerevisiae*. *Genetics, 78*, 415–437.

Edgar, R. C. (2004). MUSCLE: Multiple sequence alignment with high accuracy and high throughput. *Nucleic Acids Research, 32*, 1792–1797.

Formighieri, E. F., Tiburcio, R. A., Armas, E. D., et al. (2008). The mitochondrial genome of the phytopathogenic basidiomycete *Moniliophthora perniciosa* is 109 kb in size and contains a stable integrated plasmid. *Mycological Research, 112*, 1136–1152.

Fourie, G., van der Merwe, N. A., Wingfield, B. D., et al. (2013). Evidence for inter-specific recombination among the mitochondrial genomes of *Fusarium* species in the *Gibberella fujikuroi* complex. *BMC Genomics, 14*, 605.

Fritsch, E. S., Chabbert, C. D., Klaus, B., et al. (2015). A genome-wide map of mitochondrial DNA recombination in yeast. *Genetics, 198*, 755–771.

Gao, Q., Jin, K., Ying, S. H., et al. (2011). Genome sequencing and comparative transcriptomics of the model entomopathogenic fungi *Metarhizium anisopliae* and *M. acridum*. *PLoS Genetics, 7*(1), e1001264.

Ghikas, D. V., Kouvelis, V. N., & Typas, M. A. (2006). The complete mitochondrial genome of the entomopathogenic fungus *Metarhizium anisopliae* var. *anisopliae*: Gene order and trn gene clusters reveal a common evolutionary course for all Sordariomycetes, while intergenic regions show variation. *Archives of Microbiology, 185*, 393–401.

Ghikas, D. V., Kouvelis, V. N., & Typas, M. A. (2010). Phylogenetic and biogeographic implications inferred by mitochondrial intergenic region analyses and ITS1-5.8 S-ITS2 of the entomopathogenic fungi *Beauveria bassiana* and *B. brongniartii*. *BMC Microbiology, 10*, 174.

Gray, M. W. (2012). Mitochondrial evolution. *Cold Spring Harbor Perspectives in Biology, 4*(9).

Hausner, G. (2003). Fungal mitochondrial genomes, plasmids and introns. *Applied Mycology and Biotechnology, 3*, 101–131.

Hirose, Y., Ikeda, K. T., Noro, E., et al. (2015). Precise mapping and dynamics of tRNA-derived fragments (tRFs) in the development of *Triops cancriformis* (tadpole shrimp). *BMC Genetics, 16*, 83.

Hodgkinson, A., Idaghdour, Y., Gbeha, E., et al. (2014). High-resolution genomic analysis of human mitochondrial RNA sequence variation. *Science, 344*(6182), 413–415.

Kouvelis, V. N., Ghikas, D. V., & Typas, M. A. (2004). The analysis of the complete mitochondrial genome of *Lecanicillium muscarium* (synonym *Verticillium lecanii*) suggests a minimum common gene organization in mtDNAs of Sordariomycetes: Phylogenetic implications. *Fungal Genetics and Biology, 41*, 930–940.

Lai, Y., Liu, K., Zhang, X., et al. (2014). Comparative genomics and transcriptomics analyses reveal divergent lifestyle features of nematode endoparasitic fungus *Hirsutella minnesotensis*. *Genome Biology and Evolution, 6*, 3077–3093.

Larriba, E., Jaime, M. D., Carbonell-Caballero, J., et al. (2014). Sequencing and functional analysis of the genome of a nematode egg-parasitic fungus, *Pochonia chlamydosporia*. *Fungal Genetics and Biology, 65*, 69–80.

Lei, D., Lin, R., Yin, C., et al. (2014). Global protein–protein interaction network of rice sheath blight pathogen. *Journal of Proteome Research, 13*, 3277–3293.

Lin, R., Liu, C., Shen, B., et al. (2015). Analysis of the complete mitochondrial genome of *Pochonia chlamydosporia* suggests a close relationship to the invertebrate-pathogenic fungi in Hypocreales. *BMC Microbiology, 15*, 5.

Lin, R., He, L., He, J., et al. (2016). Comprehensive analysis of microRNA-Seq and target mRNAs of rice sheath blight pathogen provides new insights into pathogenic regulatory mechanisms. *DNA Research*. doi:10.1093/dnares/dsw024.

Lohse, M., Drechsel, O., Kahlau, S., et al. (2013). OrganellarGenomeDRAW – A suite of tools for generating physical maps of plastid and mitochondrial genomes and visualizing expression data sets. *Nucleic Acids Research, 41*(Web Server issue), W575–W581.

Lotz, C., Lin, A. J., Black, C. M., et al. (2014). Characterization, design, and function of the mitochondrial proteome: From organs to organisms. *Journal of Proteome Research, 13*, 433–446.

Manzanilla-López, R. H., Atkins, S. D., Clark, I. M., et al. (2009). Measuring abundance, diversity and parasitic ability in two populations of the nematophagous fungus *Pochonia chlamydosporia* var. *chlamydosporia*. *Biocontrol Science and Technology, 19*, 391–406.

Manzanilla-López, R. H., Esteves, I., Finetti-Sialer, M. M., et al. (2013). *Pochonia chlamydosporia*: Advances and challenges to improve its performance as a biological control agent of sedentary endo-parasitic nematodes. *Journal of Nematology, 45*, 1–7.

Matthews, L. R., Vaglio, P., Reboul, J., et al. (2001). Identification of potential interaction networks using sequence-based searches for conserved protein-protein interactions or "interologs". *Genome Research, 11*, 2120–2126.

Ng, S. K., Zhang, Z., & Tan, S. H. (2003). Integrative approach for computationally inferring protein domain interactions. *Bioinformatics, 19*, 923–929.

Oh, J., Kong, W. S., & Sung, G. H. (2015). Complete mitochondrial genome of the entomopathogenic fungus *Beauveria pseudobassiana* (Ascomycota, Cordycipitaceae). *Mitochondrial DNA, 26*, 777–778.

Rackham, O., Mercer, T. R., & Filipovska, A. (2012). The human mitochondrial transcriptome and the RNA-binding proteins that regulate its expression. *RNA, 3*, 675–695.

Ran, J. H., Gao, H., & Wang, X. Q. (2010). Fast evolution of the retroprocessed mitochondrial *rps3* gene in conifer II and further evidence for the phylogeny of gymnosperms. *Molecular Phylogenetics and Evolution, 54*, 136–149.

Schmidt, H. A., Strimmer, K., Vingron, M., et al. (2002). TREE-PUZZLE: Maximum likelihood phylogenetic analysis using quartets and parallel computing. *Bioinformatics, 18*, 502–504.

Sethuraman, J., Majer, A., Iranpour, M., et al. (2009). Molecular evolution of the mtDNA encoded *rps3* gene among filamentous ascomycetes fungi with an emphasis on the ophiostomatoid fungi. *Journal of Molecular Evolution, 69*, 372–385.

Shang, Y., Xiao, G., Zheng, P., et al. (2016). Divergent and convergent evolution of fungal pathogenicity. *Genome Biology and Evolution, 8*, 1374–1387.

Sung, G. H. (2015). Complete mitochondrial DNA genome of the medicinal mushroom *Cordyceps militaris* (Ascomycota, Cordycipitaceae). *Mitochondrial DNA, 26*, 789–790.

Tamura, K., Stecher, G., Peterson, D., et al. (2013). MEGA6: Molecular evolutionary genetics analysis version 6.0. *Molecular Biology and Evolution, 30*, 2725–2729.

Wang, N., Zhang, Y., Hussain, M., et al. (2016a). The mitochondrial genome of the nematode endoparasitic fungus *Hirsutella rhossiliensis*. *Mitochondrial DNA Part B*. doi:10.1080/23802359.2016.1143336.

Wang, G., Liu, Z., Lin, R., et al. (2016b). Biosynthesis of antibiotic leucinostatins in bio-control fungus *Purpureocillium lilacinum* and their inhibition on *Phytophthora* revealed by genome mining. *PLoS Pathogens, 12*(7), e1005685.

Wilson, A. J., & Xu, J. (2012). Mitochondrial inheritance: Diverse patterns and mechanisms with an emphasis on fungi. *Mycology, 3*, 158–166.

Wu, B., Buljic, A., & Hao, W. (2015). Extensive horizontal transfer and homologous recombination generate highly chimeric mitochondrial genomes in yeast. *Molecular Biology and Evolution, 32*, 2559–2570.

Xiao, G., Ying, S. H., Zheng, P., et al. (2012). Genomic perspectives on the evolution of fungal entomopathogenicity in *Beauveria bassiana*. *Scientific Reports, 2*, 483.

Xin, B., Lin, R., Shen, B., et al. (2015). The complete mitochondrial genome of the nematophagous fungus *Lecanicillium saksenae*. *Mitochondrial DNA*. doi:10.3109/19401736.2015.1110794.

Xu, J., Huang, B., Qin, C., et al. (2009). Sequence and phylogenetic analysis of *Beauveria bassiana* with mitochondrial genome. *Mycosystema, 28*, 718–723.

Yao, Y., Lin, R., Tian, X., et al. (2016). The complete mitochondrial genome of the nematophagous fungus *Acremonium implicatum*. *Mitochondrial DNA Part A, 27*, 3246–3247.

Yoshida, K., Schuenemann, V. J., Cano, L. M., et al. (2013). The rise and fall of the *Phytophthora infestans* lineage that triggered the Irish potato famine. *eLife, 2*, e00731.

Zhang, Y. J., Zhang, S., & Liu, X. Z. (2015a). The complete mitochondrial genome of the nematode endoparasitic fungus *Hirsutella minnesotensis*. *Mitochondrial DNA*. doi:10.3109/19401736.2015.1046126.

Zhang, Y. J., Zhang, S., Zhang, G., et al. (2015b). Comparison of mitochondrial genomes provides insights into intron 4 dynamics and evolution in the caterpillar fungus *Cordyceps militaris*. *Fungal Genetics and Biology, 77*, 95–107.

Zheng, P., Xia, Y., Xiao, G., et al. (2011). Genome sequence of the insect pathogenic fungus *Cordyceps militaris*, a valued traditional Chinese medicine. *Genome Biology, 12*(11), R116.

Chapter 10
Pochonia chlamydosporia: Multitrophic Lifestyles Explained by a Versatile Genome

Federico Lopez-Moya, Nuria Escudero, and Luis V. Lopez-Llorca

Abstract The nematophagous fungus *Pochonia chlamydosporia* (Goddard) Zare & Gams is a facultative parasite of nematode females and eggs. This fungus has a world-wide distribution and has the capacity to survive in the absence of the nematodes. *Pochonia chlamydosporia* is rhizosphere competent and can colonize the rhizosphere of crops of economic importance, such as tomato and barley. The infection of nematode eggs by *P. chlamydosporia* requires adhesion, differentiation of appressoria, and penetration of their egg shells. Since the 1980s, proteases have been described as the main group of enzymes that the fungus uses to penetrate nematode eggs. Recent studies found that these genes are also expressed when the fungus endophytically colonizes barley roots. The genome of the fungus was sequenced in 2014, this being the first publicly available genome of a fungus parasite of nematode eggs. Since then, many publications based on the fungus genome have been published. Study of the major gene families in the genome shows the tools that the fungus uses in its different lifestyles. Transcriptomic studies using DNA microarrays and RNA-seq have been done to study the molecular mechanisms that regulate endophytic colonization of *Pochonia chlamydosporia*. A general analysis of the genome allowed us to identify a large number of carbohydrate active enzymes (Cazy) which help the fungus in colonizing plants and infecting nematodes.

F. Lopez-Moya (✉) • L.V. Lopez-Llorca
Laboratory of Plant Pathology, Multidisciplinary Institute for Environment Studies (MIES) Ramón Margalef, Department of Marine Sciences and Applied Biology, University of Alicante, Alicante, Spain
e-mail: federico.lopez@ua.es; luislopezllorca1@gmail.com

N. Escudero
Laboratory of Plant Pathology, Multidisciplinary Institute for Environment Studies (MIES) Ramón Margalef, Department of Marine Sciences and Applied Biology, University of Alicante, Alicante, Spain

Departament d'Enginyeria Agroalimentària i Biotecnologia, Universitat Politècnica de Catalunya, Barcelona, Spain
e-mail: nuria.escudero@ua.es

R.H. Manzanilla-López, L.V. Lopez-Llorca (eds.), *Perspectives in Sustainable Nematode Management Through* Pochonia chlamydosporia *Applications for Root and Rhizosphere Health*, Sustainability in Plant and Crop Protection,
DOI 10.1007/978-3-319-59224-4_10

10.1 New Generation Sequencing Applied to Filamentous Fungi

Since the sequencing of the human genome in 2001 (Lander et al. 2001; Venter et al. 2001) genomics has achieved an impressive impact in science. As of 2016, more than 80,000 prokaryotic genomes and ca 4000 eukaryotic genomes have been sequenced, indicating the last decade as the Genomics Era. The first fully sequenced genome of an eukaryotic organism was that of the budding yeast *Saccharomyces cerevisiae* (Goffeau et al. 1996). Seven years later the genome of the filamentous fungus *Neurospora crassa* was completely sequenced (Galagan et al. 2003), thereby starting the massive sequencing of fungal genomes. With the development of next generation sequencing tools, more than 634 genomes from 388 species of fungi have now been sequenced (Kersey et al. 2015), among them fully sequenced genomes of nematophagous fungi.

Nematophagous fungi are natural antagonisms of plant-parasitic nematodes (PPN) of economic importance worldwide. They are responsible for cases of natural soil suppression to these pathogens in agroecosystems. The diverse biology of nematophagous fungi and the lack of molecular tools have delayed progress in understanding the molecular basis of their mode of action. The study of the genomics of this group of fungi is a key new tool for the identification of the molecular determinants of pathogenicity to PPN. This new knowledge will help development and improvement of nematophagous fungi as an alternative to fumigants and toxic chemical nematicides for the safe management of PPN.

10.2 Comparative Genomics of Nematophagous Fungi

Nematode-trapping fungi have evolved sophisticated hyphal structures such as nets, knobs, branches or rings, in which nematodes are captured mechanically or by adhesion (Nordbring-Hertz et al. 2006). The first nematophagous fungus genome to be sequenced was that of *Arthrobotrys oligospora* (Yang et al. 2011). In the latter work the genome sequence was compared with that of pathogenic and non-pathogenic fungi. The *A. oligospora* genome differs from those of *Eurotiomycetes* (*Aspergillus* spp.) or *Sordariomycetes* (*Fusarium*, *Magnaporthe* or *Neurospora* spp.), indicating that in this genome several cluster of genes that codify for proteins allow *A. oligospora* to develop its specialised strategy to catch nematodes. The *A. oligospora* genome encodes for several hydrolytic enzyme families, such as subtilisin, cellulase, cellobiohydrolase, and pectinesterases. In a subsequent study, the *Monacrosporium haptotylum* (= *Dactylellina haptotyla*) genome (Meerupati et al. 2013) was sequenced. They found a large number of orthologues differentially expressed between *M. haptotylum*, a knob-forming fungus, and *A. oligospora*, a net-forming fungus thus indicating that each fungus presents a specific strategy to catch PPN. The *M. haptotylum* genome encodes more small secreted proteins than that of *A. oligospora*. This would indicate that the capability of this fungus to degrade external barriers of nematodes is more specialised than that of *A. oligospora.*

Genome sequencing of the nematode predatory fungus *Drechslerella stenobrocha* (Liu et al. 2014) shows high similarities with *A. oligospora*. Enzymatic analysis of the *D. stenobrocha* genome shows enrichment of genes encoding degrading enzymes, mainly those for chitin, cellulose and protein. The large numbers of genes encoding carbohydrate degrading enzymes found in their genomes suggest that these fungi have capabilities for establishing within and utilizing plant roots as a natural niche in order to develop a network for trapping nematodes. Proteases and chitinases play a determinant role for degrading nematode juveniles which are the main target of this fungus. An exhaustive analysis of the *D. stenobrocha* genome demonstrates a rare occurrence of transposons and Repeat-Induced Point mutations (RIPs). This could indicate the slow evolution and primitive state of the *D. stenobrocha* genome. This analysis, using genomic data, suggests the trapping of nematodes to be a primitive strategy in comparison to that of nematophagous fungi (e.g., *P. chlamydosporia*) which parasitize female nematodes and eggs within the roots.

Endoparasitic fungi are nematophagous fungi that form spores which either adhere to the nematode cuticle or are swallowed by the host. Most are obligate parasites of nematodes and live in a close relationship with their target host (Lopez-Llorca et al. 2008). Genomes of the most important species of nematode fungal endoparasites, *Drechmeria coniospora* (Lebrigand et al. 2016) and *Hirsutella rhossiliensis*, have recently been sequenced (Lai et al. 2014).

In comparison to nematode-trapping fungi, endoparasitic nematophagous fungi show a high amount of transposable elements in their genomes together with an expansion of RIP mutation elements. These characteristics indicate high genomic evolution in this group of fungi. This may suggest that nematode-trapping fungi have strategies for catching nematodes which are more primitive than those of nematode endoparasitic fungi. The *H. rhossiliensis* genome reveals many transposable elements together with an expansion of secondary metabolite genes associated with host adaptation. However, this fungus has lost genes involved in cellulose and hemicellulose degradation, suggesting that this species does not have a close interaction with plants. Another example of this group of fungi is *Drechmeria coniospora*, which has been proposed as a model to study nematode immunity (Rouger et al. 2014). The *D. coniospora* genome reveals that this fungus shares, with other nematophagous fungi, genes encoding important putative pathogenicity factors such as subtilisin-like serine proteases and gene clusters for biosynthesis of secondary metabolites. However, *D. coniospora* also possess specific codifying enzymes not found in other nematophagous fungi that allow this fungus to infect nematodes as an endoparasite (Lebrigand et al. 2016).

Genomes of the fungal parasites of nematode eggs and sessile females, such as *Purpureocillium lilacinum* (= *Paecilomyces lilacinus*) (Prasad et al. 2015) and *Pochonia chlamydosporia* (Larriba et al. 2014), have also been sequenced recently and the genome of *P. chlamydosporia* will be discussed in this chapter. Both fungi display a multitrophic lifestyle and besides their nematophagous life they are soil saprotrophs, and endophytically colonize plant roots (Bordallo et al. 2002; Maciá-Vicente et al. 2009; Manzanilla-López et al. 2011; Castillo Lopez et al. 2014). *Purpureocillium lilacinum* genome presents a large number of genes encoding diverse hydrolytic enzymes such as proteases, glycoside hydrolases and carbohydrate esterases involved in the pathogenicity of the fungus (Prasad et al. 2015). This genome also has expanded genes involved in secondary metabolism. *Drechmeria coniospora* and *P. lilacinum* genome sequencing has also allowed their phylogenetic classification in the *Phiocordycipitaceae* (Prasad et al. 2015; Lebrigand et al. 2016).

10.3 Impact of Genomics in *Pochonia chlamydosporia* Research

Before the development of the New Generation Sequencing platforms (NGS), several other tools (e.g. proteomics, cloning and traditional sequencing) were applied to understand the molecular basis of pathogenicity of *P. chlamydosporia* to PPN (Lopez-Llorca 1990; Larriba et al. 2012). For molecular studies on the role of *P. chlamydosporia* proteases, see Chap. 6 of this book.

Recent studies have applied genomics to analyse the potential role of chitin deacetylases and chitosanases in the infection of root-knot nematodes eggs by *P. chlamydosporia* (Aranda-Martínez et al. 2016). A key factor for testing the performance of biocontrol agents such as *P. chlamydosporia* is the quantification of their inocula levels in the environment (soil/rhizosphere). Initially, this was performed using culture techniques (Lopez-Llorca and Duncan 1986; Kerry and Bourne 2002). However, this overestimates fungus inoculum, especially in the endophytic stage (Escudero and Lopez-Llorca 2012). New molecular biology tools have been developed to quantify the presence of this fungus in the environment. DNA-based quantitative PCR (qPCR) methods using species-specific primers can improve the estimation of fungal abundance of *P. chlamydosporia* var. *chlamydosporia* in soil and during plant root colonization (Mauchline et al. 2002; Atkins et al. 2009; Escudero and Lopez-Llorca 2012). These molecular approaches, together with the application of traditional culture methods, provide complementary information on fungal abundance in soil and during plant colonisation. This can facilitate precision agriculture techniques to deliver *P. chlamydosporia* inoculum at the most effective time for more efficient PPN management. Recently, the *P. chlamydosporia* genome has been sequenced and functionally analyzed (Larriba et al. 2014). This ongoing project has opened new opportunities to study the multitrophic lifestyles of *P. chlamydosporia* (Lin et al. 2015; Larriba et al. 2015; Zavala-Gonzalez et al. 2015).

A key factor in the study of the fungal pathogens is the origin and evolution of pathogenicity. In the seminal book of Barron (1977), *The Nematode-destroying Fungi*, the author had already hypothesized that the nematophagous habit in fungi probably evolved independently several times. Our sequencing of *P. chlamydosporia* genome (see next section) supports this view.

By phylogenetically analyzing the global comparisons of deduced proteomes from nematophagous fungi sequencing (Zhang et al. 2016), we can draw the following conclusions:

- *Pochonia chlamydosporia* is extremely close (same clade) to *Metarhizium* spp. entomopathogenic fungi. This may indicate that parasitism in both groups of organisms evolved together.
- However, other entomopathogens (*Beauveria bassiana*) are more distantly related to *P. chlamydosporia.*
- *Pochonia chlamydosporia* is also phylogenetically close to *Clavicipitaceae* which are endophytic on grasses such as *Epichloë* spp. and similar species. These fungi are known to be active producers of secondary metabolites.
- Endoparasitic nematophagous fungi producers of adhesive spores to infect nematodes (*D. coniospora* and *Hirsutella* spp.) are also close relatives of *P. chlamydosporia.*
- Finally, predatory nematophagous fungi are far distant from the nematode egg parasite *P. chlamydosporia.* This is for both 3D adhesive trap (*A. oligospora*) and adhesive hyphal branch (*Monacrosporium haptotylum* [*Dactylellina haptotyla*]) formers.

10.4 *Pochonia chlamydosporia* Genome Sequencing

The fungus sequenced isolate was *P. chlamydosporia* var. *chlamydosporia* (Larriba et al. 2014) isolate 123 (ATCC MYA-4875). This fungus was isolated from field infected eggs of the cereal cyst nematode (*Heterodera avenae*) in a wheat field from south-west Spain (Seville). The genome of this fungus was sequenced using the Illumina Technology by generating: pair-end (200 bp) and mate-pair (3 kb) libraries to obtain deep coverage sequencing. After sequencing, the depth of sequencing was around 136-fold coverage. The raw data was analyzed with the software Augustus. The whole assembly (N50, 225 kb) comprised 901 scaffolds and 57 additional contigs >3 kb (not assembled into scaffolds) which, after discarding ambiguous bases, yielded a final genome size of 41.2 Mb. This analysis revealed a G + C content of about 50% with 12,122 protein-coding genes of which 672 were unique proteins. Gene prediction was determined using Maker2 (Holt and Yandell 2011) and was also completed with a reannotation using Blast2Go (Conesa et al. 2005). After annotation, we could identify similarities with the entomopathogenic fungi *Metarhizium anisopliae* and *M. acridium.* These results were also confirmed after sequencing the *P. chlamydosporia* mitochondrial genome (Lin et al. 2015; Chap. 9). Using a phylogenetic analysis, based on the orthologous genes found in the genome of more than 25 fungi, these authors confirmed that the genome of *P. chlamydosporia* is close to that of *Metarhizium* spp. (entomopathogenic fungi) and far distant from the nematode-trapping fungus *A. oligospora* (Larriba et al. 2014). The *P. chlamydosporia* genome shows genes adapted to infect hosts with external barriers based on chitin/protein, such as nematode egg shells. This can explain the close similarity of *P. chlamydosporia* genome with that of the entomopathogenic *Metarhizium* spp. fungi which have to penetrate chitin enriched insect cuticles (Gao et al. 2011). Enzymes essential to penetrate nematode egg shells (proteases, chitinases or chitosanases) are highly enriched in the *P. chlamydosporia* genome (Aranda-Martínez et al. 2016; Escudero et al. 2016). These enzymes also appear to be widely expanded in insect parasitic fungi (St Leger et al. 1993; Pinto et al. 1997; Moraes et al. 2003).

After genome sequencing, our research group developed a study to determine the main gene functional annotation in the *P. chlamydosporia* genome. Using different tools such as Blast2Go (Conesa et al. 2005) a *de novo* prediction was done. In addition, and to improve the annotation, gene families were established using the InterPro (http://www.ebi.ac.uk/interpro) and Kyoto Encyclopedia of Genes and Genomes (KEGG) databases (http://www.genome.jp/kegg/pathway.html). Other biosequence analysis databases based upon the Hiden Markov Model, such as HMMR (Finn et al. 2011) including the protein family Pfam, TIGRFAM, Gene 3D and Superfamily databases, were also used. This analysis revealed 7887 protein-coding annotated genes, representing more than 60% of the *P. chlamydosporia* genes predicted in the genome. With the aim of standardizing the criteria determining functional annotation we used Gene Ontology (GO). This analysis identified more than 20% of the genes with associated GO terms. These were identified with primary metabolism in the Biological Processes category. Within this GO category highly represented gene functions, such as the cellular metabolic process or macromolecules metabolic process, were found. This analysis revealed that *P. chlamydosporia* metabolism is highly represented in its genome. This may facilitate adaptation of this fungus to a wide range of environmental conditions. The functional analysis based on Molecular Functions GO category also revealed that genes related to hydrolase activity, nucleic acid and protein binding have a large presence in the genome. *Pochonia chlamydosporia* genome re-sequencing of different strains opens new possibilities for studying the multitrophic lifestyle of *P. chlamydosporia.* Crop plants, PPN and environmental conditions could thus be analyzed.

10.5 Role of Hydrolytic Enzymes from the *Pochonia chlamydosporia* Genome

Pochonia chlamydosporia genome displays expanded protein families of hydrolytic enzymes, especially proteases and glycosidases, which provide further evidence to support the multitrophic lifestyles (saprotroph, endophyte, nematode pathogen) of this versatile biocontrol agent (Fig. 10.1; Larriba et al. 2014). *Pochonia chlamydosporia* has also expanded genes encoding membrane transporters of various types, e.g., oligopeptides, sugars, ABC (ATP-binding cassette transporter) or MFS (Major Facilitator Superfamily). Those proteins are mainly involved in substrate assimilation and detoxification. Also remarkable is the expansion in the number of genes

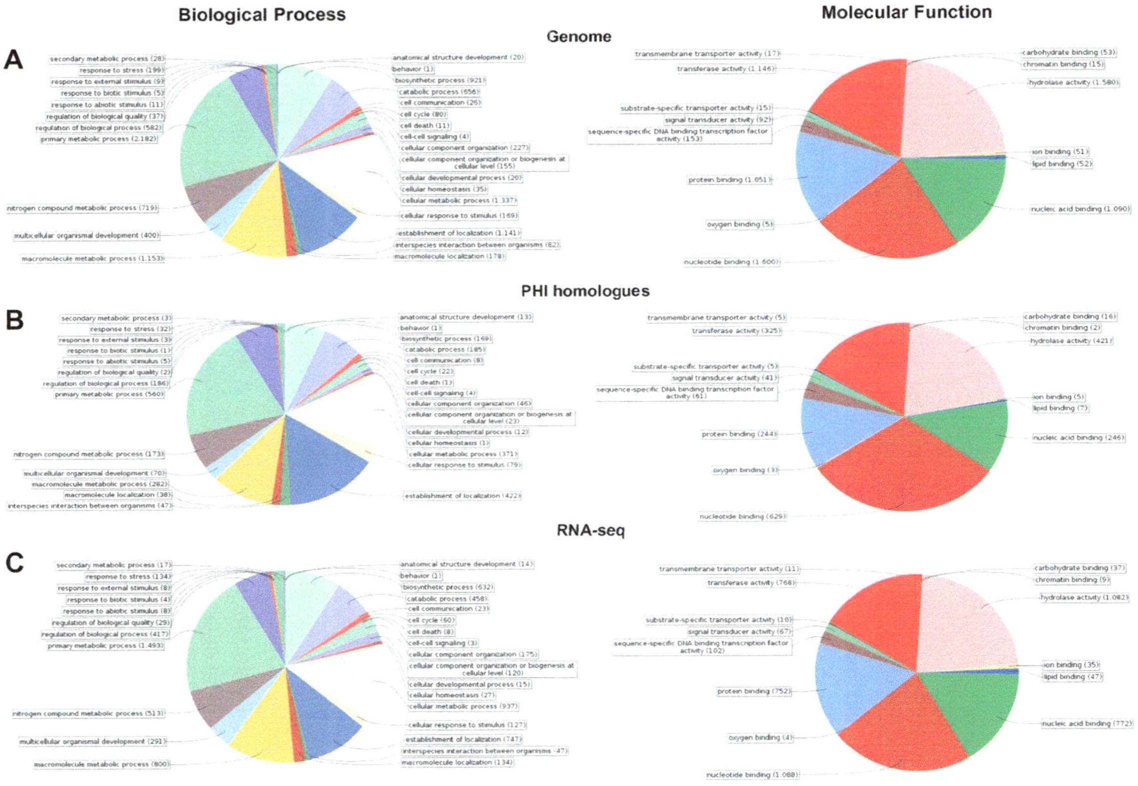

Fig. 10.1 Gene Ontology (GO) functional annotation of proteins encoded by genes in the *Pochonia chlamydosporia* genome. Gene Ontology charts were generated using generic Slim terms at Level 3 within the GO Biological Process (*left*) and Molecular Function (*right*) domains. (**A**): Protein-coding genes predicted in the *P. chlamydosporia* genome; (**B**): *P. chlamydosporia* genes with homologous counterparts in the PHI database; (**C**): Genes expressed during endophytic colonization of barley roots by *P. chlamydosporia* detected by RNA-Seq (Figure obtained from original publication Larriba et al. 2014)

that codify for transcriptions factors, mainly C2H2 zinc finger transcription factors and Zn(2)-C6 fungal-type transcription factors (Larriba et al. 2014). A large number of proteins are related with signal transduction, mainly protein kinases found in *P. chlamydosporia* genome, which probably regulates the mechanisms of pathogenicity such as appressorium differentiation and plant cell invasion. This is important in plant pathogenic fungi such as *Magnaporthe oryzae* [= *Pyricularia oryzae*] (Zhao et al. 2005; Wilson and Talbot 2009).

Pochonia chlamydosporia has an impressive expansion in the number of glycoside hydrolases in its genome in comparison with those of other nematophagous, entomopathogenic or mycoparasitic fungi. This fungus shows 292 glycoside hydrolases, more than *M. anisopliae* (192), *A. oligospora* (202), or even *Trichoderma atroviride* (255). Focusing on plant cell wall degrading enzymes that allow fungi to colonize plants, *P. chlamydosporia* presents fewer (6) cellulose degrading enzymes than *Trichoderma* spp. (7-9), and especially *A. oligospora* (31). The same occur with other glycoside hydrolases, such as xylanases and pectinases. Pectinases loosen the cell wall in plant tissues. The striking difference in encoded pectin-target enzymes between *P. chlamydosporia* (3) and *A. oligospora* (25) (Aranda-Martínez et al. 2016) can explain why the trapping fungus, unlike *P. chlamydosporia*, causes necrosis and decortication of barley roots (Bordallo et al. 2002). *Pochonia chlamydosporia* infects nematode females and eggs embedded in root tissues indicating its capability as a plant root mutualist. *Pochonia chlamydosporia* shows a highly expanded family of hydrolases related with chitin modification. This fungus encodes chitin deacetylases (CE4) and a large number of chitinases (GH18) and chitosanases (GH75). Genome comparative analyses demonstrate that this fungus has a higher number (77) of chitin modifying enzymes than entomopathogenic and mycoparasitic fungi (Aranda-Martínez et al. 2016). These enzymes could be involved in the ability of the fungus to degrade chitin and chitosan during nematode egg infection. A recent study (Aranda-Martínez et al. 2016) demonstrates that chitosanases and chitin deacetylases play a role during egg infection. They probably interact with the chitin layer in the egg shell allowing its degradation and facilitating penetration by the fungus. This study complements previous studies in which *P. chlamydosporia* proteases, together with chitinases, were found to be involved in egg infection.

Production of chitosan on the surface of the nematode egg shell could induce expression of protein degrading enzymes. Proteomics studies revealed that chitosan can induce expression of VCP1 serine protease, a putative pathogenicity factor of *P. chlamydosporia* to infect nematode eggs (Palma-Guerrero et al. 2010). Chitosan also induced expression of VCP1 and SCP1 (a serine carboxypeptidase) in appressoria of *P. chlamydosporia* infecting root-knot nematode eggs, enhancing infection (Escudero et al. 2016).

10.6 Concluding Remarks

Pochonia chlamydosporia genomics has been shown to have a very useful role in understanding the multitrophic biology of the fungus. It has clarified the evolutionary relationships of the fungus and functional analysis of the *P. chlamydosporia* genome has shown new aspects of pathogenicity, such as the role of chitosan in this process. In future, this tool will help in the development of *P. chlamydosporia* in agrobiotechnology and in other aspects, such as the industrial application of the fungus using synthetic biology approaches.

References

Aranda-Martínez, A., Lenfant, N., Escudero, N., et al. (2016). CAZyme content of *Pochonia chlamydosporia* reflects that chitin and chitosan modification are involved in nematode parasitism. *Environmental Microbiology, 18*, 4200–4215.

Atkins, S. D., Peteira, B., Clark, I. M., et al. (2009). Use of real time quantitative PCR to investigate root and gall colonisation by coinoculated isolates of the nematophagous fungus *Pochonia chlamydosporia. The Annals of Applied Biology, 155*, 143–152.

Barron, G. (1977). *The nematode-destroying fungi. Topics in mycobiology No. 1*. Guelph: Canadian Biological Publications.

Bordallo, J. J., Lopez-Llorca, L. V., Jansson, H. B., et al. (2002). Colonization of plant roots by egg-parasitic and nematode-trapping fungi. *The New Phytologist, 154*, 491–499.

Castillo Lopez, D., Zhu-Salzman, K., Ek-Ramos, M. J., et al. (2014). The entomopathogenic fungal endophytes *Purpureocillium lilacinum* (formerly *Paecilomyces lilacinus*) and *Beauveria bassiana* negatively affect cotton aphid reproduction under both greenhouse and field conditions. *PloS One, 9*(8), e103891. doi:10.1371/journal.pone.0103891.

Conesa, A., Götz, S., García-Gómez, J. M., et al. (2005). Blast2GO: A universal tool for annotation, visualization and analysis in functional genomics research. *Bioinformatics, 21*, 3674–3676.

Escudero, N., & Lopez-Llorca, L. V. (2012). Effects on plant growth and root-knot nematode infection of an endophytic GFP transformant of the nematophagous fungus *Pochonia chlamydosporia. Symbiosis, 57*, 33–42.

Escudero, N., Ferreira, S. R., Lopez-Moya, F., et al. (2016). Chitosan enhances parasitism of *Meloidogyne javanica* eggs by the nematophagous fungus *Pochonia chlamydosporia. Fungal Biology, 120*, 572–585.

Finn, R. D., Clements, J., & Eddy, S. R. (2011). HMMER web server: Interactive sequence similarity searching. *Nucleic Acids Research, 39*, 29–37.

Galagan, J. E., Calvo, S. E., Borkovich, K. A., et al. (2003). The genome sequence of the filamentous fungus *Neurospora crassa. Nature, 422*, 859–868.

Gao, Q., Jin, K., Ying, S. H., et al. (2011). Genome sequencing and comparative transcriptomics of the model entomopathogenic fungi *Metarhizium anisopliae* and *M. acridum. PLoS Genetics, 7*(1), e1001264.

Goffeau, A., Barrell, B. G., Bussey, H., et al. (1996). Life with 6000 genes. *Science, 274*, 546–567.

Holt, C., & Yandell, M. (2011). MAKER2: An annotation pipeline and genome-database management tool for second-generation genome projects. *BMC Bioinformatics, 12*, 491.

Kerry, B. R., & Bourne, J. M. (2002). *A manual for research on Verticillium chlamdosporium, as potential biological control agent for root-knot nematodes*. Ghent: IOBC/WPRS.

Kersey, P. J., Allen, J. E., Armean, I., et al. (2015). Ensembl genomes 2016: More genomes, more complexity. *Nucleic Acids Research*. doi:10.1093/nar/gkv1209.

Lai, Y., Liu, K., Zhang, X., et al. (2014). Comparative genomics and transcriptomics analyses reveal divergent lifestyle features of nematode endoparasitic fungus *Hirsutella minnesotensis*. *Genome Biology and Evolution, 6*, 3077–3093.

Lander, E. S., Linton, L. M., Birren, B., et al. (2001). Initial sequencing and analysis of the human genome. *Nature, 409*, 860–921.

Larriba, E., Martín-Nieto, J., & Lopez-Llorca, L. V. (2012). Gene cloning, molecular modeling, and phylogenetics of serine protease P32 and serine carboxypeptidase SCP1 from nematophagous fungi *Pochonia rubescens* and *Pochonia chlamydosporia*. *Canadian Journal of Microbiology, 58*, 815–827.

Larriba, E., Jaime, M. D. L. A., Carbonell-Caballero, J., et al. (2014). Sequencing and functional analysis of the genome of a nematode egg-parasitic fungus, *Pochonia chlamydosporia*. *Fungal Genetics and Biology, 65*(C), 69–80.

Larriba, E., Jaime, M. D., Nislow, C., et al. (2015). Endophytic colonization of barley (*Hordeum vulgare*) roots by the nematophagous fungus *Pochonia chlamydosporia* reveals plant growth promotion and a general defense and stress transcriptomic response. *Journal of Plant Research, 128*, 665–678.

Lebrigand, K., He, LD., Thakur, N. (2016). Comparative genomic analysis of *Drechmeria coniospora* reveals core and specific genetic requirements for fungal endoparasitism of nematodes. *PLoS Genetics*. doi:10.1371/journal.pgen.1006017.

Lin, R., Liu, C., Shen, B., Bai, M., et al. (2015). Analysis of the complete mitochondrial genome of *Pochonia chlamydosporia* suggests a close relationship to the invertebrate-pathogenic fungi in Hypocreales. *BMC Microbiology, 15*, 5.

Liu, K., Zhang, W., Lai, Y., et al. (2014). *Drechslerella stenobrocha* genome illustrates the mechanism of constricting rings and the origin of nematode predation in fungi. *BMC Genomics, 15*, 1–14.

Lopez-Llorca, L. V. (1990). Purification and properties of extracellular proteases produced by the nematophagous fungus *Verticillium suchlasporium*. *Canadian Journal of Microbiology, 36*, 530–537.

Lopez-Llorca, L. V., & Duncan, J. M. (1986). New media for the estimation of fungal infection in eggs of the cereal cyst nematode, *Heterodera avenae* Woll. *Nematologica, 32*, 486–489.

Lopez-Llorca, L. V., Maciá-Vicente, J. G., Jansson, H. B., et al. (2008). Mode of action and interactions of nematophagous fungi. In A. Ciancio & K. G. Mukerji (Eds.), *Integrated management and biocontrol of vegetable and grain crops nematodes. Integrated management of plant pests and diseases* (Vol. 2, pp. 51–76). Heidelberg: Springer.

Maciá-Vicente, J., Rosso, L., Ciancio, A., et al. (2009). Colonisation of barley roots by endophytic *Fusarium equiseti* and *Pochonia chlamydosporia*: Effects on plant growth and disease. *The Annals of Applied Biology, 155*, 391–401.

Manzanilla-López, R. H., Esteves, I., Powers, S. J., et al. (2011). Effects of crop plants on abundance of *Pochonia chlamydosporia* and other fungal parasites of root-knot and potato cyst nematodes. *The Annals of Applied Biology, 159*, 118–129.

Mauchline, T. H., Kerry, B. R., & Hirsch, P. R. (2002). Quantification in soil and the rhizosphere of the nematophagous fungus *Verticillium chlamydosporium* by competitive PCR and comparison with selective plating. *Applied and Environmental Microbiology, 68*, 1846–1853.

Meerupati, T., Andersson, K. M., & Friman, E. (2013). Genomic mechanisms accounting for the adaptation to parasitism in nematode-trapping fungi. *PLoS Genetics, 9*(11), e1003909.

Moraes, C. K., Schrank, A., & Vainstein, M. H. (2003). Regulation of extracellular chitinases and proteases in the entomopathogen and acaricide *Metarhizium anisopliae*. *Current Microbiology, 46*, 205–210.

Nordbring-Hertz, B., Jansson, H. B., & Tunlid, A. (2006). Nematophagous fungi. In *Encyclopedia of life sciences*. Chichester: Wiley. doi:10.1038/npg.els.0004293.

Palma-Guerrero, J., Gómez-Vidal, S., Tikhonov, V. E., et al. (2010). Comparative analysis of extracellular proteins from *Pochonia chlamydosporia* grown with chitosan or chitin as main carbon and nitrogen sources. *Enzyme and Microbial Technology, 46*, 568–574.

Pinto, A. D. S., Barreto, C. C., Vainstein, M. H., et al. (1997). Purification and characterization of an extracellular chitinase from the entomopathogen *Metarhizium anisopliae*. *Canadian Journal of Microbiology, 43*, 322–327.

Prasad, P., Varshney, D., Adholeya, A., et al. (2015). Whole genome annotation and comparative genomic analyses of bio-control fungus *Purpureocillium lilacinum*. *BMC Genomics, 16*, 1–14.

Rouger, V., Bordet, G., Couillault, C., et al. (2014). Independent synchronized control and visualization of interactions between living cells and organisms. *Biophysical Journal, 106*, 2096–2104.

St Leger, R. J., Staples, R. C., & Roberts, D. W. (1993). Entomopathogenic isolates of *Metarhizium anisopliae*, *Beauveria bassiana*, and *Aspergillus flavus* produce multiple extracellular chitinase isozymes. *Journal of Invertebrate Pathology, 61*, 81–84.

Venter, J. C., Adams, M. D., Myers, E. W., et al. (2001). The sequence of the human genome. *Science, 291*, 1304–1351.

Wilson, R. A., & Talbot, N. J. (2009). Under pressure: Investigating the biology of plant infection by *Magnaporthe oryzae*. *Nature Reviews. Microbiology, 7*, 185–195.

Yang, J., Wang, L., Ji, X., et al. (2011). Genomic and proteomic analyses of the fungus *Arthrobotrys oligospora* provide insights into nematode-trap formation. *PLoS Pathogens, 7*(9), e1002179.

Zavala-Gonzalez, E. A., Escudero, N., Lopez-Moya, F., et al. (2015). Some isolates of the nematophagous fungus *Pochonia chlamydosporia* promote root growth and reduce flowering time of tomato. *The Annals of Applied Biology, 166*, 472–483.

Zhang, L., Zhou, Z., Guo, Q., et al. (2016). Insights into adaptations to a near-obligate nematode endoparasitic lifestyle from the finished genome of *Drechmeria coniospora*. *Scientific Reports*. 6. doi:10.1038/srep23122.

Zhao, X., Kim, Y., & al, P. G. (2005). A mitogen-activated protein kinase cascade regulating infection-related morphogenesis in *Magnaporthe grisea*. *The Plant Cell, 17*, 1317–1329.

Part IV
Methodology

Chapter 11
Methodology Part I. *Pochonia* spp. *In Vitro* Culturing: Media, Strain Maintenance and Deposition

Rosa Navarrete-Maya, Ivânia Esteves, and Rosa H. Manzanilla-López

Abstract Culturing on agar is a common method used in fungal isolation, cultivation, preservation, microscopic examination, biochemical, and physiological characterization. The present chapter reviews research carried out on *Pochonia chlamydosporia* culturing under *in vitro* conditions, which have made possible the production of mycelium, conidia and chlamydospores in sufficient quantities to allow laboratory, glasshouse and field experiments to be carried out with different isolates and strains of the fungus. Maintenance of isolates and deposit in culture collections is also covered. Knowledge of factors that can promote the growth, sporulation and development of resting structures of the fungus is essential for planning experimental assays and development of fungal strains as potential biopesticides.

11.1 Introduction

To culture *Pochonia chlamydosporia* (formerly known as *Verticillium chlamydosporium*; see Chap. 2) and other fungi *in vitro* it is first necessary to know the basic nutrients and conditions needed for vegetative growth, reproduction and production

R. Navarrete-Maya (✉)
UNIGRAS (Unidad de Investigación en Granos y Semillas), Facultad de Estudios Superiores Cuautitlán, Universidad Nacional Autónoma de México,
Av. Dr. Jorge Jiménez Cantú s/n, Cuautitlán Izcalli, C.P. 54740, México
e-mail: rosa_navarrete@hotmail.com

I. Esteves
CFE-Centre for Functional Ecology, Department of Life Sciences, Calçada Martim de Freitas, University of Coimbra, P3000-456 Coimbra, Portugal
e-mail: iesteves@cu.pt

R.H. Manzanilla-López
Centro de Desarrollo de Productos Bióticos (Visiting Professor),
Yautepec, Morelos, C.P. 62731, México
e-mail: rosa.manzanilla@gmail.com

R.H. Manzanilla-López, L.V. Lopez-Llorca (eds.), *Perspectives in Sustainable Nematode Management Through* Pochonia chlamydosporia *Applications for Root and Rhizosphere Health*, Sustainability in Plant and Crop Protection,
DOI 10.1007/978-3-319-59224-4_11

of resting or survival structures when grown in an artificial environment (Griffin 1994; Ciancio et al. 2016). *Pochonia* spp. colonies that are grown on agar are white or whitish to ochre yellow and cottony, with the same colour on the reverse side and can reach 15–40 mm diameter in 10 days. Conidiophores are prostrate and little differentiated from vegetative hyphae, or erect and differentiated, verticillate or solitary. Conidia adhere to form globose heads or chains, and are subglobose, ellipsoidal to rod-shaped, isodiametric-polyhedric or falcate with blunt ends and 3–4 μm by 1.5–2.0 μm in size (Gams 1988). Chlamydospores 15–30 × 10–20 μm in diameter are usually formed on short stalks (Zare et al. 2001).

Pochonia chlamydosporia can be cultured not only in agar but on a range of liquid and solid artificial media. The choice of culture medium depends upon the cultivation objective, since the composition of the culture medium influences vegetative growth, colony morphology, sporulation, metabolite production, and pigments (Kerry et al. 1986; Liu and Chen 2003; Mo et al. 2005). The functioning of a medium is determined not only by the available nutrients (e.g., minerals, carbon sources, nitrogen sources, and vitamins), redox potential, pH, water activity (a_w) and potential (Ψ), and the presence of inhibitors, but also by the vitality of the inocula (Griffin 1994). In the different sections of the chapter we summarize the research carried out on *P. chlamydosporia* culturing *in vitro* as related to media composition (nutrients), culture-related physicochemical factors and the ambient conditions required to produce different inocula sources Culture preservation is also covered.

11.2 Nutrients

Nutritional elements may be divided into two major classes depending on concentration of macronutrients (10^{-3} M) and micronutrients (10^{-6} M). Vitamins are usually required at micromolar concentrations as coenzymes, whereas growth factors are usually required at higher concentrations (100 μM). The latter are also known as structural elements and include amino acids, sterols, purines and pyrimidines (Griffin 1994). Macronutrient elements include carbon, hydrogen, oxygen, phosphorus, potassium, nitrogen, sulphur, and magnesium (Griffin 1994). These elements are essential for fungal growth and development although the type and quantity required of each element varies with fungus age and environmental conditions (Maheshwari 2012). Among the different components and mineral elements for the sporulation of *P. chlamydosporia*, Gao (2011) found that optimum sporulation occurred in the presence of $ZnSO_4.7H_2O$, $MnSO_4.H_2O$, and H_3BO_4, all of them at 0.05 g/l.

11.2.1 Carbon:Nitrogen (C:N) Ratio

Organisms that decompose organic matter use carbon as a source of energy and nitrogen for building cell structure (Whatcom 2016). The rate of decay and nutrients released to the soil vary greatly. Nitrogen is one of the major products of plant decay while the undigested portion is primarily carbon. The term 'substrata' refers to a particular carbon constituent of a tissue from which a fungus is deriving its energy supply (Garrett 1981). During substrata colonization and exploitation, the most rapidly available carbon-source, the sugars, are used first by microorganisms whereas more complex substrata, such as cellulose and lignin, persist longest (Garrett 1981; Whatcom 2016). Exhaustion of available nitrogen supply is considered the commonest cause of the cessation of mycelial activity and the production of resting bodies such as oospores, ascospores, chlamydospores, undifferentiated dormant mycelium, and sclerotia (Garrett 1981). Higher organic matter levels with excess carbon can result in 'robbing' the soil of nitrogen but, if carbon is less than that required for converting available nitrogen into protein, organisms use the available carbon and dispose of excess nitrogen as ammonia (Whatcom 2016). The spectrum of carbon constituents that may be available to a particular fungal species is a primary determinant of its ecological niche in the soil economy (Garrett 1981). However, the share of a potential substrate in the native soil habitat and niche will always be smaller than the substrate proportion that the fungus can exploit in pure culture in the laboratory (Garrett 1981). Carbon and nitrogen are very important in mycelial growth and the sporulation of fungi, fungal cell dry weight estimated to constitute 50% carbon and 5% nitrogen (Maheshwari 2012). The C:N ratio is used as an indicator of the relative proportion of carbon (C) and nitrogen (N) elements available for microorganisms, either in organic substrata or media cultures. The optimum ratio in soil organic matter is a C:N ratio of about 10:1 (Whatcom 2016). Carbon:nitrogen ratios have been investigated in different media and culturing conditions, including temperature and pH, in order to optimize vegetative growth, sporulation, and ovicidal or enzymatic activity in *P. chlamydosporia*.

Sykes (1994) analysed the effect of different glucose concentrations on growth rate and conidiospore numbers using Vogel's medium (http:www.fgsc.net/methods/vogel.html) supplemented by decreasing concentrations of glucose. The medium (50 ml) was inoculated with 5% (v/v) of a mycelia/spore suspension of *P. chlamydosporia* and incubated at 25 °C on a rotary shaker at 200 rpm. The overall growth rate, biomass yield (36 h), and presence of chlamydospores (at 21 days) remained constant at each glucose concentration (1, 2, 5, 10 g/l) but conidiospore concentration increased from 0.84×10^7 (C:N 0.4:1) to 3.10×10^7 conidia/ml (C:N 4.4:1) as C:N ratio increased.

Carbon and nitrogen absorption can also vary according to nitrogen source. Mo et al. (2005) examined the effect of 21 carbon sources, 15 nitrogen compounds and pH range (2.5–8.0) on the growth and sporulation of *P. chlamydosporia* in liquid culture. The best combination for mycelia growth was sweet potato and L-tyrosine with a C:N ratio of 40:1, whereas the best sporulation source was sweet potato and casein peptone medium with a C:N ratio 10:1. The effect of C and N ratios on the efficacy of *P. chlamydosporia* to parasitise *Meloidogyne incognita* eggs *in vitro* was investigated by Luambano et al. (2015). Glucose (30 g/l water) was the source of C while ammonium nitrate (40 g/l water) provided N. All treatments were buffered using 0.1 M potassium buffer (pH 6.8) and incubated at 25.5 °C in a shaker at 150 rpm for 48 h. Among the C:N ratios tested, the highest and lowest infection were recorded in media containing C:N ratios of 5:100 (31%) and 10:5 (12.03%), respectively.

The effect of carbon and nitrogen sources on gene expression in *P. chlamydosporia* has also been investigated (see Chap. 5). The alkaline serine protease VCP1 is involved at the onset of the parasitic phase of *P. chlamydosporia* life cycle. It digests the outer proteinaceous vitelline membrane of nematode eggs, thereby facilitating penetration and infection. In the study made by Ward et al. (2012) different fungal isolates were grown in Czapek Dox broth (CDB) and incubated at 28 °C for 3 days with constant shaking at 200 rpm. Subsequently, the mycelium was harvested, washed and transferred to flasks containing 0.01 M K phosphate buffer (pH 6.8) or phosphate buffer containing one or more of the following: 2% D-glucose, 200 mM NH_4Cl, or *M. incognita* eggs (1 egg/μl). Gene expression experiments monitoring both enzyme and mRNA of VCP1 confirmed regulation by readily-metabolised carbon sources, such as glucose, at relatively low levels (10 mM). The regulation of gene expression by nitrogen sources involves GATA transcription factors (i.e., a family of transcription factors with ability to bind to DNA sequence GATA), and the presence of GATA motifs in the upstream region of *vcp1*, indicates that this gene would also be subject to regulation from nitrogen sources (Ward et al. 2012).

11.3 Water Activity

Substrates and enzymes are all in aqueous solution or colloidal suspension and no enzymatic activity can occur in the absence of water. The movement of water into and out of the hypha occurs by osmosis: a diffusion process in response to concentration gradients across the semipermeable plasmalemma (Griffin 1994). The osmotic pressure in a medium is expressed as water activity (a_w) and the range of microorganisms capable of developing on a medium decreases with decreasing a_w. Hence, media are often made selective to more osmo-tolerant organisms by the addition of polar solutes like NaCl, glucose, fructose, saccharose and glycerol (Mossel 1970).

11.3.1 Water Potential

Water potential (Ψ) refers to the ability of water molecules to move freely in solution and is expressed in megaPascal (MPa) pressure units. The partition of Ψ in turgor, osmotic and matric components facilitates studies of their individual influence on the growth and physiological functioning of organisms (Griffin 1994). In solid substrates, where solute potential is the major force, a_w is commonly used to express osmotic pressure instead of Ψ. In soil and cereal crop residue, matric potential is the major component of total Ψ (Magan and Lynch 1986) and, therefore, matric stress is thought to exert a greater influence on fungal growth and germination than osmotic potential stress (Brownell and Schneider 1985). Since moisture in soil can vary and may decrease substantially during certain seasons of the year, knowledge concerning the water requirements and limiting Ψ for fungal growth is particularly important in soil fungi such as *P. chlamydosporia* as it enables a better understanding of the fungus survival and growth strategies in the soil environment. The response of fungi to changes in Ψ can be studied in the laboratory and several solutes are used for adjusting the Ψ of growth media, including various salts, glucose, sucrose, glycerol or polyethylenglycol (PEG) (Griffin 1994). To overcome water stress, fungal cells must be able to take up water until, by turgor pressure, the Ψ inside the cells is in equilibrium with that outside (Cook and Baker 1983). During osmoregulation fungi are able to lower their intracellular Ψ by synthesizing compatible solutes that are accumulated in cell protoplasmic and vacuolar spaces (Brown and Simpson 1972). These solutes are mainly polyhydroxyalcohols (polyols), organic acids and sugars (Brown 1978; Luard 1982; Meikle et al. 1991; Van Eck et al. 1993; Hallsworth and Magan 1994a, b).

In general, fungi can grow under conditions of lower Ψ within theoretical limits that lie between Ψ of 0, and −81 MPa, where DNA becomes denatured (Griffin 1994). *Pochonia chlamydosporia* can survive in greater numbers of colony forming units (cfu) at a Ψ of −0.2 MPa than at other potentials (Bourne and Kerry 2000). According to Gao et al. (2009), the best Ψ for *P. chlamydosporia* growth was −3.9 MPa at pH 3, 24 h light and 23 °C; the best factors for sporulation included a water potential of −3.9 MPa, pH 4, 12 h light, and 23 °C temperature. However, tolerance to lower Ψ may be variable depending on type of stress (matric or osmotic), isolate, and the solute used to generate the reduction of Ψ. Esteves et al. (2009) reported that three isolates of *P. chlamydosporia* were able to grow at Ψ of −7.0 MPa on potato dextrose agar (PDA) amended with KCl, but that none was able to grow at this potential in the medium modified with glycerol. In the same study, differences in the radial growth of the fungus were detected between isolates of *P. chlamydosporia* in response to matric stress on PEG-modified media, although isolates responded similarly on osmotic-modified PDA medium amended with KCl and glycerol (Esteves et al. 2009). Tolerance to matric stress was thought to be related to the accumulation of high amounts of erythritol in PEG-modified media, which is built up by many xerophylic fungi in response to water stress as it enables enzyme systems to work more effectively. Under different types of water stress, the fungus accumulated a combination of sugar alcohols, i.e., polyols (arabitol, erythritol, glycerol, mannitol) and sugar (glucose and trehalose) known to be important in osmoregulation and the differential and proportional amounts of these compounds were dependent on the solute used to generate the water stress (Esteves et al. 2009). These results suggest that different polyols and sugars are used by *Pochonia* to tolerate osmotic and matric stress for osmoregulation and enzyme activity and that such versatility might allow the fungus to survive in soil when environmental conditions are either less favourable or highly variable (Esteves et al. 2009). According to Bourne and Kerry (2000), the fungus is able to survive to Ψ of −9.7 MPa in soil amended with NaCl at temperatures of 25–30 °C, whereas in a different study, the limiting Ψ for growth in Sabouraud dextrose agar amended with PEG was even lower and about −16.5 MPa (Olivares-Bernabeu and Lopez-Llorca 2002). Since most plants in soil reach the permanent wilting point at about −1.5 MPa (Deacon 2006), the fungus is therefore capable of tolerating Ψ values much below those that are limiting for plant growth.

As well as the effects in mycelial growth and accumulation of endogenous reserves, the *in vitro* growth of *P. chlamydosporia* cultures in water-restrictive media can influence the type, quantity and viability of spores produced. Vieira dos Santos et al. (2012) found that the production of chlamydospores by five isolates of the fungus was repressed in both osmotic and matric modified media when compared with growth in control plates where water was freely available (1.7% Corn Meal Agar [CMA]). Although the production of conidia was increased in CMA osmotic-modified medium with KCl, the germination rate of the conidia produced was lower than that of those obtained from CMA control plates. However, in some isolates a significant increase in percentage parasitism of Potato Cyst Nematode (PCN) eggs was observed after 25 days when the fungus was subjected to water-stress conditions unfavourable for growth (Vieira dos Santos et al. 2012). The results suggested that the osmoregulation mechanisms used to compensate water stress in this species play a role in *in vitro* growth, sporulation and parasitism, and thus, together with other abiotic, ecological and key intrinsic biological factors, are likely to affect the performance of the fungus as a biological control agent (BCA) when released into soil. Fungal spores resistant to drought have been obtained from media with trehalose (Gao 2015).

11.4 Hydrogen Ion (pH)

Most fungi grow well when the initial pH of the culture is adjusted to the range 4–7. However, fungi change the pH of the medium during growth (Griffin 1994). Although the extracellular pH has essentially no influence on the cytoplasmic pH, effects of pH may be indirect on the cellular surface or on extracellular components; pH also affects the ionization of weak acids or bases in the medium (Griffin 1994). Absorption of nitrogen sources, such as ammonium sulphate, and loss of H+ ions from the fungal cells can cause a rapid acidification of the medium which then stops growth (Maheshwari 2012).

Optimal fungal growth and sporulation may occur under different pH conditions and pH values can vary for the same or different species according to the media and whether media are buffered or not. Initial pH adjustment of the culture media should consider the appropriate selection of the acids or bases used for this purpose. Acids like lactic acid and acetic acid possess intrinsic antimicrobial properties, whereas citric acid and tartaric acid show no other inhibitory effects other than those resulting from reducing the pH of a medium (Mossel 1970).

Pochonia chlamydosporia var. *chlamydosporia* has been reported to grow best in solid media at pH 6.5–7.7 (Nagesh et al. 2007) but also at pH 4–7 (Karakas et al. 2012); while *P. chlamydosporia* var. *catenulata* requires a pH range of 6–7 (Karakas et al. 2012). Mo et al. (2005) obtained the greatest biomass at pH 6.8 and optimum sporulation at pH 3.7 when growing *P. chlamydosporia* in a liquid medium; it was noted that pH increased after 7 days due to fungal activity.

Pochonia chlamydosporia growth in buffered (0.05 M sodium citrate/citric acid) Vogel's medium (Vogel 1956), inoculated with a 5% (v/v) mycelia/spore suspension and incubated in flasks at 25 °C on a rotary shaker at 200 rpm at different pH, resulted in filamentous (pH 3), aggregated (pH higher than 5.4) or pelleted growth (pH below 3) (Sykes 1994). Mycelia pellets (i.e., stable, spherical mycelia aggregate) were also noticed by Ward et al. (2012) when isolates of the two varieties of *P. chlamydosporia* were cultured in a medium composed of $NaNO_3$ (3 g), KCl (0.5 g), $C_3H_7MgO_6P$ (0.5 g), $FeSO_4$ (0.01 g), K_2SO_4 (0.35 g), sucrose (30 g), 0.5 g yeast extract/l, pH 6.8, and incubated at 28 °C for 3 days with constant shaking at 200 rpm.

VCP1 production by the fungus is also regulated by ambient pH, as shown by Ward et al. (2012). Different fungal isolates were cultured in CDB and incubated for 3 days at 28 °C with constant shaking at 200 rpm. Mycelium was harvested and washed in sterile distilled water before transferring to flasks containing 0.01 M K phosphate buffer (pH 6.8) or phosphate buffer containing one or more of the following additives: 2% D-glucose, 200 mM NH_4Cl, *M. incognita* eggs (1 egg/μl). To investigate the effect of pH, 0.01 M K phosphate buffers at pH 5.8 and pH 8.0 were used. In Pc10 and Pc147 (*P. chlamydosporia* var. *chlamydosporia*) at 8 h and 24 h VCP1 (enzyme and mRNA) levels were significantly ($P = 0.05$) higher in the pH 8 media than in the other two media. In Pc392 (*P. chlamydosporia* var. *catenulata*) enzyme levels were low in all three buffers used and there were no significant differences between them. However, VCP1 mRNA levels were significantly ($P = 0.05$) higher when grown at pH 6.8 than at the other two pH levels (Ward et al. 2012).

11.5 Media Formulation

In addition to the manufacturer's quality control on media formulation, attention must also be given to the mode of medium preparation, glassware, water quality to be used in reconstitution, mode of dissolving dry ingredients, and the way labile components are dissolved and added (Mossel 1970). Any sterilization method has the potential to change media composition and functioning, media productivity or selectivity (Mossel 1970), for example in antibiotics and fungicides used in the selective-agar to culture *Pochonia* spp.

Pochonia chlamydosporia can be easily grown on various solid media. Water Agar (WA) is much less selective than a nutrient agar and mycelia must be subcultured on a nutrient agar for production of identifiable structures such as conidia and chlamydospores (Garrett 1981). Water agar, CMA, Czapek Dox Agar (CDA), modified Vogel's agar (MVA), and semi-selective growth agar were among the media tested by Sykes (1994). Media components were added to distilled water prior to autoclaving at 121 °C for 15 min and molten media (ca 50 °C) dispensed as 20 ml aliquots into 9 cm Petri dishes or as 10 ml aliquots into 25 ml plastic or glass universal bottles (Sterilin™), and dried in a sterile air flow in a laminar flow unit. Other media, such as potato carrot agar (PCA), PDA, malt extract agar (MEA), straw agar (8 g agar, 40 g chopped straw), and oat meal agar (OA), have been used to grow and produce the anatomical structures needed for identification and morphological characterization of *Pochonia* spp. (Gams 1988; Clyde 1993; Zare et al. 2000, 2001; Nonaka et al. 2013).

De Carvalho et al. (2014) tested different culture media to grow *P. chlamydosporia* in order to assess its ovicidal activity on *Parascaris equorum*. Media included 2% WA; agar-chitin (AC); YPSSA (yeast extract, K_2HPO_4, $MgSO_4 \cdot 7H_2O$, and soluble starch); AELA extract (starch + water +agar); 2% CMA; and 2% PDA. Eggs were incubated at 25 °C in the dark for 21 days. The best media was AELA where the fungus caused 44.9% egg destruction (see also Chap. 17).

11.5.1 Growth and Sporulation on Solid Media

Mycelial plugs used to inoculate solid fresh media are usually taken from the margin of actively growing colonies of *P. chlamydosporia* cultures grown on CMA plates or other media. Sykes (1994) measured the colony diameter of the fungus in CDA and CMA plates inoculated with CMA plugs. Colonies grew linearly over a 14 day period on all agar media tested and sporulation was recorded 10 days after inoculation at 25 °C. Colonies showed the best growth on CDA and CMA, but more conidiospores were formed on CDA than CMA.

11.5.1.1 Carbon and Nitrogen Sources in Culture Media

Sun and Liu (2006) tested the effects of 33 carbon sources on spore germination, hyphal growth and sporulation of different entomopathogenic, mycoparasitic and nematophagous fungi, including *P. chlamydosporia*, by varying the composition of a solid agar basal medium. Carbon sources such as soluble starch, methyl cellulose, ethanol, methanol, trehalose, agarose, dextrin, pectin, D-glucose, maltose, stearic acid, and sucrose stimulated *P. chlamydosporia* mycelial growth, whereas spore germination was favoured by sucrose, agarose, methyl cellulose, and oxalic and stearic acids. Dodecanoic, linoleic and sorbic acid inhibited germination of all tested fungi. Agarose, methanol, and stearic acid were good carbon sources for sporulation of *P. chlamydosporia*, although no information was provided on the type of spores (conidia or chlamydospores) produced in the different media. Results supported the thesis that nematophagous fungi have diverse carbon nutritional requirements but that their response to carbon sources might also be strain dependent (Sun and Liu 2006).

Gao (2011) investigated sucrose and maltose as carbon sources, and NaO_3 and urea as nitrogen source in a 'two-step' culturing method in solid media to optimize vegetative growth and sporulation. The first step or stage of the method included a basal medium made of sucrose (19 g), soy peptone (4.06 g), K_2HPO_4 (1 g), KCl (0.5 g), $MgSO_4$ (0.5 g), $FeSO_4$ (0.01 g) and agar (17 g) per litre, which was used for vegetative culturing of the fungus for 4 days. Different combinations of C and N were also tested and 8 g/l of sucrose and 0.8 g/l of soy peptone (C:N ratio of 10:1) were selected for the second culturing stage where small, fresh mycelium pieces were transferred into the culturing medium for another 4 days. The best C:N ratio 10:1 increased sporulation 30 times in comparison to traditional methods. This method had also the advantage that production of conidia could be upgraded to industrial scale (Gao 2011).

11.5.1.2 Oxidative Stress Treatment

According to Xu et al. (2009), the production of conidia in solid media can be increased by treating *P. chlamydosporia* colonies with hydrogen peroxide (H_2O_2). Numbers of conidia obtained from colonies that were sprayed with 90 mM of H_2O_2 and subsequently cultured on WA were five times greater than those that were transferred to WA alone (i.e., without H_2O_2 treatment); three times greater when colonies were treated with H_2O_2 and then transferred to PDA, and 15 times greater than the continuous culturing of the fungus on PDA plates (Xu et al. 2009). The H_2O_2 treatment affected the quantity, but not the viability, of the conidia produced. Although chlamydospore production was also enhanced using this method, the increase was not as remarkable as conidial production. Conidial production in *P. chlamydosporia* seems to be promoted in response to both oxidative and low-nutrient stresses, although a different mechanism may promote chlamydospore formation (Xu et al. 2009).

11.5.2 Semi-selective Medium

Selectivity of culture media can be achieved by the addition of one or more inhibitors of microbial growth. The activity of such inhibitors is determined by concentration, degree of inactivation during sterilization or partial inactivation during medium storage and the period it is incubated after inoculation (Mossel 1970).

The isolation of *P. chlamydosporia* from parasitized nematode eggs can be done by plating a suspension of eggs on poor nutrient medium (0.8% tap water agar) followed by plating on a richer medium to allow the fungus to grow and produce spores (Kerry and Crump 1977). However, bacterial contamination and overlap of fungal colonies can adversely affect the ease with which nematophagous fungi are isolated. Plating onto selective media overcame this limitation as chemicals restrict bacteria and other fungi growth, thus facilitating the identification and counting of *P. chlamydosporia* colonies. A selective medium containing 1% peptone dextrose agar, 1000 μl/ml Triton X-100, 50 μg/ml streptomycin sulphate, 50 μg/ml aureomycin and 50 μg/ml Rose Bengal, was developed by Lopez-Llorca and Duncan (1986) to estimate fungal infection of eggs of *Heterodera avenae* collected from cereal soils in Scotland. Plating eggs on this medium allowed the discrimination of colonies from different fungi, as fungal growth was restricted, making it easier to count and directly identify small, dense colonies on plates. In this study, *H. avenae* eggs were parasitized by more than three fungi, *P. chlamydosporia* (= *Verticillium chlamydosporium*) being the most abundant (Lopez-Llorca and Duncan 1986). Plating onto selective media is an important tool to measure the relative abundance of the fungus in soil and in the rhizosphere of roots infested with nematodes (Mauchline et al. 2002; Atkins et al. 2003). For this purpose, another semi-selective medium (sometimes also known as selective-agar) was developed by Kerry et al. (1993). To prepare 1 L of semi-selective-agar 17 g CMA, 17.5 g NaCl, 75 mg Rose Bengal are added in 800 ml distilled water. After autoclaving, antibiotics (50 mg streptomycin sulphate, 50 mg chloramphenicol, 50 mg chlortetracycline), fungicides (37.5 mg thiabendazole, 37.5 mg carbendazim) and 3 ml Triton X-100, are dissolved in 200 ml of sterile distilled water and added to the medium. Since *P. chlamydosporia* is relatively tolerant to high concentrations of sodium chloride (Ceiro et al. 2013), the incorporation of 1.75% NaCl into the medium inhibits the growth of salt-sensitive fungal species and restricts fungal spread. Rose Bengal is a colouring agent that increases selectivity as it helps in the suppression of most bacteria, restricts the overgrowth of rapidly growing fungi and provides contrast for visualization of colonies (Ottow 1972). The inhibition of bacterial growth is also prevented by the use of antibiotics, whereas the development of contaminant fungi is restricted by the addition of two benzimidazole fungicides to which *P. chlamydosporia* has a relatively high level of tolerance compared to other soil fungi (Hirsch et al. 2001). As well as reducing the rate of colony growth, the surfactant Triton X-100 also helps in the solubilization of the two fungicides. After pouring the cooled medium onto 9-cm diameter Petri dishes, plates should be kept in the dark, as light causes degradation of Rose Bengal. The inoculation of *P. chlamydosporia* onto the semi-selective medium is done by pipetting 200 μl of a 0.05% WA suspension containing the fungal propagules onto each plate. To distribute the sample, the plates are gently swirled or, alternatively, a sterile L-shaped glass or plastic spreader (e.g., Sigma or Sterilin™) can be used. After incubation at 25 °C for 10–15 days, colonies of *P. chlamydosporia* are translucent white with sparse smooth mycelium and can vary in diameter. By measuring the abundance of viable fungal propagules in terms of cfu, dilution plate counting in semi-selective medium is a useful technique to monitor changes in the abundance of the fungus. However, it cannot distinguish between the nature of these propagules (hyphal fragments, conidia, chlamydospores) or between different isolates. As a result, molecular methods should be combined with classic plating methods to provide complementary information on fungal diversity and abundance (Mauchline et al. 2002; Atkins et al. 2003, 2005).

11.5.3 Liquid Culture

The capacity to produce infective propagules *in vitro* in large quantities has been explored for *Pochonia* spp. Mycelia and conidia of *P. chlamydosporia* produced in liquid culture can remain viable for up to 12 months at 25 °C and be biologically active when introduced into soil in a granular formulation (Stirling et al. 1998). A commercial product (Xianchongbike) made from mycelia and conidia of *P. chlamydosporia* produced in liquid culture reduced *Meloidogyne* spp. root-knot nematode (RKN) in field trials in tobacco (Mo et al. 2005). Sporulation and conidial germination was increased in a basal liquid medium originally composed of 10 g maltose, 2.0 g $NaNO_3$, 0.5 g $MgSO_4{\cdot}7H_2O$, 0.5 g KH_2PO_4, 0.65 g Na_2HPO_4 and 0.5 g KC1, per litre of distilled deionised water (Mo et al. 2005). The basal liquid medium was modified for carbon:nitrogen tests where D-fructose was used as carbon source and L-glutamic acid as nitrogen source. Maximum sporulation occurred 7 days after inoculation (51.4 × 10^6 conidia/ml) in media that had a C:N ratio of 10:1 and at pH 3.7 (see also Sects. 11.2.1 and 11.3.1).

11.5.3.1 Chlamydospore Production in Liquid Culture

It has been reported that *P. chlamydosporia* does not produce chlamydospores and blastospores in liquid culture and that the fungus produces conidia directly on phialides in liquid culture that do not differ in size and shape from those produced in solid culture (Stirling et al. 1998; Mo et al. 2005). However, Sykes (1994) reported that fungal growth was prolific in shake-flask culture, stirred-tank reactor, or in an air lift fermenter in Vogel's liquid minimal medium (Vogel 1956) and that chlamydospores were found within the mycelial rind formed in shake-flask or stirred-tank reactor cultures. Chlamydospores can also be produced in a liquid medium supplemented by organic matter such as straw, barley, and rice. A wheat-straw-based liquid culture composed of 10 g of finely chopped straw (ca 2 mm in length) in 1 L distilled of water was described by Sykes (1994). An aliquot of 50 ml of the liquid culture inoculated with a 5% (v/v) mycelia/spore suspension of *P. chlamydosporia* and incubated at 25 °C on a rotary shaker at 200 rpm, produced chlamydospores after 10 days of incubation, yield reaching 4.7 × 10^4 chlamydospores/ml after 21 days. The straw water medium had a low concentration of available carbon (0.346 g/l), nitrogen (0.021 g/l), and no sulphur. Chlamydospore formation was at its greatest (1.35 × 10^5 spores/ml) when the medium contained 20 g straw/l. Barley can be also used in liquid culture to produce chlamydospores. Barley (20 g) and deionised water (40 ml) are put into 250 ml flasks, left to incubate overnight and excess water removed. Flasks and their contents are autoclaved at 121 °C for 20 min and left to cool before the fungus is inoculated into flasks, sealed with a sterile cotton plug and incubated for 4 weeks at 25 °C. Fifty ml of water are added to the flasks and blended for 1 min, the suspension is then washed through several sieves (300 μm mesh to 50 μm mesh) and the chlamydospores retained with a final rinse on a 10 μm mesh (De Leij et al. 1993; Yang et al. 2012).

11.5.3.2 Rice Biphasic Culture

A standard biphasic culture (liquid-solid) protocol for mass production of *P. chlamydosporia* chlamydospores was developed at Centro Nacional de Sanidad Agropecuaria (CENSA), Cuba (Hidalgo-Díaz 2003; see Chap. 15). This protocol was used in Rothamsted Research and other research laboratories. The protocol includes two phases: (i) preparation of inoculum; and (ii) preparation of substrate and harvesting of chlamydospores.

Preparation of Inoculum

Forty grams of rice are added to 1 l of tap water and boiled for 30 min. The rice broth is filtered and distributed into 200–250 ml conical flasks, then autoclaved (15 min at 121 °C) and left to cool. Three plugs (5 mm) of the fungus cultured on CMA (i.e., 7 days at 25 °C) are added to 200/250 ml of the rice broth. Flasks are then placed in an orbital shaker for 72 h at 130 rpm at 25 °C.

Preparation of Substrate

One kg rice is washed under tap water (three times), drained and dried in a tray placed in a flow cabinet (30 min) to reach 10–12% humidity. Three hundred ml of tap water are added to the rice, mixed and left to rest (30 min) before pre-cooking in a boiling water bath for 15 min, then distributed in conical flasks (100 g/250 ml flask), covered with a cotton plug and sterilized (15 min at 121 °C). Ten ml of rice broth/100 g substrate are added to the flask which is then shaken vigorously to mix the broth and the substrate, and incubated for 21 days (minimum of 16 days) at 25 °C.

Chlamydospore Harvesting

Twenty one days after inoculation the rice containing the chlamydospores is tipped from the flask onto a sieve (250 μm mesh), and rinsed under a spray of water to collect the substrate and chlamydospores onto a second sieve (10 μm mesh). The sieve is blotted underneath with a sponge and chlamydospores are collected from the sieve top surface mesh with a spatula. Chlamydospores are mixed with fine sand in a 10:1 w:w ratio (sand:chlamydospores) and stored at 5 °C for up to a week, although chlamydospore viability must always be checked before use (Esteves 2007).

11.6 Culture Contamination

Sources of culture contamination may include glassware, nutrients of biological origin, macronutrient salts, sugars, water supply, and inoculum sample (Mossel 1970). Sequential transfers using the control medium (i.e., semi-selective agar) may sometimes eliminate contamination. However, the success of media inoculated with pure cultures will depend on the viability and purity of the inocula. Fungal populations brought in or on media may not be axenic and became contaminated with other microorganisms either of a different or the same species. In such instances, different organisms may influence each other's development by synergism or antagonism. A fast growing organism can exhaust the supply of one or more nutrients and suppress the development of slower growing organisms that also require such factors (Mossel 1970). Puertas et al. (2006) mentioned that isolates of *P. chlamydosporia* and *Trichoderma harzianum*, when growing as saprophytes, can compete for the substrate.

11.7 Controls

Appropriate controls, including media composition, are essential to any experiment; for example, amino acids used as the source of nitrogen or sulphur also provide carbon. Therefore, depending on the objectives of experiments, it is necessary to include a control lacking the tested carbon source (Griffin 1994). This is very important when testing the effect of C:N ratios on the growth and other physiological process of *Pochonia* spp. such as the switch from the saprophytic to parasitic phase.

11.8 Controlled Environment Conditions

11.8.1 Light

Light is a very important factor in most fungi, mainly for the sporulation process, which can be started with just a flash of light. Growing the fungi under light conditions can lower glucose absorption even after sporulation has started. Light can also inhibit or stimulate secondary metabolite production (Maheshwari 2012) and growth. The optimum photoperiod for growth of *P. chlamydosporia* var. *chlamydosporia* in potato dextrose broth media occurred under a 12 h light and 12 h dark photoperiod, while for *P. chlamydosporia* var. *catenulata* growth required 12 days of continuous darkness (Karakas et al. 2012).

11.8.2 Effect of Temperature on Growth and Sporulation

Sykes (1994) assessed the optimal growth temperature for *P. chlamydosporia* by measuring the colony radial growth rate on CDA plates kept at 17–29 °C. Conidiospore and chlamydospore sporulation was noted at all tested temperatures. However, subsequent studies have shown that optimum temperatures can vary from 25–30 °C, depending on fungal isolate or biotype (Bourne and Kerry 2000; Arevalo et al. 2009). The influence of temperature on sporulation was also studied by Vieira dos Santos et al. (2013) in five biotypes of the fungus. None of the biotypes was able to grow at 10, 33 and 35 °C but spores/hyphal fragments subjected to these temperatures remained viable for nearly 1 month at 10 and 33 °C and resumed growth after incubation at 25 °C. Spore viability varied between biotypes exposed to 35 °C. Nevertheless, chlamydospore and conidia production, spore viability and nematode egg parasitism were affected by exposure to these growth limiting temperatures and differed among biotypes (Vieira dos Santos et al. 2013). The Spanish strain of *P. chlamydosporia* ATCC No. MYA-4875 can survive at 40 °C, but such a high temperature is lethal for Scottish isolates of *P. chlamydosporia* (Escudero and Lopez-Llorca 2014).

11.9 Attenuation in Fungi

Some fungi, when cultured in nutritionally deprived media, may adapt their growth resulting in morphological or physiological effects (Smith 1991). Serial subculturing methods using agar plating are known to cause loss of genetic material and/or certain important fungal characteristics (Jenkins and Grzywacz 2000). Numerous fungi that were heavily ovicidal when first isolated lost this property during cultivation on artificial media (Chalupova et al. 1977). Frequent subculturing of *P. chlamydosporia* has been reported to result in a reduction in chlamydospore production but not loss of virulence (Coosemans 1990). Sykes (1994) also noticed that periodic subculturing and freeze drying resulted in a decrease in sporulation as well as reduction in specific growth rate in liquid culture and colony radial growth rate (Kr) in solid culture. Jacobs (2000) noticed a 98.5% decline in pathogenicity of the fungus after three passages through PCN females. The stability of phenotypic strains, including virulence and sporulation, can vary in biotypes; therefore stability should be monitored in both the commercial production setting (Vandenberg and Cantone 2004) and in small scale production in research laboratories. The saprophytic ability of subcultures of the fungus was tested by Esteves (2007) to study the effects of repeated subculturing on the stability of the fungus and the possibility of the development of attenuation. The ability of three fungal biotypes to colonize the rhizosphere of maize plants in an *in vitro* bioassay and in soil pots was not affected by the continuous subculturing of the fungus after 60 weeks of repeated culture. The ability to parasitize RKN eggs *in vitro* was also maintained, the biotypes remaining stable in chlamydospore production, retaining spore viability, and maintaining similar enzyme profiles to their parent cultures. The stability of *P. chlamydosporia* var. *catenulata* Cuban strain IMISD187 related to production of hydrolytic enzymes was also maintained through subculturing (Peteira et al. 2007). Nevertheless, due to the increased risks of contamination, subculturing methods using agar are not recommended for the long term preservation of the fungus. However, for short term storage, the fungus can be kept on PDA slopes under light mineral oil, sterilized by heating at 80 °C. After cooling, the oil is poured over the slope to a depth of 5 mm and cultures can be kept at room temperature, in darkness, for several months (Jacobs 2000).

11.10 Long Term Storage of Cultures

Maintenance of original cultures must be strongly controlled and monitored to ensure that the fungus retains pathogenicity, together with other important characteristics for biocontrol performance (Ryan et al. 2002). To avoid problems encountered with serial transfer, preservation methods that reduce metabolism to an extent that it induces artificial dormancy have proved to be successful for conservation (Smith 1991). This is usually achieved by dehydration or freezing. Preservation using freeze-drying (lyophilisation) and liquid nitrogen are two examples of techniques that can suspend metabolism (Smith and Onions 1994). The choice of preservation methods depends upon many factors such as fungal species, type of fungal material available (spores or mycelia), cost of equipment, and technical experience.

11.10.1 Preservation of **Pochonia** *spp. in Media Cultures*

To preserve *P. chlamydosporia* isolates, cultures can be stored on 1% (w/v) water-agar slants at 4 °C (Verdejo-Lucas et al. 2002), kept in test tubes with 2% CMA at 4 °C in the dark for 10 days (Silva et al. 2010) or stored on PDA at 4 °C (Arevalo et al. 2009). The agar strips method uses pieces of colonized dried agar transferred to sterile ampoules, vacuum-dried, and sealed. To revive cultures, agar strips are placed on fresh medium of choice (Nakasone et al. 2004). Corn meal agar and PDA cultures that have dried out can sometimes be reactivated by adding CDB and keeping the cultures within a moisture chamber at 25 °C for up to 1 week. Whichever method is chosen, cultures must be checked periodically for contamination and desiccation (Nakasone et al. 2004).

11.10.2 Preservation in Soil and Sand Cultures

Some fungi can be preserved for many years in dry, sterile soil or sand (Nakasone et al. 2004). In one method, described by Yang et al. (2012), *P. chlamydosporia* strains were grown at 22 °C on a sterilized mixture made of sandy loam, sand and compost (2:1:1) in test tubes that were capped and Parafilm sealed. After dehydration of substrate the tubes were stored at 16 °C. For new cultures a few crumbs can be spread on PDA plates (Yang et al. 2012). Cultures can also be preserved in silica gel crystals (Smith and Onions 1994), a technique that prevents all fungal growth and metabolism. To reactivate cultures, a few silica gel crystals are scattered on an agar plate (see Nakasone et al. 2004). *Pochonia* spp. isolates had been preserved in the Rothamsted Research collection in soil and silica gel in tubes or vials. However, storage conditions in sand at 5 °C can affect inoculum viability (Clyde 1993).

11.10.3 Freeze-Drying

To freeze-dry isolates, 2 ml of autoclaved nutrient broth composed of 100 ml distilled water, 2.5 g (Oxoid Nutrient broth no. 2) and 7.5 g glucose are added to 9 mm CMA plates containing 2- to 3-week-old fungal colonies that have produced plenty of chlamydospores. A suspension of spores and mycelia is obtained by gently scrapping the fungus from plates using a sterile L-shaped glass spreader. Aliquots of 0.5–1 ml of the suspension are added into freeze-dry vials and taken to the freeze drying machine for lyophilisation. To recover the fungus, samples are re-hydrated by adding 1 ml of sterile distilled water and allowed 30 min to re-hydrate. The contents of the vial are streaked on suitable agar media, such as CMA, MEA or PDA, and incubated at an appropriate temperature for growth (Bourne and Kerry 2000).

11.10.4 Liquid Nitrogen

Long term storage of *P. chlamydosporia* cultures can be achieved by placing 0.5 cm^3 sections of colonized agar into cryovials. The cryovials are filled with 10% glycerol, sealed and allowed to slowly cool to −35 °C in a polythene box containing 400 g dry ice and 75 ml 90% ethanol, before being plunged into liquid nitrogen (Jacobs 2000). Alternatively, the fungus can be kept viable at −80 °C using 25% glycerol as storage medium (Franco-Navarro et al. 2009).

11.11 Monosporic Cultures

Single spore cultures are important not only to ensure culture purity, identity of an isolate or strain, but as a first hand source of reference to check morphological, molecular characteristics, and original biological activity with that of strains that may have changed through continuous culturing. Single spore cultures can be maintained on agar and stored at 4 °C or kept as small pieces of agar in cryo vials with 10% glycerol in liquid nitrogen, freeze drying or other methods (Choi et al. 1999). There are different techniques for producing single spore cultures of fungi (Ciancio et al. 2016) and some are described below.

11.11.1 Direct Transfer

Take a few conidia from the original culture with a sterilized needle and distribute conidia using a sterile glass triangle onto the surface of 3% WA in a Petri dish (plate). Incubate at 25 °C for 24 h and observe under a stereoscopic microscope to find germinated conidia. Transfer the germinated conidia into tubes or plates prepared with fresh media and incubate at 25 °C for 2 weeks (Sanglard et al. 2009).

11.11.2 Conidial Suspension

Make a conidial suspension in 3 ml of sterilized water, mix vigorously, take a 0.5 ml aliquot and spread it with an L-shaped glass rod onto WA in a Petri dish. Incubate at 25 °C for 24, 48 or 72 h and observe under a stereoscopic microscope to find germinated conidia. Cut the agar surrounding the conidia and transfer it to plates with fresh media and incubate at 25 °C for 2 weeks (Gilchrist-Saavedra et al. 2006).

11.11.3 Sterilized Glass Slide

Put a drop of sterilized water onto a sterilized glass slide; take a few conidia from the original culture and mix them with the water. Take an aliquot of the suspension and streak it onto a Petri dish containing a thin WA layer. Incubate at 24 °C overnight and then observe under the stereoscopic microscope to find germinated conidia. Transfer individual germinated conidia onto Petri plates with fresh media and incubate at 25 °C for 2 weeks (Waller et al. 2002).

11.11.4 Micromanipulator

Prepare an agar medium made of 8.35 g CDB and 3.0 g of Daishin agar (Brunswigchemie, Amsterdam), dissolve and autoclave in 250 ml of sterile distilled water. Pour 5–6 ml of medium per Petri dish (plate) to produce a thin transparent agar layer (2–3 mm thick) that will facilitate conidia observation under a stereo- or inverted-microscope. Immediately after pouring the medium, plates are covered and left to solidify on a level surface, care being taken that humidity within the plate is not too high so as to avoid condensation on the plate lid. Up to 100 μl of a diluted conidia suspension are spread onto the centre of the plate with a sterile glass bacteriological loop. This aliquot volume provides enough plate humidity to work for up to 3 h using a stage-mounted micromanipulator (Singer). Plates inoculated with single conidia using the micromanipulator are incubated at 25 °C until mycelium, conidiophores and conidia are produced (RH Manzanilla-López unpublished information).

11.12 Collections and Deposit of Isolates

Fungal isolates can be preserved and maintained in research collections of institutes or universities. Information on isolates collection should be kept in laboratory notebooks and a proper database. According to Nakasone et al. (2004), database records for each strain (or isolate) should include: strain number, genus, species, variety, taxonomic authority, required growth conditions, date of isolation, isolation location, isolation substrate or host, person who isolated the culture, provenance of the strain, isolate numbers in other collections that reference the same strain, synonyms, and other important comments, i.e., molecular markers used to confirm strain or variety identity. The method of preservation, number of replicate preservation vials and dates they were prepared should also be included.

11.12.1 Distribution and Exchange of Isolates

Collections are a valuable resource to scientific investigators (Nakasone et al. 2004), research institutions or private companies. However, some collections will require official permits or licences issued by government offices before releasing cultures for research purposes. Material transfer agreements between institutions may also be required. If fungal germplasm is distributed to other institutions or individuals, records should be kept in the collection database (Nakasone et al. 2004). Regulations and safety measures for the handling of nematophagous fungi were reviewed by Mubyana-John and Taylor (2015).

11.13 Description of New Species and Varieties

It is likely that morphological and molecular studies of new isolates will continue to reveal the existence of different varieties within *Pochonia* spp. (see Chaps. 2, 12, 13), necessitating specimens/samples to be properly described and kept in a formal collection. A review of taxonomy criteria, and an example on how to describe a fungal species, is provided by Seifert and Rossman (2010). In summary, the International Code of Nomenclature for algae, fungi and plants (http://www.iapt-taxon.org/nomen/main.php?page=title) requires any description of a new species to be 'effectively, legitimately and validly published' (Seifert and Rossman 2010; Minnis et al. 2016; see also Chap. 2), which means that the new species description must be published in a journal accessible to the scientific community and that any new species must have a unique binomial, as well as a designated and permanently preserved 'type' (i.e., holotype). Living cultures are allowed as holotypes as long as they are preserved in a metabolically inactive state (i.e., liophilization or liquid nitrogen) and deposited in an internationally recognized culture collection. Type specimens can also be required by law to be deposited in public reference collections in the country from where the specimens were originally isolated. It is increasingly expected that descriptions of new species should be accompanied by molecular data to integrate them into molecular phylogenies (Seifert and Rossman 2010).

11.14 Future Perspectives

The feasibility of *in vitro* culturing and production of mycelium, conidia and chlamydospores of *Pochonia* spp. at small or large scale has contributed to the increasing use of *Pochonia* spp. as a biological control agent of plant-parasitic nematodes. The ovicidal potential of *P. chlamydosporia* has also started to be explored in order to control animal-parasitic nematodes and other helminths in cattle, and new culture methods are continuously being produced. However, the pathogenicity of different propagules to nematodes, including conidia and blastospores produced under different nutritional conditions, still needs to be assessed (Mo et al. 2005). Some challenges lie ahead, including standardization of culture and production methods, enhancing the shelf life of *Pochonia* propagules, development of formulations and workable doses, and application methods, in order to ensure that available products commonly produced using some of the methods mentioned in this chapter can be used effectively as another tool within IPM of plant-parasitic nematodes and other invertebrates.

References

Arevalo, J., Hidalgo-Díaz, L., Martins, I., et al. (2009). Cultural and morphological characterization of *Pochonia chlamydosporia* and *Lecanicillium psalliotae* isolated from *Meloidogyne mayaguensis* eggs in Brazil. *Tropical Plant Pathology, 34*, 158–163.

Atkins, S. D., Hidalgo-Diaz, L., Clark, I. M., et al. (2003). Approaches for monitoring the release of *Pochonia chlamydosporia* var. *catenulata*, a biocontrol agent for root-knot nematodes. *Mycological Research, 107*, 206–212.

Atkins, S. D., Clark, I. M., Pande, S., et al. (2005). The use of real-time PCR and species-specific primers for the identification and monitoring of *Paecilomyces lilacinus*. *FEMS Microbiology Ecology, 51*, 257–264.

Bourne, J. M., & Kerry, B. R. (2000). Observations on the survival and competitive ability of the nematophagous fungus *Verticillium chlamydosporium* in soil. *International Journal of Nematology, 10*, 9–18.

Brown, A. D. (1978). Compatible solutes and extreme water stress in eukaryotic microorganisms. *Advances in Microbial Physiology, 17*, 181–242.

Brown, A. D., & Simpson, J. R. (1972). Water relations of sugar-tolerant yeasts: The role of intracellular polyols. *Journal of General Microbiology, 72*, 589–591.

Brownell, K. H., & Schneider, R. W. (1985). Roles of matric and osmotic components of water potential and their interaction with temperature in the growth of *Fusarium oxysporum* in synthetic media and soil. *Phytopathology, 75*, 53–57.

Ceiro, W. G., Arevalo, J., Puertas, A., et al. (2013). Tolerancia de *Pochonia chlamydosporia* var. catenulata (Kamyschko ex Barron y Onions) Zare y W. Gams a diferentes niveles de cloruro de sodio. *Revista de Protección Vegetal, 28*, 70–73.

Chalupova, V., Lenhart, K., & Lysek, H. (1977). Ovicidal activity of morphological mutants of *Verticillium chlamydosporium*. *Acta Universitatis Palackianae Olomucensis, 81*, 51–61.

Choi, Y. W., Hyde, K. D., & Ho, W. H. (1999). Single spore isolation of fungi. *Fungal Diversity, 3*, 29–38.

Ciancio, A., Colagiero, M., Pentimone, I., et al. (2016). Formulation of Pochonia chlamydosporia for plant and nematode management. In N.K. Arora et al. (Eds.), Bioformulations for sustainable agriculture (pp.177–197). Springer India

Clyde, J. M. F. (1993). The cyst nematode pathogen *Verticillium chlamydosporium*, PhD thesis, The University of Leeds, Department of Pure and Applied Biology, UK.

Cook, R. J., & Baker, K. F. (1983). *The nature and practice of biological control of plant pathogens*. Minnesota: The American Phytopathological Society.

Coosemans, J. (1990). Seed pellet inoculation of sugar beet with freeze dried *Verticillium chlamydosporium* culture against *Heterodera schachtii*. In *Proceedings of the international symposium on crop protection* (Vol. 55, pp. 713–718). Gent: Mededelingen van de Faculteit Landbouwwetenschappen Rijksuniversiteit.

de Carvalho, L. M., Braga, F. R., Domingues, R. R., et al. (2014). Interaction of the nematophagous fungus *Pochonia chlamydosporia* and *Parascaris equorum* eggs in different culture media. *Journal of Basic Microbiology, 54* (Suppl 1), S109–S114. doi:10.1002/jobm.201300586

Deacon, J. W. (2006). *Fungal Biology* (4th ed.). Oxford: Blackwell Publishing.

De Leij, F. A. A. M., Kerry, B. R., & Dennehy, J. A. (1993). *Verticillium chlamydosporium* as a biological control agent for *Meloidogyne incognita* and *M. hapla* in pots and micro-plot tests. *Nematologica, 39*,115–126

Escudero, N., & Lopez-Llorca, L. V. (2014). Estudio de la interacción tritrófica: tomate *Meloidogyne javanica y Pochonia chlamydosporia*. *Boletín Informativo de la Sociedad Española de Fitopatología, 87*, 73–81.

Esteves, I. (2007). Factors affecting the performance of *Pochonia chlamydosporia* as a biological control agent for nematodes, PhD thesis, Cranfield University, UK.

Esteves, I., Peteira, B., Powers, S., et al. (2009). Effects of osmotic and matric potential on growth and accumulation of endogenous reserves in three isolates of *Pochonia chlamydosporia*. *Biomedical Science and Technology, 19*, 185–199.

Franco-Navarro, F., Vilchis-Martínez, K., & Miranda-Damián, J. (2009). New records of *Pochonia chlamydosporia* from Mexico: Isolation, root colonization and parasitism of *Nacobbus aberrans* eggs. *Nematropica, 39*, 133–142.

Gams, G. (1988). A contribution to the knowledge of nematophagous species of *Verticillium*. *Netherlands Journal of Plant Pathology, 94*, 123–148.

Gao, L. (2011). Optimize culture conditions for biomass and sporulation of the nematophagous fungus *Pochonia chlamydosporia*, isolate HSY-12-14 by "two-step" cultivation. *African Journal of Microbiology Research, 5*, 4731–4738.

Gao, L. (2015) Application of two-stage cultivation for exploring the nutritional requirements for sporulation of three biocontrol fungi. *Biotechnology Research International*, Art ID 682839, 5. http://www.dx.doi.org/10.1155/2015/682839

Gao, L., Liu, X. Z., Sun, M. H., et al. (2009). Use of a novel two-stage method to determine the effects of environmental factors on the growth and sporulation of several biocontrol fungi. *Mycoscience, 50*, 317–321.

Garrett, S. D. (1981). *Soil fungi and soil fertility* (2nd ed.). Oxford: Pergamon Press.

Gilchrist-Saavedra, L., Fuentes-Dávila, G., Martínez-Cano, C., et al. (2006). *Practical guide to the identification of selected diseases of wheat and barley* (2nd ed.). Mexico: CIMMYT.

Griffin, D. H. (1994). *Fungal physiology* (2nd ed.pp. 130–157). New York: Wiley-Liss.

Hallsworth, J. E., & Magan, N. (1994a). Effect of carbohydrate type and concentration on polyhydroxy alcohol and trehalose content of conidia of three entomopathogenic fungi. *Microbiology, 140*, 2705–2713.

Hallsworth, J. E., & Magan, N. (1994b). Effects of KCl concentration on accumulation of acyclic sugar alcohols and trehalose in conidia of three entomopathogenic fungi. *Letters in Applied Microbiology, 18*, 8–11.

Hidalgo-Díaz, L. (2003). *Standard operating procedures for mass production of Pochonia chlamydosporia*. Havana: OCIC, CENSA.

Hirsch, P. R., Atkins, S. D., Mauchline, T. H., et al. (2001). Methods for studying the nematophagous fungus *Verticillium chlamydosporium* in the root environment. *Plant & Soil, 232*, 21–30.

Jacobs, H. (2000). *Development of a fungal biological control agent for potato cyst nematodes in Jersey*, PhD thesis, University of Bedfordshire, UK.

Jenkins, N. E., & Grzywacz, D. (2000). Quality control of fungal and viral biocontrol agents – assurance of product performance. *Biomedical Science and Technology, 10*, 753–777.

Karakas, M., Benli, M., & Cebesov, S. (2012). Effects of some cultural conditions on the growth of nematophagous fungus *Pochonia chlamydosporia* (fungi: Clavicipitacea) isolated from *Meloidogyne incognita* eggs. *Journal of Animal and Veterinary Advances, 11*, 4644–4647.

Kerry, B. R., & Crump, D. H. (1977). Observations on fungal parasites of females and eggs of the cereal cyst-nematode, *Heterodera avenae*, and other cyst nematodes. *Nematologica, 23*, 193–201.

Kerry, B. R., Irving, F., & Hornsey, J. C. (1986). Variation between strains of the nematophagous fungus *Verticillium chlamydosporium* Goddard I. Factors affecting growth *in vitro*. *Nematologica, 32*, 461–473.

Kerry, B. R., Kirkwood, I. A., De Leij, F. A. A. M., et al. (1993). Growth and survival of *Verticillium chlamydosporium* Goddard, a parasite of nematodes, in soil. *Biomedical Science and Technology, 3*, 355–365.

Liu, X. Z., & Chen, S. Y. (2003). Nutritional requirements of *Pochonia chlamydosporia* and ARF18, fungal parasites of nematode eggs. *Journal of Invertebrate Pathology, 83*, 10–15.

Lopez-Llorca, L. V., & Duncan, J. M. (1986). New media for the estimation of fungal infection in eggs of the cereal cyst nematode, *Heterodera avenae* Woll. *Nematologica, 32*, 486–490.

Luambano, N. D., Manzanilla-López, R. H., Kimenju, J. W., et al. (2015). Effect of temperature, pH, carbon and nitrogen ratios on the parasitic activity of *Pochonia chlamydosporia* on *Meloidogyne incognita. Biological Control, 80*, 23–29.

Luard, E. J. (1982). Effect of osmotic shock on some intracellular solutes in two filamentous fungi. *Journal of General Microbiology, 128*, 2575–2581.

Magan, N., & Lynch, J. M. (1986). Water potential, growth and cellulolysis of fungi involved in decomposition of cereal residues. *Journal of General Microbiology, 132*, 1181–1187.

Maheshwari, R. (2012). *Fungi. Experimental methods in biology* (2nd ed.). Boca Raton: CRC Press.

Mauchline, T. H., Kerry, B. R., & Hirsch, P. R. (2002). Quantification in soil and the rhizosphere of the nematophagous fungus *Verticillium chlamydosporium* by competitive PCR and comparison with selective plating. *Applied and Environmental Microbiology, 68*, 1846–1853.

Meikle, A. J., Chudek, J. A., Reed, R. H., et al. (1991). Natural abundance 13C-nuclear magnetic resonance spectroscopic analysis of acyclic polyol and trehalose accumulation by several yeast species in response to salt stress. *FEMS Microbiology Letters, 66*, 163–167.

Minnis, A. M., Seifert, K. A., Redhead, S. A., et al. (2016). *2014 nomenclatural procedures FAQ (ICN/Melbourne code): Help and checklist for authors*. Mycologia. http://www.mycologia.org/site/misc/FAQvers2.xhtml. Accessed 3 Nov 2016.

Mo, M. H., CK, X., & Zhang, K. Q. (2005). Effects of carbon and nitrogen sources, carbon-nitrogen ratio, and initial pH on the growth of nematophagous fungus *Pochonia chlamydosporia* in liquid culture. *Mycopathologia, 159*, 381–387.

Mossel, D. A. A. (1970). Microbiological culture media as ecosystems. In: Van Bragt, J., Mossel, D. A. A., Pierik, R. L. M., Veldstra, H. (Eds.), *Effects of sterilization on components in nutrient media*. Miscellaneous papers 9 (1971) Landbouwhogeschool (pp. 15–39), Wageningen, The Netherlands.

Mubyana-John, T., & Taylor, J. (2015). Nematophagous fungi: Regulations and safety. In H. T. Askary & M. PPB (Eds.), *Biocontrol agents of phytonematodes* (pp. 203–213). Wallingford: CABI.

Nagesh, M., Hussaini, S. S., Ramanujam, B., et al. (2007). Molecular identification, characterization, variability and infectivity of Indian isolates of the nematophagous fungus *Pochonia chlamydosporia. Nematologia Mediterranea, 35*, 47–56.

Nakasone, K. K., Peterson, S. W., & John, S. C. (2004). Preservation and distribution of fungal cultures. In G. M. Mueller, G. F. Bills, & M. S. Foster (Eds.), *Biodiversity of fungi. Inventory and monitoring methods* (pp. 37–47). Burlington: Elsevier Academic Press.

Nonaka, K., Omura, S., Masuma, R., et al. (2013). Three new *Pochonia* taxa (Clavicipitaceae) from soils in Japan. *Mycologia, 105*, 1202–1218.

Olivares-Bernabeu, M. C., & Lopez-Llorca, L. V. (2002). Fungal egg-parasites of plant-parasitic nematodes from Spanish soils. *Revista Iberoamericana de Micología, 19*, 104–110.

Ottow, J. C. G. (1972). Rose Bengal as a selective aid in the isolation of fungi and Actinomycetes from natural sources. *Mycologia, 64*, 304–315.

Peteira, B., Esteves, I., Montes de Oca, N., et al. (2007). Estabilidad de la cepa IMI SD 187 de *Pochonia chlamydosporia* var. *catenulata* en medio sólido. *Revista de Protección Vegetal, 22*, 124–127.

Puertas, A., Arévalo, J., Montes de Oca, N., et al. (2006). Efecto de diferentes concentraciones de inóculo de la cepa IMI SD 187 de *Pochonia chlamydosporia* var. *catenulata* para el control de *Meloidogyne incognita. Revista de Protección Vegetal, 21*, 74–79.

Ryan, M. J., Bridge, P. D., Smith, D., et al. (2002). Phenotypic degeneration occurs during sector formation in *Metarhizium anisopliae. Journal of Applied Microbiology, 93*, 163–168.

Sanglard, D. A., Balbi, B. P., Barros, E. G., et al. (2009). An efficient protocol for isolation, sporulation and maintenance of *Pseudocercospora griseola. Annual Report of the Bean Improvement Cooperative, 52*, 62–63.

Seifert, K. A., & Rossman, A. Y. (2010). How to describe a new fungal species. *IMA Fungus, 1*, 109–116.

Silva, A. R., Araújo, J. V., Braga, F. R., et al. (2010). In vitro ovicidal activity of the nematophagous fungi *Duddingtonia flagrans*, *Monacrosporium thaumasium* and *Pochonia chlamydosporia* on *Trichuris vulpis* eggs. *Veterinary Parasitology, 172*, 76–79.

Smith, D. (1991). Maintenance of filamentous fungi. In B. E. Kirsop & A. Doyle (Eds.), *Maintenance of microorganisms and cultured cells – A manual of laboratory methods* (pp. 133–159). London: Academic.

Smith, D., & Onions, A. H. S. (1994). *The preservation and maintenance of living fungi, IMI technical handbooks No. 2* (2nd ed.). Wallingford: CAB International.

Stirling, G. R., Licastro, K. A., West, L. M., et al. (1998). Development of commercially acceptable formulations of the nematophagous fungus *Verticillium chlamydosporium*. *Biological Control, 11*, 217–223.

Sun, M. H., & Liu, X. Z. (2006). Carbon requirements of some nematophagous, entomopathogenic and mycoparasitic Hyphomycetes as fungal biocontrol agents. *Mycopathologia, 161*, 295–305.

Sykes, D,. (1994). The growth and sporulation of *Verticillium chlamydosporium*, MSc thesis, University of Manchester, UK.

Van Eck, J. H., Prior, B. A., & Brandt, E. V. (1993). The water relations of growth and polyhydroxy alcohol production by ascomycetous yeasts. *Journal of General Microbiology, 139*, 1047–1054.

Vandenberg, J. D., & Cantone, F. A. (2004). Effect of serial transfer of three strains of *Paecilomyces fumosoroseus* on growth in vitro, virulence and host specificity. *Journal of Invertebrate Pathology, 85*, 40–45.

Verdejo-Lucas, S., Ornat, C., Sorribas, F. J., et al. (2002). Species of root-knot nematodes and fungal egg parasites recovered from vegetables in Almería and Barcelona, Spain. *Journal of Nematology, 34*, 405–408.

Vieira dos Santos, C., Esteves, I., Abrantes, I. (2012). In vitro water stress bioassays with the nematophagous fungus *Pochonia chlamydosporia*: Effects on growth and parasitism. *Biological Control, 63*, 310–319.

Vieira dos Santos, M. C., Esteves, I., Kerry, I., et al. (2013). Biology, growth parameters and enzymatic activity of *Pochonia chlamydosporia* isolated from potato cyst and root-knot nematodes. *Nematology, 15*, 493–504.

Vogel, H. J. (1956). A convenient growth medium for *Neurospora* (medium N). *Microbial Genetics Bulletin, 13*, 42–43.

Waller, J. M., Lenné, J. M., & Waller, S. J. (2002). *Plant pathologist's pocketbook* (3rd ed.pp. 390–391). Oxon: CABI.

Ward, E., Kerry, B. R., Manzanilla-López, R. H., et al. (2012). The *Pochonia chlamydosporia* serine protease gene*vcp1* is subject to regulation by carbon, nitrogen and pH: Implications for nematode biocontrol. *PloS One, 7*(4), e35657.

Whatcom County Station. (2016). *Compost fundamentals*, Washington State University. http://whatcom.wsu.edu/ag/compost/fundamentals/needs_carbon_nitrogen.htm. Accessed 23 May 2016.

Xu, L. L., Li, F., & al, X. H. (2009). A novel method for promoting conidial production by a nematophagous fungus, *Pochonia chlamydosporia* AS6.8. *World Journal of Microbiology and Biotechnology, 25*, 1989–1994.

Yang, J., Loffredo, A., Borneman, J., et al. (2012). Biocontrol efficacy among strains of *Pochonia chlamydosporia* obtained from a root-knot nematode suppressive soil. *Journal of Nematology, 44*, 67–71.

Zare, R., Gams, W., & Cullan, A. (2000). A revision of *Verticillium* section *Prostrata*. I. Phylogenetic studies using ITS sequences. *Nova Hedwigia, 71*, 465–480.

Zare, R., Gams, W., & Evans, H. C. (2001). A revision of *Verticillium* section *Prostrata*. V. The genus *Pochonia*, with notes on *Rotiferophthora*. *Nova Hedwigia, 73*, 51–86.

Chapter 12
Methodology Part II. *Pochonia* spp.: Screening and Isolate Selection for Managing Plant-Parasitic Nematodes

Ivânia Esteves, Rosa Navarrete-Maya, and Rosa H. Manzanilla-López

Abstract Production of *Pochonia* spp. in laboratory conditions has facilitated studies on its biology, abundance, dispersion, rhizosphere colonization, host preference, and isolate virulence. Research in biological control of nematodes requires suitable, standardized methods. In this chapter we review the commonest, non-molecular, standard *in vitro* culture methods to isolate, screen, and select isolates, some of which may eventually be produced on a larger scale for application in combination with other management strategies for plant-parasitic nematodes.

12.1 Introduction

Research in nematophagous fungi and their tritrophic interactions (fungus-plant-nematode) is continuously advancing from basic to high tech molecular studies. This trend is especially noticeable in countries and research institutions where *Pochonia* spp. studies started in the 1970s. However, studies of nematophagous fungi as potential biological control agents of nematodes are now starting in other countries, where they focus primarily in "the detection of isolates to be used either

I. Esteves (✉)
CFE-Centre for Functional Ecology, Department of Life Sciences, Calçada Martim de Freitas, University of Coimbra, Coimbra P3000-456, Portugal
e-mail: iesteves@uc.pt

R. Navarrete-Maya
UNIGRAS (Unidad de Investigación en Granos y Semillas), Facultad de Estudios Superiores Cuautitlán, Universidad Nacional Autónoma de México, Av. Dr. Jorge Jiménez Cantú s/n, Cuautitlán Izcalli C.P. 54740, Mexico
e-mail: rosa_navarrete@hotmail.com

R.H. Manzanilla-López
Centro de Desarrollo de Productos Bióticos (Visiting Professor), Yautepec, Morelos C.P. 62731, Mexico
e-mail: rosa.manzanilla@gmail.com

R.H. Manzanilla-López, L.V. Lopez-Llorca (eds.), *Perspectives in Sustainable Nematode Management Through* Pochonia chlamydosporia *Applications for Root and Rhizosphere Health*, Sustainability in Plant and Crop Protection, DOI 10.1007/978-3-319-59224-4_12

for local control of nematode pests and/or future development of biopesticide commercial products for nematode control" (Nicolay and Sikora 1991). It is worth mentioning then, that inadequate selection of methodology at an early stage of a research programme may result in 'the worldwide promotion of isolates with low levels of activity and a significant loss in research funding, manpower and progress' (Nicolay and Sikora 1991). Therefore, it is important to be familiar from the outset with at least some of the standard, basic guidelines and techniques used to study the fungus in order to detect isolates that can work in "effective biological control systems" and to avoid techniques that "favour isolation of saprophytic fungi and not parasitic species" (Nicolay and Sikora 1991; Kerry 2000). An extended version of the methods described in this chapter can be found in two bulletins from the International Organization for Biological Control and Integrated Control of Noxious Animals and Plants (Kerry 1991; Kerry and Bourne 2002). Readers are especially encouraged to refer to the *Manual for research on* Verticillium chlamydosporium, *a potential biological control for root-knot nematodes* (Kerry and Bourne 2002), as an essential introduction to *Pochonia chlamydosporia* standard working practices. The manual was an output of the European funded project FAIR5-PL97-3444 (Development of a sustainable strategy for the management of root-knot nematodes in vegetable crops in southern Europe – an alternative to the use of the methyl bromide). The many contributions made by scientists helped to produce and standardize the protocols to study this nematophagous fungus. Many of these protocols are used worldwide. Figure 12.1 is a summary of the basic steps involved in the processes involved from fungal isolation to field trials. These and other related methods are revised in the context of *Pochonia* spp. as a potential biological control agent for root-knot (*Meloidogyne* spp.), false root-knot (*Nacobbus aberrans sensu lato*) and cyst nematodes (*Heterodera* spp., *Globodera* spp.).

12.2 Sampling and Isolation from Soil, Roots and Nematodes

Populations of *P. chlamydosporia* (= *V. chlamydosporium*) are implicated in the decline of plant-parasitic nematodes in nematode suppressive soils (Kerry 2000) yet little is known of the genetic diversity of the fungus in naturally- or agroecosystem-occurring populations (Morton et al. 2003; Franco-Navarro et al. 2009; Giné et al. 2016). A nematode suppressive soil can be defined as one where a nematode population, which was once at an economic damaging level, becomes suppressed and remains at a level that causes little or no economic damage to a nematode susceptible plant cultivar (Chen and Dickson 2012). Therefore, either places where suppressiveness to nematodes and other soil-borne pathogens has been observed, or intensively cropped soils with a long history of nematode infestation are suitable environments to look for the fungus (Kerry and Bourne 2002). It is worth considering that different fungal biotypes or isolates with different molecular profiles or nematode host preferences can be found co-existing in the same host and soil sample location. *Pochonia chlamydosporia* may also be part of a complex of similar

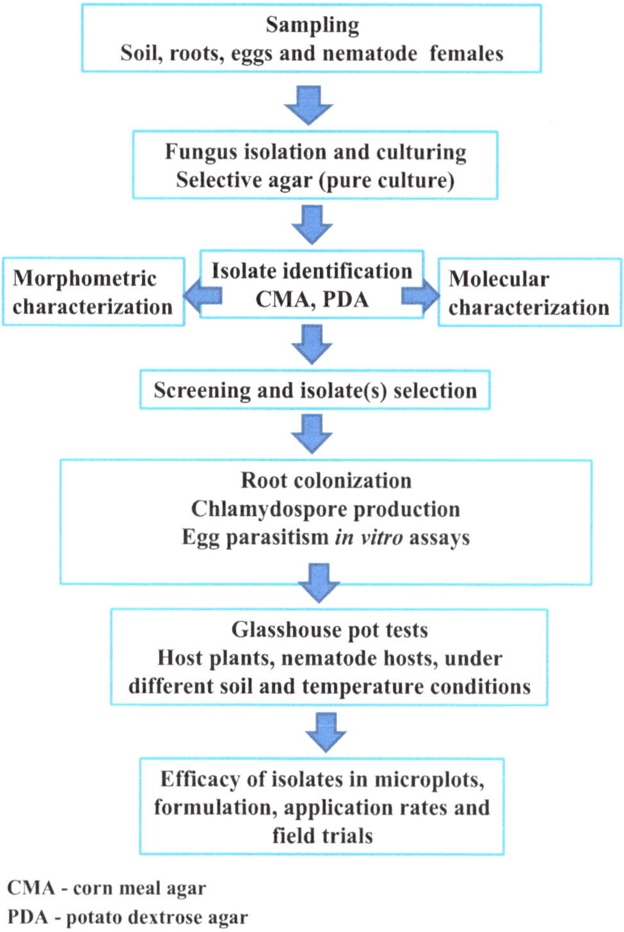

Fig. 12.1 Diagram showing *Pochonia chlamydosporia* screening processes of isolates from isolation of the fungus from soil, roots, eggs and nematode samples, to field trials of selected isolates. (After Kerry and Bourne 2002)

species and varieties with considerable variation between biotypes, especially *in vitro* growth and pathogenicity (Irving and Kerry 1986; Kerry et al. 1986; see also Chaps. 2, 3, and 4). Understanding the importance of such character variability and their interactions is crucial for the successful use of this natural microorganism as a biological control agent (Esteves 2007).

12.2.1 Where to Look?

Pochonia chlamydosporia has been isolated worldwide from soils, molluscs, nematodes, insects, and soil-borne fungi (Domsch et al. 1980; Sykes 1994; Zare et al. 2001; Chap. 2). The fungus can behave as a saprophyte, facultative parasite, hyperparasite, and endophyte (Lopez-Llorca et al. 2002a, b, 2010; Maciá-Vicente et al. 2009a, b; Schulz et al. 2015).

Despite a worldwide distribution there are few records in the literature of extensive and intensive surveys having been carried out looking specifically for *Pochonia* spp. in natural ecosystems, the search for the fungus in agroecosystems mainly being focused on crops affected by root-knot nematodes (RKN) in countries such as Brazil (Carneiro et al. 2008; Arevalo et al. 2009; Dallemole-Giaretta et al. 2012), Cuba (Hidalgo-Díaz et al. 2000), China (Sun et al. 2006; Aminuzzaman et al. 2013), Kenya (Bourne et al. 2004), Mexico (Flores-Camacho et al. 2007; Franco-Navarro et al. 2009; Medina-Canales et al. 2013), Spain (Verdejo-Lucas et al. 2002), USA (Morgan-Jones et al. 1981); and also cyst nematodes in countries such as Germany, the UK, The Netherlands, Spain (Olivares-Bernabeu and López-Llorca 2002), Norway and Poland (Chap. 16), India (Nagesh et al. 2007), Iran (Moosavi et al. 2010), and the USA (Chen and Chen 2003), among others.

Nematophagous fungi are part of dynamic natural communities where populations, community composition and function change continuously both qualitatively and quantitatively over time (Lynch 1996). Field surveys for the fungus, must consider that soil health, natural vegetation, local climate, crop, and population levels of the target nematode pest or other hosts may play an important role in fungus abundance, diversity and parasitic ability. In general, fungi sampling is carried out using methods applied to sampling macro-organisms and therefore targets macro-habitats rather than microhabitats, possibly resulting in loss of valuable information on habitat and fungi associations (Lynch 1996).

Pochonia chlamydosporia var. *chlamydosporia*, *P. chlamydosporia* var. *catenulata* and *P. suchlasporia* (= *V. suchlasporium*) have been isolated from tropical and subtropical soil samples (Hidalgo-Díaz et al. 2000; Arevalo et al. 2009; Franco-Navarro et al. 2009) whereas *P.* [*Metapochonia*] *suchlasporia* has also been isolated from temperate areas (Bourne and Kerry 2000). In a study carried out in a natural forest and secondary forest region of Los Tuxtlas (Mexico), Franco-Navarro et al. (2009) isolated *P. chlamydosporia* var. *chlamydosporia* and *P. chlamydosporia* var. *catenulata*, the later variety being isolated for the first time in Mexico from these ecosystems. The number of isolates obtained of the two *Pochonia* varieties was higher in forest soils in comparison to those occurring in soils from pasture and maize fields. In a Cuban survey of the two main coffee-growing regions, 32% of the 83 isolates obtained belonged to *P. chlamydosporia* var. *chlamydosporia*, 63% to *P. chlamydosporia* var. *catenulata*, and 1.2% to *P.* [*Metapochonia*] *suchlasporia*; the remaining 4% belonged to other fungi. In the same survey, the number of colony forming units (cfu) of *Pochonia* spp. in Cuban soils ranged from 13–575 cfu/g soil, thus showing that the fungus can be present at very low levels in the coffee agro-ecosystem (Hidalgo-Díaz et al. 2000). In Brazil, *P. chlamydosporia* var. *chlamydosporia* and *P. chlamydosporia* var. *catenulata* were isolated from soil and eggs samples taken from *M. mayaguensis* (= *M. enterolobii*) in commercial guava plantations in an area having a semiarid climate and sandy soils (Arevalo et al. 2009). In general, isolates are more often isolated from vegetable crops soil than from parasitized nematode eggs. This may be partly related to the fungus abundance, diversity, saprophytic capacity, and competition with other fungi, but also to the difficulty in finding infective propagules at an appropriate sampling time as RKN and other nematodes can decay rapidly in soil (Nagesh et al. 2007). A survey for *Pochonia* isolates parasitizing *N. aberrans s.l.* in ten different geographical locations and vegetable crops from Central Mexico (Flores-Camacho et al. 2007) produced only four *P. chlamydosporia* var. *chlamydosporia* isolates from soil, and one isolate from egg masses. A similarly low number of *Pochonia* isolates was recorded from another vegetable growing region in Central Mexico (Medina-Canales et al. 2013), where *M. arenaria*, *M. hapla* and *M. incognita* occur as important pests. Only 7 (<1%) out of 105 screened samples belonged to *P. chlamydosporia* var. *mexicana*. In another Brazilian survey in the Federal district of Brazil, this time from vegetable crops, five strains of *Purpureocillium lilacinum* (= *Paecilomyces lilacinus*), one strain of *Beauveria bassiana* and one strain of *P. chlamydosporia* were isolated from *Meloidogyne* spp. eggs; by comparison, a larger number of entomopathogenic fungi isolates were isolated from soil, including 20 strains of *B. bassiana*, 26 of *Metarhizium anisopliae* and 20 of the nematophagous fungus *P. lilacinum* (= *P. lilacinus*). Microflora isolates taken from eggs and females of *Meloidogyne* spp. in China showed that *P. lilacinum* (= *P. lilacinus*) was present in 49.3% of fungi isolates whereas only 6.9% belonged to *P. chlamydosporia* (Sun et al. 2006).

Continuous cultivation or monoculture of crops can sometimes lead to a selective increase in antagonistic organisms and the development of disease suppressive soils (Timper 2014). It is advisable, therefore, that when searching for isolates of the fungus in monoculture locations, sites where crops are already heavily infested by nematodes are not selected as high nematode populations can indicate that biological agents are unable to exert control (see also Verdejo-Lucas et al. 2002). Under such conditions, nematode eggs can escape infection due to physical inaccessibility to the fungus as eggs are lodged within large egg masses (Bourne et al. 1996). Care should be taken during the isolation process as isolates of the fungus taken from eggs are preferred for further testing rather than those isolated from soil or roots (rhizoplane) so as to ensure a parasitic rather than saprophytic activity of the isolate (Nicolay and Sikora 1991).

12.3 When to Look?

The activity of fungal egg parasites is affected by crop, specific development stages of the target nematode, and other fungi occurring as part of the natural succession process in the soil (Crump 1987a, b; Bernard et al. 1997). Fungi succession can be defined as ‘the sequential occupation of the same site by thalli (normally mycelia) either of different fungi or of different associations of fungi’ (Rayner and Todd 1979). Therefore, more than one sampling at different time of crop phenology and nematode development may be required to target parasitic isolates of *Pochonia* spp. The fungus has both endophytic and rhizosphere abilities, the latter of these abilities allowing the fungus to colonize the rhizoplane as a saprotroph until it infects nematode females (i.e., cyst nematodes) and/or eggs as a parasite. In the case of RKN, under experimental conditions, the fungus can be isolated from 5 to 9 weeks after the addition of second-stage juveniles (J2) to soil and when egg masses have been produced (Bourne and Kerry 1999; Manzanilla-López et al. 2009, 2011a). In contrast to RKN, cyst nematode eggs are not directly exposed to the rhizosphere since they remain inside the encysted female until they develop into J2. Therefore, it is considered that the time that the fungus has to infect and kill RKN eggs is shorter than for cyst nematode eggs (Kerry 1997). Ideally, local studies of the population dynamics or epidemiology of the target nematode could be useful in selecting the appropriate timing to look for the fungus in colonized egg masses or cysts.

12.4 Sampling

Plants play an important role as a host of both *P. chlamydosporia* and the nematode; the three of them being part of a tritrophic interaction. Depending on research objectives sampling methods may vary according to when and where the sampling is going to be carried out and the target (natural plant communities, annual or perennial crops, fungus species or fungal community, and nematode species). Sampling should include always both soil and root samples. Data such as sampling date, geographical location (using GPS), isolation substrate or host, and the name of the person who took the samples should always be recorded.

Nematodes are currently sampled close to the plant rhizosphere at 5–30 cm soil depth (Giné et al. 2016). The vertical distribution of the fungus occurs in the first 50 cm of soil (De Leij et al. 1993a; Kerry and Bourne 1996) and its horizontal distribution can reach 10–15 cm from the plant rhizosphere (De Leij et al. 1993a; Kerry and Bourne 1996). Arevalo et al. (2009) collected six randomized samples of soil (500 g) and *Meloidogyne* infected roots at a depth of 10 cm when sampling commercial guava plantations. The samples were placed in polythene bags and kept at 4 °C until processing for isolation of egg parasitic fungi. Franco-Navarro et al. (2009) sampled forest, pasture and maize soils following a grid pattern. Sampling points were positioned 200 m from each other and subsamples taken where grid lines crossed following two concentric circles centred around each point, i.e., four subsamples from the inner circle (3 m diameter) and another four from the outer circle (6 m diameter). Each subsample (100 g) was taken at a depth of 10–30 cm. All subsamples per locality were mixed thoroughly in polythene bags and 100 g soil subsamples taken to isolate the fungus.

Two intensive vegetable production areas of north and south Spain (Almería and Barcelona) with different cropping cycles, climatic conditions, and agronomic practices, were selected to determine the *Meloidogyne* spp. and their egg parasites occurring in plastic houses and the field. Composite samples per site included 6–10 plants that were dug from patches with RKN damage symptoms, galled roots, and rhizosphere soil (ca 2 kg soil per site). Samples were sieved to separate roots from soil, and one 500 cm^3 soil subsample per site was used for nematode extraction on

Baermann trays. Fungal parasites were found in 37% (Almería) and 45% (Barcelona) of the sites but parasitism was never greater than 5% (Verdejo-Lucas et al. 2002). In another study carried out in tunnel houses, 25 soil samples (200–500 g each) were taken to a depth of 15 cm at random along a 'W walk' within each tunnel house and the samples bulked per site (Manzanilla-López et al. 2009).

12.4.1 Soil Analysis

Soil samples should be sent for texture, pH, nutrient content, organic matter content, and other relevant soil physicochemical analysis. This information is needed to identify the original habitat conditions where the fungus, crop and nematodes co-exist. Implementation of sustainable management strategies will require such 'soil health' history/records to enhance the selected isolate's chances of success as a biological control agent.

12.4.2 Other Fungi Isolated from Samples

Different fungi may co-exist in soils as members of a microbial community. Hence, information about other fungi isolated from roots, soils or nematode samples from which *Pochonia* spp. are isolated should not be discarded or ignored but considered in producing a more holistic view of the soil microbial community and potential multitrophic interactions.

12.4.3 Identification of Nematode Hosts

Nematode host species from which isolates are taken must be identified as accurately as possible. Molecular confirmation of identification is required for *Meloidogyne* spp. (see Carneiro et al. 2008), *Heterodera* spp. and *Globodera* spp. (see Hunt and Handoo 2012; Subbotin 2012).

12.5 Isolation

The fungus can be isolated as an isolate, i.e., "the first 1-spore or pure isolation of a fungus from any place" (Kirk et al. 2008) from soil, roots, females and nematode eggs using dilution plating on a selective-agar medium (Kerry and Bourne 2002; Ciancio et al. 2016). Individual colonies of *Pochonia chlamydosporia* can be visually identified as *Pochonia* sp. with the aid of a stereo-microscope at

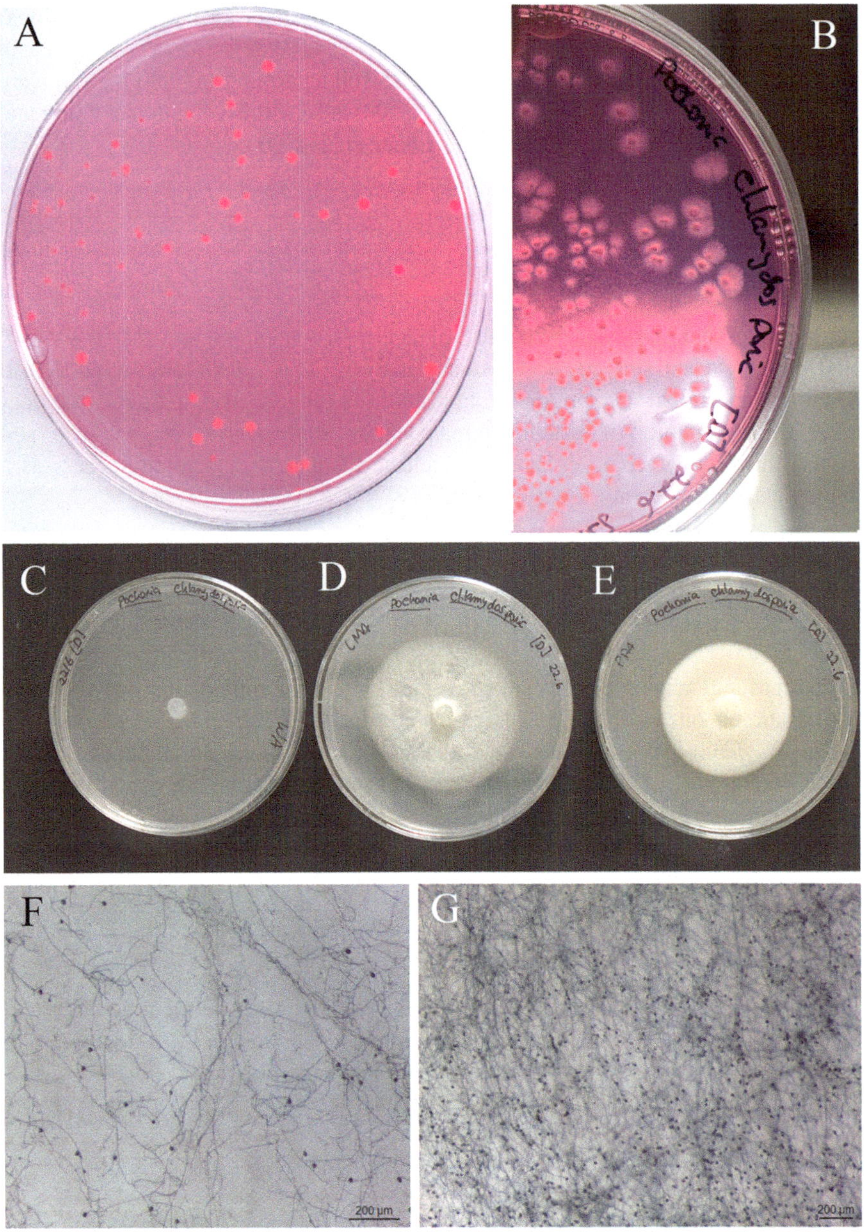

Fig. 12.2 *Pochonia chlamydosporia* colony growth in different media plates 14 days after inoculation with one pre-colonised corn meal agar plug. (**A**), (**B**) Selective agar; (**C**) Water agar; (**D**) Corn meal agar; (**E**) Potato dextrose agar; (**F**) Hyphae and chlamydospores in water agar; (**G**) Corn meal agar (Courtesy I. Esteves)

25 × magnification (Fig. 12.2). However, it is advisable to make a slide to look for morphological features of the different *Pochonia* varieties using a compound microscope. To confirm *Pochonia* spp. identification, colonies are picked from selective-agar plate (Fig. 12.2A, B) and transferred onto 0.8% water agar (WA, Fig. 12.2C), or onto 1.7% Corn Meal Agar (CMA, Fig. 12.2D), or Potato Dextrose Agar (PDA, Fig. 12.2E) to look for characteristic conidiophores (PDA) or for chlamydospore production (Fig. 12.2F, G). Isolate identity should also be confirmed molecularly (see Chap. 13). Strains or 'a group of clonally related individual or cells' (Kirk et al. 2008) can be produced from selected isolates through culturing.

12.5.1 Isolation from Soil

The fungus is a relatively poor competitor in soils (Bourne and Kerry 2000), its proliferation being greater in organic than in mineral soils (Kerry et al. 1984; De Leij et al. 1993b). It can also occur in soil in the absence of plants as conidia, mycelium, resting spores or chlamydospores.

A standard protocol to isolate the fungus from the original bulk soil requires the whole bulk sample to be thoroughly mixed so that a single subsample can be representative (Kerry and Bourne 2002). The subsample (1 g) is suspended for a few minutes in 9 ml of autoclaved and cooled 0.05% agar solution, mixed and shaken vigorously (20 s). The solution is diluted at 10^{-2} and 10^{-3} in 9 ml vials (i.e., glass universals) and 0.2 ml of each dilution are plated onto the semi selective medium (usually 3 plates/dilution). Plates are incubated at 25 °C for 10–15 days and the colonies counted to calculate the cfu/g soil. To give a measure of cfu/g dry soil a known amount of soil is weighed, dried and the amount of water/g soil calculated (Kerry and Bourne 2002). It is not recommended to leave soil samples longer than 1 h in the agar solution before plating to avoid reducing fungal propagule viability. Soil samples can be stored at 4 °C before processing if necessary (Kerry and Bourne 2002).

12.5.2 Extraction of Chlamydospores from Soil

A method to separate chlamydospores from soil involves wet sieving of soil followed by centrifugation in anhydrous magnesium sulphate solution (Specific Gravity 1.30, concentration 650 g/l). This technique has been used to quantify *Nematophthora gynophila* and *P. chlamydosporia* in soils and monitor changes in chlamydospore numbers in field trials (Crump and Kerry 1981; Crump 1991; Kerry and Bourne 2002).

12.5.3 *Isolation of* Pochonia *spp. from Roots*

Host-plant susceptibility influences the number of nematodes invading the roots, their rate of development, fecundity, and the number of eggs colonized by the fungus. The extent of fungal growth in the plant is also affected by the plant species and cultivar, which in turn affects nematode control (Kerry 1995, 2000).

To isolate the fungus from roots, select healthy-looking roots, wash them gently, cut into 1 cm sections, mix, and remove a 1 g subsample. Gently crush the roots in 9 ml of a 0.05% sterile agar solution with a sterilized mortar and pestle. Make dilution series at 10^{-2} and 10^{-3} in 9 ml vials and plate 0.2 ml of each dilution onto the selective-agar (3 plates/dilution). Incubate at 25 °C for 10–15 days, count the colonies and calculate the cfu/g root. Chlamydospores are produced after 3–5 days (0.8% WA) to 2 weeks (1.7% CMA). Make a control by incubating sections of the roots examined on water agar (WA) with antibiotics or CMA (Kerry and Bourne 2002). To prepare water agar plus antibiotics, autoclave 1 litre of 0.8% agar and let the medium cool slightly before adding 50 mg of streptomycin sulphate, 50 mg of chloramphenicol, and 50 mg of chlortetracycline. These three antibiotics are added to other culturing media described in subsequent sections of the chapter.

12.5.4 *Isolation from Nematode Eggs*

Meloidogyne egg masses are picked off the root galls surface, pooled and gently separated using a glass rod in a glass tube containing 0.5 ml sterile distilled water (sdH_2O). The percentage of parasitized eggs is assessed in 100 eggs aliquots spread onto 0.8% WA plus antibiotics plates (Sect. 12.5.3) and after 3 days of incubation at 25 °C. Infected eggs are transferred onto selective-agar and incubated at 25 °C for 10–15 days; colonies of *P. chlamydosporia* are selected and grown in CMA and PDA for microscope identification at 25 × magnification. Each colony growing out of an egg is transferred to CMA to obtain a pure culture (Kerry and Bourne 2002); also prepare monosporic cultures (see Chap. 11).

Verdejo-Lucas et al. (2002) started the isolation process from galled roots collected in *Meloidogyne* spp. infested plastic tunnels. Galled roots were chopped and mixed, and 30–40 egg masses were handpicked from one 5 g root subsample. Eggs were dispersed from egg masses with ca 0.2 ml sdH_2O using a pestle in an Eppendorf microcentrifuge tube, spread on a restrictive growth medium (Lopez-Llorca and Duncan 1986) in three replicate Petri dishes, and incubated at 25 °C. After 48 h eggs were examined for parasitism at × 50 magnification, parasitized eggs being transferred individually to CMA to establish pure cultures. In another method, egg masses are collected from galled roots and placed in a humid chamber, fungi being isolated from the eggs by culturing in PDA with 250 mg chloramphenicol/litre (Carneiro et al. 2008).

Globodera spp. and *Heterodera* spp. young cysts can be picked from roots but older cysts are extracted from soil using the Trudgill column or Fenwick can (Hooper 1986; Lopez-Llorca and Duncan 1986). Cysts are put into small excavated blocks and rinsed twice with dsH_2O; then put in a drop of sdH_2O placed on the channel (15 mm wide, 0.05 cm deep) of a sterilized channelled aluminium slide (7.5 × 2.5 cm) to release the eggs by squashing the cysts with a glass slide placed on top of them. To suspend the eggs, add dsH_2O (250 μl) and transfer the egg suspension into a 2 ml tube. Mix and transfer a 0.5 ml aliquot to another 2 ml tube. One tube (0.5 ml) is used to assess egg infection and the other tube to assess egg viability. Cysts can also be crushed and eggs plated on agar in a similar manner to RKN egg masses to isolate the fungus from eggs. Fungal infection in eggs of cereal cyst nematodes (*Heterodera* spp.) can be estimated in media containing 1% PDA, 1000 μl/ml Triton X-100, 50 μg/ml streptomycin sulphate, 50 μg/ml aureomycin, and 50 μg/mg Rose Bengal (Lopez-Llorca and Duncan 1986).

12.5.4.1 Isolation of the Fungus from Cyst Females

Pochonia chlamydosporia can parasitize cyst females soon after they have emerged through the roots and before egg production (Crump and Kerry 1987). The fungus can be isolated by culturing infected females on agar to induce sporulation and facilitate identification (Kerry and Crump 1977). This can be achieved by placing a sterile glass cover slip (24 mm square) onto nutrient agar (with or without antibiotics) in a Petri dish and placing four females that have been broken opened in a drop of water on the top of the cover slip (0.5 mm from each edge) and left to incubate at 20 °C. If females are infected, the fungus will grow away from them in search for nutrients and across the cover slip until reaching the agar where it will grow rapidly. This method lessens the need to surface sterilise the female because saprophytic fungi are less likely to reach the agar than the fungi that are feeding on the female (Crump 1987a).

12.6 Nematode Baits (Slide Mount Technique)

The bait method is based on a baiting technique developed by Lumsden (1981). Egg masses or cysts are used as a bait for the fungus in treated or untreated soil. Ten egg masses of RKN are wrapped in nylon fabric mesh which are held in place supported by plastic slide mounts or frames (24 × 36 mm) with glasses removed prior to being buried in the soil (Fig. 12.3) at a depth of about 30 mm in soil contained in 200 g capacity pots. Pots are watered as required. After 7 days, the egg masses are removed from the soil by pulling out the plastic slide mounts, placed into excavated glass blocks, and cleaned with sdH_2O. The eggs are then released from egg masses using a sterile glass cyst crusher. From the ensuing egg suspension, 200 μl is pipetted onto plates containing sorbose agar with antibiotics (Kerry and Bourne 2002, see also

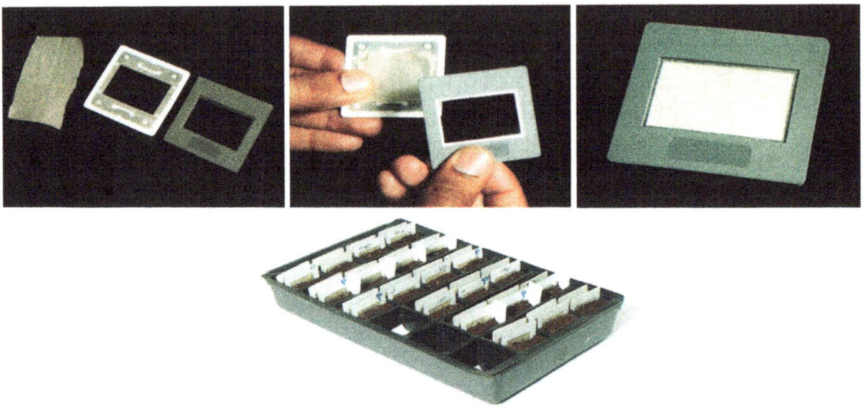

Fig. 12.3 Baiting slides. Egg masses or cysts are wrapped in nylon fabric mesh which are held in place and supported by plastic slide mounts or frames (glasses removed) prior to being buried in the soil (Image B.R. Kerry slide collection)

Sect. 12.5.4) and incubated at 25 °C for 72 h. Afterwards, 100 randomly selected eggs are examined and counted for fungal infection under a stereomicroscope (50 × magnification) to estimate the percentage of eggs infected by *P. chlamydosporia* (Luambano et al. 2015). Cysts (ca 25) can be used as bait instead of egg masses (Atkins et al. 2003).

The plastic slide mounts or frames technique has also been used, with some differences, to study the diversity of egg pathogenic fungi (Pyrowolakis et al. 1999). *Globodera pallida* cyst were placed in the soil (five cysts/slide mount). Fourteen days later the cysts were collected, crushed and the egg suspension plated onto WA with antibiotics (150 ppm of streptomycin, 150 ppm of penicillin). The plates were incubated at 23 °C for 2–5 days. The fungal egg pathogens were isolated from the eggs (stereomicroscope × 40) and each isolate was transferred onto a new agar plate. Species diversity, species richness and species evenness indexes were calculated using their respective formula (Pyrowolakis et al. 1999).

12.7 Isolate Growth on Selective-Agar

The selective-agar medium (also known as semi-selective medium) contains chemicals that selectively restrict fungal growth to facilitate counting of *P. chlamydosporia* colonies (see Chap. 11, Sect. 11.5.2). However, such inhibition could lead to underestimation of the amount of fungus quantified in the soil. It is therefore necessary to compare propagule growth on CMA with growth on selective-agar medium to establish if growth is inhibited (Bourne and Kerry 2000). The procedure is carried out with isolates obtained from infected eggs. The colony, which has been grown on CMA and incubated for 3 weeks at 25 °C, is scraped with a glass rod and washed with sdH_2O into a sterile beaker. Dilution series are prepared by adding one ml of the fungal suspension into 9 ml 0.05% WA aiming to count about 50–100 colonies/plate. Aliquots (0.2 ml) of the appropriate dilution are spread onto plates of CMA (non-selective) and selective-agar medium. Plates are incubated at 25 °C and colonies counted 3 weeks later to compare numbers of colonies on both media and estimate the inhibition (%). Corn meal agar colonies may need to be counted before those growing in the selective agar as the CMA colonies can grow too large to be able to distinguish between them (Kerry and Bourne 2002).

12.8 Isolate Screening

Host plant and environmental conditions have a significant effect on the fungus (Bourne et al. 1996). Therefore, it is preferable to identify and select native isolates with potential for colonizing the rhizosphere of local crops (Hidalgo-Díaz et al. 2000). Selection of fungal isolates (screening) depends on laboratory and glasshouse tests that initially evaluate: (i) capacity to colonize the rhizosphere; (ii) chlamydospore production and viability; and (iii) egg mortality (Kerry and Bourne 2002). After screening, the isolates selected will need to be further characterized (see Fig. 12.1) in glasshouse tests to assess efficacy of the isolates in controlling the target nematode at different densities, their ability to colonize the rhizosphere of plant species susceptible or resistant to nematodes, and the effect of different temperatures and soil organic matter content on the isolates. Organic matter added to soil at 0.5–1% w/w concentrations usually improve plant growth and provide significant nematode control (Mian and Rodríguez-Kábana 1982; Stirling 1991; McSorley 2011; Timper 2014) – hence the need for the inclusion of a substrate or food base treatment control when testing microbial species for activity against nematodes (Rodríguez-Kábana et al. 1984). Tests carried out only in sterilized soil may lead to an inaccurate prediction of the isolate's ability to perform in the field. Therefore, all pot tests should be done in unsterilized soil with active microbial communities to determine the final population density of the fungus in soil and rhizosphere in treated and untreated soil. In addition, nematode populations and reproduction rate (Pf/Pi) should also be estimated (Kerry and Bourne 2002).

12.8.1 Rhizosphere Colonization

Isolates which may proliferate in soil substrates have no significant effect on nematode multiplication rates unless they are able to colonize the rhizosphere (Bourne and Kerry 1999). Therefore, isolate(s) rhizosphere colonization tests must be carried out in both sterile (see Sects. 12.8.1.1 and 12.8.2) and non-sterile conditions (Sect. 12.8.3) using conidia inoculated seeds, colonized agar plugs or chlamydospores as inocula.

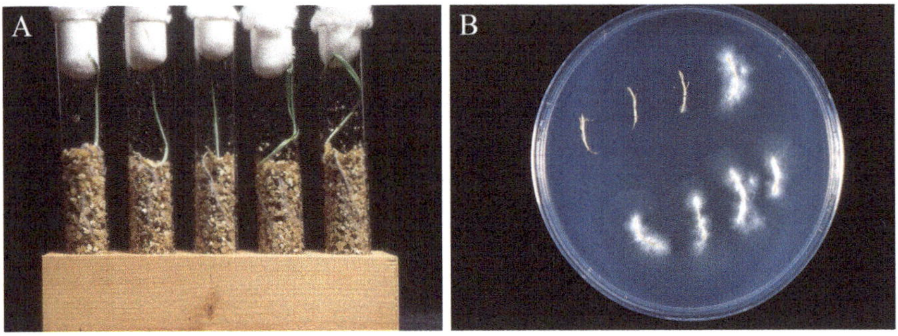

Fig. 12.4 (**A**) Barley seedlings grown in vermiculite inoculated with *Pochonia chlamydosporia*; (**B**) Colonized roots (Image B.R. Kerry slide collection)

12.8.1.1 Conidia Inoculated Seeds

Conidia can be produced in Czapek Dox broth plus salts (CDB) autoclaved and cooled (Kerry et al. 1986). Ten CMA colonized plugs (5 mm diameter) of the fungus are added to 200 ml of CDB into conical flasks that are kept 14 days at 15–20 ° C in a shaker at 120 rpm. The colonized medium is washed on a 10 μm mesh sieve and centrifuged to reduce liquid volume. Conidia are suspended and counted in 1% manucol (sodium alginate) using an haemocytometer and volume adjusted to 10^6 conidia/ml. Barley seeds are surface sterilised in 7% calcium hypochlorite for 45 min on a magnetic stirrer, washed in sdH_2O (5 times) and seeds dipped in conidia suspension (10^5 conidia/seed). Seeds are air dried overnight in a sterile cabinet and planted in sterile moist vermiculite in glass tubes (15 × 2.5 cm diameter) stoppered with cotton wool. Seeds are grown for 10 days at 25 °C with an appropriate light regime (Fig. 12.4A). Roots are cut into 1 cm portions and placed on WA, incubated, and the number of colonized root segments count after 2–3 days of incubation at 25 °C (Fig. 12.4B). If the isolate will not colonise the roots (<80%) in non-sterile conditions it can be discarded. Seed inoculum viability is checked by plating 3 treated seeds onto 0.8% WA incubated for 2–3 days at 25 °C (Kerry and Bourne 2002).

12.8.2 Corn Meal Agar Colonized Plugs

Barley seeds are surface sterilised (see Sect. 12.8.1.1) and transferred to seed germinating media composed of 10 g/l glucose, 0.1 g/l yeast extract, 0.1/g l peptone, and 12 g/l technical agar (Kerry et al. 1984; Manzanilla-López et al. 2009). Glass tubes are filled with moist vermiculite (see Sect. 12.8.1.1) and 4 × 2 mm plugs *P. chlamydosporia* grown on CMA culture are placed below the surface of the vermiculite (Bourne et al. 1994). Seedlings are placed on top of the plugs, ensuring that they are covered with the vermiculite. Cover tube up to level of vermiculite with aluminium foil, incubate for 3 weeks at 25 °C in an appropriate light regime, then cut the roots, place on WA, and count the number of colonized root segments as described in Sect. 12.8.1. Root colonization is generally quicker and more extensive than with seed dipped in conidia, but the failure rate of isolates in soil is higher (Kerry and Bourne 2002).

12.8.3 Chlamydospore Inoculum

Chlamydospores do not require additional nutrients and are thus considered the most effective inoculum for establishing the fungus in soil and rhizosphere, and a useful measure of soil and rhizosphere colonization when screening for isolates. Chlamydospores can be produced on a variety of substrates (Lopez-Llorca et al. 1999; Coutinho et al. 2009; Luambano et al. 2015; see also Chap. 11). The basic protocol (Crump and Irving 1992; Kerry and Bourne 2002) requires of 30 g of milled barley or maize grain (wheat, rice or other substrate) washed through a 53 µm mesh sieve. The sieve is blotted dry underneath and the grain mixed with an equal volume of coarse sand to provide aeration and to prevent clogging. The slightly moist mixture is put into a 250 ml flask and autoclaved. When cool, the flask is shaken to loosen the mixture and inoculated with 4–5 mm CMA or WA fungus colonized plugs. The fungus is left to grow on the agar plugs for 2–3 days (without shaking) at 25 °C until it looks like cotton wool, then dispersed throughout the medium by vigorous shaking and incubated for 3 weeks at 25 °C. To extract the chlamydospores, the colonized substrate is washed with a fine spray of water through a 250 µm mesh sieve placed over a 53 and 10 µm mesh sieves; the underneath of the 10 µm mesh sieve is blotted and chlamydospores collected using a spatula and mixed with fine sand (i.e., low iron; Fisher Scientific, Loughborough, UK) in a ratio of 10 parts of sand to one part of chlamydospores (w:w). Whichever method is chosen, chlamydospore viability should always be estimated (see Sect. 12.8.3.1).

Also, when adding chlamydospores grown in an organic substrate, the inclusion of controls, consisting of chlamydospores inoculum without organic substrate, should be considered in order to separate the effects of the fungus from its food source as cfu count can be greater when chlamydospores, or another source of inoculum, are applied without substrate than equivalent amounts of colonized bran (Rodríguez-Kábana et al. 1984; De Leij and Kerry 1991; Stirling 1991). Ayatollahy et al. (2008) found that application of chlamydospores without their colonized barley substrate appeared to be more efficient in reducing the final number of *H. schachtii* females than when the fungus was applied in infected barley substrate. A possible explanation is that application of colonized bran supports the growth of other soil-microorganisms, which then compete with the inoculated fungus and suppress its growth in the soil, whereas chlamydospores have little effect on the residual soil microflora as they tend to leak only small amounts of nutrients (De Leij and Kerry 1991).

12.8.3.1 Chlamydospore Viability

To test chlamydospore viability, suspend 1 g of sand:chlamydospore inoculum (see Sect. 12.8.3.1) in 9 ml water and make a 10^{-1} to 10^{-3} dilution series. Plate 0.2 ml of several dilutions (ca 100–500 chlamydospores/plate) on a 9 cm Petri dish onto sorbose agar (12 g technical agar, 2 g sorbose per litre) plus antibiotics (see Sect. 12.5.3). Incubate at 25 °C for 2–3 days. Count percentage (%) germination (Kerry and Bourne 2002).

12.8.3.2 Chlamydospores as Inoculum Source

Chlamydospores are mass produced as described in Chap. 11 (Sect. 11.5.3). Take a 1 g substrate sample and estimate cfu/g. Inoculate soil (peat:coarse sand 3:1 v/v) with 5000 chlamydospores/g soil. Transplant a 3–4 week-old tomato seedling and allow the plant to establish for about 1 week before adding nematodes. Inoculate soil around plant with approximately 3000 J2 of RKN. After a 6–8 week period, harvest the plants and assess for fungal colonisation of soil (Sect. 12.5.2), roots (Sect. 12.5.3) and eggs (Sect. 12.5.4) and total nematode populations (Kerry and Bourne 2002; Flores-Camacho et al. 2007). Although 5000 chlamydospores/g soil is usually sufficient, the number of chlamydospores required can be higher in order to achieve nematode control, depending on the isolate (see Ayatollahy et al. 2008; Podestá et al. 2009).

12.9 Egg Parasitism Test

Wash a slope or plate of the fungus growing on WA or CMA with 2–5 ml of sdH_2O. Scrape the surface with a sterile glass rod to suspend fungal material and count conidia or chlamydospores in a haemocytometer. Take and spread an aliquot (0.2 ml) onto a 9 cm Petri dish of WA plus antibiotics (see Sect. 12.5.3). Incubate plates at 25 °C for 2–3 days. To collect egg masses, wash roots of plants infected with RKN (i.e., 4–6 weeks after J2 inoculation) and handpick several hundred egg masses to extract eggs. It is preferable not to use sodium hypochlorite (NaClO) solution to separate the eggs from the gelatinous matrix as the bleach can affect the permeability of the egg membrane. Instead, crush the egg masses using a glass homogenizer and, in the case of cyst nematodes, a cyst crusher and glass rod (Shepherd 1986). The egg suspension can be filtered (200–250 μm mesh sieve) to separate eggs from egg mass debris and centrifuged to reduce liquid volume. Add and spread ca 200–300 RKN eggs per plate. Incubate at 25 °C for 3–4 days and then count the number of parasitized eggs (%) (Kerry and Bourne 2002; Esteves 2007; Manzanilla-López et al. 2009).

To prevent bacterial contamination, Ayatollahy et al. (2008) surface-sterilized with 0.1% NaClO for 30 min, then with a solution of 100 ppm of tetracycline, streptomycin sulphate and chloramphenicol another 30 min, and finally rinsed *H. schachtii* cysts and females in sdH_2O before being placed for 48 h on WA. One fungal pre-colonized PDA plug was placed in the centre of a fresh Petri dish containing WA (0.8%) plus the antibiotics listed above, and individual surface-sterilised female or cyst were added onto the fungus colonized plug. After 3 weeks of incubation in the dark at 20 °C, females or cysts were placed in a lactoglycerol drop onto a glass slide, crushed with a cover slip, and examined for fungal colonization at 200 × and 400 × magnification using a light microscope. Escudero and Lopez-Llorca (2012) surface-sterilized *M. javanica* eggs before placed them in Petri dishes with 1% WA, 50 μg/ml penicillin and 50 μg/ml ampicillin. Each plate was inoculated with 10 μl of 10^6 conidia/ml suspension. Plates were incubated at 25 °C in the dark. After a 4-day period, egg colonization was scored.

12.10 Temperature

Different isolates may require different optimum temperatures to grow, produce conidia and chlamydospores (De Leij et al. 1992; see also Chap. 11). Therefore, as part of the screening process, selected isolates should be grown *in vitro* in different media and/or glasshouse conditions over a range of temperatures to determine mycelia growth and optimum sporulation temperatures (Hidalgo-Díaz et al. 2000; Arevalo et al. 2009). Temperature also affects the rate of egg production, hatching and development from the juvenile to the egg laying female nematode. The rate at which nematode eggs are laid and develop also affects the rate at which the fungus can infect the eggs (De Leij et al. 1992). Isolates from a temperate region may not perform so well in warmer conditions as tropical isolates and *vice versa*.

12.11 Quantitative Assessment of *Pochonia chlamydosporia* in Rhizosphere and Soil

Although all methods to estimate the size of populations of nematophagous fungi in soil and the rhizosphere have major limitations (Kerry and Bourne 2002), quantitative enumeration of a single fungus species allows insights into its biology, ecology and significance for biological control (Lohmann et al. 1991). Measuring cfu number to monitor fungal biomass in pure culture is not straightforward as each spore usually gives rise to a colony but only a proportion of the mycelium will give rise to colonies (Lynch 1996). The use of dilution plate techniques in conjunction with a selective-agar medium has aided understanding of the proliferation and survival of *P. chlamydosporia* in the soil and rhizosphere (Kerry and Bourne 2002). However, cfu counting is time consuming and does not allow separation of propagules (hyphae, conidia or chlamydospores) which occur together in the soil so that apparent changes in abundance may have resulted from increased vegetative growth or from sporulation (Kerry and Bourne 2002). See also Chap. 13 for quantitative real-time PCR (qPCR).

Esteves (2007) compared the conventional bioassay of counting chlamydospore using a haemocytometer with a spectrophotometer to read the absorbance of chlamydospore suspensions in order to find a quick and easy method to determine changes in spore production. Serial chlamydospore suspensions were counted using a haemocytometer and the level of absorbance (600 nm) of the spore suspension was determined using a spectrophotometer (CaryWin UV). Comparisons between chlamydospore average size was analysed using linear regression. A positive, linear and significant relationship between the level of absorbance and chlamydospore concentration was obtained for five isolates when chlamydospores were suspended in polyethylene glycol (PEG) at 10^5 to 10^7 chlamydospores/ml concentrations (Esteves 2007).

12.11.1 Glasshouse Pot Test Experiments

Pot experiments provide a suitable basis for examining nematode control by isolates since there may be fewer and more easily controlled environmental variables than occur in field tests (Clyde 1993). Once the *in vitro* screening process has been completed, selected isolates are tested, this time, under selected glasshouse conditions (Fig. 12.1). Variables may include temperature range for development, soil type, crop, photoperiod, inocula levels for isolates, and nematode host(s). Further tests to assess the performance of isolates should be done in microplots and field conditions (see Sect. 12.8).

12.11.2 Nematode Cultures

To carry out *in vitro* and pot tests, it is necessary to maintain healthy, non-contaminated nematode cultures, and to ensure that the potting soil is free of *Pochonia* spp. or other nematophagous fungi. The origin of the nematode population must be recorded and RKN nematode cultures started from a single egg mass. To maintain healthy RKN cultures, inoculate new tomato plants every 2–3 months with ten fresh egg masses per plant (see Kerry and Bourne 2002). The choice of plant species or cultivar is also important to ensure that sufficient quantities of egg masses or cysts are available for experiments as and when required. The life cycle length of the nematode can vary according to glasshouse conditions and temperature. Average temperature records, degree days (dd), humidity and photoperiod used to grow host plants should also be recorded. Planning and timing in establishing enough plant cultures to reproduce the nematodes is essential so that: i) nematode eggs are going to be ready in good supply, and ii) eggs are not too old for egg parasitism tests. It is very important to achieve synchronous production of both nematode and fungus inocula so that experiments can be carried out in a timely manner.

12.11.3 Chlamydospore Soil Inoculation

Once chlamydospores/g of substratum have been quantified they can be added to a known quantity of soil in pots. Chlamydospores produced on barley substrate (see Sect. 12.8.3) or rice culture (see Chap. 11, Sect. 11.5.3.2) are mixed with fine sand in a 10:1 w:w ratio (sand:chlamydospores). One gram of inoculum is added to 9 ml of water agar (0.05%) and thoroughly mixed before the chlamydospores are counted using a haemocytometer and dilutions made to produce a final concentration of 5×10^3 chlamydospores/ml. Chlamydospore viability and germination percentage are evaluated on sorbose agar plus antibiotics (Esteves 2007; see also Sect. 12.5.3). To prepare each pot, soil is weighed and placed in a reclosable polythene plastic bag before adding the chlamydospores (5×10^3/g). Soil and chlamydospores are thoroughly mixed in the bag and then transferred to each pot (Manzanilla-López et al. 2011a).

12.11.4 Double Pots

Double pots (Fig. 12.5) allow better development of the root system for a longer growth period without diluting the fungal inoculum (De Leij et al. 1993b). One tomato seedling (i.e., cv. Tiny Tim) is transplanted into each pot and inoculated 3 days later with 2000 freshly hatched J2 of RKN. Each pot is then placed in a larger (15 cm) pot containing sterilized potting compost (double potted) and left under glasshouse conditions until eggs have been produced and colonized (Manzanilla-López et al. 2009).

12.11.5 Split Roots

There is evidence that fungal growth is stimulated by the presence of the nematodes in the roots, a seemingly partly systemic effect according to results from experiments using split-root systems where nematodes are separated from the fungus (Bourne and Kerry 1999). Monfort et al. (2005) excised the middle seminal root of wheat seedlings and transplanted the seedling into a Y-shaped plastic tube. Each half of the root system was inserted into one of a pair of plastic pots taped together. Observation chambers (Sect. 12.13.1) can also be adapted as a split root system.

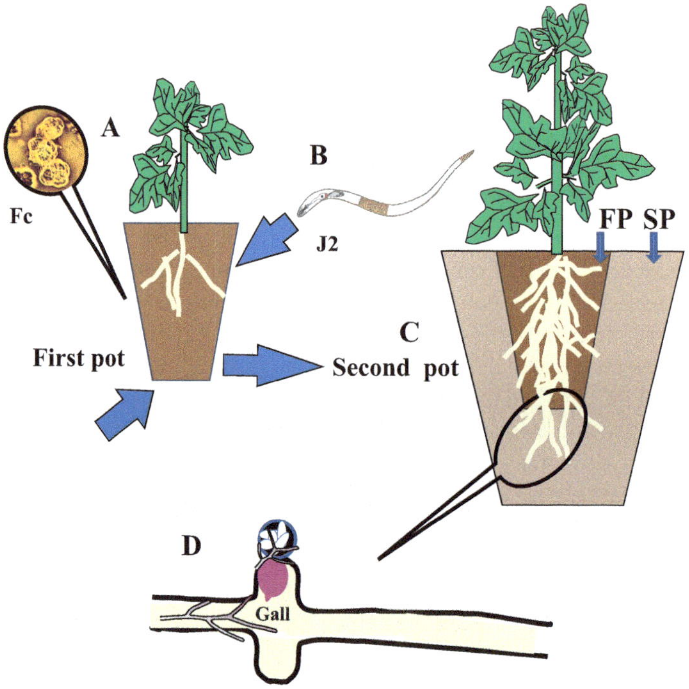

Fig. 12.5 Testing fungal colonisation of nematode-infested tomatoes in double pots. (**A**) Chlamydospores of the fungus (Fc) are mixed with soil before planting; (**B**) Second-stage juvenile (J2) nematodes are inoculated into the first pot; (**C**) The first pot is placed inside a second, larger, pot containing soil to allow the growth of roots and galled roots into another (second) pot; (**D**) Gall, nematode female and egg mass colonised by the fungus. *FP* First pot, *SP* second pot (Original courtesy P.R. Hirsch and modified by R.H. Manzanilla-López)

12.11.6 **Pochonia chlamydosporia** *Rhizosphere Competence and Population Density*

Good control of nematodes can be achieved if fungal isolates are capable of proliferating in the rhizosphere whereas nematode control will be poor if fungal isolates are incapable of colonizing the rhizosphere (Kerry 2000). Rhizosphere competence can be assessed by dilution plating of homogenized root to give cfu/g root (rhizoplane sampling), and dilution plating of soil attached to roots to give cfu/g soil (rhizosphere sampling) (Clyde 1993). Counts of colony forming units are usually higher in the rhizosphere than soil, as shown by experiments made using nutrient depleted soil (i.e., acid washed sand, pH 6) where abundance may have been related to carbon and nitrogen provided by the plant in rhizodeposits or through root leakage/exudates induced by the nematode and subsequently utilized by the fungus to support its saprophytic behaviour (Manzanilla-López et al. 2011a). Increasing the application rate can increase the amount of fungus in the soil phase, but not necessarily increase levels in the rhizosphere if the plant is a poor host of the fungus (Kerry 2000). However, root colonization by *P. chlamydosporia* cannot always be correlated with the level of nematode suppression (Kerry 1995; Bourne et al. 1996; Timper 2014).

Rhizosphere cfu count (see Sect. 12.8.1) is one of the commonest methods to estimate root colonization but, in order to estimate *P. chlamydosporia* root population density (cfu/cm^2 root), a correction for root length is needed, a procedure described by Kerry and Bourne (2002) based on a procedure adapted from Tennant's line intersect method of establishing root length (1975). A root sample (0.2 g) is spread on a Petri dish and 0.02% agar poured onto the roots. The dish is placed over a 1 cm grid to count root/grid line intersects. The equation: Root length (L) = number of intercepts × 0.7857 (length conversion factor) is used to calculate L/ g of root. The mean diameter (mm) of roots is measured to calculate radius (r) and Surface area = 2Π r x L (mm^2/g of root). Surface area is converted to cm^2/g root and cfu/cm^2 of root calculated (Tennant 1975; Bourne et al. 1996; Kerry and Bourne 2002). However, other methods, such as digital gridline intersection, to count root/grid line intersects are now available (Shen et al. 2016).

12.11.7 Soil and Compost Analyses

The crop rhizosphere or fresh organic matter may enable the fungus to proliferate yet may also prevent it from switching to its parasitic phase. However, it has not been clearly demonstrated whether significant egg parasitism occurs only after more easily metabolised energy sources have been depleted. Therefore, different rates of compost or green manures should be tested to ensure that chosen rates do not reduce egg parasitism nor have phytotoxic effects (Mian and Rodríguez-Kábana 1982; Pyrowolakis et al. 1999; Timper 2014). Also worth considering is the length of the composting process and the maturity of the composts to be added. Whenever possible, soil and compost used in experiments should be analysed, including data for nutrients and C:N ratio (see Sect. 12.4.1).

12.12 Isolation of Fungus from Egg Parasitism Tests

To isolate or re-isolate the fungus from eggs see Sects. 12.5.4 and 12.5.4.1. Plate the eggs onto WA plus antibiotics, incubate at 25 °C for 2–4 days and transfer each colony growing out of an egg to CMA in order to obtain a pure culture.

12.12.1 Conidia Egg Infection in Yeast Extract Liquid Medium

This is a quicker alternative to the WA plus antibiotics method (see Sect. 12.9). Conidia are harvested from PDA agar cultures by flooding each plate with 5 ml of sdH_2O and gently scraping the mycelium using a sterile L-shaped glass rod. A 20 μm mesh sieve, previously rinsed with 70% ethanol and UV irradiated in a flow cabinet for 20 min, is used to pour the conidia suspension. Ten ml of sdH_2O are added into the sieve and the conidia suspension collected from a Petri dish placed underneath. Conidia are counted with a haemocytometer and adjusted to a spore concentration of 5×10^3/ml (Esteves 2007). To assess egg infection 0.0125 g of yeast extract are added in 1 litre of dH_2O and autoclaved (Esteves 2007). Eight ml of medium are placed into universal glass bottles (30 ml volume) and to each of these 1 ml containing 1000 nematode eggs and 1 ml containing 5.5×10^4 conidia are mixed and incubated at 25 °C at 150 rpm for 24–48 h. After incubation, the bottles are taken out and whirl-mixed before removing a 3 ml aliquot and placing onto a counting slide. Data on parasitized eggs are collected by assessing 100 randomly selected eggs under a dissecting microscope at 50 × magnification to estimate the percentage of parasitized eggs (Esteves 2007; Luambano et al. 2015).

12.13 Soil Suppressiveness

General soil suppressiveness to soil borne pathogens includes the total microbial biomass and is not transferable between soils, whereas "specific suppression owes its activity to the effects of individual or select groups of microorganisms and is transferable" (Weller et al. 2002). The activity of most suppressive soils can continue under glasshouse conditions, thus allowing further testing of the mechanisms (see Sects. 12.13.1 and 12.13.2) and soil properties involved (Weller et al. 2002; Mitchinson 2009).

12.13.1 Observation Chambers

One standard method to visualize the effect of fungi on cyst and RKN nematodes in the rhizosphere is the use of observation chambers made of Petri dishes (9 cm diameter) that have had a piece removed (ca 1.5 cm) from the circumference of the dish (Crump 1977). The Petri dish lid and base are held together and the dish filled with soil into which germinated seedlings are planted (Fig. 12.6A). Second-stage juveniles are used to inoculate the seedlings, timing of inoculation depending on crop seedling. The method allows repeated sampling to check for females or cysts development and fungal infection at different times and to remove them from the roots by raising the lid or base of the dish. Larger observation chambers (Crump 1987a) have

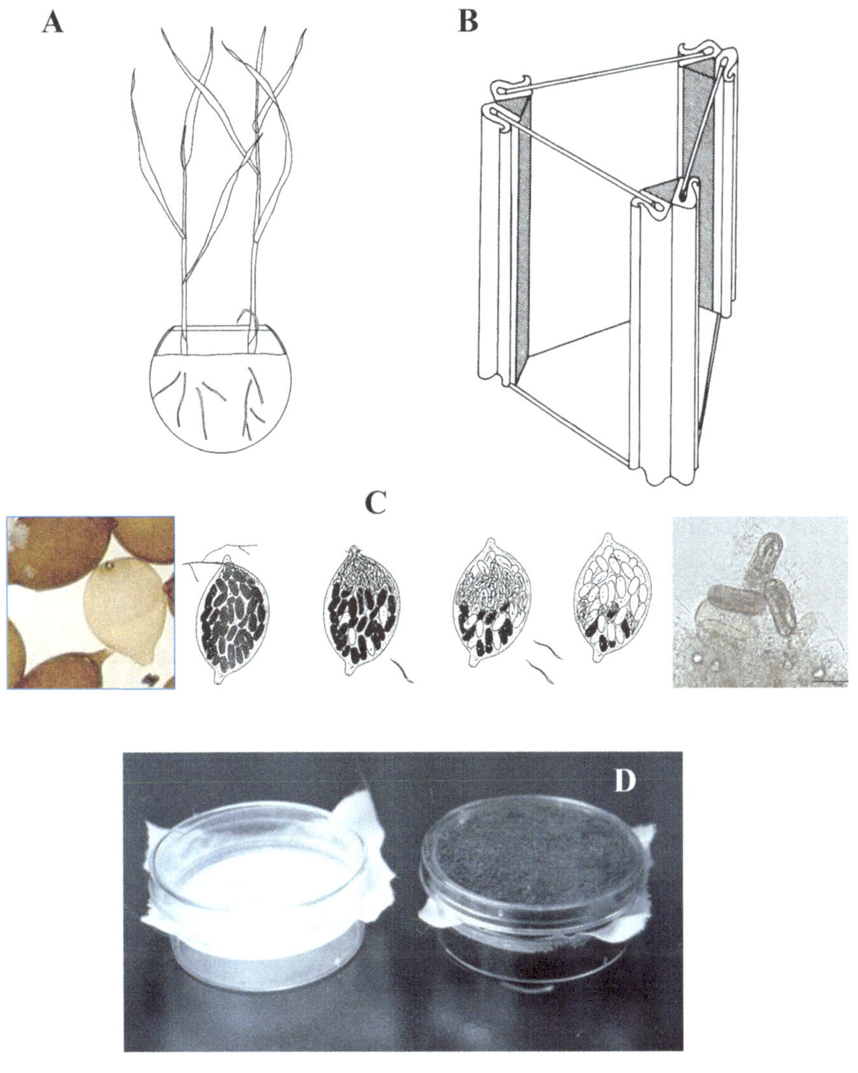

Fig. 12.6 Observation chambers. (**A**) Simple chamber; (**B**) Triangular chamber (modified after Crump 1987a; Crump and Kerry 1977); (**C**) *Heterodera schachtii* cysts colonized by the fungus. Empty eggs can either be destroyed by the fungus or the second-stage juvenile has hatched and escaped fungal parasitism (Figure after Nicolay and Sikora 1989; courtesy Richard Sikora); (**D**) Nylon membrane device to assess soil and cyst colonization (Image B.R. Kerry slide collection)

also been developed to support a sugar beet plant and development for two generations (Fig. 12.6B, C) of the beet cyst nematode (*H. schachtii*). The triangular chamber has three removable Perspex® sides that allow surface root examination under a microscope at 15 × magnification. Observation chambers help to gather information on the fungal parasites and the point in the growing season at which infection occurs. *Pochonia chlamydosporia* (= *V. chlamydosporium*), for example, was found to be most active on the first generation of *H. schachtii*, whereas *Cylindrocarpon destructans* (= *Ilyonectria destructans*) was more active on the second generation (Crump 1991). The chambers can also be used to investigate factors that may increase natural control, such as the addition of organic amendments or artificially cultured fungal inocula, in order to assess their effect on nematode multiplication (Crump 1991; de Haan and Thijssen 1991).

12.13.2 Soil Membrane Assays

A method derived from Kerry (1991) was used by Clyde (1993) to examine the extent of fungal colonization in a sandy loam soil (Fig. 12.6D). The soil was sieved, air dried, adjusted to 25% water (w/v), and used to fill 75% of individual 5 cm Petri dishes. A nylon mesh (50 μm aperture) was placed in contact with the upper surface of autoclaved or non-sterile soil, and the mesh held in place by a ring. Freshly collected chlamydospores and/or *H. schachtii* females were placed directly onto the mesh centre and covered with soil to the upper edge of the ring. A lid was placed on each Petri dish and incubated at 20 °C for 3 weeks. The soil and inoculum source were removed from the nylon mesh and the mesh pressed onto the selective medium in a flow cabinet for 30 s. Colonisation was estimated from the diameter of the colonised area that had developed after 4 days of inoculation at 25 °C. Spores did not colonise autoclaved soil to any greater extent than non-sterile soil, thus supporting the thesis that the fungus is able to compete effectively with the residual soil microflora in non-sterile soil, although hyphal growth on the nylon meshes was very slow: just 1–2 cm in 3 weeks (Clyde 1993).

Monfort et al. (2006) assessed soil receptivity to nematophagous fungi by measuring radial surface growth in a soil-membrane assay in Petri dishes containing natural or autoclaved field soil. Ten CMA colonized plugs (5 mm diameter) were incubated with 50 g of surface sterilized ryegrass (*Lolium rigidum*) seeds in flasks for 8–11 days at 25 °C. One pre-autoclaved polyvinylidene difluoride membrane (0.22 μm pore-size) was placed on top of 40 g of air-dried soil saturated with sdH_2O per Petri dish. A single seed of ryegrass pre-colonized with the nematophagous fungus under study was placed at the centre of each membrane. The Petri dishes were sealed with Parafilm® and incubated for 4 weeks at 25 °C. Soil adhering to the membrane was removed with dH_2O and the membrane dried in a laminar air flow cabinet. Mycelium on the membrane was stained with 1% trypan blue in lactic acid and incubated 12 h at room temperature. The colony radius was measured and the area of the fungal colonies calculated. Hyphal density was estimated using analySIS (Soft Imaging System) from three sampling points chosen at random between the fungal inoculum and the edge of the membrane. Fungal relative growth (RG) was determined according to: RG = Radial growth (cm) on unsterilized soil/radial growth (cm) on sterilized soil (Grünwald et al. 1997). Nematophagous fungi had higher growth rates in completely sterilized soil than in dishes where non-sterile soil was added (Monfort et al. 2006).

12.13.3 Suppressive Soil Sterilization

Elimination of specific microbial groups is one strategy to determine the microbiological basis of suppressive soils and to identify the specific microbial group (Weller et al. 2002). The first step is to determine whether suppressiveness can be destroyed by pasteurization (moist heat, 60 °C for 30 min) or selective biocides (e.g., novobiocin or chloropicrin), or whether harsher treatments (e.g., steam, methyl bromide, autoclave or gamma radiation) are required (Weller et al. 2002). Yang et al. (2012) pasteurized soil by submerging samples in plastic bags in water at 60 °C for 30 min and then quickly cooling the bags to room temperature under running tap water. Soil suppressiveness against *H. schachtii* has been proven by soil fumigation and by transferring soil suppressiveness to suitable sites (Westphal and Becker 1999, 2000). Pyrowolakis et al. (2002) rated California soils from different agricultural areas for their suppressiveness against *M. incognita* by comparing the nematode population development in methyl iodide fumigated (50 kg/ha) treated soil and re-infested soil to the equivalent non-treated soils. Mauchline et al. (2002) develop a competitive PCR (cPCR) assay to quantify *P. chlamydosporia* (= *Verticillium chlamydosporium*) in ɣ-irradiated soil seeded with different numbers of chlamydospores.

12.13.3.1 Formalin Soil-Application

Formalin has been used as a useful tool to investigate the impact of *P. chlamydosporia* and *N. gynophila* on *H. avenae* populations suppressed by fungal agents in the soil (Crump and Kerry 1977; Kerry et al. 1982). Mitchinson (2009) used formalin treated soil in pots to examining whether fungal agents were responsible for keeping cereal cyst nematodes below the economic threshold level in the UK. Soil field capacity was determined by measuring the amount of water that each soil in the 12.5 cm pot could hold without leakage out of the pot. Formalin, at a dose of 3.7 ml/12.5 cm pot, the equivalent to the applied field rate of 3000 l/ha, was then mixed with the correct amount of water for each soil, to ensure the formalin soaked into all of the soil in the pot. The formalin-treated pots were kept moist and left for 3 months in a ventilated, unheated glasshouse to ensure that all traces of formalin had disappeared from the treated pots before cereal seeds were planted (Mitchinson 2009). **Note:** health and safety procedures must always be followed when handling formaldehyde.

12.14 Determination of Fungal Egg Parasitism in Field Populations

The rate of parasitism of nematode eggs is decisive for quantitative investigations in field populations (Stirling 1979). Nicolay and Sikora (1989, 1991) investigated the rate of parasitism in newly formed eggs in observation chambers (see Sect. 12.13.1) and soil fractioning (Nikolay and Sikora 1991). The latter method consisted of extraction of all cysts from a soil sample, release of eggs and fungi from the cysts by crushing, reincorporation of the cyst contents into the original soil sample, and estimation of parasitic activity of newly formed eggs. It was concluded that the proportion of parasitized eggs reflects the parasitic activity of any particular soil sample and that the parasitism rate of newly produced *H. schachtii* eggs was not correlated with nematode densities but with fungus density.

12.15 Experimental Design

Before starting any experiment, whether in the laboratory, glasshouse pots, micro-plots or field plots, it is important to have clear 'SMART' objectives (e.g., **S**pecific, **M**easurable, **A**chievable, **R**eachable and **T**ime-bound). Good planning, appropriate experimental design, selection of variables and robust statistical analyses are of utmost importance. Readers are referred to Powers (2012) concerning the importance of statistical practice in nematology, and to the introductory statistics text of Mead et al. (2003). Stirling (1991) produced basic considerations (see list below) which, ideally, should be implemented through planning and execution of field experimentation with biological control agents (BCA).

Experimentation key points (after Stirling 1991)

1. Apply the test organism or organic amendment at feasible application rates.
2. Include appropriate treatments in addition to the untreated control such as:
   - Test organism and its substrate (native or experimental substrate).
   - Test substrate alone.
   - Test organism alone.
   - Test autoclaved, colonized substrate.
3. Look for other native *Pochonia* spp. in soils prior to introduction/inoculation of test isolate(s).
4. Quantify (cfu or qPCR) of native isolates before and after introduction of new or test isolates.
5. Monitor populations of the test organism following its introduction (i.e., classic or molecular fingerprinting methods).
6. Measure the level of nematode mortality caused by the test organism (i.e., nematode initial/final populations and reproduction rate).

7. Additional measurements and monitoring may be required to assess multitrophic interactions, in similar or different physical environments.

12.15.1 Interaction Experiments

Sikora and Carter (1987) reviewed some key factors that affect studies on plant-parasitic nematodes (PPN) interactions with other pathogens but which can equally be applied to BCA experiments. Amongst others, it was pointed out that experimental and glasshouse conditions should represent the field conditions where interactions occur in order to avoid producing artificial or forced interactions. Naturally occurring population levels of both PPN or BCA should be investigated as a prerequisite for selection of experimental population levels. Some common glasshouse failures that occur are the use of root-bound pots, abnormally high and low temperature ranges, rapid changes in moisture level (i.e., saturation or drought), low light intensity, and studies being conducted in soil immediately after steam sterilization. Another factor to consider is the predisposition that can result due to concomitant or concomitant vs sequential primary pathogen (or a BCA) inoculation, which could mask or prevent interactions from developing, and alter those affected by the rhizosphere physicochemical environmental factors that can change drastically with plant age (Sikora and Carter 1987).

12.16 Microscopy and Bioimaging

Stereo- and compound light microscopes are essential tools commonly used to identify fungal structures, fungal infection and rhizosphere colonization (Kerry and Bourne 2000). Bioimaging can be carried out with a camera and software attached to the microscope. Depending on research objectives and availability, a wide range of microscopes, microscopy tools and techniques have been used to investigate the fungus and the tritrophic interaction, including light microscopy (Lopez-Llorca et al. 2002b; Bordallo et al. 2002), confocal and fluorescence light microscopy (Maciá-Vicente et al. 2009a), scanning electron microscopy (SEM), low-temperature SEM (Lopez-Llorca et al. 2002b; Manzanilla-López et al. 2014), transmission electron microscopy (Lopez-Llorca and Duncan 1991; Lopez-Llorca and Robertson 1992) cryofracture, 'cryo-trimming', and cryoplaning (Manzanilla-López et al. 2014).

The process of fungal root colonisation and nematode infection can be investigated *in vitro* by applying a fungal isolate containing a visible marker gene (Fig. 12.7) (see Maciá-Vicente et al. 2009a; Escudero and Lopez-Llorca 2012) or the use of fungus-specific fluorescent antibodies. Immunological methods have the advantage of being potentially specific, allowing the identification of individual strains of isolates (Clyde 1993). One polyclonal antibody raised against *V. chlamydosporium* (= *P. chlamydosporium*) hyphae and conidia, allowed the visualiza-

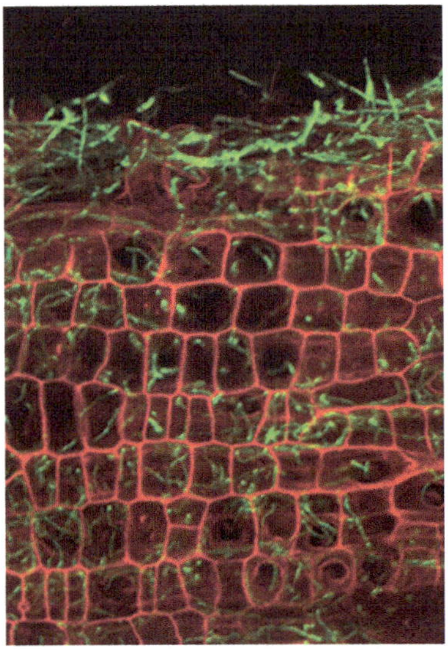

Fig. 12.7 Root colonization by *Pochonia chlamydosporia* stable transformant Pc123gfp (Courtesy N. Escudero)

tion of *P. chlamydosporia* at a microscopic level on roots and when colonizing female cyst nematodes (Clyde 1993). The incidence of fungus on roots and the surface of females was demonstrated by immunofluorescence and also enabled fungal distribution analysis. Visualisation of hyphal colonization of roots and females labelled with fluorescein isothiocyanate (FITC, Fig. 12.8) was greatly enhanced by confocal laser fluorescence microscopy and allowed the study of *H. schachtii* whole females (Clyde 1993). Hirsch et al. (2001) screened monoclonal antibodies (MABs) and identified specific MABs for the hyphae and chlamydospores (Vc3 and Vc162.2), conidia (Vc145.1), and the identification of the fungus on barley roots (Vc164.6). Hirsch et al. (2001) proposed that MABs could be useful in comparative studies with fungi transformed with visual marker genes since MABs do not alter any part of the fungus genome nor affect fitness of the fungus.

12.16.1 Microscopy Stains

Biological or histological stains may be required to detect fungal structures in the roots or nematodes, as well as viability of nematode eggs. One common fungal stain is cotton blue dissolved in lactophenol or lactoglycerol (1% cotton blue in lactic acid). Roots and nematodes within roots can be stained in acid fuchsin lactoglycerol, and New Blue "R" to differentiate between dead and alive eggs. Other procedures and commonly used stains in nematology can be found in Hooper (1986).

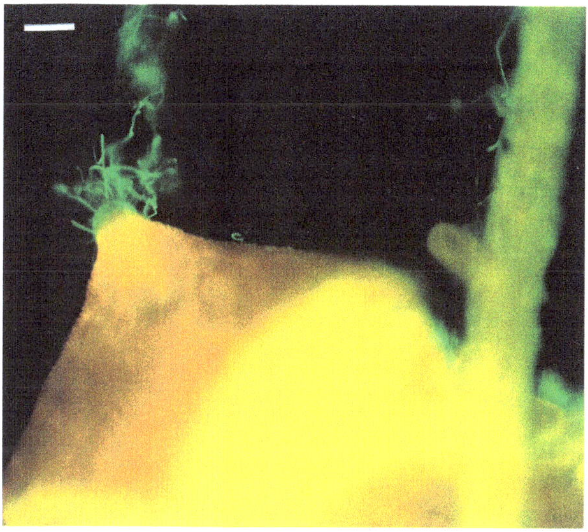

Fig. 12.8 Vulval colonization of a female *Heterodera schachtii* by *Verticillium chlamydosporium* (= *Pochonia chlamydosporia*). Female stained with fluorescein isothiocyanate (FITC) conventional epifluorescence. Scale bar 40 μm (Clyde 1993, image courtesy of H. Atkinson, Leeds University)

12.16.2 Root Staining

The endophytic nature of the fungus has been confirmed for some isolates. Hence is worth considering staining roots for locating the fungus within the roots. Lopez-Llorca et al. (2002b) studied barley roots fungal colonization in OCT embedded roots which were subsequently frozen and sectioned at −20 °C in a cryostat and stained in 0.01% W/V Tolouidine Blue O in 0.1 M potassium dihydrogen phosphate-NaOH buffer pH 6. Manzanilla-López et al. (2009) adapted the protocol described by Menéndez et al. (1997) to detect root endophytes in *P. chlamydosporia* colonized root segments using a combination of lactophenol-ethanol (1:2 v/v) and Trypan blue lactophenol (0.05%). For histochemistry and fluorescence microscopy see Bordallo et al. (2002). Some fluorescence stains such as HOE 33258 (Hua'an et al. 1991) have been used to identify fungal structures, cell walls and nuclei interchange in anastomosis process using a fluorescence microscope (Manzanilla-López et al. 2011b; Fig. 12.9). The germ tubes and appressoria was assessed microscopically (400 × magnification) by Esteves (2007) using Calcofluor white M2R after 24 h incubation at 25 °C in a moist chamber.

12.16.3 Inoculation of Tomato Seedlings Free of Contaminants

Escudero and Lopez-Llorca (2012) produce tomato seedlings free of contaminants for endophytic studies. They first surface-sterilized tomato seeds, before plating on germinating medium (see Sect. 12.8.1.1) and incubating in darkness for 7 days at 25 °C. Seedlings were plated on Petri dishes containing CMA colonised by *Pochonia* isolates (21 days) and incubated at 25 °C under a photoperiod of 16:8 h (light:dark).

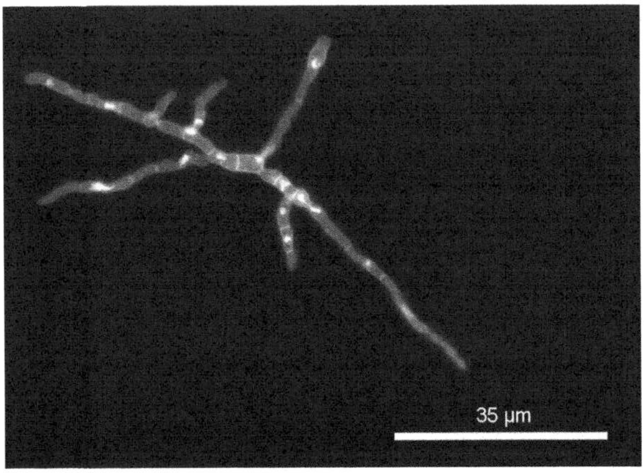

Fig. 12.9 *Pochonia chlamydosporia*. Colony stained with HOE 33258 (Copyright Rothamsted Research Ltd.)

12.17 Future Perspectives

Laboratory and glasshouse experiments provide valuable information to understand the factors that influence the relationship between *Pochonia* spp. and nematode hosts. The use of good practices and proper experiment design are essential. They are also useful to determine application rates, timing and types of inocula in field applications and whether establishment can be improved by modifying the soil environment using organic amendments or other soil treatments (Stirling 1991, 2014).

New methods to estimate the length of external arbuscular mycorrhizal fungal (AMF) hyphae in soil, such as digital gridline intersection, are now available (Shen et al. 2016) and their use may be worth considering to estimate root colonization by *Pochonia* spp. Estimates of fungal abundance that rely solely on culture-based methods may not give reliable estimates and should be used in combination with molecular methods. DNA based quantitative PCR (qPCR) with species-specific primers improves the estimation of fungal abundance but also detects DNA from non-viable and moribund material (Hirsch et al. 2000; Siddiqui et al. 2009). The use of monoclonal antibodies and ELISA (enzyme-linked immnunosorbent assays) techniques and other molecular tools may also provide suitable quantitative alternatives.

12.18 Health and Safety

All procedures should be the subject of risk analysis and recommended procedures, including the use of appropriate personal protective equipment, should be rigorously implemented and followed. Hazard data sheets for the chemicals involved should also be available to staff to minimize risk. Individual countries have their own legal requirements in these matters and it is the responsibility of the user to ensure that the appropriate legislation and procedures are followed.

References

Aminuzzaman, F. M., Xie, H. Y., Duan, W. J., et al. (2013). Isolation of nematophagous fungi from eggs and females of *Meloidogyne* spp. and evaluation of their biological control potential. *Biocontrol Science and Technology, 23*, 170–182.

Arevalo, J., Hidalgo-Díaz, L., Martins, I., et al. (2009). Cultural and morphological characterization of *Pochonia chlamydosporia* and *Lecanicillium psalliotae* isolated from *Meloidogyne mayaguensis* eggs in Brazil. *Tropical Plant Pathology, 34*, 158–163.

Atkins, S. D., Clark, I. M., Sosnowska, D., et al. (2003). Detection and quantification of *Plectosphaerella cucumerina*, a potential biological control agent of potato cyst nematodes by using conventional, PCR real-time PCR, selective media, and baiting. *Applied and Environmental Microbiology, 69*, 4788–4793.

Ayatollahy, E., Fatemy, S., & Etebarian, H. R. (2008). Potential for biological control of *Heterodera schachtii* by *Pochonia chlamydosporia* var. *chlamydosporia* on sugar beet. *Biocontrol Science and Technology, 18*, 157–167.

Bernard, E. C., Self, L. H., & Tyler, D. D. (1997). Fungal parasitism of soybean cyst nematodes, *Heterodera glycines* (Nematoda: Heteroderidae) in different cropping-tillage regimes. *Applied Soil Ecology, 5*, 57–70.

Bordallo, J. J., Lopez-Llorca, L. V., Jansson, H. B., et al. (2002). Colonization of plant roots by egg-parasitic and nematode-trapping fungi. *New Phytologist, 154*, 491–499.

Bourne, J. M., & Kerry, B. R. (1999). Effect of the host plant on the efficacy of *Verticillium chlamydosporium* as a biological control agent of root-knot nematodes at different nematode densities and fungal application rates. *Soil Biology and Biochemistry, 31*, 75–84.

Bourne, J. M., & Kerry, B. R. (2000). Observations on the survival and competitive ability of the nematophagous fungus *Verticillium chlamydosporium* in soil. *International Journal of Nematology, 10*, 9–18.

Bourne, J. M., Kerry, B. R., & De Leij, F. A. A. M. (1994). Methods for the study of *Verticillium chlamydosporium* in the rhizosphere. *Journal of Nematology, 26* (suppl), 587–591.

Bourne, J. M., Kerry, B. R., & De Leij, F. A. A. M. (1996). The importance of the host plant on the interaction between root-knot nematodes (*Meloidogyne* spp.) and the nematophagous fungus, *Verticillium chlamydosporium* Goddard. *Biocontrol Science and Technology, 6*, 539–548.

Bourne, J. M., Karanja, P. K., Kalisz, H., et al. (2004). Incidence and severity of damage caused by *Meloidogyne* spp. and isolation and screening of the nematophagous fungus *Pochonia chlamydosporia* from some of the main vegetable growing areas in Kenya. *International Journal of Nematology, 14*, 111–122.

Carneiro, R. M. D. G., Almeida, M. R. A., Martins, I., et al. (2008). Occurrence of *Meloidogyne* spp. and nematophagous fungi on vegetables in the Federal District of Brazil. *Nematologia Brasileira, 32*, 135–141.

Chen, S. Y., & Chen, F. J. (2003). Fungal parasitism of *Heterodera glycines* eggs as influenced by egg age and pre-colonization of cysts by other fungi. *Journal of Nematology, 35*, 271–277.

Chen, S., & Dickson, D. W. (2012). Biological control. In R. H. Manzanilla-López & N. Marbán-Mendoza (Eds.), *Practical plant nematology* (pp. 761–811). Mexico: BBA Biblioteca básica de agricultura, Editorial del Colegio de Postgraduados.

Ciancio, A., Colagiero, M., Pentimone, I., et al. (2016). Formulation of *Pochonia chlamydosporia* for plant and nematode management. In N.K. Arora et al. (Eds.), *Bioformulations for sustainable agriculture* (pp. 177–197). Springer India.

Clyde, J. M. F. (1993). *The cyst nematode pathogen Verticillium chlamydosporium*. PhD thesis, The University of Leeds, Department of Pure and Applied Biology, UK.

Coutinho, M. M., Freitas, L. G., Dallemolle-Giaretta, R., et al. (2009). Control of *Meloidogyna javanica* by *Pochonia chlamydosporia* and papaya seed flour. *Nematologia Brasileira, 33*, 169–175.

Crump, D. H. (1987a). A method for assessing the natural control of cyst nematode populations. *Nematologica, 33*, 232–243.

Crump, D. H. (1987b). Effect of time of sampling, method of isolation and age of the nematode on the species of fungi isolated from females of *Heterodera schachtii* and *H. avenae*. *Revue de Nématologie, 10*, 369–373.
Crump, D. H. (1991). Estimation of suppressiveness and isolation of fungal parasites of cyst nematodes. In B. R. Kerry, & D. G. Crump (Eds.). *Methods for studying nematophagous fungi*. IOBC / WPRS Bulletin XIV (2):18–22.
Crump, D. H., & Irving, F. (1992). Selection of isolates and methods of culturing *Verticillium chlamydosporium* and its efficacy as biological control agent of beet and potato cyst nematodes. *Nematologica, 38*, 367–374.
Crump, D. H., & Kerry, B. R. (1977). Maturation of females of the cereal cyst-nematode on oat roots and infection by an *Entomophthora*-like fungus in observation chambers. *Nematologica, 23*, 398–402.
Crump, D. H., & Kerry, B. R. (1981). A quantitative method for extracting resting spores of two nematode parasitic fungi, *Nematophthora gynophila* and *Verticillium chlamydosporium* from soil. *Nematologica, 27*, 330–338.
Crump, D. H., & Kerry, B. R. (1987). Studies on the population dynamics and fungal parasitism of *Heterodera schachtii* in soil from a sugar beet monoculture. *Crop Protection, 6*, 49–55.
Dallemole-Giaretta, R., Freitas, L. G., Lopes, E. A., et al. (2012). Screening of *Pochonia chlamydosporia* Brazilian isolates as biocontrol agents of *Meloidogyne javanica*. *Crop Protection, 42*, 102–107.
de Haan, J., & Thijssen, L. (1991). A bioassay for assessing the potential of facultative parasites of the potato cyst nematode. In B. R. Kerry, D. G. Crump (Eds.). *Methods for studying nematophagous fungi*. IOBC / WPRS Bulletin XIV(2):29–33.
De Leij, F. A. A. M., & Kerry, B. R. (1991). The nematophagous fungus *Verticillium chlamydosporium* as a potential biological control agent for *Meloidogyne arenaria*. *Revue de Nématologie, 14*, 157–164.
De Leij, F. A. A. M., Dennehy, J. A., & Kerry, B. R. (1992). The effect of temperature and nematode species on interactions between the nematophagous fungus *Verticillium chlamydosporium* and root-knot nematodes (*Meloidogyne* spp.) *Nematologica, 38*, 65–79.
De Leij, F. A. A. M., Kerry, B. R., & Dennehy, J. A. (1993a). Effect of watering on the distribution of *Verticillium chlamydosporium* in soil and the colonisation of egg masses of *Meloidogyne incognita* by the fungus. *Nematologica, 39*, 250–265.
De Leij, F. A. A. M., Kerry, B. R., & Dennehy, J. A. (1993b). *Verticillium chlamydosporium* as a biological control agent for *Meloidogyne incognita* and *M. Hapla* in pots and micro-plot tests. *Nematologica, 39*, 115–126.
Domsch, K. H., Gams, W., & Anderson, T.-H. (1980). *Compendium of soil fungi*. London: Academic.
Escudero, N., & Lopez-Llorca, L. V. (2012). Effects on plant growth and root-knot nematode infection of an endophytic GFP transformant of the nematophagous fungus *Pochonia chlamydosporia*. *Symbiosis, 57*, 33–42.
Esteves, I. (2007). *Factors affecting the performance of* Pochonia chlamydosporia *as a biological control agent for nematodes*. Ph.D. Thesis. Cranfield University, United Kingdom.
Flores-Camacho, R., Manzanilla-López, R. H., Cid del Prado-Vera, I., et al. (2007). Control of *Nacobbus aberrans* (Thorne, 1935) Thorne y Allen, 1944 with *Pochonia chlamydosporia* (= *Verticillium chlamydosporium*) (Goddard) Zare and W. Gams. *Revista Mexicana de Fitopatología, 25*, 26–34.
Franco-Navarro, F., Vilchis-Martínez, K., & Miranda-Damián, J. (2009). New records of *Pochonia chlamydosporia* from Mexico: Isolation, root colonization and parasitism of *Nacobbus aberrans* eggs. *Nematropica, 39*, 133–142.
Giné, A., Carrasquilla, M., Martínez-Alonso, M., et al. (2016). Characterization of soil suppressiveness to root-knot nematodes in organic horticulture in plastic greenhouse. *Frontiers in Plant Science*. doi:10.3389/fpls.2016.00164.

Grünwald, N. J., Workneh, F., Hu, S., et al. (1997). Comparison of an *in vitro* and a damping-off assay to test soils for suppressiveness to *Pythium aphanidermatum. European Journal of Plant Pathology, 103*, 55–63.

Hidalgo-Díaz, L., Bourne, J. M., Kerry, B. R., et al. (2000). Nematophagous *Verticillium* spp. in soils infested with *Meloidogyne* spp. in Cuba: Isolation and screening. *International Journal of Pest Management, 46*, 277–284.

Hirsch, P. R., Mauchline, T. H., Mendum, A., et al. (2000). Detection of the nematophagous fungus *Verticillium chlamydosporium* in nematode-infested plant roots using PCR. *Mycological Research, 104*, 435–439.

Hirsch, P. R., Atkins, D. S., Mauchline, T. H., et al. (2001). Methods for studying the nematophagous fungus *Verticillium chlamydosporium* in the root environment. *Plant and Soil, 232*, 21–30.

Hooper, D. J. (1986). Extraction of free-living stages from soil. In J. F. Southey (Ed.), *Laboratory methods for work with plant and soil nematodes* (6th ed., pp. 5–30). London: Ministry of Agriculture, Fisheries and Food.

Hua'an, Y., Sivasithamparam, K., & O'Brien, P. A. (1991). An improved technique for fluorescence staining of fungal nuclei and septa. *Australasian Plant Pathology, 20*, 119–121.

Hunt, D. J., & Handoo, Z. (2012). Root-knot nematodes. In R. H. Manzanilla-López & N. Marbán-Mendoza (Eds.), *Practical Plant Nematology* (pp. 359–409). Mexico: BBA Biblioteca Básica de Agricultura, Editorial del Colegio de Postgraduados.

Irving, F., & Kerry, B. R. (1986). Variation between strains of the nematophagous fungus, *Verticillium chlamydosporium* Goddard. II. Factors affecting parasitism of cyst nematode eggs. *Nematologica, 32*, 474–485.

Kerry, B. R. (1991). Preface. In B. R. Kerry, D. G. Crump (Eds.). *Methods for studying nematophagous fungi*. IOBC / WPRS Bulletin XIV (2):I–II.

Kerry, B. R. (1995). New strategies for the management of plant parasitic nematodes with special emphasis on biological control. *Arab Journal of Plant Protection, 13*, 47–52.

Kerry, B. R. (1997). *Biological control of nematodes: Prospects and opportunities. FAO corporate document repository*. Plant nematode problems and their control in the near east region. FAO plant production and Protection paper, 144. http://www.fao.org/docrep/v9978e/v9978e00.HTM. Accessed 14 Oct 2016.

Kerry, B. R. (2000). Rhizosphere interactions and the exploitation of microbial agents for the biological control of plant-parasitic nematodes. *Annual Review of Phytopathology, 38*, 423–441.

Kerry, B. R., & Bourne, J. R. (1996). The importance of rhizosphere interactions in the biological control of plant parasitic nematodes – A case study using *Verticillium chlamydosporium. Pesticide Science, 47*, 69–75.

Kerry, B. R., & Bourne, J. (2002). *A manual for research on Verticillium chlamydosporium a potential biocontrol agent for root-knot nematodes*. Gent: IOBC/WPRS.

Kerry, B. R., & Crump, D. H. (1977). Observations on fungal parasites of females and eggs of the cereal cyst-nematode, *Heterodera avenae*, and other cyst nematodes. *Nematologica, 23*, 193–201.

Kerry, B. R., Crump, D. H., & Mullen, L. (1982). Studies of the cereal cyst nematode, *Heterodera avenae* under continuous cereals, 1974-1978. 1. Plant growth and nematode multiplication. *The Annals of Applied Biology, 100*, 477–487.

Kerry, B. R., Simon, A., & Rovira, A. D. (1984). Observations on the introduction of *Verticillium chlamydosporium* and other parasitic fungi into soil for control of the cereal cyst-nematode *Heterodera avenae. The Annals of Applied Biology, 105*, 509–516.

Kerry, B. R., Irving, F., & Hornsey, J. C. (1986). Variation between strains of the nematophagous fungus, *Verticillium chlamydosporium* Goddard. I. Factors affecting growth in vitro. *Nematologica, 32*, 461–473.

Kirk, P. M., Cannon, P. F., Minter, D. W., et al. (2008). *Dictionary of the fungi* (10th ed.). Wallingford: CAB International.

Lohmann, U., Sikora, R. A., & van den Boogert, P. (1991). Isolation and determination of the antagonistic potential of fungal endoparasites of nematodes. In B. R Kerry, & D. G. Crump (Eds.). *Methods for studying nematophagous fungi*. IOBC / WPRS Bulletin XIV(2):1–12.

Lopez-Llorca, L. V., & Duncan, J. M. (1986). New media for the estimation of fungal infection in eggs of the cereal cyst nematode, *Heterodera avenae* Woll. *Nematologica, 32*, 486–490.

Lopez-Llorca, L. V., & Duncan, G. H. (1991). Effects of fungal parasites on cereal cyst nematode (*Heterodera avenae* Woll.) from naturally infested soil – A scanning electron microscopy study. *Canadian Journal of Microbiology, 37*, 218–225.

Lopez-Llorca, L. V., & Robertson, W. M. (1992). Ultrastructure of infection of cyst nematode eggs by the nematode fungus *Verticillium suchlasporium. Nematologica, 39*, 65–74.

Lopez-Llorca, L. V., Carbonell, T., & Salinas, J. (1999). Colonization of plant waste substrates by entomopathogenic and mycoparasitic fungi-a SEM study. *Micron, 30*, 325–333.

Lopez-Llorca, L. V., Olivares-Bernabeu, C., & Salinas, J. (2002a). Pre-penetration events in fungal parasitism of nematode eggs. *Mycological Research, 106*, 499–506.

Lopez-Llorca, L. V., Bordallo, J. J., Salinas, J., et al. (2002b). Use of light and scanning electron microscopy to examine colonisation of barley rhizosphere by the nematophagous fungus *Verticillium chlamydosporium. Micron, 33*, 61–67.

Lopez-Llorca, L. V., Gómez-Vidal, S., Monfort, E., et al. (2010). Expression of serine proteases in egg-parasitic nematophagous fungi during barley root colonization. *Fungal Genetics and Biology, 47*, 342–351.

Luambano, N. D., Manzanilla-López, R. H., Kimenju, J. W., et al. (2015). Effect of temperature, pH, carbon and nitrogen ratios on the parasitic activity of *Pochonia chlamydosporia* on *Meloidogyne incognita. Biological Control, 80*, 23–29.

Lumsden, R. D. (1981). A nylon fabric technique for studying the ecology of *Pythium aphanidermatum* and other fungi in soil. *Phytopathology, 71*, 282–285.

Lynch, J. M. (1996). Monitoring of antagonistic fungi, perspectives, needs and legislation. In D. F. Jensen, H. B. Hansson, & A. Tronsmo (Eds.), *Monitoring antagonistic fungi deliberately released into the environment, Developments in Plant Pathology* (Vol. 8, pp. 1–9). Dordrecht: Klower Academic Publishers.

Maciá-Vicente, J. G., Jansson, H. B., Talbot, N. J., et al. (2009a). Real-time PCR quantification and live-cell imaging of endophytic colonization of barley (*Hordeum vulgare*) roots by *Fusarium equiseti* and *Pochonia chlamydosporia. The New Phytologist, 182*, 213–228.

Maciá-Vicente, J. G., Rosso, L. C., Ciancio, A., et al. (2009b). Colonisation of barley roots by endophytic *Fusarium equiseti* and *Pochonia chlamydosporia*: Effects on plant growth and disease. *The Annals of Applied Biology, 155*, 391–401.

Manzanilla-López, R. H., Atkins, S. D., Clark, I. M., et al. (2009). Measuring abundance, diversity and parasitic ability in two populations of the nematophagous fungus *Pochonia chlamydosporia* var. *chlamydosporia. Biocontrol Science and Technology, 19*, 391–406.

Manzanilla-López, R. H., Esteves, I., Powers, S. J., et al. (2011a). Effects of crop plants on abundance of *Pochonia chlamydosporia* and other fungal parasites of root-knot and potato cyst nematodes. *The Annals of Applied Biology, 159*, 118–129.

Manzanilla-López, R. H., Clark, I. M., Atkins, S. D., et al. (2011b). Exploring competitiveness and variation in the nematophagous fungus *Pochonia chlamydosporia* var. *chlamydosporia* and its significance for biological control. *Bulletin OILB/SROP, 63*, 37–40.

Manzanilla-López, R. H., Devonshire, J., Ward, E., et al. (2014). Cryo-scanning electron microscopy with cryo-planning: A complementary approach for studying the infection of *Meloidogyne incognita* by *Pochonia chlamydosporia. Nematology, 16*, 1059–1067.

Mauchline, T. H., Kerry, B. R., & Hirsch, P. R. (2002). Quantification in soil and the rhizosphere of the nematophagous fungus *Verticillium chlamydosporium* by competitive PCR and comparison with selective plating. *Applied and Environmental Microbiology, 68*, 1846–1583.

McSorley, R. (2011). Overview of organic amendments for management of plant-parasitic nematodes with case studies from Florida. *Journal of Nematology, 43*, 69–81.

Mead, R., Curnow, R., & Hasted, A. M. (2003). *Statistical methods in agriculture and experimental biology* (3rd ed.). Boca Raton: Chapman and Hall.

Medina-Canales, M. G., Rodríguez-Tovar, A. V., Manzanilla-López, R. H., et al. (2013). Identification and molecular characterisation of new Mexican isolates of *Pochonia chlamydosporia* for the management of *Meloidogyne* spp. *Biocontrol Science and Technology, 24*, 1–21.

Menéndez, A., Bertoni, M. D., & Cabral, D. (1997). Endofitos fúngicos en *Juncus imbricatus* var *chamissonis*: identificación de los patrones de colonización. *Revista Iberoamericana de Micología, 14*, 125–128.

Mian, I. H., & Rodríguez-Kábana, R. (1982). Survey of the nematicidal properties of some organic materials available in Alabama as amendments to soil for control of *Meloidogyne arenaria*. *Nematropica, 12*, 235–246.

Mitchinson, S. M. (2009). *New cyst nematode threats to cereals in Southern Britain*. Reading: Reading University.

Monfort, E., Lopez-Llorca, L. V., Jansson, H. B., et al. (2005). Colonisation of seminal roots of wheat and barley by egg-parasitic nematophagous fungi and their effects on *Gaeumannomyces graminis* var. *tritici* and development of root-rot. *Soil Biology and Biochemistry, 37*, 1229–1235.

Monfort, E., Lopez-Llorca, L. V., Hans-Börje, J., et al. (2006). In vitro soil receptivity assays to egg-parasitic nematophagous fungi. *Mycological Progress, 5*, 18–23.

Moosavi, M. R., Zare, R., Zamanizadeh, H. R., et al. (2010). Pathogenicity of *Pochonia* species of *Meloidogyne javanica*. *Journal of Invertebrate Pathology, 104*, 125–133.

Morgan-Jones, G., Godoy, G., & Rodríguez-Kábana, R. (1981). *Verticillium chlamydosporium*, fungal parasite of *Meloidogyne arenaria females*. *Nematropica, 11*, 115–119.

Morton, C. O., Mauchline, T. H., Kerry, B. R., et al. (2003). PCR-based DNA fingerprinting indicates host-related genetic variation in the nematophagous fungus *Pochonia chlamydosporia*. *Mycological Research, 107*, 198–205.

Nagesh, M., Hussaini, S. S., Ramanujam, B., et al. (2007). Molecular identification, characterization, variability and infectivity of Indian isolates of the nematophagous fungus *Pochonia chlamydosporia*. *Nematologia Mediterranea, 37*, 47–56.

Nicolay, R., & Sikora, R. A. (1989). Techniques to determine the activity of fungal egg parasites of *Heterodera schachtii* in field soil. *Revue de Nématologie, 12*, 97–102.

Nicolay, R., & Sikora, R. A. (1991). Estimation of the activity of fungi parasitic on nematode eggs. In B. R. Kerry & D. G. Crump (Eds.). *Methods for studying nematophagous fungi*. IOBC / WPRS Bulletin XIV(2):23–28.

Olivares-Bernabeu, C. M., & López-Llorca, L. V. (2002). Fungal egg-parasites of plant-parasitic nematodes from Spanish soils. *Revista Iberoamericana de Micología, 19*, 104–110.

Podestá, G. S., Dallemole-Giaretta, R., Freitas, L. G., et al. (2009). Actividade nematófaga de *Pochonia chlamydosporia* em solo natural ou autoclavado sobre *Meloidogyne javanica*. *Nematologia Brasileira, 33*, 191–193.

Powers, S. J. (2012). Statistics in nematology. In R. H. Manzanilla-López & N. Marbán-Mendoza (Eds.), *Practical plant nematology* (pp. 841–870). Mexico: BBA Biblioteca básica de agricultura, Editorial del Colegio de Postgraduados.

Pyrowolakis, A., Schuster, R. F., & Sikora, R. A. (1999). Effect of cropping pattern and green manure on the antagonistic potential and the diversity of egg pathogenic fungi in fields with *Heterodera schachtii* infection. *Nematology, 1*, 165–171.

Pyrowolakis, A., Westphal, A., Sikora, R. A., et al. (2002). Identification of root-knot nematode suppressive soils. *Applied Soil Ecology, 19*, 51–56.

Rayner, A. D. M., & Todd, N. K. (1979). Population and community structures and dynamics of fungi in decaying wood. *Advances in Botanical Research, 7*, 333–420.

Rodríguez-Kábana, R., Morgan-Jones, G., Godoy, G., et al. (1984). Effectiveness of species of *Gliocladium, Paecilomyces* and *Verticillium* for control of *Meloidogyne arenaria* in field soil. *Nematropica, 14*, 155–170.

Schulz, B., Haas, S., Junker, C., et al. (2015). Fungal endophytes are involved in multiple balanced antagonisms. *Current Science, 109*, 39–45.

Shen, Q., Kirschbaum, M. U. F., Hedley, M. J., et al. (2016). Testing an alternative method for estimating the length of fungal hyphae using photomicrography and image processing. *PloS One, 11*(6), edD157017. doi:10.1371/journal.pone.0157017.

Shepherd, A. M. (1986). Extraction and estimation of cyst nematodes. In J. F. Southey (Ed.), *Laboratory methods for work with plant and soil nematodes* (6th ed., pp. 31–49). London: Ministry of Agriculture, Fisheries and Food.

Siddiqui, A., Atkins, S. D., & Kerry, B. R. (2009). Relationship between saprotrophic growth in soil of different biotypes of *Pochonia chlamydosporia* and the infection of nematode eggs. *The Annals of Applied Biology, 155*, 131–141.
Sikora, R. A., & Carter, W. W. (1987). Nematode interactions with fungal and bacterial plant pathogens –fact or fantasy. In J. A. Veech & D. W. Dickson (Eds.), *Vistas on Nematology: A commemoration of the twenty-fifth anniversary of the society of nematologists* (pp. 307–312). De Leon Springs: E.O. Painter Printing Co..
Stirling, G. R. (1979). Techniques for detecting *Dactylella oviparasitica* and evaluating its significance in field soils. *Journal of Nematology, 11*, 99–100.
Stirling, G. R. (1991). *Biological control of plant-parasitic nematodes*. Wallingford: CAB International.
Stirling, G. R. (2014). *Biological control of plant-parasitic nematodes: Soil ecosystem management in sustainable agriculture* (2nd ed.). Croydon: CABI.
Subbotin, S. (2012). Cyst nematodes. In R. H. Manzanilla-López & N. Marbán-Mendoza (Eds.), *Practical Plant Nematology* (pp. 299–357). Mexico: BBA Biblioteca básica de agricultura, Editorial del Colegio de Postgraduados.
Sun, M., Gao, L., Shi, Y., et al. (2006). Fungi and actinomycetes associated with *Meloidogyne* spp. eggs and females in China and their biocontrol potential. *Journal of Invertebrate Pathology, 93*, 22–28.
Sykes, D. (1994). *The growth and sporulation of Verticillium chlamydosporium*. MSc thesis, University of Manchester, UK.
Tennant, D. (1975). A test of modified line intersect method of establishing root length. *Journal of Ecology, 63*, 995–1001.
Timper, P. (2014). Conserving and enhancing biological control of nematodes. *Journal of Nematology, 46*, 75–89.
Verdejo-Lucas, S., Ornat, C., Sorribas, F. J., et al. (2002). Species of root-knot nematodes and fungal egg parasites recovered from vegetables in Almería and Barcelona, Spain. *Journal of Nematology, 34*, 405–408.
Weller, D. M., Raaijmakers, J. M., Gardener, B. B. M., et al. (2002). Microbial populations responsible for specific soil suppressiveness to plant pathogens. *Annual Review of Physiology, 40*, 309–348.
Westphal, A., & Becker, J. O. (1999). Biological suppression and natural population decline of *Heterodera schachtii* in a California field. *Phytopathology, 89*, 434–440.
Westphal, A., & Becker, J. O. (2000). Transfer of biological soil suppressiveness against *Heterodera schachtii*. *Phytopathology, 90*, 401–406.
Yang, J., Loffredo, A., Borneman, J., et al. (2012). Biocontrol efficacy among strains of *Pochonia chlamydosporia* obtained from a root-knot nematode suppressive soil. *Journal of Nematology, 44*, 67–71.
Zare, R., Gams, W., & Evans, H. C. (2001). A revision of *Verticillium* section Prostrata. V. The genus *Pochonia*, with notes on *Rotiferophthora*. *Nova Hedwigia, 73*, 51–86.

Chapter 13
Molecular Diagnostics of *Pochonia chlamydosporia*

María Gabriela Medina-Canales and Aída V. Rodríguez-Tovar

Abstract Molecular tools have been increasingly used in identification and taxonomy studies of *Pochonia* (*Clavicipitaceae*). In recent years there have been important advances in *Clavicipitaceae* taxonomy, including resolving the differences between *Verticillium* and *Pochonia* through ITS sequences molecular analysis. Current molecular identification of *Pochonia* spp. relies on different sets of molecular markers to differentiate species, varieties, and host-preference biotypes. The diversity of *Pochonia* spp. populations has been studied through several genomic fingerprinting techniques, such as RAPD, SCAR, ERIC and REP. Quantification of the fungus in soil and roots can be made using quantitative PCR in combination with other classic microbiological techniques. It is well known that the genus is complex and the current classification is mainly based on the phylogenetic analysis of several genes such as *β-tubulin*, ITS, *nrSSU*, *nrLSU*, *rpb1*, *rpb2* and *EF1-α*. Molecular advances and recent genome sequencing of *Pochonia chlamydosporia* has opened a new era in the study of this important fungus, thus broadening the possibilities for studying the molecular mechanisms of differentiation, pathogenesis and diagnostics within *Pochonia* spp.

13.1 *Clavicipitaceae* Classification and Molecular Approaches

The interest in studying the classification and taxonomy of this fungal family is partly due to the capacity shown by several of its members to infect invertebrate species that belong to different phyla, including insects and nematodes (Kepler et al. 2012). Currently the family *Clavicipitaceae* is placed within the phylum *Ascomycota*, subphylum *Pezizomycotina* in the class *Sordariomycetes*, order

M.G. Medina-Canales (✉) • A.V. Rodríguez-Tovar
Escuela Nacional de Ciencias Biológicas del Instituto Politécnico Nacional, Ciudad de México, Mexico
e-mail: magameca@yahoo.com.mx; avrodriguez@hotmail.com

R.H. Manzanilla-López, L.V. Lopez-Llorca (eds.), *Perspectives in Sustainable Nematode Management Through* Pochonia chlamydosporia *Applications for Root and Rhizosphere Health*, Sustainability in Plant and Crop Protection, DOI 10.1007/978-3-319-59224-4_13

Hypocreales. Many species belonging to this family play important roles, not only as biological control agents, but also as crop pathogens (Scholte et al. 2005; Kanzok and Jacobs-Lorena 2006; Sung et al. 2007; Haarmann et al. 2009). Molecular taxonomy has helped to reveal new phylogenetic relationships as well as diverse nutritional pathways for these fungi. Initially, this group of fungi was considered to be a single diverse superfamily. However, phylogenetic studies using molecular markers have revealed the existence of three different families: *Clavicipitaceae*, *Cordycipitaceae* and *Ophiocordycipitaceae* (Sung et al. 2007).

Clavicipitaceae is a very large family of at least 33 genera and approximately 800 species (Rogerson 1970; Eriksson and Hawksworth 1985; Sung et al. 2007), which include pathogens, parasites and symbionts of plants, invertebrate animals or other fungi, which can also produce different types of neurotropic alkaloids (Schardl et al. 2013). They have a cosmopolitan distribution. This family is easily recognized because of the presence of ascospores in the sexual phase (Spatafora and Blackwell 1993; Sung et al. 2007).

Gäumann (1926) was the first to attempt to classify the Clavicipitaceae and partitioned the family into three groups: group 1 *Oomycetes-Ascopolyporus*; group 2 *Epichloë-Claviceps*; and group 3 *Cordyceps* (Sung et al. 2007). However, this classification was based on sexual reproduction structures and the subsequent development and availability of molecular tools has resulted in many phylogenetic analyses of the *Clavicipitaceae* using DNA ribosomal sequences (Spatafora and Blackwell 1993; Glenn et al. 1996; Gams 1988; Sung et al. 2001, 2007). Taxon monophyly was rejected on the grounds of the limited number of samples that had been examined which was not considered to be representative of the morphological diversity and ecology within the group (Sung et al. 2007). Sung et al. (2007) made a phylogenetic study that included 91 samples of *Hypocreales* that were representative of families *Bionectriaceae*, *Nectriaceae*, *Hypocreaceae* and *Clavicipitaceae*. Using this approach, they were able to evaluate ecological diversity and family relationships. The phylogenetic analysis was done using five nuclear genes: the nuclear ribosomal small and large subunits (*nrSSU*, *nrLSU*), *β-tubulin*, the elongation factor 1α (*EF-1α*), the largest and the second largest subunits of RNA polymerase II (*rpb1* and *rpb2*) and one mitochondrial gene ATP synthase (*mtATP6*). The results showed that the family *Clavicipitaceae* has three clades, the genus *Pochonia* being placed in sub-clade A3 together with *Metarhizium* and *Rotiferophthora* (Sung et al. 2007).

Many clavicipitaceous anamorphs produce bioactive molecules, some of which have industrial and agricultural applications, e.g., *Claviceps purpurea* produces ergot alkaloids and *Cordyceps subsessilis* synthesizes cyclosporine. Within the *Clavicipitaceae* different genera are arthropod pathogens (i.e., *Cordyceps*, *Hypocrella* and *Torrubiella*), although others (i.e., *Claviceps*, *Balansia*) are grass endophytes of *Neotyphodium* spp. (Rogerson 1970; Shiba and Sugawara 2005; Sung et al. 2007). Entomopathogenic fungi, such as *Metarhizium anisopliae* and *Beauveria bassiana*, are also used as microbial insecticides (Feng et al. 1994; Bidochka et al. 2001). *Purpureocillium lilacinum* (= *Paecilomyces lilacinus*) and *Pochonia chlamydosporia* (= *Verticillium chlamydosporium*) also have a great potential as biocontrol agents of cyst and root-knot nematodes (Kerry et al. 1984; Olivares-Bernabeu and Lopez-Llorca 2002).

The broad diversity of species found within the genus *Pochonia*, has made it difficult to classify this genus within the *Clavicipitaceae* (see also Chap. 2). Classification was initially based exclusively on morphological characteristics but these were not enough to discriminate between very similar species. Therefore, other characteristics were used in the diagnosis, such as vegetative compatibility, virulence differences and biochemical characteristics (Joaquim and Rowe 1990), the latter being also used in nematophagous fungi identification. Although biochemical characteristics can differentiate between species, their use was limited, in part because of complexity in the methods employed. Molecular techniques, by comparison, have advantages over the other methods, including greater sensitivity, quicker identification and fingerprinting of isolates (Morton et al. 1995).

13.2 *Verticillium* and *Pochonia* ITS Sequencing

In the 1990s *Pochonia chlamydosporia* was known as *Verticillium chlamydosporium* and for a long time *V. chlamydosporium* and *V. catenulatum* were both considered as varieties of the same species: *Verticillium chlamydosporium* (Gams 1988). The first attempt to separate *Verticillium* sect. *Prostrata* using molecular methods was made by Sung et al. 2001. In this work they amplified *nrLSU* and *nrSSU* genes and placed *V. chlamydosporium* in the *V.* sect. *Prostrata* group D2. This group contained other species, including *V. suchlasporium* (= *Pochonia suchlasporia*), *V. gonioides* (= *P. gonioides*). For a long time the genera *Pochonia* and *Lecanicillium* were placed in *Verticillium* because of their similar morphology, except for the presence of resistant structures or chlamydospores produced by *Pochonia*. However, distinguishing these genera was still difficult, and Gams and Zare (2001) recommended the use of techniques based on phylogenetic inference of the Internal Transcriber Space (ITS) region (ITS1–5.8S–ITS2) for identification of species within the genera *Pochonia*, *Lecanicillium* and *Verticillium.*

13.3 The Genus *Pochonia* from PCR Fingerprinting to Complete Genome

The molecular identification of *P. chlamydosporia* has been mainly focused on identifying which variety isolates belong to, their host preference (i.e., nematode biotype), variation and detection in soil and rhizosphere through fingerprinting and quantitative PCR (qPCR) of specific isolates. It has been a long journey from the beginning of PCR fingerprinting to the present sequencing and genome deciphering of the fungus (Table 13.1). Taking into account all the molecular information that

Table 13.1 PCR-based tools and techniques used in molecular diagnostics and analyses of *Pochonia chlamydosporia* varieties

Molecular technique	MT use and *P. chlamydosporia* variety	References
ERIC-PCR	*P. chlamydosporia* var. *chlamydosporia* genetic variation of isolates	Arora et al. (1996), Morton et al. (2003a) and Flores-Camacho et al. (2008)
	P. chlamydosporia var. *chlamydosporia* biotypes fingerprinting	Manzanilla-López et al. (2009a)
PCR-RAPD	*P. chlamydosporia* var. *chlamydosporia* fingerprinting of different isolates soil and rhizosphere	Mauchline et al. (2004) and Flores-Camacho et al. (2008)
ITS-PCR	Molecular identification of the genus *Pochonia*	Gams and Zare (2001)
	Identification of *P. chlamydosporia* var. *catenulata* biotype 'A'	Atkins et al. (2003)
	Identification of *P. chlamydosporia* var. *mexicana*	Medina-Canales et al. (2014)
PCR-*β-tubulin* gene	Identification *P. chlamydosporia* var. *chlamydosporia*	Hirsch et al. (2001)
ITS1	Identification *P. chlamydosporia* var. *catenulata*	Atkins et al. (2003)
Genes *SSU*, *LSU*, *rpb1*, *rpb2*, *EF-1α*	Identification of *P. chlamydosporia* var. *spinulospora* and *P. chlamydosporia* var. *ellipsospora*	Nonaka et al. (2013)
vcp1 gene	Host preference test	Morton et al. (2003b) and Siddiqui et al. (2009)
Transcriptome	Gene expression profiles in the saprophytic and parasitic phases of *P. chlamydosporia*	Rosso et al. (2011)
Complete genome	*P. chlamydosporia* endophytic, parasitic and saprophytic genes	Larriba et al. (2014)
Mitogenome	*P. chlamydosporia* conserved genes (e.g., cytochrome c oxidase, apocytochrome b, ATP synthase subunits, NADH dehydrogenase subunits)	Lin et al. (2015)

MT molecular technique

has been generated, we should now be in a better position to correctly identify *Pochonia* species and varieties, and better understand its diversity and ecology.

13.3.1 Fingerprinting Markers

Arora et al. (1996) used PCR fingerprinting to identify genetic variations between different isolates of *V. chlamydosporium* (= *P. chlamydosporia*). They demonstrated the potential of restriction fragment analyses (RFLP) of PCR-ITS and PCR-IGS

(Intergenic spacer) rDNA for identification. They also proposed the use of repetitive extragenic palindromic (REP) and Enterobacterial Repetitive Intergenic Consensus (ERIC) sequences to differentiate strains of *P. chlamydosporia*. This was the first step in using molecular tools for *P. chlamydosporia* identification.

The next step was to detect the fungus on roots (i.e., in the rhizosphere) using molecular tools. Hirsch et al. (2000) attempted to detect *P. chlamydosporia* (Vcp10 strain) from DNA extracted from nematode-infested tomato roots using ERIC-PCR, ITS-PCR and primers designed to amplify the *β-tubulin* gene to identify fungi grown in pure culture. They noticed that ERIC and ITS-PCR did not detect the fungus on DNA extracted from infected tomato roots, but that specific primers designed to amplify *β-tubulin* of *Pochonia chlamydosporia* var. *chlamydosporia* did. This technique was considered to be both specific and sensitive, and allowed fungal identification up to *P. chlamydosporia* variety (Hirsch et al. 2001). Subsequently, Atkins et al. (2003) designed a new set of specific primers to identify *Pochonia chlamydosporia* var. *catenulata*. They carried out restriction fragment length polymorphic analyses of the ITS region (RFLP-ITS) with *Hin*FI, found that both varieties had different restriction patterns, and proceeded to analyze the ITS sequences. They found that the ITS1 was the most variable fragment of the ITS1–5.8S–ITS2 region and designed *P. chlamydosporia* var. *catenulata*-specific primers to be used for the molecular identification of this variety.

The intraspecific variation of *P. chlamydosporia* isolates and nematode host preference was also studied by Morton et al. (2003a). They used ERIC and ITS-PCR to analyze isolates from different geographical origins and nematode hosts (biotype) from soils infested with *Heterodera* spp. and *Meloidogyne* spp. The analysis revealed that isolates from soils infested with *Heterodera* spp. formed a different group to the isolates from soils infested with *Meloidogyne* spp., thereby suggesting that genetic variation also relates to host.

Random Amplification of Polymorphic DNA (RAPD-PCR) was used by Mauchline et al. (2004) to detect and fingerprint different isolates of *P. chlamydosporia* in the soil and rhizosphere. The technique was successful in identifying differences between isolates, but could not distinguish between varieties. Flores-Camacho et al. (2008) also used ERIC-PCR, ITS-RFLP, RAPD-PCR and specific primers (β-tubulin) to confirm the identity of Mexican isolates of *P. chlamydosporia*. ERIC-PCR showed that these isolates had two different profiles, but when they were examined with RAPD-PCR, three profiles were found instead, thus demonstrating the existence of intra-specific differences that may or may not be detected depending on the fingerprinting protocol used (Fig. 13.1A, B). In another study, ERIC-PCR separated three different profiles of Portuguese isolates of *P. chlamydosporia* var. *chlamydosporia* (Manzanilla-López et al. 2009a). Yang et al. (2012) used a multiple phylotyping of *P. chlamydosporia* based on analysis of β-tubulin, rRNA ITS, rRNA SSU, and ERIC to discriminate between three strains of isolates from a *Meloidogyne incognita*-suppressive soil.

The RAPD technique was also used in designing Sequence Characterized Amplified Regions (SCAR) specific primers to differentiate between Chinese isolates and strains of *P. chlamydosporia* var. *chlamydosporia* and *Paecilomyces*

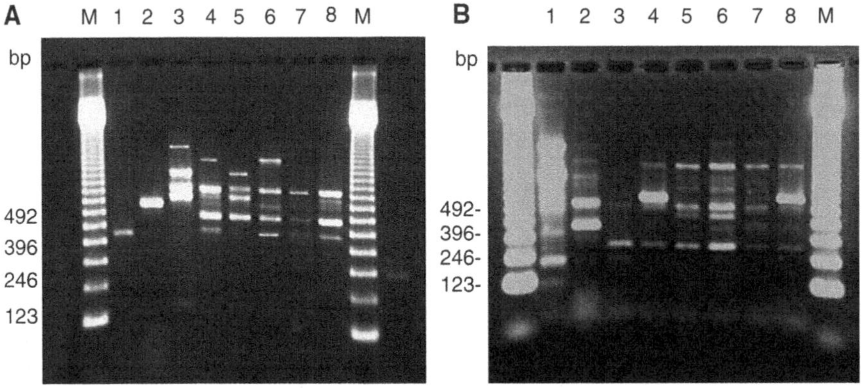

Fig. 13.1 Agarose gel electrophoresis of ERIC and RAPD-PCR amplification products of different isolates of *Pochonia chlamydosporia* var. *chlamydosporia* and other fungi. (**A**): ERIC-PCR profiles; (**B**): RAPD-PCR profiles of different isolates. M 123 base pair ladder marker, lane 1 *Purpureocillium* sp., lane 2 *Acremonium* sp., lane 3 *P. chlamydosporia* var. *chlamydosporia* isolate Pc10, lane 4 *P. chlamydosporia* isolate SMB3, lane 5 *P. chlamydosporia* isolate SC1, lane 6 *P. chlamydosporia* isolate SMB3A, lane 7 *P. chlamydosporia* isolate SM4, lane 8 *P. chlamydosporia* isolate MHCH (after Flores-Camacho et al. 2008)

lilacinus (= *Purpureocillium lilacinum*). The selection of specific fragments of 1.2 kb (Vc1200) and 2.0 kb (Vc2000) specific for strain ZK7 (*P. chlamydosporia* var. *chlamydosporia*), 1.4 kb (P1400) and 0.85 kb (P850) specific for IPC *Paecilomyces lilacinus* (= *Purpureocillium lilacinum*) was done using the random primers OPL-02, OPD-05, OPD-05 and OPC-11. These fragments were cloned and sequenced, SCAR specific primers being designed for the two strains. These four primers specifically recognized the corresponding fragments in the DNA template of the two fungal strains. The SCAR-PCR was a sensitive technique, detecting 10–1000 pg of DNA and differentiating between native and introduced *P. chlamydosporia* var. *chlamydosporia* isolates. It was concluded that SCAR-PCR has more potential in the future detection and ecological studies (Zhu et al. 2006).

Fingerprinting has proved to be useful in assessing variation within *P. chlamydosporia* isolates, but proper extraction of high quality DNA is indispensable to carry out reproducible fingerprinting as shown by Manzanilla-López et al. (2009b). In order to select the best and reliable method they tested different DNA extractions protocols, including Alkaline lysis (Klimyuk et al. 1993), MicroLYSIS® PLUS, FTA® Cards, and DNeasy® in order to perform PCR-β-tubulin and PCR-ERIC protocols. All DNA samples amplified a 270 bp band using the β-tubulin primers, but for ERIC-PCR the best extraction methods were Alkaline lysis and DNeasy®, which obtained more and longer amplicons. These results need to be considered as reproducibility of fingerprinting depends on DNA extraction quality.

13.3.2 Quantitative PCR (qPCR) and RNA Analysis

Colony forming unit (cfu) counts were initially used as the only procedure to measure *P. chlamydosporia* soil abundance but this method is very time consuming and eventually new and faster methods to measure abundance, such as qPCR, were explored. Real time PCR or qPCR is based on self-fluorescent probes, including fluorochromes (intercalating agents) or probes. Specific probes also provide highly reliable results. Ciancio et al. (2005) used qPCR to detect *P. chlamydosporia* for the first time in soil samples using Scorpion probes (DNA-based fluorescent molecules) from ITS-2 rDNA gene. The test is very sensitive and only needs 0.2 pg of DNA to detect *P. chlamydosporia* in the soil and rhizosphere.

Peteira et al. (2005) designed Taqman® PCR primers and probes for *P. chlamydosporia* var. *catenulata* ITS sequences. The probe was designed over the same region as the reverse primer described by Atkins et al. (2003). The primers CATrF and CATrtR generated a 97 bp PCR product for all isolates of *P. chlamydosporia* var. *catenulata* tested. Colony forming unit counts *vs* qPCR efficiency were compared and it was demonstrated that both methods were effective in measuring soil abundance, although qPCR was faster.

13.4 Transcriptome

Pochonia chlamydosporia transcriptome studies were done by Rosso et al. (2011). They studied expression profiles of the saprophytic and parasitic phases of the fungus and found that differential gene expression occurred when the fungus was in contact with root-knot and cyst nematode eggs (see Chap. 5). Some genes showed a high expression (i.e., up-regulated) and were involved in eggs parasitism while other genes were shown to be constitutive (i.e., a gene transcribed at a relatively constant level regardless of environmental conditions). This was an important breakthrough as knowledge of constitutive genes and up-regulated genes in the host-parasite interaction between *Pochonia* spp. and nematodes eggs was still limited. It became apparent from this study that *P. chlamydosporia* reprograms genes when switching from the saprophytic to the parasitic phase. Switching from saprophytism to parasitism showed that up-regulated genes were involved in detoxification, metabolism and enzymatic activities. Some up-regulated genes of the parasitic phase could not only be used in qPCR to assess *P. chlamydosporia* abundance in soil and roots, but also to obtain more information about the parasitic capacity of the fungus in soil.

13.5 Complete Genome Sequencing

The study of any microorganism can be facilitated if its complete genome has been sequenced. Knowledge about different functions and mechanisms to survive and infect hosts is indispensable if microorganisms are to be used as biological control agents. Ward et al. (unpublished data) sequenced the genome of *P. chlamydosporia* Pc10 isolate (*P. chlamydosporia* var. *chlamydosporia*). The genome size was ca 45 Mb and supported similarities between *P. chlamydosporia* (nematophagous fungus) and the entomopathogenic fungus *Metarhizium* (Manzanilla-López et al. 2013). In 2014, Larriba et al. obtained similar results related to *P. chlamydosporia* genome size and relationships and included more information, not only about saprophytic and parasitic activities, but also the endophytic behaviour of the fungus in some plant hosts. The genes involved in the endophytic behaviour/phase of the fungus are the most abundant and obvious as this phase requires high enzymatic, signalling transductions and metabolic activities. An example of these genes are those encoding G proteins α-subunits (Larriba et al. 2014). Some of these genes will probably be used to quantify abundance of *P. chlamydosporia* on roots. The genome study also identified the presence of many serine proteases. The most abundant serine protease found was prolyl aminopeptidase (S33). This protease could play a crucial role as a nematode egg shell disruptor (Jia et al. 2000; Newport et al. 2003; Hung et al. 2005; Soloviev et al. 2011; Larriba et al. 2014), the eggs shells of root-knot and cyst nematodes having proline residues (Bird and McClure 1976), the S33 protease recognizing the proline residues in the egg shell (Larriba et al. 2014) and may even be involved in starting the parasitic phase. For these reasons this serine protease could be a candidate to select and monitor isolates of *P. chlamydosporia* with high potential for parasitism of nematode eggs.

The *P. chlamydosporia* genome has also revealed genes for the synthesis of a large number of secondary metabolites (see Chapter 7), including aurovertins, which are involved in the parasitism of the fungus on the nematode (Larriba et al. 2014; Wang et al. 2015) and the pochonins (i.e., genes involved in the biosynthesis of radicicol) (Reeves et al. 2008). Aurovertin is a secondary metabolite yellow pigment produced by this fungus but is not produced by all *P. chlamydosporia* isolates. There are four types of aurovertins (D, E, F and J) of which the D type has the most nematode-toxic activity (Wang et al. 2015). The aurovertins have a nematicidal activity on eggs and juvenile nematode stages (Niu et al. 2010; Wang et al. 2015). Some aurovertin genes could be suitable to target isolates having a more efficient egg parasitism.

Mining *P. chlamydosporia* genome data could also help in selecting and studying more genes to provide a quicker and better diagnostic resolution to identify *Pochonia* spp. and varieties (Hirsch et al. 2000; Atkins et al. 2003; Nonaka et al. 2013; Medina-Canales et al. 2014).

13.6 Molecular Identification of *Pochonia chlamydosporia* Varieties

The importance of correctly identifying and fingerprinting isolates has recently increased for several reasons, among them the need, because of environmental regulations, to monitor commercial, non-native isolates that are added to soil. Although several varieties of *P. chlamydosporia* have been characterized both morphologically and molecularly, it has increasingly been shown that morphology on its own is not enough, and that the use of molecular tools is a 'must'. In the following section we review the molecular methodologies with which *Pochonia* varieties have been successfully identified.

13.6.1 Pochonia chlamydosporia *var.* chlamydosporia

Pochonia chlamydosporia var. *chlamydosporia* was the first variety identified using molecular methods (Hirsch et al. 2000). They designed universal primers for the *β-tubulin* gene from a range of ascomycetes. They used the *β-tubulin* sequence of *Pochonia* to design new and more specific primers (tub 1f 5'TTT GCA GTA TCT CAG TGT TC 3′ tub 1r 5'ATG CAA GAA AGC CTT GCG AC 3′), and in so doing found a region with an intron (64 bp) that was specific to *P. chlamydosporia* var. *chlamydosporia*. The PCR product was 270 bp long. However, when these primers were tested using isolates of the fungus from Cuba, the primers did not produce an amplicon. The Cuban isolates were later identified as belonging to another variety: *Pochonia chlamydosporia* var. *catenulata*. A similar result was obtained by Medina-Canales et al. (2014) using *β-tubulin* primers when testing the *P. chlamydosporia* var. *chlamydosporia* primers with isolates from Mexico, which also failed to produce an amplicon. The Mexican isolates were subsequently identified as *Pochonia chlamydosporia* var. *mexicana* (*P. chlamydosporia* var. *mexicana*) (Fig. 13.2A).

13.6.2 Pochonia chlamydosporia *var.* catenulata

Although *P. chlamydosporia* var. *chlamydosporia* had been successfully identified with molecular markers (Hirsch et al. 2000), another variety of the fungus, *P. chlamydosporia* var. *catenulata*, was still being identified using only morphological characters until Atkins et al. (2003) carried out molecular analyses of *P. chlamydosporia* var. *catenulata* Cuban isolates. They first amplified the ITS region of different isolates of *P. chlamydosporia* var. *catenulata* and *P. chlamydosporia* var. *chlamydosporia* followed by ITS digestion with *Hin*FI. They found that both varieties had different restriction patterns and designed specific primers for *P. chlamydosporia* var. *catenulata*: PcatF 5'GTG AAC TTA TAC CAT TTT TTG and PcatR

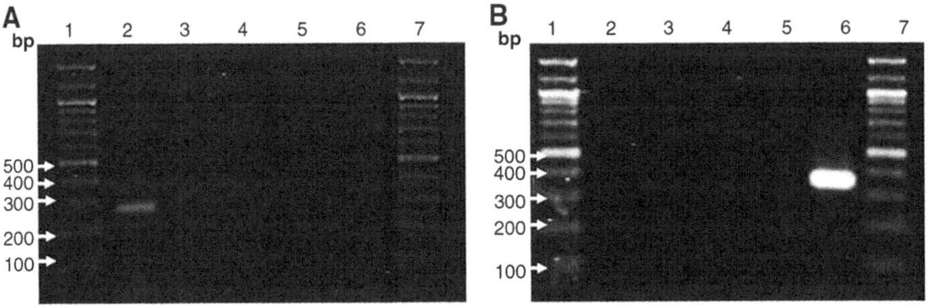

Fig. 13.2 Agarose gel electrophoresis of PCR amplification products using specific primers for diagnostic discrimination of three varieties of *Pochonia chlamydosporia*. (**A**): β-tubulin primers specific for *P. chlamydosporia* var. *chlamydosporia*; (**B**): ITS1 primers specific for *P. chlamydosporia* var. *catenulata*. Lane 1, 6 100 bp ladder marker, lane 2 isolate Pcp2, lane 3 isolate Pcp21, lane 4 isolate Pcp31 (Mexican isolates of *P. chlamydosporia* var. *mexicana*), lane 5 Pc10, of *P. chlamydosporia* var. *chlamydosporia*, lane 6 isolate Pc392 of *P. chlamydosporia* var. *catenulata* (after Medina-Canales et al. 2014)

3'CAC AAG TCC CCA TCC GC. These primers were tested using isolates from varieties, *P. chlamydosporia* var. *chlamydosporia* and *P. chlamydosporia* var. *catenulata*, the results showing that only *P. chlamydosporia* var. *catenulata* produced an amplicon of 450 bp, thus demonstrating the specificity of this new set of primers for the latter variety. In 2014 Medina-Canales et al. corroborated *P. chlamydosporia* var. *catenulata* primers specificity when *P. chlamydosporia* var. *mexicana* did not produce an amplicon (Fig. 13.2B).

13.6.3 **Pochonia chlamydosporia** ***var.*** **spinulospora** ***and*** **P. chlamydosporia** ***var.*** **ellipsospora**

Nonaka et al. (2013) used a multigene locus to identify different *Pochonia* isolates from Japan. They amplified the *nrSSU*, *nrLSU*, and ITS regions, including 5.8S rDNA, *EF1-α*, *rpb1* and *rpb2*, and using these molecular markers (except ITS) made a phylogenetic analysis based on the maximum parsimony (MP) and maximum likelihood (ML) methods. Their results showed the differentiation of *P. chlamydosporia* var. *chlamydosporia* from *P. chlamydosporia* var. *catenulata*. They also described two new *P. chlamydosporia* varieties (*P. chlamydosporia* var. *spinulospora* and *P. chlamydosporia* var. *ellipsospora*). They noted that all varieties of *P. chlamydosporia* formed a monophyletic clade and concluded that, for correct identification of *P. chlamydosporia* varieties, it is necessary to implement a procedure including the use of the six molecular markers tested (Nonaka et al. 2013), followed by a phylogenetic reconstruction, and which correlates the molecular analysis with morphological characteristics.

13.6.4 Pochonia chlamydosporia *var.* mexicana

Medina-Canales et al. (2014) isolated and identified *P. chlamydosporia* var. *mexicana* as a new variety of *P. chlamydosporia*. This variety was identified using a set of different molecular protocols starting with the primers discriminating *P. chlamydosporia* var. *chlamydosporia* and *P. chlamydosporia* var. *catenulata* (Hirsch et al. 2000; Atkins et al. 2003), but which did not produce an amplicon with the Mexican isolates. The ITS region of the Mexican isolates was amplified (White et al. 1990) and a molecular reconstruction was made based on the ML method. It was clear that the isolates were very close to *P. chlamydosporia* varieties but were not the same. Finally, they carried out a 'molecular host preference' test by amplifying a fragment of the *vcp1* gene as described by Morton et al. (2003b). These molecular-based results showed a root-knot 'host preference', although one of the isolates tested, the Pcp21 isolate, showed a putative host preference for both root-knot and cyst nematodes. Molecular evidence and morphological features of the Mexican isolates were used to propose a new variety (*P. chlamydosporia* var. *mexicana*) with the recommendation to carry out more molecular investigations using other genes and molecular markers.

13.7 Molecular Identification of *Pochonia chlamydosporia* Biotypes

Varieties of *P. chlamydosporia* var. *chlamydosporia* and *P. chlamydosporia* var. *catenulata* have shown intra-specific variation, also referred to as molecular biotypes, as shown by ERIC-PCR (Morton et al. 2003a; Mauchline et al. 2004). At least three molecular biotypes of *Pochonia* var. *chlamydosporia* have been identified using ERIC profiles (Manzanilla-López et al. 2009a, 2011) (Fig. 13.3). *Pochonia chlamydosporia* var. *catenulata* also has biotypes (Hidalgo-Díaz et al. 2000). Biotype A of *P. chlamydosporia* var. *catenulata* was initially identified only morphologically (Hidalgo-Díaz et al. 2000) and it was not until 2003 that it was molecularly characterized by ITS region amplification and *Hin*FI digestion. Biotype A has a similar profile to that of *P. chlamydosporia* var. *chlamydosporia*, but did not produce a PCR product with the *β-tubulin* region-specific primers for *P. chlamydosporia* var. *chlamydosporia* (Atkins et al. 2003).

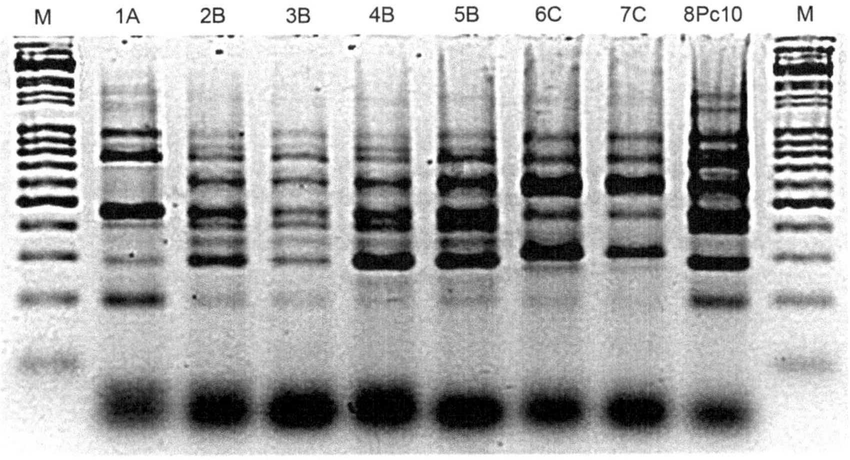

Fig. 13.3 ERIC-PCR DNA profiles from *Pochonia chlamydosporia* var. *chlamydosporia* isolates from soil and nematode eggs. *M* ladder marker (3 µl 100bp ladder + 3 µl λ*Hind*III/*Eco*RI). Eggs: *1A*, *5B*; soil *2B*, *3B*, *4B*, *6C*, *7C*, *8Pc10* (after Manzanilla-López et al. 2011). Copyright® Rothamsted Research Ltd.

13.8 Molecular Identification of *Pochonia chlamydosporia* Host Preference

In *P. chlamydosporia* the non-molecular 'biotype' term refers to nematode host from which a fungal isolate was originally isolated and it is also used to denote host preference (e.g., root-knot or cyst nematode). This host preference has been in part attributed to the *P. chlamydosporia* VCP1 serine protease enzyme. This enzyme takes part in the initial stages of nematode egg parasitism at least in *P. chlamydosporia* var. *chlamydosporia*. The amino acid sequence of the VCP1 enzyme shows polymorphism which has been linked to the capacity to infect root-knot or cyst nematodes eggs (Morton et al. 2003b; Siddiqui et al. 2009). The amino acid VCP1 polymorphisms are located in protein sequence positions number 171 and 208. VCP1 polymorphism in sequence position 171 is shown by the replacement of glutamic acid in root-knot nematode VCP1, and by glutamine in cyst nematode VCP1. Polymorphism in the VCP1 root-knot nematode sequence position 208 is shown by a replacement of glycine in the former by alanine in cyst nematodes VCP1. In order to identify the fungus host preference, Siddiqui et al. (2009) designed a molecular test using primers derived from the *vcp1* gene. They used two sets of primers according to host preference or biotype (i.e., root-knot or cyst nematode). The VCP1 polymorphism found by Morton et al. (2003b) was used to design the primers. They tested ten isolates on eggs of root-knot and cyst nematodes. Results showed that primers could differentiate host preference for each fungal isolate. This type of molecular host preference test associated to VCP1 polymorphism is yet another tool that could be used to improve screening and selection processes of isolates for biological control purposes. Host preference can be putatively identified for those isolates without a known nematode host and which were originally isolated from soil samples (Manzanilla-López et al. 2009a; Siddiqui et al. 2009; Medina-Canales et al. 2014). Although some isolates can parasitize eggs from both root-knot and cyst nematodes, fungal egg parasitism has shown to be higher in the original biotype or preferred host than in other potential hosts. *Pochonia chlamydosporia* VCP1 double polymorphism also supports that parasitic behaviour (Siddiqui et al. 2009).

13.8.1 Pochonia chlamydosporia *var.* catenulata *VCP1 Activity*

Differences in VCP1 activity have been reported for *P. chlamydosporia* (Peteira 2005; Atkins et al. 2009; Esteves et al. 2009; Peteira et al. 2009; Ward et al. 2012). VCP1 low levels observed in *Pochonia chlamydosporia* var. *catenulata* could be related to VCP1 polymorphisms or other molecular differences. Ward et al. (2012) isolated and sequenced the upstream flanking regions of *vcp1* from *P. chlamydosporia* var. *chlamydosporia* isolates using a combination of random and gene specific primers. An alignment of these sequences was then used to design new specific PCR primers (VCPuff5 and VCP1 SFrev) to amplify and sequence the region of both varieties. All sequenced isolates were ≥97% identical except for Pc392 (*P. chlamydosporia* var. *catenulata*) which was only 91–92% identical to the others. Putative regulatory motifs CREA and CREB (C-repression), GATA (N-repression) and PacC (pH regulation) showed that isolate Pc392 shared the same CREA and CREB motif profiles as the other *P. chlamydosporia* var. *chlamydosporia* isolates but lacked the putative GATA sites at 314 and 61 nt upstream, and it also had an extra putative PacC site 381 nt upstream of the transcription start. Gene expression changes in Pc392 and other *P. chlamydosporia* var. *chlamydosporia* isolates were possible to monitor due to real-time PCR greater sensitivity and important differences between *vcp1* expression were found between the two varieties. VCP1 enzyme levels in Pc392 were very low, whether glucose was present or not while lower concentrations of glucose (50 mM and 10 mM) were effective at repression of VCP1 (enzyme and mRNA) in Pc10 and Pc280 (*P. chlamydosporia* var. *chlamydosporia*). VCP1 mRNA levels in the presence of ammonium chloride were significantly (*P* 0.05) lower in Pc147 and Pc392 and higher (*P* 0.05) in Pc10. The expression of *vcp1* was also affected by the pH of the medium; *vcp1* expression being higher in Pc392 when grown at pH 6.8. Ward et al. (2012) pointed out the importance of further work to better understand the effects of different combinations of pH and carbon and nitrogen sources and ratios on *vcp1* expression and pathogenicity.

13.9 *Pochonia chlamydosporia* Detection of Viable Propagules

Measuring diversity and abundance of fungi in soil and roots is very difficult (Bridge and Spooner 2001). Nowadays, molecular methods and direct DNA extraction from soil and plant samples allows the monitoring of individual species, genera or varieties of fungi in these habitats (Anderson and Cairney 2004; Giné et al. 2016). Although qPCR measures fungal abundance based on gene copies rather than biomass (Atkins et al. 2009), it has the disadvantage that it is not possible to distinguish between viable and dead material.

13.9.1 **Pochonia chlamydosporia** *Detection in Soil*

The measurement of *P. chlamydosporia* abundance in soil is very important to confirm establishment when the fungus has been introduced as a non-native strain or to assess whether populations of native isolates have increased. Soil survival is also an indirect way to evaluate the success of a biological control agent (Kerry and Hirsch 2011).

Pochonia chlamydosporia detection in soil was initially done using the cfu (colony forming unit) count method, but more accurate and faster quantitative methods, such as qPCR, have been implemented to quantify the fungus. However, qPCR disadvantages (see 11.3.2) still require the use of both viable count (CFU) and qPCR in concert in order to obtain a more realistic assessment of fungal soil abundance (Atkins et al. 2009; Manzanilla-López et al. 2013).

One quantitative molecular assays was carried out by Mauchline et al. (2002) to develop a competitive PCR (cPCR) assay to quantify *P. chlamydosporia* (= *Verticillium chlamydosporium*) in γ-irradiated soil seeded with different numbers of chlamydospores and sampled at time intervals up to 8 weeks. Samples were analysed by cPCR and by plating onto selective medium. The results suggested that both methods detected three different fungal growth phases or stages. The first stage of growth (DNA replication) was demonstrated by the rapid increase in cPCR estimates 1 week after incubation, when the presumed carrying capacity of the soil (PCC) was reached. The second stage (increase in fungal propagules) was related to fungus cell division, sporulation and hyphal fragmentation, as indicated by a less rapid increase in cfu; the third stage was related to reach the PCC, which required 3 weeks. It was also shown that saprophytic fungal growth was limited probably by soil microflora competition and that the PCR results were less variable than the equivalent plate counts. This was the first time a PCR-based quantification had been made of a fungal biological control agent of nematodes in the soil and rhizosphere. The study also provided evidence for host specificity by the fungus.

Other quantitative PCR assays have been done using a specific set of *P. chlamydosporia* var. *chlamydosporia* molecular beacons, which allowed the quantification of *P. chlamydosporia* var. *chlamydosporia* from soil and rhizosphere-extracted

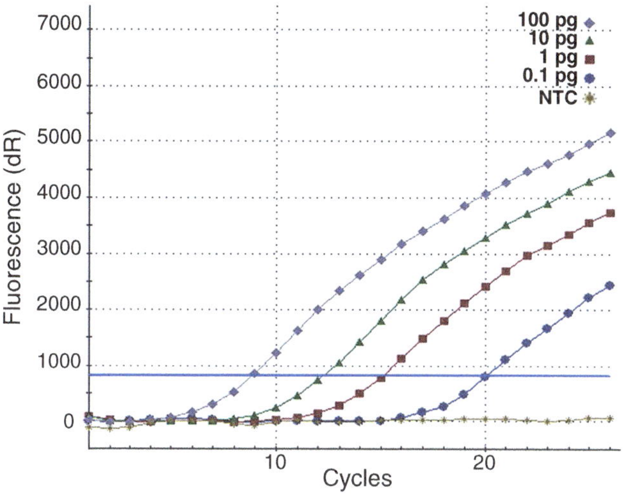

Fig. 13.4 Detection of *Pochonia chlamydosporia* by nested RT-PCR of DNA extracted with magnetic beads. Fluorescence corresponds to different amounts of DNA from treated soil (5000, 1000, 100, 10 spores/g). DNA from untreated soil (NTC) was used as control (courtesy of A. Ciancio)

DNA. In 2007, Rosso et al. tested methods of DNA soil extraction for qPCR and two VCP1 beacons to detect *P. chlamydosporia* in soil. They concluded that the use of magnetic beads was the best DNA soil extraction technique and qPCR using VCP1 beacons was very specific and could be used to quantify fungi in soil (Fig. 13.4).

Peteira et al. (2005) used qPCR with Taqman® beacons and carried out an experiment using soil, fungus grown on rice (colonized substrate) or fungus alone, root-knot nematode and tomato. Four months later they extracted DNA from soil, performed qPCR and counted the number of cfu colonies grown in a semi-selective medium. Two months after inoculation of colonized rice, the results showed that both the DNA (qPCR) (Fig. 13.5A) and the cfu increased (Fig. 13.5B). With these results it was demonstrated that qPCR and the cfu had a positive correlation under their experimental conditions. The comparative advantage of qPCR is the short time required to deliver results (i.e., three hours) while cfu counting can take up to 2 weeks. However, there are other reports that have not shown a correlation between cfu and qPCR (Manzanilla-López et al. 2009a). This could be due to problems with the yield and quality of DNA extracted from soil samples or related to the distribution of *P. chlamydosporia* in soil and the sampling procedures employed.

The Denaturing Gradient Gel Electrophoresis (DGGE) is a fingerprinting technique to determinate the total microbial community (including uncultured microorganisms) and genetic diversity, allowing analyses of the total microbial structure of the soil (Muyzer et al. 2004; Smalla and Heuer 2006). Giné et al. (2016) made a DGGE analysis using the ITS region as a molecular marker and DNA extracted from suppressive and non-suppressive soils to root-knot nematodes. They determined the fungal genetic profiles and found little fungal diversity. The fungal genetic patterns were different in suppressive and non-suppressive soils. The *Ascomycota*

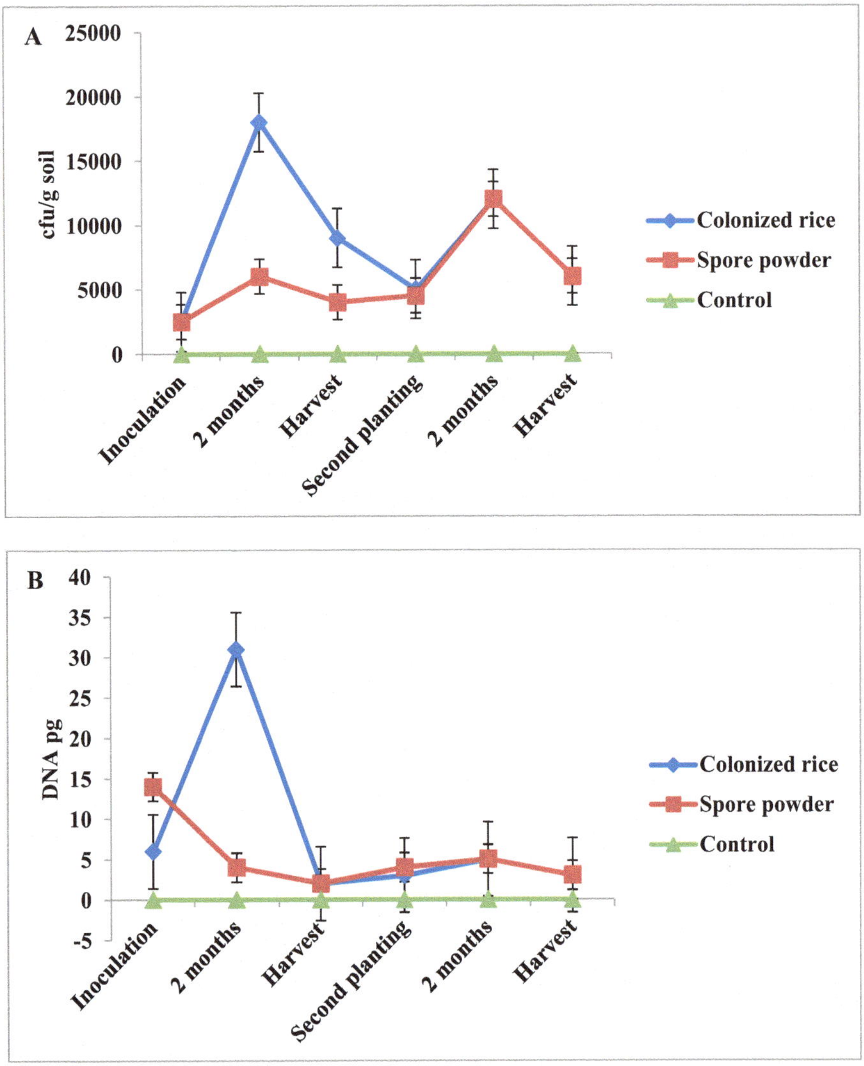

Fig. 13.5 Soil abundance of *Pochonia chlamydosporia* (cfu/g soil vs qPCR DNA pg). (**A**): *P. chlamydosporia* var. *catenulata* measured as cfu/g soil; (**B**): *P. chlamydosporia* var. *catenulata* measured as DNA pg. All treatments data were obtained over a 6 month period in a time course experiment (courtesy of B. Peteira Delgado-Oramas)

fungi, for example *Fusarium equiseti* (= *Gibberella intricans*), *Graphiopsis chlorocephala* and other uncultured fungi, were the most abundant in both soils. Unfortunately, the authors did not find a suitable sequence of *P. chlamydosporia* to be used with this technique because molecular information for this fungus in the gene bank is still limited. However, they could isolate *P. chlamydosporia* from suppressive soil. The DGGE technique could become an alternative tool to measure

P. chlamydosporia diversity in soil fungi communities, but it is necessary to first select the correct molecular marker. Probably the *vcp1*, aurovertins or pochonins genes could improve detection with the DGGE technique.

13.9.2 **Pochonia chlamydosporia** *Detection in Roots*

Pochonia chlamydosporia isolates need to be able to colonize roots to be considered as a potential biological control agent of root-knot and cyst nematode eggs (Kerry 2000). Hence, the importance of quantifying fungal abundance on rhizosphere and roots. The usual method to measure abundance is the cfu count (cfu/g root or cfu/cm^2 root) (Bourne et al. 1996; see also Chap. 12). However, the cfu count method is not sensitive enough and can result in inaccurate measurements of abundance. In contrast, molecular techniques can deliver quicker and better abundance measurements. Therefore molecular techniques used in combination with microbiological methods could be an alternative to measure the real abundance of *Pochonia* in soil and roots (Mauchline et al. 2004; Peteira et al. 2005; Atkins et al. 2009; Manzanilla-López et al. 2009a; Escudero and Lopez-Llorca 2012).

Mauchline et al. (2004) measured *P. chlamydosporia* abundance in soil and roots by making a comparative saprophytic competence test with three isolates of *P. chlamydosporia* in soil, with and without tomato plants. In order to evaluate fungal abundance they made a cfu count and RAPD-PCR using OPV-17 specific primers to detect these three isolates. They observed that the fungal abundance had an accumulative effect in mixed treatments (i.e., using two isolates) that was confirmed by RAPD-PCR (Fig. 13.2). They suggested that the use of RAPD-PCR could help to identify mixed *P. chlamydosporia* isolates in soil and root samples.

As previously mentioned, molecular differentiation between *P. chlamydosporia* var. *chlamydosporia* and *P. chlamydosporia* var. *catenulata* was made possible by using the *β-tubulin* gene and the ITS region. Atkins et al. (2009) designed specific primers to detect these two varieties in the rhizosphere by qPCR. They inoculated three isolates of *P. chlamydosporia* var. *chlamydosporia* (isolates 200 and 309) and *P. chlamydosporia* var. *catenulata* (isolate SD187), either on its own or together, into soil containing root-knot nematode eggs and planted with tomato seedlings. Eight weeks after inoculation the roots and galls were removed and DNA extracted. The qPCR was carried out and results showed that isolate SD187 was the most abundant in roots and galls and also produced the greatest amount of DNA when compared to the other two isolates. It was the first investigation that was able to detect and quantify the DNA from different isolates in the same habitat at the same time. Due to the high specificity of the primers, it is possible with this technique to measure abundance of these two *Pochonia* varieties on roots.

13.10 Future Perspectives to Detect Different Varieties of *Pochonia chlamydosporia*

The search for molecular markers to identify varieties of *P. chlamydosporia* has become an important topic. In China, Lin et al. (2015) sequenced the complete mitochondrial (mt) genome of *P. chlamydosporia*. The mt genome has a length of 25,615 bp and Lin et al. (2015) were able to determine that the genome encodes a set of conserved genes, such as cytochrome c oxidase, apocytochrome b, ATP synthase subunits, NADH dehydrogenase subunits, etc. They also found the transfer RNA genes (*trn*) and that genes could be mobile elements involved in gene rearrangement (see Chap. 9). These genes could probably be used as molecular markers to identify the varieties of *P. chlamydosporia* although it will be necessary to study the genes in more detail.

The elongation factor 1-α is a much conserved protein involved in translation. For this reason it has been considered to have properties suitable for a molecular marker (Roger et al. 1999). Recently, *EF1-α* has been used as molecular maker to discriminate between very closely related fungus species, such as *Fusarium* spp. (Scauflaire et al. 2011) and *Metarhizium* spp. (Bischoff et al. 2006). Nonaka et al. (2013) identified two varieties of *P. chlamydosporia* using different molecular markers, including *EF1-α*. However, they did not find that this region of *EF1-α* could discriminate between the varieties of *P. chlamydosporia*. In 2015 Muñoz-Saucedo et al. used the *β-tubulin*, *EF-1α*, *nrLSU*, *nrSSU*, *rpb1* and *rpb2* as molecular markers to identify the varieties of *P. chlamydosporia* and noticed that the *EF-1α* could discriminate between three varieties (*P. chlamydosporia* var. *chlamydosporia*, *P. chlamydosporia* var. *catenulata*, *P. chlamydosporia* var. *mexicana*). Discrimination between *P. chlamydosporia* varieties was possible because of the use of a different partial *EF1-α* region, which was apparently more informative, to the partial *EF1-α* region used by Nonaka et al. (2013). This molecular marker could potentially be used for correct and rapid identification of *P. chlamydosporia* varieties although it is necessary to test it with more isolates and to include the other two varieties of the fungus in future studies (i.e., *P. chlamydosporia* var. *spinulospora* and *P. chlamydosporia* var. *ellipsospora*).

Sequence Characterized Amplified Regions (SCAR) has been developed by converting RAPD fingerprints into specific molecular markers (Koveza et al. 2001; Arnedo-Andrés et al. 2002; Zhu et al. 2006; Feng et al. 2011). The SCAR have been are used to monitor *Trichoderma* cf. *harzianum* in experimental fungal communities (Pérez et al. 2014) and therefore this molecular marker could be worth trying for estimating the relative abundance of *P. chlamydosporia* in soil, roots or galls. It could be also used to design specific primers for accurate identification of varieties.

Nowadays, with more affordable equipment and reagents in molecular biology studies it is possible to sequence the complete genome of the five varieties of *P. chlamydosporia*. Using this approach, we could not only search for molecular markers to allow us to identify varieties in a faster, easier and more reliable manner, but also carry out other studies in ecology of the fungus. We might, therefore, be able to identify more efficient isolates and thereby improve parasitism and the biological control of root-knot and cyst nematode eggs.

References

Anderson, I. C., & Cairney, J. W. G. (2004). Diversity and ecology of soil fungal communities: Increased understanding through the application of molecular techniques. *Environmental Microbiology, 6*, 769–779.

Arnedo-Andrés, M., Gil-Ortega, R., Luis-Arteaga, M., et al. (2002). Development of RAPD and SCAR markers linked to the Pvr4 locus for resistance to PVY in pepper (*Capsicum annuum* L.) *Theoretical and Applied Genetics, 105*, 1067–1074.

Arora, D. K., Hirsch, P. R., & Kerry, B. R. (1996). PCR-based molecular discrimination of *Verticillium chlamydosporium* isolates. *Mycological Research, 100*, 801–809.

Atkins, S. D., Hidalgo-Díaz, L., Clark, I. M., et al. (2003). Approaches for monitoring the release of *Pochonia chlamydosporia* var. *catenulata*, a biocontrol agent of root-knot nematodes. *Mycological Research, 107*, 206–212.

Atkins, S. D., Peteira, B., Clark, I. M., et al. (2009). Use of real-time quantitative PCR to investigate root and gall colonisation by co-inoculated isolates of the nematophagous fungus *Pochonia chlamydosporia*. *The Annals of Applied Biology, 155*, 143–152.

Bidochka, M. J., Kamp, A. M., Lavender, T. M., et al. (2001). Habitat association in two genetic groups of the insect-pathogenic fungus *Metarhizium anisopliae*: Uncovering cryptic species? *Applied and Environmental Microbiology, 67*, 1335–1342.

Bird, A. F., & McClure, M. A. (1976). The tylenchid (Nematoda) egg shell: Structure, composition and permeability. *Parasitology, 72*, 19–28.

Bischoff, J. F., Rehner, S. A., & Humber, R. A. (2006). *Metarhizium frigidum* sp. nov.: A cryptic species of *M. anisopliae* and a member of the *M. flavoviride* complex. *Mycologia, 98*, 737–745.

Bourne, J. M., Kerry, B. R., & De Leij, F. A. A. M. (1996). The importance of the host plant on the interaction between root-knot nematodes (*Meloidogyne* spp.) and the nematophagous fungus, *Verticillium chlamydosporium* Goddard. *Biocontrol Science and Technology, 6*, 539–548.

Bridge, P., & Spooner, B. (2001). Soil fungi: Diversity and detection. *Plant and Soil, 232*, 147–154.

Ciancio, A., Loffredo, A., Paradies, F., et al. (2005). Detection of *Meloidogyne incognita* and *Pochonia chlamydosporia* by fluorogenic molecular probes. *EPPO Bulletin, 35*, 157–164.

Eriksson, O. E., & Hawksworth, D. L. (1985). Outline of the ascomycetes. *Systema Ascomycetum, 4*, 1–79.

Escudero, N., & Lopez-Llorca, L. V. (2012). Effects on plant growth and root-knot nematode infection of an endophytic GFP transformant of the nematophagous fungus *Pochonia chlamydosporia*. *Symbiosis, 57*, 33–42.

Esteves, I., Peteira, B., Atkins, S., et al. (2009). Production of extracellular enzymes by different isolates of *Pochonia chlamydosporia*. *Mycological Research, 113*, 867–876.

Feng, M. G., Poprawski, T. J., & Khachatourians, G. G. (1994). Production, formulation and application of the entomopathogenic fungus *Beauveria bassiana* for insect control: Current status. *Biocontrol Science and Technology, 4*, 3–34.

Feng, X. M., Holmberg, A. I. J., Sundh, I., et al. (2011). Specific SCAR markers and multiplex real-time PCR for quantification of two *Trichoderma* biocontrol strains in environmental samples. *BioControl, 56*, 903–913.

Flores-Camacho, R., Atkins, S. D., Manzanilla-López, R. H., et al. (2008). Characterization of Mexican isolates of *Pochonia chlamydosporia* var. *chlamydosporia* (Goddard) Gams and Zare for biological control of *Nacobbus aberrans* (Thorne) Thorne and Allen. *Revista Mexicana de Fitopatología, 26*, 93–104.

Gams, W. (1988). A contribution to the knowledge of nematophagous species of Verticillium. *Netherlands Journal of Plant Pathology, 94*, 123–148.

Gams, W., & Zare, R. (2001). A revision of *Verticillium* sect. *Prostrata*. III. Generic classification. *Nova Hedwigia, 72*, 329–337.

Gäumann E. (1926). *Vergleichende Morphologie der pilze publisher in Fisher*, Jena.

Giné, A., Carrasquilla, M., Martínez-Alonso, M., et al. (2016). Characterization of soil suppressiveness to root-knot nematodes in organic horticulture in plastic greenhouse. *Frontiers in Plant Science*. doi:10.3389/fpls.2016.00164.

Glenn, A. E., Bacon, C. W., Price, R., et al. (1996). Molecular phylogeny of *Acremonium* and its taxonomic implicatioins. *Mycologia, 88*, 369–383.

Haarmann, T., Rolke, Y., Giesbert, S., et al. (2009). Ergot: From witchcraft to biotechnology. *Molecular Plant Pathology, 10*, 563–577.

Hidalgo-Diaz, L., Bourne, J. M., Kerry, B. R., et al. (2000). Nematophagous *Verticillium* spp. in soils infested with *Meloidogyne* spp. in Cuba: Isolation and screening. *International Journal of Pest Management, 46*, 277–284.

Hirsch, P. R., Mauchline, T. H., Mendum, T. A., et al. (2000). Detection of the nematophagous fungus *Verticillium chlamydosporium* in nematode-infested plant roots using PCR. *Mycological Research, 104*, 435–439.

Hirsch, P. R., Atkins, S. D., Mauchline, T. H., et al. (2001). Methods for studying the nematophagous fungus *Verticillium chlamydosporium* in the root environment. *Plant and Soil, 232*, 21–30.

Hung, C. Y., Seshan, K. R., Yu, J. J., et al. (2005). A metalloproteinase of *Coccidioides posadasii* contributes to evasion of host detection. *Infection and Immunity, 73*, 6689–6703.

Jia, Y., McAdams, S. A., Bryan, G. T., et al. (2000). Direct interaction of resistance gene and avirulence gene products confers rice blast resistance. *The EMBO Journal, 19*, 4004–4014.

Joaquim, T. R., & Rowe, R. C. (1990). Reassessment of vegetative compatibility relationships among strains of *Verticillium dahliae* using nitrate-nonutilizing mutants. *Phytopathology, 80*, 1160–1166.

Kanzok, S. M., & Jacobs-Lorena, M. (2006). Entomopathogenic fungi as biological insecticides to control malaria. *Trends in Parasitology, 22*, 49–51.

Kepler, R. M., Sung, G. H., Harada, Y., et al. (2012). Host jumping onto close relatives and across kingdoms by Tyrannicordyceps (Clavicipitaceae) gen. nov. and Ustilaginoidea (Clavicipitaceae). *American Journal of Botany, 99*, 552–561.

Kerry, B. R., Atkins, S., & Rovira, D. A. (1984). Observations on the introduction of Verticillium chlamydosporium and other parasitic fungi into soil for control of the cereal cyst-nematode *Heterodera avenae. The Annals of Applied Biology, 105*, 509–516.

Kerry, B. R., & Hirsch, P. R. (2011). Ecology of *Pochonia chlamydosporia* in the rhizosphere at the population, whole organism and molecular scales. In K. Davies & Y. Spiegel (Eds.), *Biological control of plant-parasitic nematodes* (pp. 171–182). Netherlands: Springer.

Klimyuk, V. I., Carroll, B. J., Thomas, C. M., et al. (1993). Alkali treatment for rapid preparation of plant material for reliable PCR analysis. *The Plant Journal, 3*, 493–494.

Koveza, O. V., Kokaeva, Z. G., Gostimsky, S. A., et al. (2001). Creation of a SCAR marker in pea (*Pisum sativum* L.) using RAPD analysis. *Russian Journal of Genetics, 37*, 464–466.

Larriba, E., Jaime, M. D. L. A., Carbonell-Caballero, J., et al. (2014). Sequencing and functional analysis of the genome of a nematode egg-parasitic fungus, *Pochonia chlamydosporia. Fungal Genetics and Biology, 65*, 69–80.

Lin, R., Liu, C., Shen, B., et al. (2015). Analysis of the complete mitochondrial genome of *Pochonia chlamydosporia* suggests a close relationship to the invertebrate-pathogenic fungi in Hypocreales. *BMC Microbiology, 15, 15*(1).

Manzanilla-López, R. H., Atkins, S. D., Clark, I. M., et al. (2009a). Measuring abundance, diversity and parasitic ability in two populations of the nematophagous fungus *Pochonia chlamydosporia* var. *chlamydosporia. Biocontrol Science and Technology, 19*, 391–406.

Manzanilla-López, R. H., Clark, I. M., Atkins, S. D., et al. (2009b). Rapid and reliable DNA extraction and PCR fingerprinting methods to discriminate multiple biotypes of the nematophagous fungus *Pochonia chlamydosporia* isolated from plant rhizospheres. *Letters in Applied Microbiology, 48*, 71–76.

Manzanilla-López, R. H., Clark, I. M., Atkins, S. D., et al. (2011). Exploring competitiveness and variation in the nematophagous fungus *Pochonia chlamydosporia* var. *chlamydosporia* and its significance for biological control. *Bulletin OILB/SROP, 63*, 37–40.

Manzanilla-López, R. H., Esteves, I., Finetti-Sialer, M. M., et al. (2013). *Pochonia chlamydosporia*: Advances and challenges to improve its performance as a biological control agent of sedentary endo-parasitic nematodes. *Journal of Nematology, 45*, 1–7.

Mauchline, T. H., Kerry, B. R., & Hirsch, P. R. (2002). Quantification in soil and the rhizosphere of the nematophagous fungus *Verticillium chlamydosporium* by competitive PCR and comparison with selective plating. *Applied and Environmental Microbiology, 68*, 1846–1583.

Mauchline, T. M., Kerry, B. R., & Hirsch, P. R. (2004). The biocontrol fungus *Pochonia chlamydosporia* shows nematode host preference at the infraspecific level. *Mycological Research, 108*, 161–169.

Medina-Canales, M. G., Rodríguez-Tovar, A. V., Manzanilla-López, R. H., et al. (2014). Identification and molecular characterisation of new Mexican isolates of *Pochonia chlamydosporia* for the management of *Meloidogyne* spp. *Biocontrol Science and Technology, 24*, 1–21.

Morton, A., Fabrett, A. M., Carder, J. H., et al. (1995). Sub-repeat sequences in ribosomal RNA intergenic regions of *Verticillium albo-atrum* and *V. dahlie*. *Mycological Research, 99*, 257–266.

Morton, C. O., Mauchline, T. H., Kerry, B. R., et al. (2003a). PCR-based DNA fingerprinting indicates host-related genetic variation in the nematophagous fungus *Pochonia chlamydosporia*. *Mycological Research, 107*, 198–205.

Morton, C. O., Hirsch, P. R., Peberdy, J., et al. (2003b). Cloning of a genetic variation in protease VCP1 from the nematophagus fungus *Pochonia chlamydosporia*. *Mycological Research, 107*, 38–46.

Muñoz-Saucedo LA, Tovar-Soto A, Rodríguez-Tovar AV et al. (2015). Phylogenetic analysis of native Mexican isolates of the nematophagous fungus *Pochonia chlamydosporia*. In: Abstracts of the XI Mexican congress of molecular and cell biology of fungi, Puebla, Mexico, October 25–29 2015.

Muyzer, G., Brinkhoff, T., Nübel, U., et al. (2004). Denaturing gradient gel electrophoresis (DGGE) in microbial ecology. In G. A. Kowalchuk, F. J. de Bruijn, I. M. Head, et al. (Eds.), *Molecular Microbial Ecology Manual* (Vol. 2, pp. 743–769). Dordrecht: Springer.

Newport, G., Kuo, A., Flattery, A., et al. (2003). Inactivation of Kex2 diminishes the virulence of *Candida albicans*. *The Journal of Biological Chemistry, 278*, 1713–1720.

Niu, X. M., Wang, Y. L., Chu, Y. S., et al. (2010). Nematodetoxic aurovertin-type metabolites from a root-knot nematode parasitic fungus *Pochonia chlamydosporia*. *Journal of Agricultural and Food Chemistry, 58*, 828–834.

Nonaka, K., Ōmura, S., Masuma, R., et al. (2013). Three new *Pochonia* taxa (Clavicipitaceae) from soils in Japan. *Mycologia, 105*, 1202–1218.

Olivares-Bernabeu, C. M., & López-Llorca, L. V. (2002). Fungal egg parasites of plant-parasitic nematodes from Spanish soils. *Revista Iberoamericana de Micología, 19*, 104–110.

Pérez, G., Verdejo, V., Gondim-Porto, C., et al. (2014). Designing a SCAR molecular marker for monitoring *Trichoderma* cf. *harzianum* in experimental communities. *Journal of Zhejiang University. Science. B, 15*, 966–978.

Peteira B. (2005). Caracterización del hongo nematofago cepa IMI SD 187 de *Pochonia chlamydosporia* var. *catenulata* (Kamischo ex barron y Onions) Zare y Gams. Thesis Universidad Agraria de La Habana, Cuba.

Peteira, B., Puertas, A., Hidalgo-Díaz, L., et al. (2005). Real-time PCR to monitor and assess the efficacy of two types of inoculum of the nematophagous fungus *Pochonia chlamydosporia* var. *catenulata* against root-knot nematode populations in the field. *Biotecnología Aplicada, 22*, 261–266.

Reeves, C. D., Hu, Z., Reid, R., et al. (2008). Genes for the biosynthesis of the fungal polyketides hypothemycin from *Hypomyces subiculosus* and radicicol from *Pochonia chlamydosporia*. *Applied and Environmental Microbiology, 74*, 5121–5129.

Rogerson, C. T. (1970). The hypocrealean fungi (Ascomycetes, Hypocreales). *Mycologia, 62*, 865–910.

Roger, A. J., Sandblom, O., Doolittle, W. F., et al. (1999). An evaluation of elongation factor 1 alpha as a phylogenetic marker for eukaryotes. *Molecular Biology and Evolution, 16*, 218–233.

Rosso, L., Ciancio, A., & Finetti-Sialer, M. (2007). Application of molecular methods for detection of *Pochonia chlamydosporia* from soil: *Nematropica, 37*, 1–8.

Rosso, L. C., Finetti-Sialer, M. M., Hirsch, P. R., et al. (2011). Transcriptome analysis shows differential gene expression in the saprotrophic to parasitic transition of *Pochonia chlamydosporia*. *Applied Microbiology and Biotechnology, 90*, 1981–1994.

Smalla, K., & Heuer, H. (2006). How to assess the abundance and diversity of mobile genetic elements in soil bacterial communities? In P. Nannipieriand & K. Smalla (Eds.), *Nucleic acids and proteins in soil* (Vol. 8, pp. 313–330). Heidelberg: Springer.

Scauflaire, J., Gourgue, M., & Munaut, F. (2011). *Fusarium temperatum* sp. nov. from maize, an emergent species closely related to *Fusarium subglutinans*. *Mycologia, 103*, 586–597.

Schardl, C. L., Young, C. A., Hesse, U., et al. (2013). Plant-symbiotic fungi as chemical engineers: Multi-genome analysis of the clavicipitaceae reveals dynamics of alkaloid loci. *PLoS Genetics, 9*, e1003323.

Shiba, T., & Sugawara, K. (2005). Resistance to the rice leaf bug, *Trigonotylus caelestialium*, is conferred by *Neotyphodium endophyte* infection of perennial ryegrass, *Lolium perenne*. *Entomologia Experimentalis et Applicata, 115*, 387–392.

Scholte, E. J., Ng'habi, K., Kihonda, J., et al. (2005). An entomopathogenic fungus for control of adult African malaria mosquitoes. *Science, 308*, 1641–1642.

Siddiqui, A., Atkins, S. D., & Kerry, B. R. (2009). Relationship between saprotrophic growth in soil of different biotypes of *Pochonia chlamydosporia* and the infection of nematode eggs. *The Annals of Applied Biology, 155*, 131–141.

Soloviev, D. A., Jawhara, S., & Fonzi, W. A. (2011). Regulation of innate immune response to *Candida albicans* infections by aMb2-Pra1p interaction. *Infection and Immunity, 79*, 1546–1558.

Spatafora, J. W., & Blackwell, M. (1993). Molecular systematics of unitunicate perithecial ascomycetes: The Clavicipitales-Hypocreales connection. *Mycologia, 85*, 912–922.

Sung, G. H., Spatafora, J. W., Zare, R., et al. (2001). A revision of *Verticillium* sect. Prostrata. II. Phylogenetic analyses of SSU and LSU nuclear rDNA sequences from anamorphs and teleomorphs of the Clavicipitaceae. *Nova Hedwigia, 72*, 311–328.

Sung, G. H., Sung, J. M., Nigel, L. J., et al. (2007). A multi-gene phylogeny of Clavicipitaceae (Ascomycota, Fungi): Identification of localized incongruence using a combinational bootstrap approach. *Molecular Phylogenetics and Evolution, 44*, 1204–1223.

Wang, Y., Li, L., Li, D., et al. (2015). Yellow pigment aurovertins mediate interactions between the pathogenic fungus *Pochonia chlamydosporia* and its nematode host. *Journal of Agricultural and Food Chemistry, 63*, 6577–6587.

Ward, E., Kerry, B. R., Manzanilla-López, R. H., et al. (2012). The *Pochonia chlamydosporia* serine protease gene *vcp1* is subject to regulation by carbon, nitrogen and pH: Implications for nematode biocontrol. *PLoS One, 7*(4), e35657.

White, T. J., Bruns, T., Lee, S., et al. (1990). Amplification and direct sequencing of fungal ribosomal RNA genes for phylogenetics. In M. A. Innes, D. H. Gelfand, J. J. Sninsky, & T. J. White (Eds.), *PCR Protocols: A Guide to Methods and Applications* (pp. 315–322). New York: Academic Press, Inc.

Yang, J., Loffredo, A., Bonerman, J., et al. (2012). Biocontrol efficacy among strains of *Pochonia chlamydosporia* ibtained from a root-knot nematode suppressive soil. *Journal of Nematology, 44*, 67–71.

Zhu, M. L., Mo, M. H., Xia, Z. Y., et al. (2006). Detection of fungal biocontrol agents against root-knot nematodes by RAPD markers. *Mycophatologia, 161*, 307–316.

Part V
Practical Approaches and Management Studies

Chapter 14
Priming Plant Defence Responses Can Enhance the Biological Control of *Pochonia chlamydosporia* Against Root-Knot Nematodes

Maria Clara Vieira dos Santos, Isabel Abrantes, and Rosane H.C. Curtis

Abstract Improved crop protection strategies are required to address the need for efficient use of resources and ensure food security. The soil-dwelling plant-parasitic nematodes are a major pest causing crop disease and yield loss. Root-knot nematodes, *Meloidogyne* spp., are among the most economically damaging soil dwelling plant parasites of agricultural crops. Current management practices are not economically or environmentally attractive and often rely on toxic nematicide applications. Therefore, research on sustainable strategies for nematode control is a priority. Successful management strategies could potentially be achieved by the combined use of natural enemies of nematodes, such as the biological control agent *Pochonia chlamydosporia*, a facultative egg parasite of plant-parasitic nematodes, and induction of natural defence mechanisms in plants. The fungus can saprophytically proliferate in the rhizosphere using nutrients present in root exudates and infect nematode eggs but is not able to prevent the initial infection of roots by nematode infective juveniles which can itself be achieved by priming plant immune defences. The application of *P. chlamydosporia*, combined with the exploitation of the natural mechanisms of plant defence, is a biomanagement alternative under study. The activation of specific plant biochemical defence pathways might produce changes that promote the fungal transition from the saprophytic to the parasitic phase. The main goal of this chapter is to unravel the role of induced plant defence responses as a putative resource to enhance the potential of *P. chlamydosporia* as a biocontrol agent against root-knot nematodes.

M.C. Vieira dos Santos (✉) • I. Abrantes
Centre for Functional Ecology (CFE), University of Coimbra,
Calçada Martim de Freitas, 3000-456 Coimbra, Portugal
e-mail: mcvs@sapo.pt; isabel.abrantes@uc.pt

R.H.C. Curtis
Bionemax UK Ltd, Rothamsted Centre for Research and Enterprise,
Harpenden, Herts AL5 2JQ, UK
e-mail: r.curtis@bionemax.co.uk

R.H. Manzanilla-López, L.V. Lopez-Llorca (eds.), *Perspectives in Sustainable Nematode Management Through* Pochonia chlamydosporia *Applications for Root and Rhizosphere Health*, Sustainability in Plant and Crop Protection,
DOI 10.1007/978-3-319-59224-4_14

14.1 Introduction

A major global challenge in the twenty-first century is to ensure food security and to feed the increasing human population, mainly in resource-poor areas of the world. Crop production in Europe is estimated to be only ca 70% of the optimal yield potential. The yield gap is largely affected by abiotic factors, but pests and diseases also contribute significantly to reduce crop production, having pronounced effects on yield and quality, and decreasing resource-use efficiency. Improved crop protection strategies able to prevent such damage and loss can increase food production and security. The development of efficient environmentally-friendly and durable crop protection strategies are therefore a priority. The broad range of limitations on crop production needs to be considered, those caused by plant-parasitic nematodes often being overlooked.

Plant-parasitic nematodes are important soil-dwelling pests that affect a wide range of crop plants, facilitate fungal and bacterial plant pathogens, and act as a major cause of preventable crop disease and yield loss. In addition, several plant-parasitic nematodes are major pests of quarantine importance and interfere with the free trade of agriculture products (Chitwood 2003; Lucas 2010; Nicol et al. 2011). Plant-parasitic nematodes have many natural enemies, including bacteria, fungi, protozoa, predatory nematodes, insects and mites. The efficient exploitation of these enemies as biological control agents of plant-parasitic nematodes depends on the knowledge of their biology as well as the interactions between them (Kerry 1987).

Natural control has been detected in several soils where the build-up of nematophagous fungi and bacteria, under some perennial crops and those grown in monoculture, leading to the control of cyst nematodes, *Globodera* spp. and *Heterodera* spp., and root-knot nematodes, *Meloidogyne* spp. (reviewed in Stirling 2011). Research on the manipulation of natural control has been restricted partly due to difficulties in mass production of biological control agents and manipulation of soil activity, and has mainly been concentrated on the introduction of specific agents to provide more rapid control that might be commercially exploitable (Kerry and Hominick 2002). The majority of the natural enemies found to parasitise nematode hosts in suppressive soils, i.e., soils that contain microbial communities able to prevent the increase of nematode populations on susceptible crops, are bacteria and nematophagous fungi (Viaene et al. 2006).

The characteristics of facultative fungal parasites of sedentary nematode eggs and females give them the highest potential as biological control agents. They can be rapidly grown *in vitro* and their survival in soil is not dependent on the presence of the nematodes. Among the fungal parasites, *Pochonia chlamydosporia* has been developed as a biological control agent. The fungus is able to proliferate saprophytically in the rhizosphere by using nutrients released in root exudates and can switch to a parasitic stage that feeds on nematode eggs (Kerry and Hirsch 2011). Intricate relationships between the nematode host and nutrient availability play important roles in the saprophytic-parasitic transition of the fungus (Kerry and Hirsch 2011; Ward et al. 2012).

Plants are able differentially to activate inducible broad-spectrum defence mechanisms, depending on the type of attacker encountered. A signal, caused either by or following a response to local attacks, spreads throughout the plant inducing changes in gene expression that cause a systemic defence response. In contrast, plant-parasitic nematodes not only have the ability to produce plant hormones themselves, such as cytokinins, which interfere with the plant hormonal changes, but also use a number of effector proteins to suppress plant defence genes (De Meutter et al. 2003; Siddique et al. 2015). The nematode effector proteins, the fatty-acid and retinol-binding protein, might play a role in manipulating the lipid based jasmonates signalling *in planta* to enhance nematode development *in planta* (Iberkleid et al. 2013). Therefore, it seems that the action of plant resistance or susceptibility is likely dictated by the hormonal balance that regulates and modulates complex signalling pathways. Induced plant resistance pathways can also be chemically activated by compounds that mimic the mode of action of natural elicitors. In order to promote plant resistance, natural or synthetic elicitors need to create the hormonal balance that will trigger the signalling pathway able to counteract the ability of the nematode to manipulate plant immunity to its own benefit (Maule and Curtis 2011). The broad-spectrum nature of the induced resistance provides additional options for use in integrated pest biomanagement but, to fully access the huge potential of artificially-induced resistance, studies should take into account the effects of different concentrations and application methods (Lucas 2010).

The potential of a Portuguese *P. chlamydosporia* isolate combined with benzothiadiazole (BTH) or cis-jasmone against *M. incognita* has already been demonstrated. (Vieira dos Santos et al. 2013a). Activation of the biochemical defence pathways might produce changes in the plant and/or the rhizosphere that prompt the fungal transition from the saprophytic to the parasitic phase. The putative modulation of the interactions among host plant, nematode and fungus by the activation of natural defence pathways raises new questions as to how these organisms interact. More importantly, it represents a potential new biomanagement strategy to control root-knot nematodes. The main goal of this chapter is to tease out the role of the induced plant defence responses as a significant resource to enhance the potential of *P. chlamydosporia* as a biocontrol agent against root-knot nematodes.

14.2 Root-Knot Nematodes, *Meloidogyne* spp.: Importance and Control Approaches

One of the most economically damaging plant-parasitic nematodes to horticultural and field crops are the root-knot nematodes, *Meloidogyne* spp. (Jones et al. 2013). These important, highly adapted, obligate plant pathogens are one of the oldest known causes of nematode diseases of plants. They can parasitize the roots of thousands of plant species, from monocots to dicots and herbaceous to woody plants (Eisenback and Tryantaphyllou 1991). Interacting with their hosts in a remarkable way they establish permanent polyploid, multinucleate and enlarged giant feeding cells, leading to the formation of root galls that are the first visible symptom of infection. The subsequent damage to the root system decreases plant nutrient and water uptake and root-shoot translocation, resulting in weak plants, low yields and poor quality produce. The above-ground symptoms are similar to those produced in plants that have a damaged and malfunctioning root system but are related to infestation levels, becoming more severe in high infestations (Karssen and Moens 2006).

Current control strategies include planting with resistant or tolerant cultivars, long rotations, and fertilization with high doses of organic matter. Chemical (fumigants or granular systemic compounds) and physical methods, such as solarization, have also been used to reduce the populations or damage caused by these nematodes (Nicol et al. 2011). However, the above-mentioned strategies are not economically attractive approaches. The use of resistant or tolerant cultivars may further increase selection pressure for virulent populations or limit the durability of resistance in some cultivars. Although the efficiency of nematicides against these nematodes has been demonstrated, their use is being banned or reduced due to environmental concerns (Trudgill and Blok 2001). The withdrawal of many nematicides and soil fumigants further limits control options, and urges the need for alternative control strategies (Chitwood 2003). The regulatory effects of natural enemies of nematodes and activation of natural plant defence mechanisms by elicitors have been the subject of research to develop sustainable control measures against these nematodes. Nevertheless, all of the above methods need further investigation, as control levels can be variable and are often limited.

14.3 *Pochonia chlamydosporia* – A Potential Biological Control Agent

Pochonia chlamydosporia has been associated with the decline of *H. avenae* population densities below the damage threshold under monocultures of susceptible cereals in northern Europe (Kerry 1982; Kerry et al. 1982). This facultative egg parasite of cyst and root-knot nematodes has been reported in many countries in almost all continents as an important antagonist in the regulation of cereal cyst nematodes, *Heterodera* spp., potato-cyst nematodes, *Globodera* spp., and root-knot nematodes, *Meloidogyne* spp. (reviewed in Kerry and Hirsch 2011). A recent survey conducted in Spain to assess the biodiversity and frequency of infection of fungal egg parasites of *Meloidogyne* spp. revealed that *P. chlamydosporia* prevalence was correlated with percentages of egg parasitism higher than 40% in two commercial organic vegetable production sites (Giné et al. 2016). Moreover, an isolate of this fungus has been successfully developed under the name KlamiC® as a biological control agent against root-knot nematodes in Cuba (Kerry 1987; Hernández and Hidalgo-Díaz 2008; see also Chap. 15). *Pochonia chlamydosporia* isolates differ in such factors as their ability to colonize the root surface, virulence and chlamydospore production, these being advantageous characteristics for its use as a biological control agent (Kerry and Bourne 2002). The intricate trophic interactions plant-nematode-fungus are regulated by several key factors associated with fungal isolate variability, plant species susceptibility and tolerance, application rate, and nematode species and egg maturity (Kerry 2000). The induction of natural plant defence mechanisms could significantly influence the factors closely linked to the plant-fungus and plant-nematode dynamics, thus opening the possibility of enhancing the ability of *P. chlamydosporia* as a biocontrol agent of root-knot nematodes.

The rhizosphere colonization ability of the fungus is strongly correlated with nematode control. The fungus can remain saprophytic in soil in the absence of both plant and nematode hosts and survival and proliferation in the soil is greatly influenced by abiotic factors, such as temperature and water availability (Bourne and Kerry 2000; Magan 2001; see also Chaps. 3 and 4). Nonetheless, the impact on nematode reproduction is more affected by the fungal ability to colonize the rhizosphere than the ability to proliferate in soil (De Leij and Kerry 1991; Bourne et al. 1994, 1996; Kerry and Bourne 1996). When the ability of two *P. chlamydosporia* isolates to survive and proliferate in soil and in the rhizosphere was compared in pot assays using soil from fields infected with *M. chitwoodi* following a crop rotation scheme including potato and maize (*Zea mays*), low levels of egg parasitism by these two fungal isolates was obtained. Therefore, successful egg parasitism seem to be correlated with *P. chlamydosporia* host preference, plant compatibility and tolerance to abiotic factors (Vieira dos Santos et al. 2014).

Although there seems to be limited growth of *P. chlamydosporia* in soil, the fungus is rhizosphere competent (De Leij et al. 1993; Kerry et al. 1993; Bourne and Kerry 2000). In the rhizosphere limited endophytic growth (within the roots) has been detected (De Leij and Kerry 1991; Lopez-Llorca et al. 2002) and the fungus behaves as an endophyte (Bordallo et al. 2002; Monfort et al. 2005; Mácia-Vicente et al. 2009a, b; Escudero and Lopez-Llorca 2012). Structural and functional analysis of the *P. chlamydosporia* genome revealed the production of hydrolytic enzymes, transporters, proteases, chitinases and a large number of secondary metabolites related to its endophytic behaviour (Larriba et al. 2014). Although plant species differ in their ability to support *P. chlamydosporia* growth in their rhizosphere (Bourne et al. 1996; Kerry and Bourne 1996; Bourne and Kerry 1999), selected isolates of *P. chlamydosporia* have been considered as plant-growth promoters of barley (Mácia-Vicente et al. 2009b), lettuce (Dias-Arieira et al. 2011; Dallemole-Giaretta et al. 2015), tomato (Escudero and Lopez-Llorca 2012; Dallemole-Giaretta et al. 2015) and wheat (Monfort et al. 2005). Transcriptomic analysis of the growth-promoting effect observed in barley roots endophytically colonized by *P. chlamydosporia*, revealed an up-regulation of genes implicated in the biosynthesis of the plant hormones auxin, ethylene and jasmonic acid (JA) and of other genes involved in natural plant defence mechanisms (Larriba et al. 2015). Colonization of *Arabidopsis thaliana* by *P. chlamydosporia* reduces flowering time, stimulates root growth and improves seed yield. Changes in JA signalling are strongly linked with fungal colonization (Zavala-Gonzalez et al. 2017). *Pochonia chlamydosporia* is able to saprophytically proliferate in the rhizosphere by using nutrients released in root exudates and then switch to a parasitic stage that feeds on nematode eggs. When the colonizing *P. chlamydosporia* contacts a nematode egg mass in the rhizosphere, the fungus ceases its saprophytic stage and switches to a parasitic stage. Appressoria formed in response to the contact with the surface of a nematode egg enable both the attachment to, and penetration of, the nematode egg, which involves mechanical and enzymatic action (see Chap. 4). Application of large numbers of fungal spores to the soil in agricultural systems may not result in effective nematode control because fungal establishment and rhizosphere colonisation are linked to intricate interactions of nematode-host preference, root physiological status or changes in nutrient availability in the microenvironment of the rhizosphere surrounding the egg masses (Atkins et al. 2003; Kerry and Hirsch 2011; Mauchline et al. 2004; Ward et al. 2012). Sequence analysis of the upstream regulatory region of the fungal virulence-related gene *vcp1* showed that gene expression can be influenced by carbon and nitrogen sources and pH (Ward et al. 2012), suggesting that nutrient quality and availability in the rhizosphere may induce the saprophytic-parasitic transition.

Plant susceptibility and tolerance to nematode attack is another key factor that influences the performance of *P. chlamydosporia* isolates (Kerry 2000). The fungus is not able to prevent the initial infection of roots by root-knot nematode infective juveniles (i.e., the second-stage juvenile) and is ineffective when the nematode eggs remain protected from fungal attack inside larger root galls which are formed in soil with highly susceptible hosts or with large nematode population densities (Bourne et al. 1996). Consequently, its biocontrol efficacy should be enhanced by integration with control measures that prevent or decrease initial infestations, such as crop rotation with poor hosts for the nematode or the application of plant defence activators (Kerry and Bourne 1996; Vieira dos Santos et al. 2013b).

14.4 Plant Defence Mechanisms

Plants have evolved an immune system to recognise and respond to pathogen attack by modifying the relative abundance of plant hormones and the expression of their defence genes by activating a complex response of cross-talk between signalling pathways at key regulatory steps, thereby resulting in an efficient defence response allowing for plant survival (Maule and Curtis 2011). Induced systemic resistance (ISR) of plants against pathogens has been widely studied with respect to the basic signalling pathways and to its potential use in plant protection. The plant hormones salicylic acid (SA), JA and ethylene are major players in the network of defence signalling pathways (Pieterse et al. 2001). Two essential mechanisms regulate induced resistance in plants: i) systemic acquired resistance mediated by the accumulation of endogenous SA; and ii) induced systemic resistance related with JA and ethylene pathways (Kuć 2001).

Plants are able to differentially activate inducible broad-spectrum defence mechanisms depending on the type of attacker encountered. The response to local attack by a given pathogen involves production of compounds that reduce or inhibit either its performance or further attack. Responses include an oxidative burst, which can lead to cell death, changes in cell wall composition that can inhibit penetration by the pathogen, and *de novo* synthesis of antimicrobial compounds such as phytoalexins and pathogenesis related (PR) proteins. Caused either by, or following, these local responses, a signal spreads through the plant, inducing subtle changes in gene expression in uninfected plant parts and leading to a systemic response with the possible involvement of PR proteins and phytoalexins. The functional role of both groups of compounds is still unknown, but two groups of PR proteins have been detected. Acidic PR proteins are predominantly located in the intercellular spaces, whereas basic PR proteins are mainly located intracellularly, the latter being functionally similar to the former but having different molecular weights and amino acid sequences (Heil and Bostock 2002). Some of the compounds associated with ISR are expressed in uninfected tissue in response to the initial attack, whereas other biochemical changes of ISR-expressing plants become obvious only in plant parts where effective resistance is required and only in response to a further attack. This physiological condition leading to induced resistance is associated with an enhanced capacity to mobilise plant cellular defence responses a process known as 'priming' (Conrath et al. 2001). The molecular mechanisms underlying priming are not fully understood, although chemical ISR activators such as, for example, β-aminobutyric acid (BABA), can elicit priming effects (Jakab et al. 2001). To fully access the huge potential of artificially-induced resistance, studies should take into account the effects of different concentrations and application methods of natural and synthetic compounds elicitors of plant immunity (Lucas 2010).

14.5 Plant Defence Activators

The discovery of induced resistance pathways in plants opened the possibility of chemically activating one or more of these pathways and the development of protective chemicals mimicking the mode of action of natural elicitors to provide useful tools for the development of new strategies for crop protection (Pieterse et al. 2001). The enhancement of plant defence mechanisms can be accomplished by application of plant defence activators and, when these are integrated into good agricultural practices, both productivity and resistance to disease can be enhanced. On the other hand, the performance of plant activators is often variable: they may not consistently provide the same level of disease control as nematicides and can have side effects on crop growth and development. Nonetheless, there are a number of attractive aspects for using natural plant defence systems in nematode management, with the broad spectrum nature of the induced resistance being the most appealing feature and providing additional options for their use in integrated disease and pest control programmes (Kuć 2001).

The identification of SA as an essential endogenous signal in ISR led to the synthesis of chemical mimics such as BTH (Gorlach et al. 1996). This plant activator is commercialised in Europe (Bion®) and in the USA (Actigard® and Boost®)

slowing root-knot nematode development and decreasing egg deposition on glasshouse-grown Cabernet Sauvignon grapevines (Owen et al. 2002), and reducing nematode infection in sugarcane (Berry et al. 2011). Foliar sprays of BTH were effective in reducing nematode population densities of *Pratylenchus* spp. and *M. chitwoodi* in potato crops under field conditions, increasing the nematode population densities of non-target nematodes (Collins et al. 2006). Exogenous application of SA by foliar sprays reduced nematode infection by 30% in cowpea and 77% in okra whereas reproduction was reduced 52% and 69% respectively whilst promoting plant growth (Nandi et al. 2003). On the other hand, variation in the efficacy of plant defence activators has been reported. Foliar sprays of SA had no effect on nematode reproduction in pineapple (*Ananas comosus*), where they also reduced plant growth (Chinnasri et al. 2006). When applied to eggplant, pepper and tomato, foliar sprays of SA and BTH were ineffective against *M. incognita* whereas soil drenches of SA and root dip application of SA and BTH were effective in reducing nematode reproduction, at specific dosage ranges, without affecting plant growth (Molinari 2016).

The non-protein amino acid BABA is an effective plant defence inducer not only against nematodes but also against a large number of plant-parasites. Additionally, post-infection activity has been demonstrated in tomato, tobacco, grapevine and cauliflower against several oomycetes. Unlike other plant activators, BABA has not been formulated as a commercial product due to the lack of toxicological data and biodegradation studies (Cohen 2002). Previous research has revealed that foliar sprays of BABA affected *M. javanica* reproduction in pineapple (Chinnasri et al. 2006) and infection, development and reproduction in tomato (Oka et al. 1999) and *M. incognita* infection in sugarcane (Berry et al. 2011). Soil drenches and foliar sprays of this plant defence activator were effective in reducing reproduction of *Heterodera* spp. and *Meloidogyne* spp. populations on barley and wheat (Oka and Cohen 2001).

Jasmonic acid and its methyl ester are implicated in plant defence responses triggered by wounding and insect feeding, as well as in resistance against plant pathogens (Gozzo 2003). Nematode reproduction was reduced in tomato plants treated with foliar sprays of JA or methyl-jasmonate (Cooper et al. 2005; Curtis et al. 2009; Fujimoto et al. 2011). Cowpea (*Vigna unguiculata*) and tomato seed treatment with jasmonates induced a systemic defence response by reducing infection, development and reproduction of *M. incognita* and *G. pallida* (Pankaj et al. 2013). Nonetheless, it has been suggested that the SA pathway of systemic acquired resistance has a stronger effect on root-knot nematode development and reproduction in tomato and potato plants (Vieira dos Santos et al. 2013a). Some of the above mentioned studies also reported undesirable effects on seed germination and plant growth (Oka et al. 1999; Oka and Cohen 2001; Pankaj et al. 2013; Molinari 2016). These contrasting data are likely due to differences in the method of delivering the treatment (e.g., foliar spray, drench application, root dipping or seed treatment) and the concentrations used. It has been demonstrated that SA applied as soil-drench and BTH applied as root dips in tomato plants were, against *M. incognita*, only effective plant defence activators at suitable concentrations (Molinari and Baser

2010). Many of the side effects caused by exogenous application of plant defence inducers are strongly dosage dependent and have not been directly linked to fitness costs (Heil and Baldwin 2002). Dosage application of methyl jasmonate has been directly related to the decrease in the number of root-knot nematodes and egg masses in roots of tomato plants (Fujimoto et al. 2011). A linear relationship between dosage application and nematode reproduction was also observed in pineapple plants treated with BTH (Chinnasri et al. 2006). Although dependent on the quantity and method of application, levels of root uptake, obtained through a soil drench, for example, can be influenced by a range of factors, such as cultivar, age and health of treated plants, and the environmental conditions under which the trials are set (Vallad and Goodman 2004). Nonetheless, plant defence activators can indirectly influence plant growth by their effects on a reduction in root-knot nematode infection, development and reproduction (Nandi et al. 2003).

The performance of plant defence activators may not provide the same level of disease control as conventional nematicides. Furthermore, time of application is an issue, as they need to be applied before nematode attack (Lucas 2010). Plant defence activators act through the physiology of the plant and may, therefore, have side effects on crop growth and development (Heil and Baldwin 2002). However, there are several attractive features of priming the plant immune responses using natural plant defence activators for nematode management. Plant defence activators may require fewer inputs than current management based on chemicals. In addition, the broad-spectrum nature of the induced resistance provides additional options for their use in integrated disease and pest control programmes (Lucas 2010). The success of management strategies based solely on chemical plant defence activators may be compromised by the complex and diverse range of pathogens potentially present in the rhizosphere, organisms that can be either positively or negatively affected by the immunity response induced by a given plant elicitor (Vallad and Goodman 2004; Lucas 2010). Integrated control regimes that combine the use of plant activators to strengthen the endogenous defence of plants with resistant crop genotypes and management of natural suppressive processes may be a practical alternative to current methods of root-knot nematode control.

14.6 The Combined Use of *Pochonia chlamydosporia* with Plant Defence Activators

The application of plant defence activators that reduce the overall susceptibility of crop plants to root-feeding nematodes could be a key strategy for boosting the control potential of *P. chlamydosporia*. The observed effects on the determent of nematode penetration are a highly desirable feature of the application of plant defence activators that can enhance *P. chlamydosporia* performance, as this egg parasite cannot prevent the initial infestation of roots by the mobile nematode juveniles.

The efficacy of the combined application of a Portuguese *P. chlamydosporia* isolate and two plant defence activators, representatives of the SA and JA biochemical defence pathways, BTH and *cis*-jasmone, was assessed in pot trials, against *M. chitwoodi* in potato, in controlled conditions (Vieira dos Santos et al. 2013b). The results of this study revealed that nematode penetration was significantly affected by both plant defence activators. Only BTH treatments (alone or combined with fungus) and *cis*-jasmone combined with the fungus significantly reduced nematode reproduction when compared to the untreated plants. Foliar sprays of BTH did not seem to affect fungal parasitism but this was significantly increased in the *cis*-jasmone treatment, suggesting that the biocontrol efficacy of the *P. chlamydosporia* isolate was enhanced by the application of this plant defence activator. The presence of the fungus did not affect the efficacy of the plant activators, but the application of the plant activator *cis*-jasmone may have created the conditions for a fungal switch from saprophytic to parasitic, thus increasing nematode parasitism that deserves further investigation. The effects of plant activators on nematode reproduction were also enhanced by the presence of the fungus. Egg deposition was lower in BTH and *cis*-jasmone treatments in the presence of the fungus which may be indicative of an additive effect of both strategies (activator and fungus). Rhizosphere proliferation by the fungus was lower in treatments where the plant defence activators were applied, suggesting that the plant actively responds to the endophytic colonization of its roots by the fungus (Vieira dos Santos et al. 2013b). The application of foliar sprays of plant defence activators has been reported to interfere with rhizosphere colonisation of *P. chlamydosporia* and to decrease the ability of fungus to trigger local induced resistance (Medeiros et al. 2015). Nonetheless, under controlled conditions, *P. chlamydosporia* was found to be a poorer rhizosphere coloniser but a more efficient nematode parasite (Vieira dos Santos et al. 2013b). Activation of the JA pathway may produce changes in the plant and/or the rhizosphere inducing changes in nutrients released that prompt the fungal transition from the saprophytic to the parasitic phase. Molecular and genetic evidence suggests that root colonisation is modulated by JA signalling, probably due to the effects on carbohydrate partitioning (Zavala-Gonzalez et al. 2017).

A sustainable, natural and effective strategy of nematode control may rely on the activation of the JA pathway by plant defence activators that reduce root-knot nematode penetration and promote egg parasitism by a selected *P. chlamydosporia* isolate (with biological control potential) which also acts to improve the efficacy of the plant defence activators.

14.7 Future Perspectives

The putative modulation of the interactions among the host plant, nematode and *P. chlamydosporia* by the activation of the biochemical defence pathways represents not only a potential application as a new control strategy but also poses new questions on how the organisms interact and how these interactions can be further exploited to promote nematode control. Therefore, trials should be conducted in conditions closer to those found in agricultural systems to assess how these interactions can be affected by both biotic (competition by native microbiota, effects of other herbivores and pathogens in the activation of plant defence mechanisms) and abiotic factors. Also, *in vitro* assays under controlled conditions, which focus on selected key components of all these interactions, need to be performed to better understand the biochemical and molecular mechanisms involved.

To access fully the enormous potential of chemically-induced resistance to plant-parasitic nematodes, further studies should take into account a number of variables, such as the effects of different chemical concentrations and application methods, along with the effects of inducing plant defence pathways on the complex and diverse naturally occurring interactions between plants and their associated soil microbial diversity (Molinari 2011). It has also been suggested that induced plant defence mechanisms would incur fitness costs for the plant (Heil and Baldwin 2002). Research is needed on the evaluation of the potential trade-offs in investment in plant defence and susceptibility to nematode attack to determine the most successful combinations of different forms of induced defence in different plant species/cultivars under different growing conditions. This can be experimentally tested in studies that integrate various aspects of biology: molecular analyses, plant physiology and phenotyping and ecological studies of direct and indirect effects of plant defence activators.

The involvement and regulation of various PR proteins of the defence pathways can be assessed at different stages and time points of the previous suggested assays through current proteomic tools. While gel-based protein separation techniques have led to the identification of many pathogenicity and defence-related genes and proteins expressed during plant-pathogen interactions, key proteins are often expressed in low abundance and can either be overlooked by the analyses or be difficult to extract and identify. Several emerging techniques, such as gel-free high-throughput screening technologies, could overcome the setbacks of gel-based protein separation techniques (Afroz et al. 2011). Also, studies on gene expression would provide a more detailed understanding of the underlying mechanisms responsible for specific plant-nematode-fungus interactions. This will help to clarify the molecular mechanisms behind the improved nematode control obtained under laboratory conditions.

Several aspects of the factors affecting the performance of *P. chlamydosporia*, including the use of plant elicitors to strengthen the endogenous defence of plants and the development and application of alternative integrated biocontrol strategies against plant-parasitic nematodes, must continue to be addressed.

References

Afroz, A., Ali, G. M., Mir, A., et al. (2011). Application of proteomics to investigate stress-induced proteins for improvement in crop protection. *Plant Cell Reports, 30*, 745–763.

Atkins, S. D., Hidalgo-Diaz, L., Kalisz, H., et al. (2003). Development of a new management strategy for the control of root-knot nematodes (*Meloidogyne* spp.) in organic vegetable production. *Pest Management Science, 59*, 183–189.

Berry, S. D., Rutherford, R. S., & Curtis, R. H. (2011). Preliminary investigations into inducing resistance in sugarcane against *Meloidogyne incognita*. *South African Journal of Plant and Soil, 28*, 272.

Bordallo, J. J., Lopez-Llorca, L. V., Jansson, H.-B., et al. (2002). Colonization of plant roots by egg-parasitic and nematode-trapping fungi. *The New Phytologist, 154*, 491–499.

Bourne, J. M., & Kerry, B. R. (1999). Effect of the host plant on the efficacy of *Verticillium chlamydosporium* as a biological control agent of root-knot nematodes at different nematode densities and fungal application rates. *Soil Biology and Biochemistry, 31*, 75–84.

Bourne, J. M., & Kerry, B. R. (2000). Observations on the survival and competitive ability of the nematophagous fungus *Verticillium chlamydosporium* in soil. *International Journal of Nematology, 10*, 9–18.

Bourne, J. M., Kerry, B. R., & De Leij, F. A. A. M. (1994). Methods for the study of *Verticillium chlamydosporium* in the rhizosphere. *Supplement Journal of Nematology, 26*, 587–591.

Bourne, J. M., Kerry, B. R., & De Leij, F. A. A. M. (1996). The importance of the host plant on the interaction between root-knot nematodes (*Meloidogyne* spp.) and the nematophagous fungus, *Verticillium chlamydosporium* Goddard. *Biocontrol Science and Technology, 6*, 539–548.

Chinnasri, B., Sipes, B. S., & Schmitt, D. P. (2006). Effects of inducers of systemic acquired resistance on reproduction of *Meloidogyne javanica* and *Rotylenchulus reniformis* in pineapple. *Journal of Nematology, 38*, 319–325.

Chitwood, D. J. (2003). Nematicides. In J. R. Plimmer (Ed.), *Encyclopedia of agrochemicals* (pp. 1104–1115). New York: Wiley.

Cohen, Y. R. (2002). β-aminobutyric acid-induced resistance against plant pathogens. *Plant Disease, 86*, 448–457.

Collins, H. P., Navare, D. A., Riga, E., et al. (2006). Effect of foliar applied plant elicitors on microbial and nematode populations in the root zone of potato. *Communications in Soil Science and Plant Analysis, 37*, 1747–1759.

Conrath, U., Thulke, O., Katz, V., et al. (2001). Priming as a mechanism in induced systemic resistance in plants. *European Journal of Plant Pathology, 107*, 113–119.

Cooper, W. R., Jia, L., & Goggin, L. (2005). Effects of jasmonate-induced defenses on root-knot nematode infection of resistant and susceptible tomato cultivars. *Journal of Chemical Ecology, 31*, 1953–1967.

Curtis, R. H., Birkett, M., & Pye, B., et al. (2009). *Aspects of the plant-nematode interactions: Host recognition and plant signalling molecules*. In: Abstracts of the 2nd international congress of nematology (40th ONTA and 28th SBN Meetings), 4–9 October 2009, Maceió, Brasil.

Dallemole-Giaretta, R., Freitas, L. G., Lopes, E. A., et al. (2015). *Pochonia chlamydosporia* promotes the growth of tomato and lettuce plants. *Acta Scientiarum Agronomy, 37*, 417–423.

De Leij, F. A. A. M., & Kerry, B. R. (1991). The nematophagous fungus *Verticillium chlamydosporium* as a potential biological control agent for *Meloidogyne arenaria*. *Revue de Nématologie, 14*, 157–164.

De Leij, F. A. A. M., Kerry, B. R., & Dennehy, J. A. (1993). *Verticillium chlamydosporium* as a biological control agent for *Meloidogyne incognita* and *M. hapla* in pot and micro-plot tests. *Nematologica, 38*, 115–126.

De Meutter, J., Tytgat, T., Witters, E., et al. (2003). Identification of cytokinins produced by the plant parasitic nematodes *Heterodera schachtii* and *Meloidogyne incognita*. *Molecular Plant Pathology, 4*, 271–277.

Dias-Arieira, C. R., Santana S de, M., Freitas, L. G., et al. (2011). Efficiency of *Pochonia chlamydosporia* in *Meloidogyne incognita* control in lettuce crop (*Lactuca sativa* L.) *Journal of Food, Agriculture and Environment, 9*, 561–563.

Eisenback, J. D., & Triantaphyllou, H.H. (1991). Root-knot nematodes: *Meloidogyne* species and races. In W. R. Nickle (Ed.), *Manual of agricultural nematology* (pp. 191–274). New York: Marcel Dekker.

Escudero, N., & Lopez-Llorca, L. V. (2012). Effects on plant growth and root-knot nematode infection of an endophytic GFP transformant of the nematophagous fungus *Pochonia chlamydosporia*. *Symbiosis, 57*, 33–42.

Fujimoto, T., Tomitaka, Y., Abe, S., et al. (2011). Expression profile of jasmonic acid-induced genes and the induced resistance against the root-knot nematode (*Meloidogyne incognita*) in tomato plants (*Solanum lycopersicum*) after foliar treatment with methyl-jasmonate. *Journal of Plant Physiology, 168*, 1084–1097.
Giné, A., Carrasquilla, M., Martínez-Alonso, M., et al. (2016). Characterization of soil suppressiveness to root-knot nematodes in organic horticulture in plastic greenhouse. *Frontiers in Plant Science, 7*, 164. doi:10.3389/fpls.2016.00164.
Gorlach, J., Volrath, S., Knauf-Beiter, G., et al. (1996). Benzothiadiazole, a novel class of inducers of systemic acquired resistance, activates gene expression and disease resistance in wheat. *Plant Cell, 8*, 629–643.
Gozzo, F. (2003). Systemic acquired resistance in crop protection: From nature to a chemical approach. *Journal of Agricultural and Food Chemistry, 51*, 4487–4503.
Heil, M., & Baldwin, I. T. (2002). Fitness costs of induced resistance: Emerging experimental support for a slippery concept. *Trends in Plant Science, 7*, 61–67.
Heil, M., & Bostock, R. M. (2002). Induced Systemic Resistance (IRS) against pathogens in the context of induced plant defences. *Annals of Botany, 89*, 503–512.
Hérnandez, M. A., & Hidalgo-Díaz, L. (2008). KlamiC®: Bionematicida agrícola producido a partir del hongo *Pochonia chlamydosporia* var. *catenulata*. *Revista de Protección Vegetal, 23*, 131–134.
Iberkleid, I., Vieira, P., de Almeida Engler, J., et al. (2013). Fatty acid- and retinol-binding protein, Mj-FAR-1 induces tomato host susceptibility to root-not nematodes. *PloS One, 8*(5), e64586. doi:10.1371/journal.pone.0064586.
Jakab, G., Cottier, V., Toquin, V., et al. (2001). β-aminobutyric acid-induced resistance in plants. *European Journal of Plant Pathology, 107*, 29–37.
Jones, J. T., Haegeman, A., Danchin, E. G. J., et al. (2013). Top 10 plant-parasitic nematodes in molecular plant pathology. *Molecular Plant Pathology, 14*, 946–961.
Karssen, G., & Moens, M. (2006). Root-knot nematodes. In R. N. Perry & M. Moens (Eds.), *Plant nematology* (pp. 59–90). Wallingford: CABI Publishing.
Kerry, B. R. (1982). The decline of *Heterodera avenae* populations. *EPPO Bulletin, 12*, 491–496.
Kerry, B. R. (1987). Biological control. In R. H. Brown & B. R. Kerry (Eds.), *Principles and practise of nematode control in crops* (pp. 233–257). Sidney: Academic.
Kerry, B. R. (2000). Rhizosphere interactions and the exploitation of microbial agents for the biological control of plant-parasitic nematodes. *Annual Review of Phytopathology, 38*, 423–441.
Kerry, B. R., & Bourne, J. M. (1996). The importance of rhizosphere interactions in the biological control of plant parasitic nematodes - A case study using *Verticillium chlamydosporium*. *Pesticide Science, 47*, 69–75.
Kerry, B. R., & Bourne, J. (Eds.). (2002). A manual for research on *Verticillium chlamydosporium*, a potential biological control agent for root-knot nematodes. International Organization for Biological and Integrated Control of Noxious Animals and Plants/West Palearctic Regional Section (IOBC/WPRS), Gent.
Kerry, B. R., & Hirsch, P. R. (2011). Ecology of *Pochonia chlamydosporia* in the rhizosphere at the population, whole organism and molecular scales. In K. G. Davies & Y. Spiegel (Eds.), *Biological control of plant-parasitic nematodes: Building coherence between microbial ecology and molecular mechanism* (pp. 171–182). London: Springer.
Kerry, B. R., & Hominick, W. M. (2002). Biological control. In D. L. Lee (Ed.), *The biology of nematodes* (pp. 483–509). London: Taylor and Francis.
Kerry, B. R., Crump, D. H., & Mullen, L. A. (1982). Studies of the cereal cyst-nematode, *Heterodera avenae* under continuous cereals, 1975-1978. II. Fungal parasitism of nematode female and eggs. *The Annals of Applied Biology, 100*, 489–499.
Kerry, B. R., Kirkwood, I. A., De Leij, F. A. A. M., et al. (1993). Growth and survival of *Verticillium chlamydosporium* Goddard, a parasite of nematodes, in soil. *Biocontrol Science and Technology, 3*, 355–365.
Kuć, J. (2001). Concepts and direction of induced systemic resistance in plants and its application. *European Journal of Plant Pathology, 107*, 7–12.

Larriba, E., Jaime, M. D., Carbonell-Caballero, J., et al. (2014). Sequencing and functional analysis of the genome of a nematode egg-parasitic fungus, *Pochonia chlamydosporia*. *Fungal Genetics and Biology, 65*, 69–80.

Larriba, E., Jaime, M. D., Nislow, C., et al. (2015). Endophytic colonization of barley (*Hordeum vulgare*) roots by the nematophagous fungus *Pochonia chlamydosporia* reveals plant growth promotion and a general defense and stress transcriptomic response. *Journal of Plant Research, 128*, 665–678.

Lopez-Llorca, L. V., Bordallo, J. J., Salinas, J., et al. (2002). Use of light and scanning electron microscopy to examine colonization of barley rhizosphere by the nematophagous fungus *Verticillium chlamydosporium*. *Micron, 33*, 61–67.

Lucas, J. (2010). Advances in plant disease and pest management. *The Journal of Agricultural Science, 149*, 91–114.

Mácia-Vicente, J. G., Jansson, H.-B., Talbot, N. J., et al. (2009a). Real-time PCR quantification and live-cell imaging of endophytic colonization of barley (*Hordeum vulgare*) roots by *Fusarium equiseti* and *Pochonia chlamydosporia*. *The New Phytologist, 182*, 213–228.

Mácia-Vicente, J. G., Rosso, L. C., Ciancio, A., et al. (2009b). Colonisation of barley roots by endophytic *Fusarium equiseti* and *Pochonia chlamydosporia*: Effects on plant growth and disease. *The Annals of Applied Biology, 155*, 391–401.

Magan, N. (2001). Physiological approaches to improving the ecological fitness of fungal biocontrol agents in fungi as biocontrol agents. In T. Butt, C. Jackson, & N. Magan (Eds.), *Fungi as biocontrol agents – Progress, problems and potential* (pp. 239–251). Wallingford: CABI.

Mauchline, T. H., Kerry, B. R., & Hirsch, P. R. (2004). The biocontrol fungus *Pochonia chlamydosporia* shows nematode host preference at the infraspecific level. *Mycological Research, 108*, 161–169.

Maule, A., & Curtis, R. (2011). Exploiting genomics and molecular biology to understand plant-nematode interaction. In J. Jones, L. Gheysen, & C. Fenoll (Eds.), *Genomics and molecular genetics of plant-nematode interactions* (pp. 221–251). London: Springer.

Medeiros, H. A., Resende, R. S., Ferreira, F. C., et al. (2015). Induction of resistance in tomato against *Meloidogyne javanica* by *Pochonia chlamydosporia*. *Nematoda, 2*, e10015. http://www.dx.doi.org/10.4322/nematoda.10015.

Molinari, S. (2011). Natural genetic and induced plant resistance, as a control strategy to plant-parasitic nematodes alternative to pesticides. *Plant Cell Reports, 30*, 311–323.

Molinari, S. (2016). Systemic acquired resistance activation in solanaceous crops as a management strategy against root-knot nematodes. *Pest Management Science, 72*, 888–896.

Molinari, S., & Baser, N. (2010). Induction of resistance to root-knot nematodes by SAR elicitors in tomato. *Crop Protection, 29*, 1354–1362.

Monfort, E., Lopez-Llorca, L. V., Jansson, H.-B., et al. (2005). Colonisation of seminal roots of wheat and barley by egg-parasitic nematophagous fungi and their effects on *Gaeumannomyces graminis* var. *tritici* and development of root-rot. *Soil Biology and Biochemistry, 37*, 1229–1235.

Nandi, B., Kundu, K., Banerjee, N., et al. (2003). Salicylic acid-induced suppression of *Meloidogyne incognita* infestation of okra and cowpea. *Nematology, 5*, 747–752.

Nicol, J. M., Turner, S. J., Coyne, D. L., et al. (2011). Current nematode threats to world agriculture. In J. Jones, G. Gheysen, & C. Fenoll (Eds.), *Genomics and molecular genetics of plant-nematode interactions* (pp. 21–43). London: Springer.

Oka, Y., & Cohen, Y. (2001). Induced resistance to cyst and root-knot nematodes in cereals by DL-β-amino-*n*-butyric acid. *European Journal of Plant Pathology, 107*, 219–227.

Oka, Y., Cohen, Y., & Spiegel, Y. (1999). Local and systemic induced resistance to the root-knot nematode in tomato by DL-β-amino-*n*-butyric acid. *Phytopathology, 89*, 1138–1143.

Owen, K. J., Green, C. D., & Deverall, B. J. (2002). A benzothiadiazole applied to foliage reduces development and egg deposition by *Meloidogyne* spp. in glasshouse-grown grapevine roots. *Australasian Plant Pathology, 31*, 47–53.

Pankaj, N. M., Powers, S. J., Gaur, H.-S., et al. (2013). Differential defence response due to jasmonate seed treatment in cowpea and tomato against root-knot and potato cyst nematodes. *Nematology, 15*, 15–21.

Pieterse, C. M. J., Ton, J., & Van Loon, L. C. (2001). Cross-talk between plant defence signalling pathways: Boost or burden? *AgBiotechNet, 3*, 1–6.

Siddique, S., Radkovic, Z. S., De La Torre, C. M., et al. (2015). A parasitic nematode releases cytokinin that controls cell division and orchestrates feeding site formation in host plants. *Proceedings of the National Academy of Sciences, 112*, 12669–12674.

Stirling, G. R. (2011). Biological control of plant-parasitic nematodes: An ecological perspective, a review of progress and opportunities for further research. In K. G. Davies & Y. Spiegel (Eds.), *Biological control of plant-parasitic nematodes: Building coherence between microbial ecology and molecular mechanisms* (pp. 1–38). London: Springer.

Trudgill, D. L., & Blok, V. C. (2001). Apomictic, polyphagous root-knot nematodes: Exceptionally successful and damaging biotrophic root pathogens. *Annual Review of Phytopathology, 39*, 53–77.

Vallad, G. E., & Goodman, R. M. (2004). Systemic acquired resistance and induced systemic resistance in conventional agriculture. *Crop Science, 44*, 1920–1934.

Viaene, N., Coyle, D. L., & Kerry, B. R. (2006). Biological and cultural management. In R. N. Perry & M. Moens (Eds.), *Plant nematology* (pp. 346–369). Wallingford: CABI Publishing.

Vieira dos Santos, M. C., Curtis, R., & Abrantes, I. (2013a). Effect of plant elicitors on the reproduction of the root-knot nematode *Meloidogyne chitwoodi* on susceptible hosts. *European Journal of Plant Pathology, 136*, 193–202.

Vieira dos Santos, M. C., Curtis, R., & Abrantes, I. (2013b). The combined use of *Pochonia chlamydosporia* with plant defence activators – A potential sustainable control strategy for *Meloidogyne chitwoodi*. *Phytopathologia Mediterranea, 53*, 66–74.

Vieira dos Santos, M. C., Esteves, I., Kerry, B., et al. (2014). Interactions between *Pochonia chlamydosporia* and *Meloidogyne chitwoodi* in a crop rotation scheme. *Nematropica, 44*, 37–46.

Ward, E., Kerry, B. R., Manzanilla-López, R. H., et al. (2012). The *Pochonia chlamydosporia* serine protease gene *vcp1* is subjected to regulation by carbon, nitrogen and pH: implications for nematode biocontrol. *PloS One, 7*(4), e35637. doi:10.1371/journal.pone.0035657.

Zavala-Gonzalez, E. A., Rodríguez-Cazorla, E., Escudero, N., et al. (2017). *Arabidopsis thaliana* root colonization by the nematophagous fungus *Pochonia chlamydosporia* is modulated by jasmonate signalling and leads to accelerated flowering and improved yield. *The New Phytologist, 213*, 351–364. doi:10.1111/nph.14106.

Chapter 15
Pochonia chlamydosporia Microbial Products to Manage Plant-Parasitic Nematodes: Case Studies from Cuba, Mexico and Brazil

Leopoldo Hidalgo-Díaz, Francisco Franco-Navarro, and Leandro Grassi de Freitas

Abstract Plant-parasitic nematodes are important pests of many cultivated plants. Vegetable crops account for the greatest proportion of nematicide use because of infection by root-knot nematodes (*Meloidogyne* spp.). However, the negative impact of most chemical nematicides on the environment has led to a total ban or restricted use of these chemicals. The research discussed in this chapter concerns the development, innovation and good manufacturing practices of selected native isolates of *Pochonia chlamydosporia* as potential microbial control agents for the management of root-knot and false root-knot nematodes (*Nacobbus aberrans sensu lato*), as part of research conducted by multidisciplinary groups in Brazil (Federal University of Viçosa), Cuba (CENSA, Mayabeque) and Mexico (Colegio de Posgraduados-Montecillo). The studies referred to in this chapter deal with the basic approach towards implementation of a bio-management strategy for specific agro-economic production systems in Brazil, Cuba and Mexico.

L. Hidalgo-Díaz (✉)
Dirección de Sanidad Vegetal, Centro Nacional de Sanidad Agropecuaria, San José de las Lajas, Mayabeque, Cuba
e-mail: lhidalgo@censa.edu.cu

F. Franco-Navarro
Programa de Fitopatología, Colegio de Posgraduados-Campus Montecillo, Texcoco, Estado de México, Mexico
e-mail: ffranco@colpos.mx

L. Grassi de Freitas
Departamento de Fitopatología, Universidade Federal de Viçosa, Minas Gerais, Brazil
e-mail: leandro@ufv.br

R.H. Manzanilla-López, L.V. Lopez-Llorca (eds.), *Perspectives in Sustainable Nematode Management Through* Pochonia chlamydosporia *Applications for Root and Rhizosphere Health*, Sustainability in Plant and Crop Protection, DOI 10.1007/978-3-319-59224-4_15

15.1 Introduction

Plant-parasitic nematodes (PPN) are important pests of many cultivated plants. Losses due to nematode damage average 13.5% for some 40 crops, equivalent to an annual monetary loss of US$358.24 billion according to estimates based on 2010–2013 production figures and prices (Abd-Elgawad 2014). Nematicides continue to be an important part of nematode management programmes, whether they are used as part of an integrated management approach or as the sole control component. Vegetable crops account for the greatest proportion of nematicide use and *Meloidogyne* spp., known as the root-knot nematodes (RKN), are the target for approximately half of the world's nematicide usage. Root-knot nematodes are the major pest to be considered when developing new nematicidal products (Haydock et al. 2006). However, the negative impact of most chemical nematicides on both non-target organisms and the environment has led to a total ban or restricted use of these chemicals (Wesemael et al. 2011). Therefore, the development of new control approaches, which can be integrated with other environmentally friendly pest and crop management strategies, is essential. Among these management approaches are agrotechnical measures such as: (i) crop rotation with plant species that are immune to PPN; (ii) the use of PPN-free planting material, antagonistic crops and resistant varieties; (iii) soil treatments to reduce nematode populations, including biofumigation, soil solarization and the addition of organic amendments; and (iv) the use of microbial control, which has emerged as an important alternative to chemical control.

Fungal antagonists are among the most promising microbial control agents (MCA) for nematodes (Stirling 2011; Lamovsek et al. 2013). The research discussed in this chapter concerns the development and innovation of selected native isolates of *Pochonia chlamydosporia* (Goddard) Gams & Zare isolated in different countries of the Americas as potential MCA for the management of RKN and *Nacobbus aberrans sensu lato* (false root-knot nematode), as part of research conducted by multidisciplinary groups in Brazil (Federal University of Viçosa), Cuba (CENSA, Mayabeque) and Mexico (Colegio de Posgraduados-Montecillo).

The successful development of a microbial pest control product (MPCP) requires many steps to be carried out between concept and final product. A well-targeted project strategy is necessary to succeed and a roadmap to the successful development and commercialisation of MPCP for the control of arthropods has been proposed by Ravensberg (2011). The general pathway that he proposed is divided into several process stages: (i) selection of the microbial agent; (ii) production and product development; (iii) quality control; (iv) registration; (v) implementation; and (vi) commercialisation. This chapter deals with key topics related to these processes and considers the complex interactions present in the practical application of bio-management strategies for PPN on tropical soils (Kerry 2000).

15.2 Prospecting, Identification and Selection of the Microbial Agent

The first stage in the development of a MPCP consists of the isolation and selection of a bacterium, fungus, virus or nematode able to interfere with the biological cycle of the agricultural pest. The appropriate selection of the sampling sites significantly increases the probability of obtaining successful biocontrol agents (Montesinos 2003). In the specific case of nematode MCA, suppressive soils or soil with a long history of nematode infestation are recommended as prime targets for prospecting (Kerry et al. 1982).

In Cuba, Hidalgo-Díaz (2000) carried out a survey of MCA in the two main coffee regions of the country: a drier and warm region to the east (Santiago de Cuba) and another more humid and cooler region located in the centre of the country (Villa Clara). The soil samples from coffee rhizosphere were collected in 12 different places with a history of infestation by a complex of RKN species: *M. incognita*, *M. enterolobii*, and *M. arenaria* (Rodríguez et al. 1995). Three species and two varieties of *Pochonia* spp. were identified from a total of 83 isolates of facultative parasites of nematode eggs, amongst them: *Pochonia chlamydosporia* var. *chlamydosporia* (Goddard) Zare & Gams; *P. chlamydosporia* var. *catenulata* (Kamyschko ex Barron & Onions) Zare & Gams; *Lecanicillium psalliotae* (Treschew) Zare & Gams; and *Pochonia* [*Metapochonia*] *suchlasporia* (Gams & Dackman) Zare & Gams. *Pochonia chlamydosporia* was the most frequent species (90%) and the variety *P. chlamydosporia* var. *catenulata* was the commonest (62%) of the two varieties identified (Hidalgo-Díaz et al. 2000).

In Mexico, some native isolates of *P. chlamydosporia* var. *chlamydosporia* have been isolated from three states in Central Mexico (Mexico, Tlaxcala and Puebla) in soils naturally infested with *Nacobbus aberrans s.l.* (Flores-Camacho et al. 2007). Isolates of the false root-knot nematode (false RKN) were molecularly characterised and subsequently tested for the management of this nematode on crops such as tomato (*Solanum lycopersicum*) and chilli pepper (*Capsicum annuum*) (Pérez-Rodríguez et al. 2007). During another study, carried out in the buffer zone of the Los Tuxtlas Biosphere Reserve, Veracruz, 106 soil samples were taken and examined for the presence of native isolates of *P. chlamydosporia*. Samples were collected from locations with different land use (natural forest, secondary forest, pasture fields, and maize fields) and two varieties of *P. chlamydosporia* were found in 30 soil samples (either alone or together in the same soil sample). The variety *P. chlamydosporia* var. *chlamydosporia* was more frequent (83%) than *P. chlamydosporia* var. *catenulata* (33%). More than 70% of the isolates were found in non-disturbed soils from natural and secondary forest (Franco-Navarro et al. 2009). Another Mexican isolate of *P. chlamydosporia* var. *chlamydosporia* was obtained from naturally infested soil with the false RKN at Pozo de Gamboa, Zacatecas (named IZ1) (Franco-Navarro et al. 2013). This nematode is the most important plant-parasite in several bean growing areas of that region.

In Brazil, several screenings for nematophagous fungi have been done. The egg parasitic fungi *P. chlamydosporia* var. *chlamydosporia*, *P. chlamydosporia* var. *catenulata* and *L. psalliotae* were isolated from *M. mayaguensis* (= *M. enterolobii*) eggs (Arévalo et al. 2009). In this case, six randomised samples of 500 g of soil and infected roots were collected up to a depth of 10 cm from commercial guava (*Psidium guajava*) plantations in Petrolina, Pernambuco, Brazil. This area is characterised by a semiarid climate (high temperature, low humidity and sandy soil) which is also known to be heavily infested with *M. enterolobii* (Carneiro et al. 2001). In addition, 29 isolates of *P. chlamydosporia* were recovered from *Meloidogyne*-infested soils from three states in Brazil (Minas Gerais, Espirito Santo and Paraná). Of these, 65.5% were identified as *P. chlamydosporia* var. *chlamydosporia* and 34.5% as *P. chlamydosporia* var. *catenulata* (Dallemole-Giaretta et al. 2012).

Much variation within the same fungus species (*P. chlamydosporia*) and the varieties *P. chlamydosporia* var. *chlamydosporia* or *P. chlamydosporia* var. *catenulata* has been observed in different isolates from the same tropical soil (Hidalgo-Díaz et al. 2000; Flores-Camacho et al. 2008; Arévalo et al. 2009). According to Kerry and Hirsch (2011), such variation "may be correlated with changes in the host nematode in accordance with Red Queen dynamics but such interrelationships are unproven".

Due to the variability observed in the fungus, the use of accurate protocols for selection of one or several isolates to develop a strain (e.g., a group of clonally related individuals or cells – see Kirk and Ansell 1992) or strains with potential as a biological (microbial) control agent is essential. The screening of isolates must consider, at least, three key issues: (i) their pathogenicity to RKN eggs; (ii) their ability to grow in the rhizosphere; and (iii) the production of chlamydospores, which are the preferred source of inoculum (Kerry and Bourne 2002). Thus, it is imperative to identify those fungal isolates most suited to local conditions and also those that extensively colonize the rhizosphere of locally grown crop cultivars (see 15.4.2).

15.2.1 Selection

The screening done in Mexico for fungal isolates focused not only on *Meloidogyne* spp. (Medina-Canales et al. 2013), but also on *N. aberrans s.l.* The first studies on the potential use of *P. chlamydosporia* as a biological control were focused on *N. aberrans s.l.* as a target pest, this nematode causing severe damage to crops such as tomato, chilli pepper, and bean (*Phaseolus vulgaris*). Flores-Camacho et al. (2007) evaluated five native isolates of *P. chlamydosporia* and compared isolate Vc-10 (= Pc10 Rothamsted Research (RRES) collection) in growth chamber conditions as potential biocontrol agents of three populations of *N. aberrans s.l.* (two on tomato and one on bean). All fungal isolates demonstrated high egg parasitism (60–96%) on the three populations of *N. aberrans s.l.* studied. The fungus proliferated on roots and substrate (peat moss-quartz sand 3:1 V:V) producing over 5×10^3 chlamydospores per gram of root and substrate after 12 weeks. The most efficient Mexican isolates for control the false RKN were those originally labelled as SMB3 and SC1 in the RRES collection.

Pérez-Rodríguez et al. (2007) evaluated under *in vitro* conditions the pathogenicity of the same five Mexican isolates (Flores-Camacho et al. 2007, 2008). Only three isolates, re-labelled in the Mexican collection as MPc1 (= SMB3), MPc2 (= SMB3A) and MPc3 (= SC1), parasitized more than 70% of two populations of false RKN and only the MPc3 isolate showed good results when it was applied into naturally infested soil under glasshouse conditions. This isolate, when applied at a high concentration (15×10^3 chlamydospores per gram of soil) with vermicompost as the source of organic matter, resulted in a lower galling index, greater leaf dry weight, and fewer egg masses, juveniles and mature females/g of roots in comparison to control plants. Isolate MPc3 was the only one that was re-isolated from egg masses, soil and roots at the end of the experiment.

Subsequently, Franco-Navarro et al. (2008) tested 35 isolates of both varieties of *P. chlamydosporia* for their *in vitro* ability to colonize maize roots (*Zea mays*) and parasitize eggs of *N. aberrans s.l.* Bioassays showed that 7 of the isolates produced >80% egg parasitism, 16 between 70–80%, and 12 isolates parasitized <70%; 32 of the isolates colonized more than 80–100% of root segments. Two of the most efficient egg parasites completely colonized the roots (i.e., LM17 isolate of *P. chlamydosporia* var. *catenulata* from secondary forest and LM37 isolate of *P. chlamydosporia* var. *chlamydosporia*). By studying the pathosystem Bean-*N. aberrans s.l.*, one isolate (labelled as IZ1) of *P. chlamydosporia* var. *chlamydosporia* was obtained from soil naturally infested by the nematode in the location of Pozo de Gamboa (Panuco, Zacatecas). Isolate IZ1 was evaluated as a potential biological control agent against the false RKN according to egg parasitism *in vitro* and root colonization abilities on crops used in rotation schemes in Zacatecas state, crops that included squash (*Cucurbita pepo*), tomatillo (*Physalis ixocarpa*), cabbage (*Brassica oleracea* var. *capitata*), broccoli (*Brassica oleracea* var. *italica*), and maize, as well as on different bean (*Phaseolus vulgaris*) varieties (Franco-Navarro et al. 2013). The IZ1 native isolate parasitized 82% of false RKN eggs under *in vitro* conditions and colonised 100% of the roots of the plants tested. During root colonisation tests, chlamydospores were present 7–12 days after incubating root segments with the fungus. Chlamydospores were found after 7 days of incubation on squash roots, earlier than in other crops where chlamydospores appeared 9 or 10 days after incubation. Roots of all bean varieties tested showed chlamydospores at 12 days (Franco-Navarro et al. 2013). Subsequently, in glasshouse tests, bean plants inoculated with two different doses of the native isolate (5×10^3 and 7×10^3 chlamydospores per gram of soil) showed increased foliage and root weight over the control plants and those that were treated chemically, root galling and nematode numbers inside the plants being lower after fungus application (Franco-Navarro et al. 2013).

In Cuba, 27 strains were screened and deposited in the collections of CENSA and RRES (23 *P. chlamydosporia*; 3 *L. psalliotae* and 1 *Pochonia* [*Metapochonia*] *suchlasporia* (see Chap. 2 for updated *Pochonia* spp. taxonomy). The isolates were compared with the reference strain RRES-10 (= Vc-10) of *P. chlamydosporia* var. *chlamydosporia* (Hidalgo-Diaz 2000). The selection protocol included an initial trial on the effect of temperature on the radial mycelial growth and sporulation of the three species under study, which showed that *P. chlamydosporia* strains have an optimum temperature range for mycelial growth and sporulation of 20-25 °C. The study also showed that only the strains of *P. chlamydosporia* produced resting spores (chlamydospores). In the next screening test, Hidalgo-Díaz et al. (2000) evaluated, under laboratory conditions, the ability of 23 strains of *P. chlamydosporia* (10 var. *chlamydosporia*; 8 *P. chlamydosporia* var. *catenulata*; and 5 *P. chlamydosporia* var. *catenulata* Biotype A) to colonize the rhizosphere, parasitize eggs of *M. incognita*, and produce chlamydospores. As a result of these studies, the five strains of *P. chlamydosporia* var. *catenulata* biotype A were found to have a low chlamydospore production and poor colonization of the rhizosphere. Thus, only 18 strains were studied under controlled conditions in a glasshouse and compared with the reference strain (RRES-10). The existence of a high intra-specific variability for each of the indicators that were evaluated allowed for the selection of strains with better potential as MCA for RKN. In this sense, the status of the strains of *P. chlamydosporia* was established based on its saprophytic activity (rhizosphere colonisation), parasitic activity (parasitism of eggs) and chlamydospore production (Hidalgo-Díaz 2000). At this point, a ranking order was given to each strain (Table 15.1) starting from 1 (lowest value) up to 19 (highest value).

The first ranking position was occupied by the strain Cvc 108 (RRES 392) of *P. chlamydosporia* var. *catenulata*, followed by the reference strain RRES-10 (Vc 10) of *P. chlamydosporia* var. *chlamydosporia*. The latter was isolated from *M. incognita* eggs in Brazil and proved to be a potential agent for MCA of the main species of the genus *Meloidogyne* (De Leij and Kerry 1991; De Leij et al. 1993; Bourne and Kerry 1999). This was the first strain of *P. chlamydosporia* var. *catenulata* selected for mass production and evaluation of its effectiveness under field conditions as a potential MCA of RKN (Hidalgo-Díaz 2013). The selected strain was deposited in the CABI Bioscience Collection under the code IMI SD 187.

In Brazil, among the various natural enemies of nematodes, the fungus *P. chlamydosporia* var. *chlamydosporia* isolate Pc-10 was selected as a potential MCA for the management of RKN (Dallemole-Giaretta et al. 2012). For the glasshouse screening tests, fungal chlamydospores were incorporated into the soil (5000/g soil) and one tomato seedling was transplanted into a 300 ml pot. After 1 week, each plant was inoculated with 1000 eggs of *M. javanica*. The isolates Pc-1, Pc-2, Pc-3, Pc-4, Pc-9, Pc-10, Pc-21, Pc-24, and Pc-28 were the most efficient in reducing the number of eggs of the nematode. These isolates were re-evaluated in another glasshouse experiment but this time the number of eggs was increased to 3000, both eggs and chlamydospores being inoculated simultaneously. One week after inoculation, a tomato seedling was transplanted into each pot. In this experiment, isolate Pc-10 stood out as being the most efficient nematode antagonist among the other tested Brazilian isolates. A commercial bionematicide consisting of chlamydospores from isolate Pc-10 was developed as a wettable powder formulation, with an average of 5.2×10^7 chlamydospores/g of product Rizotec® Rizoflora Biotecnologia, Viçosa, Minas Gerais state (Brazil).

Table 15.1 Ranking order of *Pochonia chlamydosporia* strains from Cuba and RRES-10. Ranking was made according to saprophytic activity (rhizosphere colonization), parasitic activity (egg parasitism) and chlamydospore production

Strain	Parasitism of eggs in vitro	Rhizosphere colonization in vitro	Chlamydospore production in vitro	Rhizosphere colonization in vivo	Egg parasitism in vitro	Mean
Cvc 108	15	15	13	15	19	15.40
RRES 10	13	13	14	18	17	15.00
Cvc 81	10	16	16	18	13	14.60
Cvc 74	18	16	14	7	18	14.60
Cvc 19	12	10	19	11	16	13.60
Cvc 107	14	5	17	16	15	13.40
Cvc 44	19	12	10	13	9	12.60
Cvc 52	10	19	12	9	11	12.20
Cvc 20	15	9	8	17	10	11.80
Cvc 65	10	11	10	19	8	11.60
Cvc 53	17	8	14	14	4	11.40
Cvc 59	9	11	18	10	7	11.00
Cvc 75	11	14	13	12	3	10.60
Cvc 49	8	14	13	4	12	10.20
Cvc 43	6	18	10	8	7	9.80
Cvc 46	7	17	11	6	5	9.20
Cvc 54	16	7	9	5	7	8.80
Cvc 57	15	4	15	4	6	8.80
Cvc 22	7	6	12	4	14	8.60

Mean values (n = 4 replicates) refer to the ranking position (1 = lowest value; 19 = highest value) of each strain calculated as the average of the order of importance for each indicator or activity

15.3 Mass Production Technology and Quality Control

Many industries implement Quality Management Systems to maintain their competitiveness in the market place. In the case of regulated industries such as pharmaceutical and food, a Good Manufacturing Practices (GMP) protocol has been implemented to obtain products that are of consistent quality and safe. In the manufacturing industry in particular, MPCP experiences are very limited and information on production details is usually protected by the intellectual property rights of the biocontrol industry (reviewed in Ravensberg 2011). Villoch et al. (2003) proposed a GMP guide for manufacturing of MPCP based on experiences of regulated industries and a study of the elements involved in its production. These guidelines include all aspects directly related to production processes, post-production, control, and other support processes for obtaining a MPCP of consistent quality.

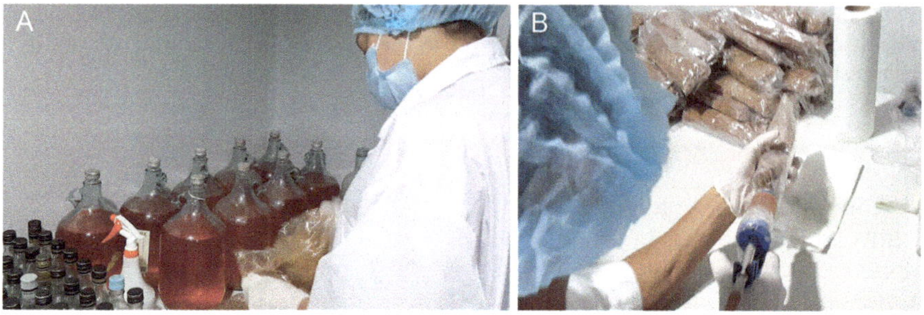

Fig. 15.1 Production process of KlamiC® at BiotorLabs (Matagalpa, Nicaragua) according to manufacturing practices developed at Centro Nacional de Sanidad Agropecuaria (CENSA), Cuba. (**A**): Preparation of pre-inoculum liquid; (**B**): Inoculation of solid substrate (courtesy of Eng. Luis F. Torres Montenegro, BiotorLabs Manager)

15.3.1 Production Process in Cuba

The selected Cuban strain IMI SD 187 of *P. chlamydosporia* var. *catenulata* is produced in solid state fermentation in a polypropylene bag with filter (Fig. 15.1A, B). This technology is implemented in a pilot-plant scale at CENSA, Cuba, following a GMP Guide which allows the specific requirements for obtaining the bionematicide KlamiC® to be accomplished (Montes de Oca et al. 2009). The main results of the implementation of the GMP Guide can be summarized as follows:

- Adaptation of an installation for production with unidirectional flow, adequate segregation of operations, minimum risk, and prevention of cross contamination.
- Establishing the production structure, responsibilities and interrelationships of personnel, and training of each employee.
- Design of a documentation system that registers all the information generated during the production process of each batch, which allows for traceability, releasing of the final products, and their post-market monitoring.
- Definition of quality control for the identification, establishment, and protection of the production strain.
- Characterisation of each process and the environment from the physical, microbiological, and waste point of view.
- Definition of specifications and assessment methods for all raw materials, intermediates, and final products involved in the production.
- Identification points and process controls, with statistical tools, ranges or values that allow production to normalise.
- Standardisation of evaluation tests and sampling for process control and final product.

KlamiC® has been produced and commercialised in Cuba by CENSA since 2009 for small farms from Mayabeque and Havana Province. It is also under large-scale production, under CENSA license, by BiotorLabs, Matagalpa (Nicaragua) with registration for commercial use in Nicaragua, Honduras and Panama in crops such as coffee, rice and banana (Fig. 15.2).

Fig. 15.2 BiotorLabs, Matagalpa, Nicaragua. Production plant with capacity to produce 3000 kg/month of formulated KlamiC® under strict quality controls (courtesy of Eng. Luis F. Torres Montenegro, BiotorLabs Manager)

15.3.2 Production Process in Brazil

In the Rizoflora bioplant the inoculum of the isolate Pc-10 (active ingredient of Rizotec®) is kept growing on solid culture medium in Petri-dishes by successive replications, and material from this colony are used to inoculate the liquid culture medium. The mass production is made through biphasic fermentation in which the isolate Pc-10 is first grown in liquid medium which is then spread on autoclaved rice in plastic bags. The bags are then transferred to incubation rooms for 21 days to allow the fungus to colonise the rice and produce chlamydospores. After the incubation period, the material is taken to drying rooms with controlled temperature and humidity. After drying, the chlamydospores are separated from the rice. The final product (Rizotec®) is packed in 2 kg plastic bags, labelled, and stored at low temperature before being sent to market in refrigerated trucks.

The Rizoflora Biotecnologia company is a spin-off of the Nematode Biological Control Laboratory, of the Plant Pathology Department. The laboratory is located in the campus of the Federal University of Viçosa and produces all the research that generated and improved the Rizotec® product, although the mass production occurs in the Rizoflora building, located at the Technology Park of Viçosa (tecnoPARQ: http://www.centev.ufv.br/en-US). The actual building has floor space of 430 m^2 and is capable of producing 30,000 kg of Rizotec®/year. Due to the good results of Rizotec® in controlling nematodes in the field, Stoller Brazil, a company that works with plant physiology and nutrition, has recently acquired most of the Rizoflora stock, so the bioplant will increase its production area to 1200m^2 in 2017. The increased production of Rizotec® is predicted to reach 130 tons in 2018. The main markets for the product in Brazil are soybean, sugar-cane, cotton, banana, coffee, and vegetables (Fig. 15.3).

Fig. 15.3 Rhizoflora, Viçosa (Brazil). Pilot Production plant capable of producing 30,000 kg of Rizotec®/year (courtesy L.G. Freitas)

15.3.3 Production Process in Mexico

In Mexico, unlike the Cuban and Brazilian experience, the registration of the technology implemented for the industrial production of *P. chlamydosporia* Mexican isolates or a registered brand for the Mexican strains has yet to be approved, production remaining at a small scale. However, the process to increase the inoculum of the fungus has evolved. Laboratory studies indicate that a good alternative to improve the growth of *P. chlamydosporia* is the use of cracked maize (Pérez-Rodríguez et al. 2007), which is cheaper in Mexico than rice, the conventional and most common substrate used for the growth of *P. chlamydosporia* in a combination of liquid and solid fermentation phases (Hidalgo-Díaz et al. 2000; Pérez-Rodríguez 2004). When the biphasic method with cracked maize (broth and grain) is used, most of the chlamydospores are produced 21 days after inoculation (Fig. 15.4), which is similar to the inoculum period estimated by other researchers (Pérez-Rodríguez 2004; Montes de Oca et al. 2009). Moreover, biphasic fermentation in cracked maize achieves greater chlamydospore yields and larger numbers of colony forming units (cfu) per gram of substrate than those produced by the fermentation process in rice (Pérez-Rodríguez et al. 2007).

15.3.4 Inoculum Stability

The main element involved in the manufacturing process of a MPCP is the strain, which is the equivalent to the active ingredient of the final product. Hence, it is necessary to demonstrate its stability and establish storage methods in such a way that it keeps its relevant MCA characteristics. Furthermore, the risk of

Fig. 15.4 Cracked maize. An alternative substrate to rice to produce chlamydospores using the biphasic (broth and grain) culture method (courtesy F. Franco-Navarro)

contamination must be avoided. Every production batch should result in propagules of the same quality and the inoculum for the production process is clearly a crucial parameter (Ravensberg 2011).

Montes de Oca et al. (2005) developed a set of original protocols covering colony level studies, enzymes, chlamydospore production, and biological activity both *in vitro* and *in vivo*. The implementation of these protocols was aimed at demonstrating whether the original characteristics of the strain, for which it was selected, were retained after repeated subculturing in Potato Agar-50 preserved at 4 °C. The results showed no differences in morphological, cultural, enzymatic characteristics, chlamydospore production or pathogenicity when subcultures (up to 100 subcultures) were compared with the original strain. The minimum values of chlamydospore production were in excess of 10^7 chlamydospores/g colonised substrate, with increased viability of 93%. Application of products for each solid fermentation subculture in pot experiments, under controlled conditions, showed levels of tomato root colonisation and soil of 1.13×10^4 and 1.49×10^5 cfu/g of root and soil, respectively. Parasitism of *M. incognita* eggs ranged from 63–67%. In general, these results agree with those obtained by Hidalgo-Diaz et al. (2000) in their original selection studies.

15.3.5 Product Specifications

During the development of a product its characteristics need to be established as part of the product standard. Every product batch must conform to these specifications and batches that do not meet the specification must either be improved or discarded (Ravensberg 2011). Testing methods and quality control protocols have been designed for the production of *P. chlamydosporia* var. *catenulata* IMI SD 187 (Montes de Oca et al. 2009) and those for KlamiC® are given in Table 15.2.

Table 15.2 Testing methods and quality control protocols criteria for the production of *Pochonia chlamydosporia* var. *catenulata* IMI SD 187 and KlamiC® (Montes de Oca et al. 2009)

Quality requirements	Assay method	Quality control criteria
Organoleptic characteristics	Visual observation	Polypropylene bags containing an ochre colour granulated product
Average mass	USP XXIII, 1995	500 g or 1Kg ±10%
Identity	Chlamydospore viability and count	Conidia in chains and chlamydospore presence
Active ingredient concentration	Chlamydospore count in Neubauer haemocytometer (Kerry and Bourne 2002)	$\geq 10^7$ chlamydospores/g substrate
Biological activity	Chlamydospore viability and egg parasitism (Kerry and Bourne 2002)	≥85% chlamydospore viability ≥70% *M. incognita* egg parasitism in every one of 4 batches
Microbial contamination	Microbial limit USP XXV, 2002	$<10^5$ colony forming unit/g product. No presence of bacteria nor pathogenic fungi
Water content	USP XXV, 2002 or Infrared Scale	< or equal to 5%
Airtightness	FEUM, 1988	Polypropylene airtight bags

15.3.6 Safety

To demonstrate the safe use of the selected strain to non-target organisms, toxicity studies were made in Cuba to assess infectivity-pathogenicity (Level I) of the Cuban strain to mammals, birds, beneficial invertebrates, and terrestrial plants. The methods used were based on the guidelines and criteria established by national and international regulatory agencies (EPA 1996; OECD 1996; Official Gazette 2007). In studies involving vertebrate animals (rats, rabbits and birds) no toxicity or infectivity was detected (García et al. 2004a, 2009). Studies of skin and eye irritation in rabbits classified the fungus as non-irritating to the skin and eyes (Garcia et al. 2004b). However, in the sensitization study, 30% of treated animals (albino guinea pigs Harley) presented allergic contact dermatitis and therefore the fungus is categorized as potentially allergenic (García 2005). Evaluation in non-target invertebrates found that the fungus was not toxic to earthworms (*Eisenia andrei*), bees (*Apis mellifera*) or lacewings (*Chrysopa exterior*), nor did it produce phytotoxicity in maize and common bean crops (García et al. 2008a, b). As a consequence, strain IMI SD 187 of *P. chlamydosporia* var. *catenulata* is classified as non-toxic, non-pathogenic, non-infective, and non-irritant to mammals. Similarly, it was evident that, according to the tests regulated for this purpose, it had no adverse ecological effects, indicating its minimum risk to animal and human health and the ecosystem. However, all mitosporic fungi are potential allergens and, in most cases, if exposure to the agent is minimised this is not a problem (Goettle et al. 2001). The use of barriers such as dust masks in production processes where there is greater exposure to spore aerosols is therefore recommended (Fig. 15.1A).

The Brazilian isolate Pc-10 passed similar toxicological and eco-toxicological tests in Brazil, as required by the environment and human health government agencies, but showed different results in not being allergenic, but causing subtle eye irritation to one of the three rabbits used in the test, resulting in the "class III – moderately toxic – eye irritation" warning to the Rizotec® product. This effect may have been because in Brazil the product should be applied to the eye of the rabbit as it is present in its formulation. Since Rizotec® is a wettable powder, it could not be suspended in water before application and it might be expected that any powder, even inert, could cause eye irritation.

15.4 Implementation of a Bio-management Strategy

The understanding of quantitative relationships between the host and its natural enemy is essential for the use of MPCP (Waage and Greathead 1988). In the case of microbial control of nematodes it is necessary to consider the tritrophic interaction established between: PPN-MCA-Plant Host. This is complex and subject to the influence of external factors such as the physical, chemical and biological properties of the soil, and environmental conditions (Kerry 2000; reviewed in Moosavi and Zare 2015). A better understanding of these complex interactions can lead to a successful MPCP.

Therefore, the development of a pre-defined bio-management strategy where the fungus is integrated with other agricultural practices to enhance their effectiveness is essential (reviewed in Hidalgo and Kerry 2008). The studies referred to in this chapter discuss the basic approach to define a bio-management strategy for specific agro-economic production systems in Brazil, Cuba and Mexico.

15.4.1 Concentration of Inoculum and Application Timing

The effect of different application rates of *P. chlamydosporia* var. *catenulata* strain IMI SD 187 on soil and root colonisation and of the parasitic activity of this fungus on *M. incognita* were studied in pots under glasshouse conditions (Puertas et al. 2006a). The levels of fungal colonisation in soil and tomato roots increased with increasing fungal inoculum concentration, as opposed to the parasitic activity that, although showing a rapid increase, tended to stabilize at an inoculum concentration of 5000 chlamydospores/g of soil. According to the results obtained, the effective minimum inoculum concentration for controlling RKN by *P. chlamydosporia* var. *catenulata* strain IMI SD 187 is 5000 chlamydospores/g of soil.

Although chlamydospores are considered the best inoculum for the application of *P. chlamydosporia*, the fungus controlled *M. javanica* on tomato plants in Brazil even when applied to the soil as colonised-rice grains, i.e., without chlamydospores. However, high dosages, such as 5 g/kg soil or more, may be needed and this is not economically feasible (Dallemole-Giaretta et al. 2014). It has been observed in several experiments that there is an optimal dose for the fungus in the soil. Two kilograms of Rizotec® per hectare is usually more effective than double the quantity. The same result occurs when 5000 and 10,000 chlamydospores are applied in pot tests, the former being more effective than the latter. It seems that the fungus has a self-regulation mechanism in the soil. It may be that toxins produced by the fungus to defend itself from other microorganisms in the soil may also act against itself when its concentration is above the ideal carrying capacity for *P. chlamydosporia* in the root or soil (Mauchline et al. 2002).

A similar conclusion was obtained by Peteira et al. (2005), who demonstrated that the ability of the Cuban selected strain (IMI SD 187) to become established within the soil was affected by the inoculum type. When the fungus was added on rice both the cfu data and the real-time PCR data indicated that it became readily established and proliferated. When the fungus was added as chlamydospores there was little proliferation. By the time the second tomato crop had been planted, the amount of fungus in the colonised rice treatment had declined to the same density as that in the other inoculum treatment. In the second crop, the fungus in both treatments fluctuated in a similar manner and detection of the fungus was the same in both diagnostic methods. An increase in the fungal population was demonstrated by both monitoring methods during the second month of the second crop due to the proliferation of the fungus in the rhizosphere, after which the levels returned to post-first harvest densities. This may also indicate that there is a threshold support level for the fungus in the soil.

15.4.2 Effect of Host Plant

Plant species differ in their root exudates and rhizodeposits, as well as in their ability to support *P. chlamydosporia* growth in their rhizosphere and in their susceptibility to infection by nematodes (Bourne et al. 1996). In Cuba, the ability of *P. chlamydosporia* var. *catenulata* IMI SD 187 was evaluated for colonization of the rhizosphere of different vegetable species commonly grown in urban organic farm (Puertas and Hidalgo-Díaz 2007). According to these studies, cabbage, cauliflower (*B. oleracea* var. *botrytis*), pak choi (*B. rapa* subsp. *chinensis*), tomato, cucumber (*Cucumis sativus*), and chilli pepper support more than 200 cfu/cm^2 of root and are considered to be good hosts for the fungus. A moderate growth of the fungus was presented on green beans, broccoli, eggplant (*Solanum melongena*), and spinach (*Spinacia oleracea*) with values of 100–200 cfu/cm^2 of root, whilst okra (*Abelmoschus esculentus*), parsley (*Petroselinum crispum*), and celery (*Apium graveolens*) supported fewer than 100 cfu/cm^2 of root (see Hidalgo-Díaz and Kerry 2008). Considering that cabbage, cauliflower and pak choi are not susceptible to *M. incognita* (Fernandez et al. 1998) but are a good host for the fungus (Puertas and

Hidalgo-Díaz 2007), it is highly recommended in the Cuban Organic Vegetable Urban Production System to use these crops in a rotation with susceptible crops such as tomato, cucumber and pepper, a strategy that combines the use of the fungus with crop rotation as previously suggested by Atkins et al. (2003).

On the other hand, the Brazilian isolates Pc-3, Pc-10 and Pc-19 of *P. chlamydosporia* var. *chlamydosporia* promote the growth of tomato and lettuce seedlings. Isolate Pc-19 colonised the rhizoplane of tomato seedlings in only 15 days and produced a large quantity of chlamydospores. This isolate was able to use cellulose as a carbon source, in addition to glucose and sucrose. Scanning electron microscopy (SEM) revealed that hyphae of the Brazilian *P. chlamydosporia* isolate Pc-10 penetrated the epidermal cells of the tomato roots, suggesting an endophytic behaviour for this isolate (Dallemole-Giaretta et al. 2015). A similar result to that for the Brazilian strain was obtained with the Cuban isolate IMI SD 187 in growth promotion of sweet pepper, cucumber, melon (*Cucumis melo*) and tomato due its endophytic behaviour (Ceiro 2015).

15.4.3 Seed and Seedling Treatment Application

In order to make *P. chlamydosporia* economically viable on soybean (*Glycine max*), the use of 250–300 g of the commercial product Rizotec® as a seed treatment, instead of a soil treatment with 3 kg/ha, would be a feasible alternative. Therefore, the efficiency of the seed treatment was investigated in sandy and clay soils for the control of *M. incognita* (Nasu 2013). For this study we used PVC tubing 55 cm in height and 10 cm in diameter divided into four sections per tube (0.0–12.5; 12.5–25.0, 25.0–37.5, 37.5–50.0 cm) and joined by adhesive tape. Two types of soils were tested, one being classified as a sandy soil and the other as a clay. Both soils had previously been treated with methyl bromide at a dose of 80 cm^3/m^3. The soils in the tubes received three different treatments: soybean seed treated with the fungus *P. chlamydosporia* (2 g of product based on chlamydospores per 100 g of seeds); soybean seed treated with *P. chlamydosporia* + 3000 eggs of *M. incognita*; soybean seed untreated + 3000 eggs of *M. incognita*. The experiment was evaluated 60 days after sowing. To do this, each tube had their four PVC sections separated using a sharp knife to cut through the adhesive tape. To do this, the tubes were placed on a table and the tapes, soil and roots cut through to separate the different soil depths. To assess the vertical distribution of *P. chlamydosporia* in both types of soil, 1 g soil was collected from each compartment to determine the population of the fungus in the soil by quantification of cfu. The plant in each compartment was evaluated as the mass of fresh roots, number of eggs per root system, number of eggs/g root, number of galls per root system and number of galls/g root. The experiment was conducted twice. Sixty days after sowing, we found soil colonization by *P. chlamydosporia* throughout the soil profile for both of the soils tested. The fungus was effective in controlling *M. incognita* in both clay and sandy soil. In sandy soil, fungus showed higher colonization, penetration and nematode control. Applying *P. chlamydosporia* by treating soybean seeds is feasible as it results in the colonization of the entire root system and promotes control of *M. incognita*. *Pochonia chlamydosporia* used in the treatment of soybean seeds can penetrate clay and sandy soil for at least 50 cm. The fungus grows best in sandy soil and thus is more effective in controlling nematodes when applied this way. Eleven field tests were conducted in Brazil from October 2015 and March 2016 in the states of São Paulo, Paraná, Mato Grosso, Goiás, Minas Gerais, and Bahia, in areas infested to varying degrees with *M. javanica*, *M. incognita*, *Heterodera glycines* and *Pratylenchus brachyurus*, in order to evaluate Pc-10 applied in the furrow or as a seed treatment. In 10 out of the 11 fields the fungus gave an increased productivity of 4.3 bags of soybean per hectare (83 kg/ha on average), showing a clear effect of growth promotion both in the presence or absence of the nematodes. This is a fundamental attribute of a product for its success in the market.

To apply the fungus as a soybean seed treatment, the farmer can choose the chemicals to use together with *P. chlamydosporia* according to a compatibility table (Table 15.3). Unfortunately, in Brazil, soybean seeds are usually sold to the farmers by the chemical companies dominating the market as pre-treated with a large amount of very toxic fungicides and insecticides, and the farmer has no alternative. One way around this problem is to apply the fungus to seeds of the crops planted after the soybean harvest, these usually being cultivated for the green mass production necessary for the next soybean no-tillage cultivation, and include millet (*Pennisetum glaucum*), oat (*Hordeum vulgare*), sorghum (*Sorghum bicolor*), *Brachiaria* grass, or *Crotalaria* spp. These seeds, therefore, do not receive any pesticide and allow the *Pochonia* to germinate together with the seeds, colonising the root system and infecting any nematode eggs remaining in the soil or in the soybean root residues and thereby diminishing the initial nematode inoculum for the next season. The desiccation of the shoots of these plants also creates a mat of straw that retains soil moisture and allows the fungus to proliferate, so increasing its concentration in the soil when the soybean crop is planted (Atkins et al. 2003). This practice has been successfully adopted by farmers in the west of Bahia for the last 4 years, the localised symptoms of yellowed and stunted plants and the losses due to nematodes diminishing year after year.

The application of *P. chlamydosporia* during seedling production of vegetable crops can be an efficient approach to control RKN since plants that are already protected are transplanted to the field. In Brazil, a powder formulation of the isolate Pc-10 containing 1×10^8 chlamydospores/g was suspended in water and applied when watering the seedlings of lettuce and cucumber (0, 4.5, 9.0, 13.5, and 18.0 g/l). The decrease in the number of eggs/g of roots was proportional to the increase of isolate Pc-10 applied to seedlings of lettuce with maximum reduction of 43.5% at the 18 g/l dose (Viggiano et al. 2015). In two experiments with cucumber seedlings, the dose of 18 g of Pc-10/l reduced the number of galls/g of roots by 46.04% and 49.44%, and the number of eggs/g of roots by 48.32% and 40.58% in the first and second experiments, respectively. Drenching the seedling substrate with Pc-10 at 18 g/l controls *M. javanica* in lettuce and cucumber crops and does not require additional fungus application in the field soil (Viggiano et al. 2014).

Table 15.3 Compatibility of *Pochonia chlamydosporia* and agrochemical products used in seed or furrow treatments (Brazil)

Pesticide Trade Name and formulation type	Active ingredient	Compatibility
(1) Avicta 500 FS	Abamectin	Compatible
(2) Avicta complete	Abamectin + cruiser + maxim xl	Compatible
(3) Azamax	Azadirachtin	Moderately compatible
(4) Bendazol	Carbendazim	Compatible
(5) Boral 500 SC	Sulfentrazone	Incompatible
(6) Bravonil ultrex WG	Chlorothalonil	Compatible
(7) Cabrio top WG	Metiram + pyraclostrobin	Compatible
(8) Cantus WG	Boscalid	Compatible
(9) Captan SC	Captan	Compatible
(10) Cercobim WG	Thiophanate-methyl	Compatible
(11) Cobre atar BR	Cuprous oxide	Compatible
(12) Confidor supra WG	Imidacloprid + beta-cyfluthrin*	Compatible
(13) Cropstar SL or EC	Imidacloprid + thiodicarb	Moderately compatible
(14) Cruiser 350 FS	Thiamethoxam	Compatible
(15) Derosal plus® FS	Carbemdazim + thiram	Moderately compatible
(16) Dipel WP	*Bacillus thuringiensis*	Incompatible
(17). Ecotrich WP	*Trichoderma harzianum*	Compatible
(18). Engeo pleno SC	Thiamethoxam + Lambda-Cyhalothrin	Compatible
(19). Evidence 700 WG	Imidacloprid	Compatible
(20) Folicur 200 EC	Tebuconazole	Incompatible
(21) Frowncide 500 SC	Fluazinam	Incompatible
(22) Gamit 360 CS	Clomazone	Compatible
(23) Gesagard SC	Triazine	Moderately compatible
(24) Infinito SC	Propamocarb hydrochloride + fluopicolide	Compatible
(25) Kraft 36 SC	EC Abamectin	Compatible
(26) Maxim XL SC	Fludioxonil + metalaxyl	Compatible
(27) Maxim FS	Fludioxonil	Moderately compatible
(28) Monseren WP	Pencycuron	Compatible
(29) Nomolt 150 SC	Teflubenzuron	Moderately compatible
(30) Permit SC	0,0-diethyl-0-phenyl phosphorothioate	Compatible
(31) Persist SC	Mancozeb concentrate	Incompatible
(32) Prisma EC	Difenoconazole	Compatible
(33) Protreat SC	Carbendazim + thiram	Incompatible
(34) Quality	*Trichoderma asperellum*	Compatible
(35) Regent WG	Fipronil	Compatible

(continued)

Table 15.3 (continued)

Pesticide Trade Name and formulation type	Active ingredient	Compatibility
(36) Revus SC	Mandipropamid	Compatible
(37) Ridomil gold MZ	Metalaxyl-m + mancozeb	Moderately compatible
(38) Roundup transorb	Glyphosate	Compatible
(39) Rovral SC	Iprodiona	Moderately compatible
(40) Rugby CS	Cadusafos*	Moderately compatible
(41) Spectro	Difenoconazol	Compatible
(42) Standak top	Pyraclostrobin + Thiophanate-methyl + Fipronil	Moderately compatible
(43) Standak	Fipronil	Moderately compatible
(44) Sumilex 500 WP	Procimidona	Compatible
(45) Talstar 100 EC	Bifentrina	Compatible
(46) Tecto SC	Thiabendazole concentrated	Compatible
(47) Vitavax SC	Carboxamide + dimethyldithiocarbamate	Incompatible

BR briquette, *CS* capsule suspension, *EC* emulsifiable concentrate, *FS* flowable concentrate for seed treatment, *SC* suspension concentrate, *SL* soluble liquid, *WG* water dispersible granules, *WP* wettable powder

Note: Beta-cyfluthrin is a pesticide of banned or restricted use (Stockholm convention on persistent Organic pollutants). Cadusafos is also a pesticide of restricted or banned use. It is classified as 1b or highly hazardous by the World Health Organization. Restrictions may differ according to country

In order to optimize the application of *P. chlamydosporia* and improve the control of *N. aberrans s.l.* in glasshouse and field conditions in Mexico, different concentrations of the fungus: 5×10^3, 15×10^3, 30×10^3 and 40×10^3 chlamydospores per gram of substrate were applied and tested at seedbed stage. The fungus was applied by mixing fungus colonized rice and, according to each treatment, with 1000 g of soil-peat moss substrate mixture (1:1 w/w) where tomato seeds were sown. Before planting fresh tomato seedlings into *N. aberrans s.l.* naturally infested soil, four older plants growing at each of the four fungal doses being tested were sampled and examined to estimate the protection (i.e., colonization) level given by the fungus to the plants by counting the cfu/g of roots and in soil (Table 15.4). In addition to these four treatments, another treatment that consisted of applying the fungus at planting only at the standard dose of 5×10^3 chlamydospores per gram of soil was included. After 45 days from planting, gall numbers and cfu/g of roots and soil were evaluated (Table 15.5). By applying 30×10^3 and 40×10^3 chlamydospores per gram of seedbed substrate, and without any additional application of the fungus, a significant presence of the fungus both in soil and on roots was guaranteed before planting tomato seedlings taken from the inoculated seedbed. However,

Table 15.4 Bioassay to estimate *Pochonia chlamydosporia* isolate MPc3 potential plant protection against *Nacobbus aberrans sensu lato* using bait tomato plants inoculated at seedbed stage and planted in soil naturally infested with the nematode. Colony forming units (cfu) per gram of root and cfu/g soil were counted to estimate initial colony forming units before planting fresh tomato seedlings inoculated with the fungus

Chlamydospore treatment (seedbed plants)	Isolate MPc3 before planting fresh tomato seedlings	
	cfu/g roots	cfu/g soil
5000	1.3×10^3 b	8.4×10^3 c
15,000	1.7×10^3 b	2.5×10^3 c
30,000	4.1×10^3 b	76×10^3 b
40,000	53×10^3 a	130×10^3 a
MSD	5.6×10^3	21×10^3

Values are means of four replicates. Numbers with similar letters are not different significantly (Tukey $\alpha < 0.01$)
MSD Minimum Significant Difference

Table 15.5 Number of galls and colony forming units (cfu) of *Pochonia chlamydosporia* isolate MPc3 per gram of tomato roots and soil 45 days after planting tomato plants to soil naturally infested by *Nacobbus aberrans sensu lato*

Treatments	45 days after planting		
	Galls/g roots	cfu/g roots	cfu/g soil
Control	7 a	0	0
P. chlamydosporia applied at planting (no previous exposure of plants to the fungus at seedbed stage)	4 ab	2.5×10^3 a	0.4×10^3 b
5000 chlamydospores applied at seedbed before planting	6 a	0.3×10^3 b	1.4×10^3 b
15,000 chlamydospores at seedbed before planting	4 ab	1.2×10^3 ab	7.6×10^3 a
30,000 chlamydospores at seedbed	5 ab	1.8×10^3 ab	2.2×10^3 b
40,000 chlamydospores at seedbed	2 b	1.9×10^3 ab	1.9×10^3 b
MSD	1.9	1.0×10^3	1.4×10^3

Values are means of four replicates. Numbers with similar letters are not different significantly (Tukey $\alpha < 0.01$)
MSD Minimum Significant Difference

45 days after planting there was a substantial reduction in the number of cfu in soil and on roots from both treatments (Franco-Navarro et al. 2008).

In Cuba, different cultivars of banana and plantain *vitro* plants were inoculated with KlamiC® at a concentration of 5.6×10^5 chlamydospores per *vitro* plant in the *ex vitro* adaptation area, in polypropylene trays and black polyethylene bags, containing a bovine compost substrate. A significant increment was obtained in all the growth variables of the *vitro* plants treated with KlamiC® (Fig. 15.5A) in comparison with the *vitro* plants not treated (Fig. 15.5B). For all the cultivars evaluated, colonization by the fungus of the substrate and rhizosphere was confirmed in all treatments using KlamiC®. These results demonstrate the potential of KlamiC® as

Fig. 15.5 Banana and plantain vitro plants. (**A**): General aspect of plants inoculated with KlamiC® (*left*) and non-inoculated plants (*right*); (**B**): Shoot and root system of plants inoculated with KlamiC® (*left*) and non-inoculated plants (*right*) (courtesy L. Hidalgo-Díaz)

a growth promoter product and suggest the evaluation of frequency of application during the production of banana and plantain *vitro* plants and its behaviour under field conditions (Hernández et al. 2016).

15.4.4 Effect of Organic Material

The use of organic manure can influence the proliferation of *P. chlamydosporia* in soil, organic soils being usually more favourable than mineral soils (Kerry 2000). Nevertheless, the application of organic manure stimulates the growth of soil micro-organisms active in decomposing the organic supplements (Rodríguez-Kábana 1997) and this may not be compatible with the MPCP introduced, either through competition or by release of toxic metabolites.

In Cuba, a glasshouse study was developed to determinate the effect of cattle manure and vermicompost mixed with natural soil at different proportions (1:1, 2:1, 3:1 v/v) on the establishment of the IMI SD 187 strain in the soil and rhizosphere of tomato. In all treatments, including natural soil without application of organic manure, good colonization of the fungus was achieved in a range of $4–10^4$ - 10^5 cfu/g soil and $6–9 \times 10^3$ cfu/g root tomato at 54 days after transplanting (Puertas and Hidalgo-Díaz 2009). The results showed that those sources of organic manure did not affect the establishment of the fungus in the soil due to its frequent use in the usual vegetable organic production system. Hence, these organic manures are recommended as a carrier for the application of the commercial product KlamiC® in the proportion of 1:10 V:V (Hernández and Hidalgo-Díaz 2008).

Furthermore, the effect of applying different doses of mineral fertilization, organic manure (vermicompost) or arbuscular mycorrhizal fungi (*Glomus hoi*-like), separately or in combination, on the establishment of the strain IMI SD 187 was studied in an experiment carried out using the tomato hybrid HA-3108 in a glasshouse at Los 3 Picos farm in Havana, Cuba. The number of cfu in soil and roots was directly related to the application of arbuscular mycorrhizal fungi and vermicompost according to Principal Component Analysis and the biplot graph. A direct relationship between the mineral fertilization and *P. chlamydosporia* colonization of roots was found, while the inverse occurred with respect to soil colonization. These results indicated that the nutritional treatments favoured *P. chlamydosporia* (IMI SD 187) growth, with more significant effects when the arbuscular mycorrhizal fungi were used combined with the vermicompost (Charles et al. 2015).

In the northwest Paraná State, in the southern region of Brazil, two experiments were conducted in areas naturally infested with the nematode. The fungus, produced on rice grain, was applied to the soil at concentrations of 0, 10, 20 and 30 g/m^2 of bed. Each gram of the product contained 1×10^8 chlamydospores. Cattle manure was applied on the top of the bed at 100 g/m^2 immediately prior to fungus application, and both product and manure were incorporated together into the soil. After 15 days, seedlings of lettuce cv. Babá de Verão were planted and grown for 45 days. In one of the experimental areas, the fungus was found to reduce the number of nematode eggs irrespective of the dosage, while in the other area the treatments reduced the final population of J2 in the soil. An increase in vegetative parameters was observed in the area of lower soil fertility when doses of 10 and 30 g of the product were used (Dias-Arieira et al. 2011).

15.4.5 Interaction with Other Beneficial Organisms

In Cuba, the interaction of the selected strain of *P. chlamydosporia* var. *catenulata* (IMI SD 187) was studied with the three most widely used microorganisms acting at rhizosphere level: the nitrogen fixing bacteria *Rhizobium* spp., the arbuscular mycorrhizal fungi *Glomus clarum* (= *Rhizophagus clarus*) and, the antagonist fungus *Trichoderma harzianum*, under glasshouse conditions (Puertas et al. 2006b). The results demonstrated the compatibility of *P. chlamydosporia* var. *catenulata* with all microorganisms. The lowest numbers of nematode infective juveniles in the substrate were found in treatments that included the nematophagous fungus. However, a significant reduction in nematode populations was also observed in the joint treatment with *T. harzianum* and *G. clarum* (= *Rhizophagus clarus*). These results suggest the possibility of combining the use of these beneficial microorganisms.

In Brazil, a glasshouse study of the interaction between the fungus *P. chlamydosporia* (Pc-12) and the rhizobacterium *Gracilibacillus dipsosauri* (MIC 14) was positive for nematode control, even though *G. dipsosauri* inhibited *P. chlamydosporia* growth by up to 30% under *in vitro* tests (Podestá et al. 2013).

15.4.6 *Biomanagement Strategy: Field Trial in Different Agricultural Systems*

15.4.6.1 Brazil

The application of the fungus *P. chlamydosporia* has been an alternative to the use of chemical nematicides on several crops in Brazil. For the management of the root-knot *M. incognita* on carrot in Minas Gerais state, Bontempo et al. (2014) found that the application of Pc-10 at 3 kg/ha increased the production of commercial roots by 41.7% and reduced the production of discarded roots due to nematode incited galls and bifurcations by 48.7%, The reproduction factor of the nematode was reduced by 61.4%, in comparison to the untreated control. However, they observed that incorporation of Pc-10 in the soil was more effective than application on the soil surface.

In the north of Minas Gerais, sandy soils, high temperatures and water availability throughout the year are ideal conditions for the proliferation of nematodes on banana. Infected plants are stunted, with yellowing leaves and malformation of the bunch, reflecting low production and reduced longevity of plantation. Plants tend to fall or topple due to damage to the roots caused by *Radopholus similis* and, interestingly, *Meloidogyne* sp., due to rotting of the root tissue (Fig. 15.6A) which begins around the RKN female and spreads to the entire gall before breaking when stressed by the weight of the bunch (Fig. 15.6B).

In addition to direct damage, nematodes predispose plants to attack by other microbes, such as the fungus *Fusarium oxysporum* f. sp. *cubense*, causing Panama disease, which leads to plant death. Banana is a perennial crop, making the use of rotation with other crops for nematode control more difficult. Thus, the most widely used control method is the use of pesticides. The nematicides organophosphates and organocarbamates used in banana are systemic, classified respectively by the Ministry of Agriculture in toxicological classes I and II and toxicological classes I and III (highly toxic), some with a grace period of up to 90 days. Government actions in this region through the PFI project Integrated Program for Fruit advocate the preservation of the environment and high quality fruit harvest in order to reach more demanding markets such as Europe, and to provide conditions for the incorporation of new technologies by farmers, such as biological control. The Pc-10 isolate of *P. chlamydosporia* var. *chlamydosporia* was taken to Janaúba, Minas Gerais, for application in the field. The area was chosen because of the high incidence of nematodes in commercial planting of banana cultivar Galil 18. One hundred and eighty-two banana mats (consisting of parent plant, daughter and granddaughter) were

Fig. 15.6 Roots infected with *Meloidogyne* spp. in banana plants. (**A**): Roots showing necrotic spots at the site where the nematode females are located. The increase in necrotic area causes breakage of roots and rotting of tissue around nematode females; (**B**): Root system affected by the nematodes (courtesy L.G. Freitas)

inoculated with the fungus, which was applied around the granddaughter plant on the soil surface in a half-moon pattern, and covered with dry leaves for protection, at the rate of 500 g per mat in June 2007. Composite samples of soil and roots were collected in May 2007 to determine the initial nematode population, and in August and November 2007 nematode populations after the application of *P. chlamydosporia* were assessed to evaluate the effect of the fungus on the population dynamics of the nematodes. There was a gradual reduction in the populations of *Meloidogyne* spp., *R. similis*, *Helicotylenchus* sp. and *Pratylenchus* sp. in the soil in the two evaluations, and the population of *Pratylenchus* sp. fell to an undetectable level by August. In the roots, there was no decline of *R. similis* in both evaluations, but there was a drastic decrease in the populations of the three other nematodes. *Helicotylenchus* sp. fell 71.6% in August, but rose to the initial level in November. *Meloidogyne* spp. 90.45% fell in August rising slightly in November, but *Pratylenchus* sp., which was undetected in the soil, was not found in the roots either (Freitas et al. 2009).

To verify the field data, *P. chlamydosporia* was tested for the management of *M. javanica* and growth promotion in banana cv. Silver Anã in an experiment conducted in a glasshouse between the spring and summer of 2015, in Viçosa, Brazil (data not published). Pots with 3 litres of soil received 5000 chlamydospores of each of seven isolates of *P. chlamydosporia* and were infested with 3000 eggs of *M. javanica*. After 10 days of soil infestation banana seedlings were transplanted. The number of nematode eggs per plant was significantly lower for the isolated Pc-10, Pc 64, Pc 123, and Pc 18 showing a reduction of 65.85%, 43.53%, 26.59%, and 13.52%, respectively, compared to the control. For the variable number of galls, only Pc-10 showed a statistical difference when compared to the other isolates. Isolates Pc-10, Pc-54, Pc-18, Pc-20, and Pc-42 provided greater shoot dry mass means with an average gain of 50.37%. It was observed that the isolates found to be present in higher concentration in the soil at the end of the experiment also provided promoted plant growth.

15.4.6.2 Cuba

A preliminary evaluation of the impact of the selected strain on the control of RKN was conducted at the 'Organico-Vivero-Alamar' in Havana, Cuba (Atkins et al. 2003). A red beet crop on the test site immediately before the experiment commenced had been severely damaged due to infestation of the soil by J2 of *M incognita* (300 juveniles/g soil). The fungus was applied (10 g m^2 of colonized rice, 6×10^6 chlamydospore/g) in a mixture with cattle manure to four plots (2 m^2), selected at random, and the mixture was incorporated to a depth of 15 cm. Control plots received the cattle manure alone. After the single application of the fungus, two poor hosts for the nematode (beans, cabbage) were grown before the fully susceptible tomato crop was planted 96 days after the application of the fungus. The bean and cabbage crop in the cropping cycle significantly reduced the nematode infestation in soil following the red beet crop. The application of the fungus before the bean crop had a small additional effect in reducing the population of the nematode after the two poor nematode hosts in the cropping cycle and prevented the population increasing on the subsequent tomato crop. In untreated soil, the nematode population increased significantly to the level that had occurred after the damaged red beet crop. The proportions of nematode egg masses and eggs colonized by the fungus were >70% and significantly greater than the numbers of nematodes colonized by the fungus in the untreated soil. *Pochonia chlamydosporia* var. *catenulata* was found in most nematode eggs from the treated soil, but was absent from the untreated soil in which <7% of the eggs were colonised by fungi and contained *Fusarium* spp. and *Purpureocillium lilacinum*. The survival of the fungus in soil was also examined under controlled conditions in which it remained in soil in densities significantly greater than its original application rate for at least 5 months. Hence, it seems that populations of this fungus may be built up in soil and have significant effects on the regulation of RKN populations.

In another field experiment located at CENSA (National Animal and Plant Health Center) was evaluated the establishment of the selected strain IMI SD 187 in soil infested with *M. incognita* from a single application of colonized rice grain or as a chlamydospore powder at rate of 5000 chlamydospores/g of soil (Peteira et al. 2005, Puertas 2007). The total area of the site was 720 m^2, divided into plots of 55.50 m^2 consisting of three beds of 12 m^2. The fungus was mixed with cattle manure prior to application. The control received only organic matter at the stated application rate. The fungus remained active throughout a cropping cycle of eight crops over a 30 month period. In one cropping cycle of tomato – tomato – cabbage – green bean – tomato – tomato – green bean – tomato, the infection of RKN eggs increased from 30% to >80% in the initial and final tomato crops, respectively and the numbers of J2 in soil decreased by >90% in treated soil compared to untreated soil.

15.4.6.3 Mexico

In a field naturally infested with *N. aberrans s.l.*, a rotation scheme was carried out by first sowing cabbage as a non-host crop during winter and then chilli pepper as a highly susceptible crop to the false RKN during the spring-summer crop season. In both cycles *P. chlamydosporia* var. *chlamydosporia* (MPc3) was inoculated to soil before planting by applying colonized rice grain (at a rate of 7.5×10^3 chlamydospores per gram of soil) mixed with vermicompost as vehicle (at a rate of 10 tonnes/ha) (Franco-Navarro et al. 2007). During the cycle with cabbage, the fungus showed a high number of cfu in soil and roots at the end of the crop (52,000/g and 15,000/g respectively), thus indicating its potential to be applied in the field and not rely upon repeated applications. The cfu in soil 90 days after planting (dap) chilli pepper, increased from 13,800 to 45,500/g; moreover, the numbers of nematodes and galls were lower where the fungus was applied. During the cycle with chilli pepper, parasitism of nematode eggs at 60 and 90 dap was 37% and 58%, respectively (Franco-Navarro et al. 2007).

In another field trial, four management strategies were tested against *N. aberrans s.l.* on chilli pepper, including application of the fungus alone or in combination. At 100 dap the incorporation of cabbage fragments, the application of vermicompost alone, and the inoculation of *P. chlamydosporia* var. *chlamydosporia* increased the foliage dry weight by 57%, 60% and 65%, respectively, and reduced the galling indices by 48%, 50% and 75%, respectively, in comparison with the absolute control. At 60, 80 and 100 dap, the number of juveniles and mature females g/ roots were lower in these treatments as compared to the control (Pérez-Rodríguez et al. 2011). Colonization of the egg masses by the fungus was 23% at 60 dap, 27% at 80 dap and 23% at 100 dap. Moreover, a higher yield of chilli pepper was obtained when treatments such as the fungus, vermicompost + cabbage fragments, and vermicompost alone, were applied (Pérez-Rodríguez et al. 2011).

Biofumigation of soil in plastic glasshouses or shadehouses prior to application of the fungus (i.e., colonized rice substrate) has been attempted to control

Meloidogyne spp. and *N. aberrans s.l.* (Cid del Prado et al. 2010). Biofumigation was carried out by preparing a mixture composed of 80 tonnes/ha of sorghum (*Sorghum bicolor*) and 40 tonnes/ha chicken manure, which was incorporated into the soil at a depth of 30 cm, watered to field capacity, covered with a black plastic sheet, and left for 4 weeks. After the organic matter had decomposed, the plastic sheet was removed and the soil left to aerate and eliminate toxic compounds for 3 weeks (Cid del Prado et al. 2010). Incorporation of the fungus at 5×10^8 chlamydospores/g colonized rice, mixed with chicken manure, vermicompost or ground lucerne (*Medicago sativa*) has also been investigated to control *M. arenaria* in guava (Torres-López et al. 2013).

15.5 Future Perspectives

Although important contributions have been made to the knowledge of *P. chlamydosporia* in Mexico (e.g., distribution, screening and identification of new potential isolates, laboratory and glasshouse management, etc.), the present situation is very different to that of Cuba or Brazil or Italy (Sellito et al. 2016), where standardized production methods and regulations to manufacture, distribute and use the fungus as a registered biocontrol product are established. As has been illustrated in this chapter, the process from isolate screening to registered product usually takes several years. Many efforts have been made in Mexico to do *P. chlamydosporia* research, but only a few of these efforts have ensured the proper use of the fungus as an efficient and effective alternative for the control and/or management of root-knot and false root-knot nematodes in medium and large scale farming systems. There is not enough staff or qualified specialized laboratories for producing the fungus on a commercial scale, and efforts made by public funded research institutions to test its biological effectiveness under standardized field and large glasshouses conditions are geographically scattered and infrequent. This scenario has been complicated during the last 5 years by the arrival of private companies claiming to have ownership of isolates of the fungus which are ready for use. Coincidentally, most of these companies are located in the highly technical, well-funded, agricultural areas of the country, where there is an increasing demand for fast and effective solutions to control the root-knot nematodes (*Meloidogyne* spp.). In most cases, there is clearly a complete lack of knowledge on fungus mode of action, the companies advertising their *Pochonia*-based products just like that of a *Trichoderma*, *Metarhizium* or *Beauveria*. As a consequence, results in nematode control have not necessarily been what the farmers have expected, although the idea of a 'silver bullet product' is still sold to them at the expense of jeopardizing the real potential and credibility of the use of *P. chlamydosporia* as a biocontrol agent. These circumstances require the development of national and objective regulatory programmes and quality protocols for BCA products. Examples of potential actions at basic and applied level include the following:

- To expand the screening of isolates with potential for biological control of root-knot and false root-knot nematodes in Mexico, through surveys in different agricultural areas of the country.
- To test under standard operation procedures the biological effectiveness of Mexican isolates of *P. chlamydosporia* var. *catenulata* or other varieties (*P. chlamydosporia* var. *mexicana*) in order to see if any has a greater potential than those already known.
- To certify all laboratories dedicated to the production of biological control organisms.
- To revise and update the Project of Official Mexican Norm (NOM-032-FITO-1995[1]) to ensure quality standards of BCA-based products used in the control of plant-parasitic nematodes.
- *Pochonia*-based products should be properly fingerprinted to allow monitoring once released in different agricultural areas of the country.
- To improve the formulation of the fungus to adapt their use to different farming and production conditions.
- To optimize the use of the fungus via more tests to confirm the feasibility of applying the fungus at the seedbed stage in order to protect plants in earlier stages of development.
- To carry out more trials for testing compatibility and effectiveness of the fungus, in combination with other management strategies for the control of root-knot and false root-knot nematodes, and even other pathogens, in both glasshouse and field conditions.

In Brazil, *Pochonia chlamydosporia* isolate Pc-10 has been tested in the field for at least 4 years, mainly with good results, in soybean, guava, carrots and beets. This has helped us to learn about how the product should or not be applied, its resilience, and its plant growth promotion effect in the absence of the nematodes. However, the commercial use of a product made from *P. chlamydosporia* has only started this year, and represents a big challenge because of the large number of farmers testing it, the large areas treated, and the high diversity of crops, soils, climates and nematode species. There is still much to be learned in respect to this fungus and its interactions with plants, nematodes, organic matter, and other microorganisms in the soil, whether pathogenic or not to the plants. A comprehensive research to investigate the compatibility of this fungus with a number of different types of agrochemicals will make its use more feasible in modern agriculture. Molecular techniques that allow the identification and quantification of the isolate in both the field soil and plants will help researchers to understand the epidemiology and ecology of the fungus. Improvements in formulation are needed to increase the shelf life of the fungal product, its persistence in the soil, and to improve delivery in low volumes of water

---

[1]The Mexican NOM-032-Fito-1995 can be consulted at: http://dof.gob.mx/nota_detalle.php?codigo=4881515&fecha=25/04/1996; and recent modifications at: http://sagarpa.gob.mx/normateca/Normateca/RESPUESTA%20a%20los%20comentarios%20y%20modificaciones%20a%20la%20NOM%20032%20FITO%201995.pdf

for in-furrow application. The next few years will be decisive in proving whether the biological control of nematodes is possible in the real world.

In Cuba there is a large amount of experience in the commercial application of *P. chlamydosporia* var. *catenulata* IMI SD 187 (KlamiC®) in vegetable production (both organic systems and protected crops). However, there is a need to reduce application dosages, optimize production and formulation methods, and conduct more validation field tests in different agricultural systems in crops such as coffee, guava, and banana. Furthermore, the plant promotion effect demonstrated for the Cuban selected strain, which is due the endophytic behaviour of the fungus, should encourage research in this area and open new uses for the product KlamiC®, even in the absence of nematodes. Hence, formulations and methods for seed and seedling treatment should be studied.

We share the working hypothesis that proper management of agro-ecosystems can enable the suppression of nematodes for long periods, as happens under natural conditions in a wide range of soils worldwide. The successful use of microbial control agents for nematodes depends greatly on the knowledge of their mechanisms of action and the interactions between the fungus and nematodes as well as with other microorganisms in the rhizosphere of the plant, even leading to the control of some phytopathogenic fungi or acting as plant growth promoters. The successful use of agents of microbial control nematodes depends greatly on the knowledge of their mechanisms of action and the interactions between them and with other microorganisms in the rhizosphere of the plant, potentially leading to their use in controlling phytopathogenic fungi or as plant growth promoters.

However, in practice, maximizing the efficacy of biological control agents through selection procedures or production and formulation methods of selected strains, will still require the support of other control measures for the sustainable management of most PPN. In addition, another challenge needs to be met for the successful exploitation of microbial control nematodes, namely the essential support services to local farmers through agricultural extension programmes.

Acknowledgements *In memoriam* of Brian R. Kerry. We would not arrive here without his guidance, his motivation and commitment to the science of microbial control of nematodes, and in particular to *Pochonia*. His memory keep us motivated to achieve bigger challenges. We also would like to acknowledge colleagues, especially all graduate and undergraduate students, who have accompanied us on this long but fascinating journey.

References

Abd-Elgawad, M. M. M. (2014). Plant-parasitic nematodes threats to global food security. *Journal of Nematology, 46*, 130–260.

Arévalo, J., Hidalgo-Díaz, L., Martins, et al. (2009). Cultural and morphological characterization of *Pochonia chlamydosporia* and *Lecanicillium psalliotae* isolated from *Meloidogyne mayaguensis* eggs in Brazil. *Tropical Plant Pathology, 34*, 158–163.

Atkins, S. D., Hidalgo-Diaz, L., Kalisz, H., et al. (2003). Development of a new management strategy for the control of root-knot nematodes (*Meloidogyne* spp.) in organic vegetable production. *Pest Management Science, 59*, 183–198.

Bontempo, A. F., Fernandes, R. H., Lopes, J., et al. (2014). *Pochonia chlamydosporia* controls *Meloidogyne incognita* on carrot. *Australasian Plant Pathology, 43*, 421–424.

Bourne, J. M., & Kerry, B. R. (1999). Effect of the host plant on the efficacy of *Verticillium chlamydosporium* as a biological control agent of root-knot nematodes at different nematodes densities and fungal application rates. *Soil Biology and Biochemistry, 31*, 75–84.

Bourne, J. M., Kerry, B. R., & De Leij, F. A. A. M. (1996). The importance of the host plant on the interaction between root-knot nematodes (*Meloidogyne* spp.) and the nemathophagous fungus, *Verticillium chlamydosporium* (Goddard). *Biocontrol Science and Technology, 6*, 539–548.

Carneiro, R. M. D. G., Moreira, W. A., Almeida, M. R. A., et al. (2001). Primeiro registro de *Meloidogyne mayaguensis* em goiabeira no Brasil. *Nematologia Brasileira, 25*, 223–228.

Ceiro, W. G. (2015). *Aportes a las bases cientifico-técnicas para el establecimiento de Pochonia chlamydosporia var. catenulata en el manejo de Meloidogyne spp.* en sistemas de producción protegida de hortalizas. Tesis en Opción al Grado de Dr. En Ciencias Agrícolas, Universidad Agraria de la Habana, Mayabeque, Cuba.

Charles, N. J., Arévalo, J., Hernández, A., et al. (2015). Effects of mineral, organic, and biological fertilization on the establishment of *Pochonia chlamydosporia* var. *catenulata* (Kamyschko ex. Barron and Onions) Zare & Gams in a protected crop. *Revista de Protección Vegetal, 30*, 239–244.

Cid del Prado-Vera, I., Lucero Pallares, M. A., Pérez-Rodríguez, I., et al. (2010). Biofumigation a very efficient alternative for the control of root-knot nematode *Meloidogyne* spp. in vegetables produced in house shadows. *Nematropica, 40*, 128. [Abstract].

Dallemole-Giaretta, R., Freitas, L. G., Lopes, E. A., et al. (2012). Screening of *Pochonia chlamydosporia* Brazilian isolates as biocontrol agentes of *Meloidogyne javanica. Crop Protection, 42*, 102–107.

Dallemole-Giaretta, R., Freitas, L. G., Xavier, D. M., et al. (2014). Soil amendment with substrate containing mycelium and conidia of *Pochonia chlamydosporia* for the management of *Meloidogyne javanica. Ciência Rural, 44*, 629–633.

Dallemole-Giaretta, R., Freitas, L. G., & Lopes, E. A., et al. (2015). *Pochonia chlamydosporia* promotes the growth of tomato and lettuce plants. *Acta Scientiarum. Agronomy*. no.4 Maringá Oct./Dec. 2015. https://doi.org/10.4025/actasciagron.v37i4.25042 . Accessed 3 Oct 2016.

De Leij, F. A. A. M., & Kerry, B. R. (1991). The nematophagous fungus *Verticillum chlamydosporium* Goddard, as a potencial biological control agent for *Meloidogyne arenaria* (Neal) Chitwood. *Revue de Nématologie, 14*, 157–164.

De Leij, F. A. A. M., Dennehy, J. A., & Kerry, B. R. (1993). Effect of catering on the distribution of *Verticillium chlamydosporium* in soil and the colonization of egg masses of *Meloidogyne incognita* by the fungus. *Nematologica, 39*, 250–265.

Dias-Arieira, C. R., Santana, S. M., Freitas, L. G., et al. (2011). Efficiency of *Pochonia chlamydosporia* in *Meloidogyne incognita* control in lettuce crop (*Lactuca sativa* L.) *Journal of Food, Agriculture and Environment, 9*, 561–563.

EPA. (1996). Prevention, pesticides and toxic substances. Series 885-Microbial Pesticide Test Guidelines.

Fernández, E., Pérez, M., Gandarilla, H., et al. (1998). Guía para disminuir infestaciones de *Meloidogyne* spp. mediante el empleo de cultivos no susceptibles. Boletín Técnico. *Sanidad Vegetal, 4*, 1–18.

Flores-Camacho, R., Manzanilla-López, R. H., Cid del Prado-Vera, I., et al. (2007). Control of *Nacobbus aberrans* (Thorne, 1935) Thorne and Allen, 1944 with *Pochonia chlamydosporia* (= *Verticillium chlamydosporium*) (Goddard) Zare and W. Gams. *Revista Mexicana de Fitopatología, 25*, 26–34.

Flores-Camacho, R., Atkins, S. D., Manzanilla-López, R. H., et al. (2008). Characterization of Mexican isolates of *Pochonia chlamydosporia* var. *chlamydosporia* (Goddard) Gams and Zare for biological control of *Nacobbus aberrans* (Thorne) Thorne and Allen. *Revista Mexicana de Fitopatología, 26*, 93–104.

Franco-Navarro, F., Pérez-Rodríguez, I., & Acosta-Hernández, R. (2007). Control microbiológico del nematodo falso nodulador, *Nacobbus aberrans*, en condiciones de campo. In: Pérez, V. A, & Ávila, R. C. (Eds.) Memoria de Resúmenes del IX Simposio Internacional y IV Congreso Nacional de Agricultura Sostenible, México, p 262.

Franco-Navarro, F., Pérez-Rodríguez, I., & Doroteo-Mendoza, et al. (2008). Estado del conocimiento de *Pochonia chlamydosporia* en México. In: Stefanova, M., Martínez, B., & Hidalgo, L. (Eds.) Memoria de Resúmenes del VI Seminario Científico Internacional de Sanidad Vegetal / II Taller Latinoamericano de Biocontrol de Fitopatógenos, pp 183–184.

Franco-Navarro, F., Vilchis-Martínez, K., & Miranda-Damián, J. (2009). New records of *Pochonia chlamydosporia* from Mexico: Isolation, root colonization and parasitism of *Nacobbus aberrans*. *Nematropica, 39*, 133–142.

Franco-Navarro, F., Cid del Prado-Vera, I., & Romero-Tejeda, M. L. (2013). Aislamiento y Potencial Parasítico de un aislamiento nativo de *Pochonia chlamydosporia* en contra de *Nacobbus aberrans* en Frijol. *Revista Mexicana de Fitopatología, 30*, 101–114.

Freitas, L. G., Giaretta-Dallemole, R., Ferraz, S., et al. (2009). Controle Biológico de Nematóides: Estudo de Casos. In L. Zambolim & M. Picanço (Eds.), *Controle Biológico de Pragas e Doenças – Exemplos Práticos* (Vol. 1, pp. 41–82). São Carlos: Suprema Gráfica e Editora Ltda.

García, L. (2005). Evaluación toxicológica y ecotoxicológica de un bionematicida constituido por la cepa IMI SD 187 de *Pochonia chlamydosporia* var. *catenulata*. Tesis en Opción al Grado de Doctor en Medicina Veterinaria. Universidad Agraria de la Habana, Mayabeque, Cuba.

García, L., Bulnes, C., Melchor, G., et al. (2004a). Safety of *Pochonia chlamydosporia* var. *catenulata* in acute oral and dermal toxicity/pathogenicity evaluations in rats and rabbits. *Veterinary and Human Toxicology, 46*, 248–250.

García, L., Melchor, G., Montes de Oca, N., et al. (2004b). Estudio de la irritación ocular y dérmica de *Pochonia chlamydosporia* var. *catenulata*. *Revista de Toxicología Española, 21*, 103–107.

García, L., Melchor, G., Arévalo, J., et al. (2008a). Evaluación de la fitotoxicidad de la cepa IMI SD 187 de *Pochonia chlamydosporia* var. *catenulata* sobre *Zea mays* L. y *Phaseolus vulgaris* L. *Revista de Protección Vegetal, 23*, 38–42.

García, L., Melchor, G., & Domínguez, Y., et al. (2008b). Ecotoxicological evaluation of *Pochonia chlamydosporia* var. *catenulata* in terrestrial invertebrates. *RETEL Revista de Toxicología en Línea*. No. 14. www.sertox.com.ar/modules.php . Accessed 11 Apr 2008.

Goettel, M. S., Hajek, A. E., Siegel, J. P., et al. (2001). Safety of fungal biocontrol agents. In T. M. Butt, C. Jackson, & N. Magan (Eds.), *Fungi as biocontrol agents. Progress, problems and potential* (pp. 347–375). Wallingford: CABI Publishing.

Gracia, L., Bules, C., Melchor, G., et al. (2009). Safety of *Pochonia chlamydosporia* var. *catenulata* inquails. *Revista de Salud Animal, 31*, 40–45.

Haydock, P. P. J., Woods, S. R., Grove, I. G., et al. (2006). Chemical control of nematode. In R. Perry & M. Moens (Eds.), *Plant nematology* (pp. 392–408). Wallingford: CABI Publishing.

Hernández, M. A., & Hidalgo-Díaz, L. (2008). KlamiC: Bionematicida agrícola producido a partir del hongo *Pochonia chlamydosporia* var. *catenulata*. *Revista de Protección Vegetal, 23*, 131–134.

Hernández, M. A., Arévalo, J., & Marrero, D., et al. (2016). Effect of KlamiC® on growth stimulation of plantain and banana vitro plants. *Cultivos Tropicales 37*, 168–172. doi: 10.13140/RG.2.2.25696.69120. Accessed 3 Oct 2016.

Hidalgo-Díaz, L. (2000). *Potencialidades de cepas autóctonas de Verticillium chlamydosporium (Goddard) como agente de control biológico de Meloidogyne spp*. Tesis en Opción al Grado de Doctor en Ciencias Agrícolas. Universidad Central de las Villas, Las Villas, Cuba.

Hidalgo-Díaz, L. (2013). Investigación, desarrollo e innovación de *Pochonia chlamydosporia* var. *catenulata*: Agente de control microbiano de nematodos formadores de agallas. Tesis en opción al grado de Doctor en Ciencias. Universidad Agraria de la Habana, Mayabeque, Cuba.

Hidalgo-Díaz, L., & Kerry, B. R. (2008). Integration of biological control with other methods of nematode management. In A. Ciancio & K. G. Mukerji (Eds.), *Integrated management and biocontrol of vegetable and grain crops nematode* (pp. 29–49). Dordrecht: Springer.

Hidalgo-Díaz, L., Bourne, J. M., Kerry, B. R., et al. (2000). Nematophagous *Verticillium* spp. in soils infested with *Meloidogyne* spp. in Cuba: Isolation and screening. *International Journal of Pest Management, 46*, 277–284.

Kerry, B. R. (2000). Rhizosphere interactions and the exploitation of microbial agents for the biological control of plant parasitic nematodes. *Annual Review of Phytopathology, 38*, 423–441.

Kerry, B. R., & Bourne, J. M. (Eds.). (2002). *A manual for research on Verticillium chlamydosporium, a potential biological control agent for rootknot nematodes*. Gent: IOBC/WPRS.

Kerry, B. R., & Hirsch, P. R. (2011). Ecology of *Pochonia chlamydosporia* in the rhizosphere at the population, whole organism and molecular scales. In K. Davies & Y. Spiegel (Eds.), *Biological control of plant-parasitic nematodes: Building coherence between microbial ecology and molecular mechanisms* (pp. 171–182). Dordrecht: Springer.

Kerry, B. R., Crump, D. H., & Mullen, L. A. (1982). Studies of the cereal cyst nematodes, *Heterodera avenae*, under continuous cereal, 1975-1978. II. Fungal parasitism of nematodes females and eggs. *The Annals of Applied Biology, 100*, 489–499.

Kirk, P. M., & Ansell, A. E. (1992). *Authors of fungal names*. Wallingford: CAB International.

Lamovsek, J., Urek, G., & Trdan, S. (2013). Biological control of rootknot nematodes (*Meloidogyne* spp.): Microbes against pests. *Acta Agriculture Slovenia, 101-2*, 263–275.

Mauchline, T. H., Kerry, B. R., & Hirsch, P. R. (2002). Quantification in soil and the rhizosphere of the nematophagous fungus *Verticillium chlamydosporium* by competitive PCR and comparison with selective plating. *Applied and Environmental Microbiology, 68*, 1846–1583.

Medina-Canales, M. G., Rodríguez-Tovar, A. V., Manzanilla-López, R. H., et al. (2013). Identification and molecular characterisation of new Mexican isolates of *Pochonia chlamydosporia* for the management of *Meloidogyne* spp. *Biocontrol Science and Technology, 24*, 1–21.

Montes de Oca, N., Arévalo, J., Acosta, N., et al. (2005). Estabilidad de la cepa IMI SD 187 de *Pochonia chlamydosporia* var. *catenulata* (Kamyscho ex Barron y Onions) Zare y W. Gams. Parte I. Indicadores morfológicos, productivos y patogénicos. *Revista de Protección Vegetal, 20*, 93–101.

Montes de Oca, N., Arévalo, J., Núñez, A., et al. (2009). KlamiC: ExperienciaTécnica-Productiva. *Revista de Protección Vegetal, 24*, 62–65.

Montesinos, E. (2003). Development, registration and commercialization of microbial pesticides for plant protection. *International Microbiology, 6*, 245–252.

Moosavi, M. R., & Zare, R. (2015). Factor affecting commercial success of biological agents of phytonematodes. In T. H. Askary & P. R. P. Martinelli (Eds.), *Biological control of phytonematodes* (pp. 423–445). Wallingford: CAB International.

Nasum, E. G. C. (2013). Tratamento de sementes de soja e algodão com *Pochonia chlamydosporia* no controle de *Meloidogyne incognita* e histopatologia da interação tritrófica. Doctoral Thesis, Universidade Federal de Viçosa, MG, Brazil.

OECD. (1996). *Environment Monograph No. 106. Data requirements for registration of biopesticides in OECD member countries: Survey results*. Paris: Organisation for Economic Co-operation and Development.

Official Gazette. (2007). Gaceta Oficial de la República de Cuba, Ministerio de Justicia: Resolución Conjunta del MINSAP y MINAGRI. No 16:77–84 (16 de abril del 2007).

Pérez-Rodríguez, I. (2004). Eficiencia de cinco aislamientos del hongo nematófago *Pochonia chlamydosporia* Goddard para el control de *Nacobbus aberrans* Thorne, 1935) Thorne and Allen 1944, en tomate (*Lycopersicon esculentum* Mill). MSci Dissertation, ITA 29. Tlaxcala, México.

Pérez-Rodríguez, I., Doroteo-Mendoza, A., Franco-Navarro, F., et al. (2007). Isolates of *Pochonia chlamydosporia* var. *chlamydosporia* from Mexico as potencial biological control agents of *Nacobbus aberrans*. *Nematropica, 37*, 127–134.

Pérez-Rodríguez, I., Franco-Navarro, F., Cid del Prado-Vera, I., et al. (2011). Control de *Nacobbus aberrans* en chile ancho (*Capsicum annuum* L.) mediante el uso combinado de enmiendas orgánicas, hongos nematófagos y nematicidas. *Nematropica, 41*, 122–129.

Peteira, B., Puertas, A., Hidalgo-Díaz, L., et al. (2005). Real-time PCR to monitor and assess the efficacy of the nematophagous fungus *Pochonia chlamydosporia* var. *catenulata* against rootknot nematode populations in the field. *Biotecnologia Aplicada, 22*, 261–266.

Podestá, G. S., Freitas, L. G., Dallemole-Giaretta, R., et al. (2013). *Meloidogyne javanica* control by *Pochonia chlamydosporia, Gracilibacillus dipsosauri* and soil conditioner in tomato. *Summa Phytopathologica, 39*, 122–125.

Puertas, A. (2007). Uso de *Pochonia chlamydosporia* var. *catenulata* (Kamyscho ex Barron y Onions) Zare y Gams como agente de control biológico de *Meloidogyne incognita* (Kofoid y White) Chitwood en cultivos hortícolas. Tesis en opción al grado científico de Doctor en Ciencias Agrícolas. Universidad Agraria de la Habana, Mayabeque, Cuba.

Puertas, A., & Hidalgo-Díaz, L. (2007). Influencia de la planta hospedante y su interacción con *Meloidogyne incognita* sobre la efectividad de *Pochonia chlamydosporia* var. *catenulata*. *Revista de Protección Vegetal, 22*, 104–109.

Puertas, A., & Hidalgo-Díaz, L. (2009). Efecto de diferentes abonos orgánicos sobre el establecimiento de *Pochonia chlamydosporia* var. *catenulata* en el sustrato y la rizosfera de plantas de tomate. *Revista de Protección Vegetal, 24*, 162–165.

Puertas, A., Arévalo, J., Montes de Oca, N., et al. (2006a). Efecto de diferentes concentraciones de inóculo de la cepa IMI SD 187 de *Pochonia chlamydosporia* var. *catenulata* en el control de *Meloidogyne incognita*. *Revista de Protección Vegetal, 21*, 74–79.

Puertas, A., Noval de la, B. M., Martínez, B., et al. (2006b). Interacción de *Pochonia chlamydosporia* var. *catenulata* con *Rhizobium* sp., *Trichoderma harzianum* y *Glomus clarum* en el control de *Meloidogyne incognita*. *Revista de Protección Vegetal, 21*, 80–89.

Ravensberg, W. J. (2011). *A Roadmap to the successful development and commercialization of microbial pest control products for control of arthropods*. Dordrecht: Springer.

Rodríguez, M. G., Rodríguez, I., & Sánchez, L. (1995). Especies del género *Meloidogyne* que parasitan el cafeto en Cuba: Distribución geográfica y sintomatología. *Revista de Protección Vegetal, 10*, 123–128.

Rodríguez-Kábana, R. (1997). Alternatives to MeBr soil fumigation. In A. Bello, J. A. González, M. Arias, & R. Rodríguez-Kábana (Eds.), *Alternatives to methyl bromide for the Southern European Countries* (pp. 17–33). Valencia: CSIC, Gráficas Papallona S.C.V.

Sellitto, V. M., Curto, G., & DallaValle, E. et al. (2016). Effect of *Pochonia chlamydosporia*-based formulations and the regulation of root-knot nematodes and plant growth response. *Frontiers in Life Sciences*. http://www.dx.doi.org/10.1080/21553769.2016.11193827. Accessed 20 Oct 2016.

Stirling, G. R. (2011). Biological control of plant-parasitic nematodes: An ecological perspective, a review of progress and opportunities for further research. In K. G. Davies & Y. Spiegel (Eds.), *Biological control of plant-parasitic nematodes: Building coherence between microbial ecology and molecular mechanisms, Progress in Biological Control 11* (pp. 1–38). Dordrecht: Springer.

Torres-López, J., Prado-Vera, C. d., Rosas-Alatorre, R., et al. (2013). Soil biodisinfection and use of *Pochonia chlamydosporia* in management of *Meloidogyne arenaria* on guava. *Nematropica, 43*, 327. [Abstract].

Viggiano, J. R., Freitas, L. G., & Lopes, E. A. (2014). Use of *Pochonia chlamydosporia* to control *Meloidogyne javanica* in cucumber. *Biological Control, 69*, 72–77.

Viggiano, J. R., Freitas, L. G., & Lopes, E. A. (2015). *Pochonia chlamydosporia* var. *chlamydosporia* (Goddard) Zare & W. Gams for the management of lettuce infected with *Meloidogyne javanica* (Treub, 1885). *Chilean Journal of Agricultural Research, 75*, 255–258.

Villoch, A., Montes de Oca, N., & Hidalgo-Díaz, L. (2003). Elaboración de una guía de buenas prácticas para la producción de biocontroles. *Revista de Protección Vegetal, 18*, 92–97.

Waage, J. K., & Greathead, D. J. (1988). Biological control: Challenges and opportunities. *Philosophical Transactions. Royal Society of London, 318*, 111–128.

Wesemael, W. M. L., Viaene, N., & Moens, M. (2011). Root-knot nematodes (*Meloidogyne* spp.) in Europe. *Nematology, 13*, 3–16.

Chapter 16
Management Strategies: *Pochonia chlamydosporia* and IPM of Plant-Parasitic Nematodes in Temperate Crops

Matthew A. Back, Danuta Sosnowska, and Ricardo Holgado

Abstract Few studies have reported findings on the use of *Pochonia chlamydosporia* for the management of plant-parasitic nematodes under field conditions. In this chapter we describe experiences of *P. chlamydosporia* application in temperate crops grown in the UK, Norway and Poland. To date, the fungus has been recovered from different endoparasitic nematodes from a range of locations across Europe. *Pochonia chlamydosporia* is an egg parasite as well as a saprophyte and plant endophyte and is primarily applied as a biological control agent to reduce nematode multiplication. In the UK, several field and micro-plot experiments have demonstrated that the fungus is capable of causing ca 50% reductions in the multiplication of *Globodera pallida* in potatoes. Further work was undertaken to evaluate the compatibility between *P. chlamydosporia* applications and the fungicide azoxystrobin which is used for managing the soil borne fungus *Rhizoctonia solani*. Although *P. chlamydosporia* is sensitive to azoxystrobin, there is evidence to suggest that it may not affect its efficacy as a biological control agent. In Norway, the fungus has been isolated from various cyst nematodes (*Heterodera* spp. and *Globodera* spp.), however, under *in vitro* conditions it was found to lose pathogenicity. Work undertaken in Poland has shown that strains of *P. chlamydosporia* can reduce populations of *H. schachtii* in sugar beet. Sugar beet grown in a 3 year rotation in combination

M.A. Back (✉)
Centre for Integrated Pest Management, Harper Adams University,
Newport, Shropshire TF10 8NB, UK
e-mail: mback@harper-adams.ac.uk

D. Sosnowska
Institute of Plant Protection – National Research Institute,
Wladyslawa Wegorka 20, Poznan, Poland
e-mail: d.sosnowska@iorpib.poznan.pl

R. Holgado
NIBIO, Norsk institutt for bioøkonomi /Norwegian Institute of Bioeconomy Research,
Pb 115, NO-1431, Ås, Akershus, Norway
e-mail: ricardo.holgado@nibio.no

R.H. Manzanilla-López, L.V. Lopez-Llorca (eds.), *Perspectives in Sustainable Nematode Management Through* Pochonia chlamydosporia *Applications for Root and Rhizosphere Health*, Sustainability in Plant and Crop Protection,
DOI 10.1007/978-3-319-59224-4_16

with a mustard green manure increased egg parasitism by *P. chlamydosporia* in comparison to other treatments which included the addition of straw or manure. Further work is discussed on the ability of strains of *P. chlamydosporia* to parasitize eggs of *Meloidogyne incognita*, *M. hapla* and *M. arenaria* at a range of temperatures.

16.1 Introduction

The nematophagous fungus, *Pochonia chlamydosporia* (Goddard) Zare and Gams, is known to reduce the multiplication of root-knot and cyst nematodes through direct parasitism of eggs and females. Its facultative nature means that it can be cultured *in vitro* on a range of substrates (see Chap. 11) where it produces thick walled chlamydospores. Chlamydospores enable the fungus to persist for long periods in the soil and tolerate environmental extremes (Chap. 3). The fact that *P. chlamydosporia* can be cultured relatively easily means that there is potential for field scale applications. Typically, the fungus is applied as chlamydospores on solid substrates such as straw or rice kernels (Davies et al. 1991). As indicated in Chap. 12, isolate selection is critical because certain isolates are more efficacious on specific nematode species. This view is supported by Kerry and Hirsch (2011) who highlight that isolates originating from egg masses of root-knot nematodes are less likely to parasitize females and eggs of cyst nematodes. Moreover, there are differences in the ability of isolates to colonise roots and produce chlamydospores (Viaene et al. 2013), a factor which again has implications for application.

Management of certain temperate plant-parasitic nematodes is challenging; varietal resistance is not always available for certain markets/produce and, more importantly, nematicide options are becoming increasingly limited. For example, the granular nematicide fosthiazate is presently (in 2016) the only option for treating potato cyst nematodes (*Globodera* spp.) in the UK. A greater pressure stems from tightening pesticide legislation such as Regulation (EC) No. 1107/2009 which governs the registration of all plant protection products in the EU. Regulation 1107/2009 is based upon hazard-based cut-off criteria, which means that active ingredients could be rejected during the evaluation process based upon their classification (Williams 2011). Consequently, there is an urgent need for non-chemical alternative crop protection strategies that provide a longer term solution to nematode pests and which have a limited impact on the environment.

Various reports have demonstrated the strong potential of *P. chlamydosporia* under glasshouse conditions, whilst a smaller number of studies have investigated field scale application (Stirling and Smith 1998; Verdejo-Lucas et al. 2003; Kerry et al. 2008; Tobin et al. 2008a; Giné et al. 2016). To date, *Pochonia chlamydosporia* has been isolated from a range of sedentary endoparasitic nematodes, including species belonging to the genera *Heterodera*, *Globodera*, *Meloidogyne*, *Nacobbus* and *Rotylenchulus* (Manzanilla-López et al. 2013). If the application of *P. chlamydosporia* is to be adopted by crop managers, it is essential that sufficient field based data is available to support methods of application, formulation, and long term nematode management. In this chapter, we will present and discuss findings of research undertaken on a field-scale in the UK, Poland, and Norway. Additionally, we will consider opportunities for integrating the use of *P. chlamydosporia* into integrated control strategies.

16.2 Experience with *Pochonia chlamydosporia* in the UK

16.2.1 Introduction

According to the Department for Environment, Food and Rural affairs (Defra 2015), the total Utilised Agricultural Area (UAA) in the UK, as of June 2015, was approximately 17.2 million hectares. A third of this area (6.3 million hectares) is occupied by crops. The main crops produced in the UK are cereals (50% of cropping area), temporary grass (22%), oilseeds (11%), horticultural crops (3%), potatoes (2%) and other crops e.g., peas, beans, and sugar beet (10%). Soil type and climatic conditions dictate where crops are grown within the country. For instance, potato production is largely concentrated in particular regions such as Norfolk, Lincolnshire, East Yorkshire, and Eastern Scotland.

Crops infected by sedentary endoparasitic nematode pests include potatoes (potato cyst nematodes; *Globodera pallida* and *G. rostochiensis*), cereals (cereal cyst nematodes; *Heterodera avenae* and *H. filipjevi*), sugar beet (beet cyst nematode; *Heterodera schachtii*), oilseed rape (*H. schachtii*), and peas (pea cyst nematode; *Heterodera goettingiana*). In addition, a number of root-knot nematodes (*Meloidogyne* spp.) are also present within the UK. For example, *Meloidogyne naasi* infects wheat crops, *M. hapla* infects various crops (CABI 2016 – see data sheet), *M. minor* has been recorded on sports turf and golf courses (Fera 2007), and a limited number of cases of *M. fallax* have also been recorded (CABI 2015).

Management of the aforementioned nematode species is largely dictated by their economic significance and prevalence within crop types. For example, potato cyst nematodes (PCN) are managed through an Integrated Pest Management (IPM) programme which employs legislative restrictions (PCN directive 2007/33/EC), extended rotations, resistant varieties, good hygiene (i.e., management of soil movement and volunteers), novel strategies such as biofumigation, and the application of nematicides at planting. Currently, there are no commercially available biological control agents for treating sedentary endoparasitic nematode pests in UK crops.

Whilst UK research on *P. chlamydosporia* has mainly focussed on its application to potato cyst nematodes (e.g., Crump and Flynn 1995; Jacobs et al. 2003; Kerry et al. 2008; Tobin 2008; Manzanilla-López et al. 2011), the fungus has been isolated from a range of cyst nematode species including *Heterodera avenae* (Crump and Kerry 1977), *H. filipjevi* (Mitchinson 2009), *H. carotae* (Crump and Kerry 1977), *H. trifolii* (Crump and Kerry 1977), and *H. schachtii* (Tribe 1979).

The following sections summarise research undertaken by Dr. John Tobin at Harper Adams University, Shropshire, UK during 2005–2008 on the application of *P. chlamydosporia* (isolate B1357 – originating from field soil collected in Cambridgeshire, UK) for the management of *G. pallida* in potatoes (Tobin 2008). The aims of this research were to: i) develop an effective method of mass production of *P. chlamydosporia*; ii) evaluate the efficacy of *P. chlamydosporia* against the potato cyst nematode *G. pallida* under field conditions; and iii) assess the factors governing the efficacy of *P. chlamydosporia* as a biological control agent (BCA).

16.2.2 Materials and Methods

16.2.2.1 Micro-plot Studies

Two micro-plot studies were conducted using 50 litre pots containing *G. pallida* infested field soil that were sunk into the ground. The first experiment was a 3-year study that was conducted to evaluate the effect of repeated applications of *P. chlamydosporia* on PCN population densities and potato (cv. Estima) development. Figure 16.1 provides the details of the treatments over the period of the experiment. A second experiment was conducted to compare different inoculum types including barley/compost, rice, and a chlamydospore suspension.

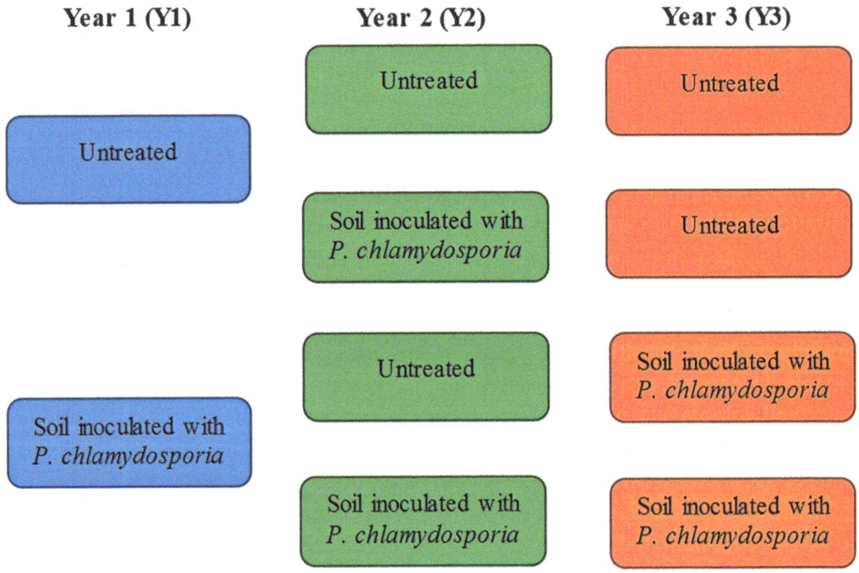

Fig. 16.1 Schematic representation of treatments used in a micro-plot experiment investigating the effect of repeated applications of *Pochonia chlamydosporia* on potato growth (cv. Estima) and *Globodera pallida* population dynamics

16.2.2.2 Field Experiments

Two field experiments were conducted in potato fields near Newport (Shropshire) during 2006 and 2007. The aim of the experiments was to assess the efficacy of *P. chlamydosporia* alone or in combination with the nematicide fosthiazate on *G. pallida* multiplication Population final/Population initial (*Pf*/*Pi*) and the invasion of potato roots. Additionally, assessments of potato growth and *P. chlamydosporia* root colonisation were also undertaken. For a greater description of the methodology used see Tobin et al. (2008a).

16.2.2.3 *In vitro* and Glasshouse Experiments

Azoxystrobin is a fungicide used in UK potato crops to reduce the incidence of soil borne pathogens of potato, including *Rhizoctonia solani* (black scurf), *Colletotrichum coccodes* (black dot) and *Helminthosporium solani* (silver scurf). The aim of these experiments was to determine the sensitivity of *P. chlamydosporia* to azoxystrobin under *in vitro* conditions and when applied at field rates under glasshouse conditions. These experiments were considered important because applications of *P. chlamydosporia* could coincide with soil treatment with azoxystrobin. The effect of azoxystrobin on the radial growth of *P. chlamydosporia* was examined in a Petri-dish experiment whereby the fungus was grown on Corn Meal Agar (CMA) amended with a range of concentrations (1–1000 μg/ml) of azoxystrobin. The radial growth of *P. chlamydosporia* colonies was monitored daily until a colony reached the edge of the Petri-dish. A further experiment assessed the effect of the same concentration range of azoxystrobin on *P. chlamydosporia* chlamydospore germination.

In order to determine the effect of azoxystrobin applied to soil, a microcosm experiment was conducted using 500 ml polypropylene containers with 200 g of sterilized loam. Treatments followed the same concentration range used in the *in vitro* experiments and chlamydospores of *P. chlamydosporia* were applied at ca 5000 chlamydospores/g substrate. The microcosms were incubated at 25 °C for 12, 20 and 40 days. After each time point chlamydospores were extracted from the microcosms and plated on CMA to determine the number of colony forming units (cfu).

A glasshouse experiment was conducted to assess the effect of azoxystrobin applied at 1, 3.2, 10 and 100 mg/kg soil on *P. chlamydosporia* colonisation of potato (cv. Estima) roots and persistence in soil. One gram samples of root and soil were collected and *P. chlamydosporia* was isolated using a semi-selective media at a 1:100 dilution (De Leij and Kerry 1991) to determine cfu.

A final experiment was conducted in the micro-plots that were briefly described in Sect. 16.2.2.1. The aim of this experiment was to determine if azoxystrobin application (1, 3 and 6 l/ha) suppressed *P. chlamydosporia* infection of *G. pallida* females by measuring *Pf*/*Pi* after a potato crop (cv. Estima). For further details of the methodology used in these experiments the reader is referred to Tobin et al. (2008b).

Table 16.1 Population dynamics of potato cyst nematodes (*Globodera pallida*) in micro-plots treated and untreated with *Pochonia chlamydosporia* in the second year of a 3 year micro-plot study conducted by Tobin (2008)

Treatment	Pi	Pf	Pf/Pi	% control
Untreated (Y1) + Untreated (Y2)	83.4	79.3	1.1	0
Untreated (Y1) + Treated (Y2)	80.5	40.8	0.6	50
Treated (Y1) + Untreated (Y2)	109.9	76.3	0.7	33
Treated (Y1) + Treated (Y2)	116.6	49.9	0.5	55
CV%	34.3	49.6	58.6	–
$SED^{(25\ d.f.)}$	16.6	14.3	0.21	–
P value	0.089	0.047	0.036	–

CV coefficient of variation, *d.f.* degrees of freedom

16.2.3 Results

16.2.3.1 Micro-plot Studies

In years 1 and 3 of Experiment 1 (Fig. 16.1), no significant differences were found between the *Pf/Pi* of *G. pallida* populations in *P. chlamydosporia* treated and untreated micro-plots. However, in year 2 (Fig. 16.1), soil inoculation with *P. chlamydosporia* was found to cause a reduction in the *Pf* and the *Pf/Pi* (see Table 16.1). Analysis of the complete 3 year dataset revealed that overall there was a significant difference in the *Pf* between *P. chlamydosporia* treated and untreated micro-plots ($P = 0.029$) and that there was a significant reduction in the *Pf* observed over the 3 years of the study. No significant differences were observed in potato growth or yield between the treatments in any of the years of assessment.

In Experiment 2, each inoculum preparation of *P. chlamydosporia* was found to cause a significant reduction ($P < 0.01$) in *Pf* and *Pf/Pi* when compared to the control, although there was no significant difference between the inoculum preparations.

16.2.3.2 Field Experiments

Two years of field experiments confirmed the findings obtained from the micro-plot experiments. The *Pf* and the *Pf/Pi* of *G. pallida* were significantly reduced following treatments of *P. chlamydosporia* applied at planting. The mean percentage reduction of *Pf* relative to the control was comparable to the previous results with reductions of 48 and 51% observed in 2006 and 2007 respectively. Interestingly, there were no significant differences between the *Pf* and the *Pf/Pi* of plots treated with *P. chlamydosporia*, the nematicide fosthiazate and *P. chlamydosporia* and fosthiazate in both 2006 and 2007. No significant differences were found in *G. pallida* fecundity (the number of eggs per cyst) between treatments in 2006 or 2007.

Analysis of *G. pallida* juveniles recovered from potato roots 49 DAP (days after planting) in 2006 indicated that there was a significant reduction ($P < 0.05$) in adult females in plots treated with *P. chlamydosporia* when compared to the control. However, this effect was not seen at 63 DAP or at either assessment time in the experiment conducted in 2007.

In both 2006 and 2007, total tuber yield of potatoes was not significantly affected by field application of *P. chlamydosporia* on its own. In contrast, fosthiazate consistently increased total tuber yield.

16.2.3.3 Experiments Examining the Compatibility of Azoxystrobin with *Pochonia chlamydosporia*

Increasing the dose of azoxystrobin had a significant ($P < 0.001$) effect on reducing radial growth and chlamydospore germination of *P. chlamydosporia*. The EC_{50} values for radial (hyphal) growth and chlamydospore germination were determined as 7 and 14 µg/ml respectively. The microcosm experiment confirmed these findings with significantly ($P < 0.001$) reduced numbers of *P. chlamydosporia* cfu found with increasing doses of azoxystrobin. The EC_{50} values were 1.25, 1, and 3 mg/kg at 12, 20, and 40 days after azoxystrobin application respectively.

The glasshouse experiment again showed the ability of azoxystrobin to reduce *P. chlamydosporia* in soil. A dose of 1 mg/kg was found to reduce *P. chlamydosporia* cfu in soil 35 days after planting. Similarly *P. chlamydosporia* cfu were significantly ($P > 0.05$) reduced on potato root tissue at this time. At 49 days after planting, a similar but less pronounced reduction was seen on *P. chlamydosporia* cfu recovered from soil and potato roots treated with azoxystrobin.

Azoxystrobin applied in the micro-plot experiment did not significantly affect the ability of *P. chlamydosporia* to reduce the multiplication rate *Pf/Pi* of *G. pallida*. Additionally, there was no significant difference between treatments in the number of *P. chlamydosporia* cfu recovered from the soil after the potatoes were harvested.

16.2.4 Discussion

Research conducted by Tobin (2008) and Tobin et al. (2008a, b) has provided a substantial amount of evidence to support the use of *P. chlamydosporia* for the management of PCN infestations on potatoes under temperate conditions. Clearly, reducing the number of females, and therefore the multiplication of the nematodes, is the main advantage of using *P. chlamydosporia* in this crop. Two years of field experiments, and to some extent the micro-plot study, have shown that applications of *P. chlamydosporia* can reduce the *Pf/Pi* of *G. pallida* in the region of 50%. It is well known that *P. chlamydosporia* is a parasite that infects the eggs of endoparasitic nematodes. Consequently, it is not particularly surprising that potato yield was not altered by soil treatment with *P. chlamydosporia* in the work conducted by Tobin (2008), because potato damage is most likely to have been inflicted earlier when second-stage juveniles (J2) invaded the roots and established syncytia. One might argue that further root damage is also caused when young females emerge by rupturing through the root cortex. However, assessment of *G. pallida* invasion of potato roots 49 and 63 DAP did not consistently show a reduction in young females (adults) in field plots treated with *P. chlamydosporia*. In the first field experiment (2006) there was a reduction in the number of adult females 49 DAP but this was not seen at 63 DAP. Kerry (1981) states that *P. chlamydosporia* is predominantly an egg parasite of PCN, although virgin females can become infected after they have emerged from the root cortex. In this respect, the preservation of potato yield through *P. chlamydosporia* application appears highly unlikely.

In terms of the IPM of PCN, *P. chlamydosporia* may have an important role to play through longer term management of the pest by suppressing multiplication. For instance, *P. chlamydosporia* application could be used in conjunction with extended rotations and/or biofumigant cover crops such as Indian mustard (*Brassica juncea*) (Ngala et al. 2015). In terms of preserving potato yield, the use of tolerant cultivars could be considered when using *P. chlamydosporia*. The field experiments conducted by Tobin et al. (2008a) demonstrated that nematicide (fosthiazate) applications improved the final yield of potatoes and reduced *Pf/Pi*. However, there was no additional benefit gained from the combination of *P. chlamydosporia* and fosthiazate applications. As well as suppressing PCN, the application of *P. chlamydosporia* may have plant growth promoting benefits on monocot and dicot crop species in the rotation. For example, Maciá-Vicente et al. (2009) observed endophytic colonisation of barley (*Hordeum vulgare* L. var. *distichum*) roots by *P. chlamydosporia*. Inoculation of barley with isolates Pc123 and Pc21 resulted in a 1.6- and 1.9-fold increase in fresh weight at 63 days after inoculation, respectively. Zavala-Gonzalez et al. (2015) recorded similar results with tomatoes (*Solanum lycopersicum* cv. Marglobe) where *P. chlamydosporia* inoculation was found to increase root growth, reduce flowering and fruiting times and increase fruit yield. Further investigations highlighted that *P. chlamydosporia* can produce Indole-3-acetic acid (IAA) and solubilise mineral phosphate, which help explain the plant growth promoting effects observed.

The work that Tobin et al. (2008b) conducted on azoxystrobin and *P. chlamydosporia* application provides useful information on the compatibility of these two crop protection inputs, particularly in the light of their shared application timing. *In vitro* experiments highlighted the sensitivity of hyphae and chlamydospores of *P. chlamydosporia* B1357 to azoxystrobin. This was confirmed in subsequent microcosm and glasshouse experiments which assessed *P. chlamydosporia* cfu in soil and potato roots treated with azoxystrobin. However, in both of these experiments it was clear that there was a decrease in the sensitivity of *P. chlamydosporia* to azoxystrobin over time. Tobin (2008) draws attention to the fact that the EC_{50} values recorded at 49 DAP the glasshouse experiment are greater than the equivalent field rate of azoxystrobin applied as Amistar® at 6 l/ha. Whilst *P. chlamydosporia* may be reduced by an application of azoxystrobin initially, the fungus may re-establish in time to infect PCN cysts emerging from the roots of potatoes. Coincidently, the final experiment (micro-plots) would appear to support this hypothesis, since *G. pallida Pf/Pi* was not significantly affected in plots inoculated with *P. chlamydosporia* and treated with azoxystrobin.

16.3 Experience with *Pochonia chlamydosporia* in Norway

16.3.1 Introduction

Agriculture in Norway is strongly influenced by natural conditions and geographic location and therefore relatively small areas are suitable for agricultural production. The Gulf Stream current makes it possible to farm even in the northernmost areas, but the conditions for agricultural production vary considerably between the different regions from south to north, and farming is also influenced by the length of the growing season.

The total area for farming is about 0.98 million ha. However, only 3% of the total land area is suitable for agriculture, 20% is productive forestland, and 70% is unproductive (Bye et al. 2015). From the total of 3% farming land, the major part is suitable for grass cultivation (67% used on silage and grassing), followed by cereals (used for forage 25%, and for food 6%), potato (1%) and vegetables (1%) (Bye et al. 2015).

During the last decade, the Norwegian government has been encouraging farmers to practise integrated pest management of pest and diseases. The Norwegian agricultural extension service, dispersed throughout the country, plays an important role in advising farmers to apply sound agricultural practices. For potato, vegetable and berry and fruits crops, pest and disease monitoring is recommended and application of pesticides is only undertaken after thresholds are exceeded or if they are needed according to established guidelines for use of pesticides.

No chemical fumigants or nematicides have been used for the management of plant-parasitic nematodes in crops such as potato or vegetables in Norway (Holgado and Magnusson 2010; Wesemael et al. 2014). Instead, Norwegian nematode management relies strongly on crop rotation using non-host crops alternating with resistant cultivars. Additionally, it is important that farmers are made aware of the current nematode management options, including phytosanitary measures and regulations (Holgado and Magnusson 2010). Sometimes crop rotations are not easy to implement due to the restricted acreage suitable for long rotations. Therefore, increasing attention has been placed on alternatives such as the use of natural enemies including predators, parasites and pathogens.

Earlier surveys from fields with the *Globodera rostochiensis* and *Heterodera avenae* complex have shown the presence of several nematode fungal antagonists (Holgado and Crump 2003). The presence of *Purpureocillium lilacinum* (= *Paecilomyces lilacinus*) and *P. chlamydosporia* nematode antagonistic fungi as natural enemies to plant-parasitic nematodes indicates that there is reason to suspect that plant-parasitic nematodes in Norwegian fields are living under various degrees of natural control. Unfortunately, for nematode management in Norway none of the nematode antagonistic fungi can be mass produced and stored for long periods of time, as our experience indicate that *P. lilacinum* (= *Paecilomyces lilacinus*) and *P. chlamydosporia* nematode antagonistic fungi isolated from Norwegian fields lost their nematode parasitical fitness after 2–3 generations in growth media and to retain their parasitic ability they must be raised continuously on living nematode cultures. Additionally, the successful use of these fungi as a BCA requires a greater understanding of the biology of both the nematode pest present in the field and the fungus.

16.3.2 Materials and Methods

The experiments described here were conducted in 2004 and 2014. Their purpose was to isolate parasitic fungi from females of *G. rostochiensis*, *Heterodera* spp., *H. avenae*, and *H. filipjevi*. Additionally, pathogenicity tests were performed in the course of the last experiment.

16.3.2.1 Occurrence of Microbial Nematode Antagonists

Previous studies investigating antagonistic fungi in fields infected with *Heterodera avenae*, *H. filipjevi* and *G. rostochiensis* have shown different levels of natural infestation by nematode antagonistic fungi. Nematophagous fungi *Pochonia chlamydosporia* and *P. lilacinus* (= *Purpureocillium lilacinum*) were isolated from cereal cyst nematodes and potato cyst nematodes (Holgado and Crump 2003; Holgado et al. 2010). Further studies were carried out in 2004 and 2014 to examine the occurrence of parasitic fungi and microbial antagonists such as *Pasteuria penetrans*.

16.3.2.2 Soil Samples from Cereals with *Heterodera* spp. and Potato Fields with *Globodera* spp.

Soil samples from eight cereal fields were collected in 2004. Using information from previous surveys on the *H. avenae* complex (Holgado et al. 2003) samples were collected from Vestfold, Østfold, Sør Trøndelag, and Nord Trøndelag counties. Similarly, using previous survey information but this time for *Globodera* (Holgado et al. 2011, 2012, 2013; Magnusson et al. 2011) a total of ten soil samples were collected during 2012–2013 from the counties of Vestfold, Nord Trøndelag, Rogaland, Vest Agder, and Østfold. In both studies, samples from each field were mixed to give a composite sample and a subsample of approximately 2 kg was taken per field and kept at 4 °C until used. The soil samples were air-dried and sieved (5 mm mesh sieve) and cysts were extracted from a 250 g subsample by means of a fluidising column (Trudgill et al. 1972). The number of *Heterodera* spp. and *Globodera* spp. cysts, viable eggs, and J2 were estimated by standard methods (Shepherd 1986).

16.3.2.3 Assessment of Fungal Parasites from Soil with *Heterodera* spp. and *Globodera rostochiensis*

After extraction of cysts belonging to *Heterodera* spp. and *Globodera rostochiensis*, eggs from each species were extracted by standard techniques (Shepherd 1986), taking care not to break the eggshells when crushing cysts.

Procedures to isolate the fungi were used as described by Kerry and Crump (1977): under sterile conditions, with the aid of a vacuum pump, egg suspensions were passed through a sintered glass filter (250 μm pore) to remove cyst wall fragments of and other debris. The eggs were collected on a second filter (20 μm pore), washed five times with sterile distilled water and made up to a known volume. Using a sterile pipette, a 0.5 ml aliquot was transferred into Petri-dishes (9 cm diam.) containing 0.8% distilled water agar (WA) and 100 ppm each of streptomycin sulphate, chloramphenicol, and chlortetracycline hydrochloride. The plates were then incubated at 19 °C for 24–48 h, infected eggs and J2 transferred on to 9 cm diam. Petri-dishes containing potato dextrose agar (PDA), and kept at room temperature (approx. 24 °C) for further fungal growth and examination.

16.3.2.4 Estimation of the Numbers of *Pochonia chlamydosporia* Colony Forming Units

Soil samples were passed through a 1 mm aperture sieve and mixed. A 2 gram subsample was taken and added to 18 ml of 0.05% water agar solution, and shaken for 15 seconds; then a 2 ml subsample was transferred into a similar volume of agar solution as part of a dilution series and this process was repeated for each dilution (Kerry 1991). Selective media for *Pochonia* isolation were prepared according to Kerry et al. (1993, see also Chaps. 11 and 12). From each 10^{-2} and 10^{-3} dilution, three Petri-dishes (9-cm diam.) containing a *P. chlamydosporia* selective medium were inoculated with 0.2 ml of the suspension under sterile conditions. The colony forming units (cfu) numbers were counted after incubation at 18 °C in the dark for 2 weeks (Kerry 1991).

The data obtained from the assessment of fungal parasites in eggs and from the numbers of cfu are only useful for comparative purposes and no statistical analyses were performed, as the main purpose of the work was the isolation of parasitic fungi from cyst nematodes (*G. rostochiensis*, *Heterodera* spp., *H. avenae*, and *H. filipjevi*).

16.3.2.5 Assessment of Fungal Parasites from Cysts of *Heterodera* spp. and *Globodera rostochiensis*

An observation chamber was used to study the infection of live females on a living root system. Cutting a section of a plastic Petri-dish (14 cm diam.) as shown in Fig. 16.2 formed the observation chamber (Crump 1987a). The soil samples were passed through a 5 mm mesh sieve before adding approximately 300 g to each chamber. Four growth chambers were filled with each test soil and fertilizer. Twenty cysts of each nematode species were inoculated into two chambers. For *G. rostochiensis*, each chamber was planted with two potato chits of the susceptible potato cultivar Beate. For *Heterodera* spp., three seeds of the susceptible barley cultivar Varde were planted. After 3 weeks females developing on the surface roots were examined weekly for 1 month (four occasions), under a microscope (× 20 magnification) with the lids removed. The infected females were removed for nematophagous fungi isolation.

Fig. 16.2 Observation chamber used for the assessment of fungal parasites from cysts of *Globodera rostochiensis* and *Heterodera* spp. infecting the roots of potato (*centre*) and barley (*right*), respectively (courtesy R. Holgado)

16.3.2.6 Fungal Isolation

The females that were considered to be infected by fungi were removed from the roots, placed onto moist filter paper in a Petri-dish, and left undisturbed for the fungus to emerge. The number of females that were infected with fungi were counted. Fungi that had sporulated, and could be identified as known nematophagous fungi, were isolated from infected females by aseptically transferring spores onto nutrient agar.

16.3.2.7 Assessment of *Pochonia chlamydosporia* Parasitism

To compare egg parasitism of *P. chlamydosporia*, isolates that were obtained in 2014 from *G. rostochiensis* eggs were used for further studies. The parasitism tests were carried out in a glasshouse using *P. chlamydosporia* (10 samples × 2 plates = 20 isolates in total). *Pochonia chlamydosporia* was grown on selective media (Kerry et al. 1993). Observation chambers were filled with sterile sand (ca 300 g) and inoculated with *P. chlamydosporia* (9 cm diam. Petri-dish) and 20 cysts of *G. rostochiensis* in nylon mesh bags. Two observation chambers per isolate were used (total 40 chambers). Two potato chits of the susceptible potato cultivar Beate were planted in each chamber.

After 50–60 days the lids of the observation chambers were removed for examination, new cysts were collected, ten new cysts were crushed in a drop of sterile water, and then eggs and a J2 suspension (ca 100 eggs and J2) were plated onto a Petri-dish containing selective medium (Kerry et al. 1993). The plates were incubated at room temperature (24 °C) and checked every second day. The 40 plates showed infection on eggs but the infection of eggs fluctuated from 39–95%.

Where eggs presented an infection rate of 95%, a further three *P. chlamydosporia* isolates were selected for additional tests. These selected isolates were plated onto Petri-dishes containing selective medium (Kerry et al. 1993) and incubated at room temperature 24 °C. Three replicates per isolate (nine plates) were used. To standardise the experiment, only colonies with 2 mm of growth from the central disk were chosen. *Globodera rostochiensis* cysts, produced in cultures reared in sterile sand, were crushed and ca 100 eggs and J2 placed on each growing *P. chlamydosporia* colony before the Petri-dishes were arranged randomly and incubated at room temperature (24 °C) in darkness. After 1 week, the eggs and J2 were inspected every second day using a light microscope (×20 and ×40 magnification) for a period of 3 weeks. Hyphae isolated from infected eggs were transferred to CMA, amended with 200 ppm streptomycin sulphate.

16.3.3 Results

16.3.3.1 Assessment of Fungal Egg Parasitism

From cysts collected from the 18 soil samples (8 from 2004 and 10 from 2014). Nematode eggs from *G. rostochiensis*, *Heterodera* sp., *H. avenae*, and *H. filipjevi*, were found to be parasitized by the fungus *P. chlamydosporia* (Fig. 16.3). No clear

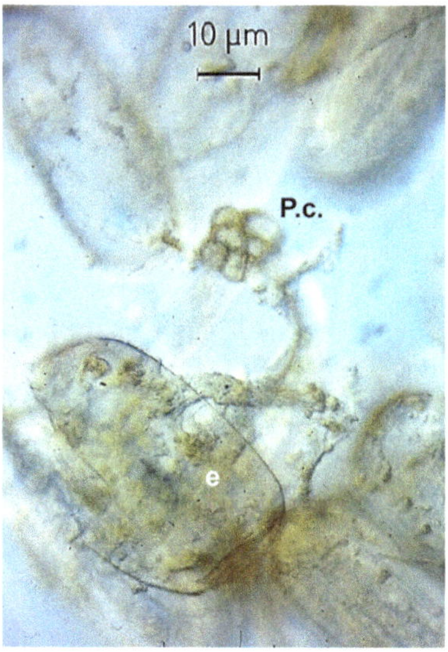

Fig. 16.3 Egg (e) of a cyst nematode infected with *Pochonia chlamydosporia* (P.c.)

differences were found between the ten soils infected with *G. rostochiensis* or the eight soil samples infected with *Heterodera* spp. populations. However, a difference in infection was observed between the counties in samples with *Heterodera* spp. cysts; Vestfold had the highest egg infection (25%) while populations from Sør Trøndelag had the lowest (5%).

16.3.3.2 Estimation of the Numbers of *P. chlamydosporia* Colony Forming Units from Soil Samples

Colony forming units of *P. chlamydosporia* were found in soils from seven fields infected with *H. avenae* and *Heterodera* spp., and eight fields infected with *Globodera* spp. A variation in the number of cfu were also found in both experiments; the majority of *P. chlamydosporia* cfu were recorded in populations from the county of Vestfold although no cfu were recorded in a soil sample from a field with *H. filipjevi* or two soil samples from fields with *G. rostochiensis*. Additionally, *P. chlamydosporia* was not detected in a sample from North Trøndelag county or another from Vest Agder county.

16.3.3.3 Assessment of Fungal Parasites from the Observation Chamber

Purpureocillium lilacinum was the only other known nematophagous fungus isolated from the 15 soils (7 soils from *Heterodera* spp. and 8 soils from *Globodera* spp.). Differences in the rate of female infection in the observation chamber plates were observed within isolates.

16.3.3.4 *Pasteuria penetrans* Infecting Second-stage Juveniles of *Globodera rostochiensis*, *Heterodera avenae*, and *Heterodera filipjevi*

In both experiments (2004 and 2014), identification of cyst nematodes species was done using a binocular Leica M 10 connected to a computer with the program Leica Q500MC 'Image Processing Analysis System'. When morphological and morphometric studies were performed on several J2 belonging to *G. rostochiensis*, *H. avenae*, and *H. filipjevi*, the bacterium *Pasteuria penetrans* was found attached to their cuticles (Fig. 16.4).

16.3.3.5 Potential for Using *Pochonia chlamydosporia* as a Biological Control Agent to Suppress Plant-Parasitic Nematodes

Among the fungal species found in our surveys, the fungus *P. chlamydosporia* is considered to be the most promising for application as a biological control agent against plant-parasitic nematodes (see Jafee 1992; Kerry 2000; Kerry and Hirsch 2011).

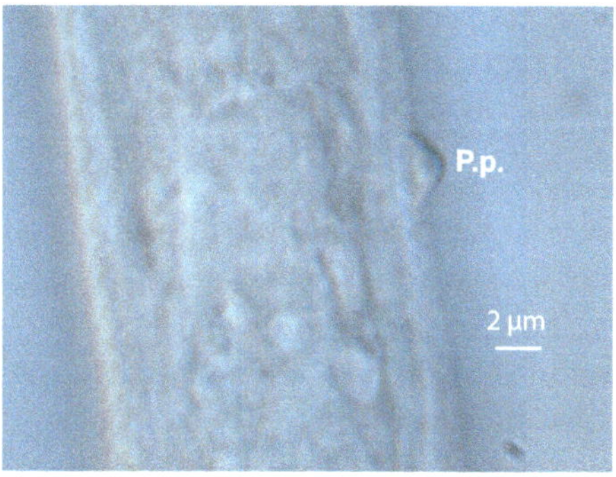

Fig. 16.4 *Globodera rostochiensis* second-stage juvenile with the bacterium *Pasteuria penetrans* (P.p.) attached to the cuticle (courtesy R. Holgado)

16.3.3.6 Assessment of *Pochonia chlamydosporia* Parasitism

Few colonies of *P. chlamydosporia* were isolated from the eggs of *Globodera* spp. after transfer to CMA. The effect of three different isolates did not show any parallel with the previous results. It appeared that *P. chlamydosporia* was not capable of continued parasitism of *Globodera* spp. eggs. On the other hand, *P. chlamydosporia* from previous isolates taken from the ten populations of *Globodera* spp. still grew on the selective media, thus indicating that they had become saprophytic.

16.3.4 Discussion

Pochonia chlamydosporia and *Purpureocillium lilacinum* have been reported as potentially effective BCAs of cereal cyst and potato cyst nematodes (Kerry and Crump 1980, 1998; Morgan-Jones et al. 1986; Rodriguez-Kabana and Morgan-Jones 1988). The rate of infection of cyst nematodes by *P. lilacinum* of native isolates showed the existence of natural control as *P. lilacinum* was isolated from soil and cyst nematodes in naturally infested Norway soils. *Purpureocillium lilacinum* is a typical soil borne fungus that has been reported from various parts of the world, in addition to its occurrence as a parasite of cyst and root-knot nematodes (Domsch et al. 1980; Kerry et al. 1982; Dackman and Nordbring-Hertz 1985; Crump 1987b). Estimating the rate of infection of cyst nematodes by *P. chlamydosporia* in the 18 samples also supported the existence of different levels of natural control on field infected by *Heterodera* spp. and *Globodera* spp. In our studies, *P. chlamydosporia* was found to be of common occurrence in *Heterodera* spp. and *Globodera* spp., in contrast to Denmark and Sweden where *Pochonia* [*Metapochonia*] *suchlasporia* was reported as a common species infecting *Heterodera* sp. with *P. chlamydosporia* more restricted to young cysts (Dackman and Nordbring-Hertz 1985; Gams 1988; Dackman et al. 1989).

It has been reported that *P. chlamydosporia* isolates differ in their virulence to cyst nematode species (Kerry 2000), their ability to colonise the root epidermis and cortex (Macia-Vicente et al. 2009), and chlamydospore production (Kerry and Hirsch 2001). These features are considered important for the development of these fungi as BCAs. It is well known that *P. chlamydosporia* is a facultative parasite. The strains that we isolated from cyst nematodes and grown on selective media, and which were re-isolated several times on selective media, apparently became attenuated or lost their parasitic capability. Carbendazim, one of the compounds added to selective media, may affect fungal mitosis and mutations. Further studies are needed concerning the facultative parasitic phase of *P. chlamydosporia*.

The results of our studies may help to provide an explanation for fluctuations in the parasitic capabilities of *P. chlamydosporia* observed by several authors (Irving and Kerry 1986; Kerry 1990, 2000). They indicate that isolates of *P. chlamydosporia*, even those collected from similar soils, differ significantly in their parasitic capabilities *in vitro* and in their efficacy in glasshouse tests (Irving and Kerry 1986; Kerry 1989, 1990, 2000; Kerry and Hirsch 2011).

There is reason to suspect that plant-parasitic nematodes in many Norwegian fields are under natural control due to parasitism of nematophagous fungi, although this has not yet been proven conclusively. It has been considered that, where nematode populations decline after several years of monoculture on susceptible hosts or where PPN populations are maintained at low levels in spite of favourable conditions, nematophagous fungi are most likely to be operating, parasitising females, eggs and J2 as reported for other countries in Europe. Natural control of cereal cyst nematodes has been reported to occur after a number of years with cereal cultivation in monoculture (Kerry et al. 1982; Kerry and Crump 1998). In contrast to these reports, damage caused by cereal cyst nematodes is increasing in fields with cereals as monoculture in Norway (Holgado et al. 2003, 2009).

It is necessary to continue extensive studies in Norway to select and determine the effectiveness of isolates for use in a wide variety of field conditions and horticultural crops and to test them in long-term field trials.

16.4 Experience with *Pochonia chlamydosporia* in Poland

16.4.1 Introduction

The area of agricultural land occupied by Polish farms is approximately 15 million ha, with the total sown area for harvest reaching about 10 million ha. Among the crops grown, wheat occupies about 16%, potato 13% and sugar beet 3% of the total hectarage.

Biological control methods are not used practically in field crops. Whilst biological control products are not registered, the Polish accession to the European Union has introduced a number of legislative acts. One of them is Directive 2009/128/EC of the European Parliament and the Council of 20 October 2009 establishing a framework for the Community action to achieve the sustainable use of pesticides. Article 14 of this Directive states that member states shall take all necessary measures to promote low pesticide input pest management, giving wherever possible priority to non-chemical methods. From 1 January 2014, professional users of pesticides should have adopted practices and products with the lowest risk to human health and the environment. Farmers have to use integrated pest control only.

The general principles of Integrated Plant Protection (IPP), which is part of Integrated Pest Management, include, among others: crop rotation, balanced fertilization and protection, and enhancement of beneficial organisms (for example nematophagous fungi).

The beet cyst nematode (*Heterodera schachtii*) is the most significant pest of sugar beet. Its direct and indirect harmfulness is not noticeable in every field in every year. In Poland, during the last century, this nematode has caused reductions of 50% in sugar beet root yield and 40% in sugar yield. Control of nematodes by chemicals, crop rotation, resistant cultivars and cultural management practices is impracticable, and biological control with nematophagous fungi is an alternative control method of this pest (Sosnowska 1996).

In Poland, the role of nematophagous fungi in natural conditions in fields where sugar beet is cultivated with modern agricultural technology (i.e., plant rotation and organic fertilizers) has been investigated. The current strategies used not only enable the reduction of population densities of *H. schachtii* but also stimulate the activity of fungal parasites of nematode eggs.

In Polish field crops, the root-knot nematode *Meloidogyne hapla* is known to have a very wide range of host plants. *Meloidogyne* species are the most economically important quarantine and non-quarantine pest of various vegetable crops, mostly under glasshouse conditions, where they cause significant yield losses. This nematode infects vegetable field crops in Poland including carrot, parsley, tomato, potato, peas, and sugar beet. The use of chemical nematicides and resistant cultivars in control of these nematodes has proved to be impractical. Biological control for the management of root-knot nematodes in vegetable crops is the most attractive from an environmental and economic standpoint. Therefore, Polish strains of the nematophagous fungus *P. chlamydosporia* were investigated for their potential to control these nematodes.

16.4.2 Materials and Methods

16.4.2.1 Field Experiments on Sugar Beet Infested with the Nematode *Heterodera schachtii*

Field experiments were carried out in the west of Poland at an Agriculture Experimental Station of the University of Mikolaj Kopernik, Koniczynka. Soil samples were taken from 100 m^2 plots that were replicated four times. Sugar beet was cultivated in different crop rotations for 2 and 3 years. In the 2-year rotation, sugar beet and spring barley were cultivated. In the 3-year rotation sugar beet, spring barley and winter wheat were cultivated. Different organic fertilizers were used, including pig manure applied at a rate of 40 tonnes/ha, Mustard cv. Salvo sown at a rate of 20 kg/ha, and straw fertiliser which consisted of winter wheat straw from a previous crop which was cut with a special machine and then fertilised with urea.

Soil samples (0.5 kg) to a depth of 25 cm (5 points per plot) were taken from fields cultivated with sugar beet and with spring barley as a post-crop after sugar beet. From each sample, 200 g of soil was weighed, washed with water on a sieve and filtered to collect cysts. Cysts were washed several times in sterile water and then transferred to 0.05% of streptomycin sulphate in sterile distilled water. Samples were incubated for 2–3 days at 22 °C and ten randomly chosen cysts from each sample (40 cysts) were crushed and fungal parasitism estimated using a microscope. To facilitate identification, some eggs were transferred to Petri-dishes with PDA to allow fungi to grow and sporulate. Analysis of variance with Tukey's multiple comparison test was used to determine differences.

16.4.2.2 The Potential of Polish Strains of *Pochonia chlamydosporia* as Biological Control Agents for Root-Knot Nematodes (*Meloidogyne* spp.)

Three Polish isolates of *P. chlamydosporia* (Vc-1, Vc-2, Vc-3) were collected from eggs of beet cyst nematode (*H. schachtii*). Chlamydospores were produced on a solid medium and extracted using standard techniques (see Chaps. 11 and 12). The chlamydospore suspension so obtained was mixed with fine sand (1:10, w:w) for storage at 4 °C. Each isolate of the fungus was washed in 2–5 ml of sterile distilled water and 0.2 ml of fungal suspension spread onto a Petri-dish (9 cm diameter) of 2% WA amended with antibiotics (streptomycin sulphate, chloramphenicol, chlortetracycline: 50 PPM). Petri-dishes were incubated at 25 °C for 2 days. Roots of plants infected with root-knot nematodes were washed. Several hundred egg masses were hand-picked and crushed using a cyst crusher and glass rod. About 200 root-knot nematode eggs were added to each plate and spread around the plate colonised by the fungus. The Petri-dishes were incubated at 25 °C for 3 days and then the number of parasitized eggs counted. Petri-dishes with eggs of root-knot nematode without fungus acted as controls.

16.4.3 Results and Discussion

16.4.3.1 Field Experiments on Sugar Beet Infested with the Nematode *Heterodera schachtii*

The data showed, that domestic strains of the nematophagous fungi *P. chlamydosporia* and *P. lilacinum* are effective bioagents, which can reduce the development of economically important plant-parasitic nematode species present in fields under

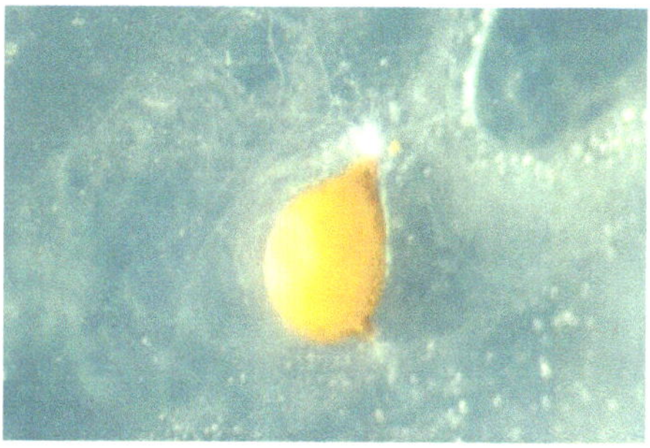

Fig. 16.5 A cyst of *Heterodera schachtii* infected with *Pochonia chlamydosporia* (courtesy D. Sosnowska)

Fig. 16.6 SEM of chlamydospores of *Pochonia chlamydosporia* – the fungus resting stage. (courtesy D. Sosnowska)

sugar beet cultivation. *Pochonia chlamydosporia* was the most frequently occurring parasite (Figs. 16.5 and 16.6).

When sugar beet was cultivated in the 2-year rotation, we observed fewer parasitized eggs inside *H. schachtii* cysts. The highest percentage of egg parasitism reached 52% in spring when sugar beet fields were fertilised with mustard green manure, and 55% in summer on the field fertilised with straw, and 45% in autumn on fields fertilised with mustard. When the fields were fertilised with pig manure we observed the lowest incidence of egg parasitism (35% in summer and 30% in autumn) (Fig. 16.7).

Observations were conducted on microplots where sugar beet plants were fertilized with manure, straw, and mustard on fields with 2- and 3-year rotations. The results showed that the highest parasitism of *H. schachtii* eggs by nematophagous fungi was observed on fields with a 3-year rotation. In spring and summer, parasitism of eggs inside the cysts of nematodes amounted to more than 80% on the fields fertilised with mustard green manure. On the fields fertilised with pig manure, egg parasitism reached 50% in spring and 75% in the summer. The lowest parasitism was observed in the autumn (Fig. 16.8).

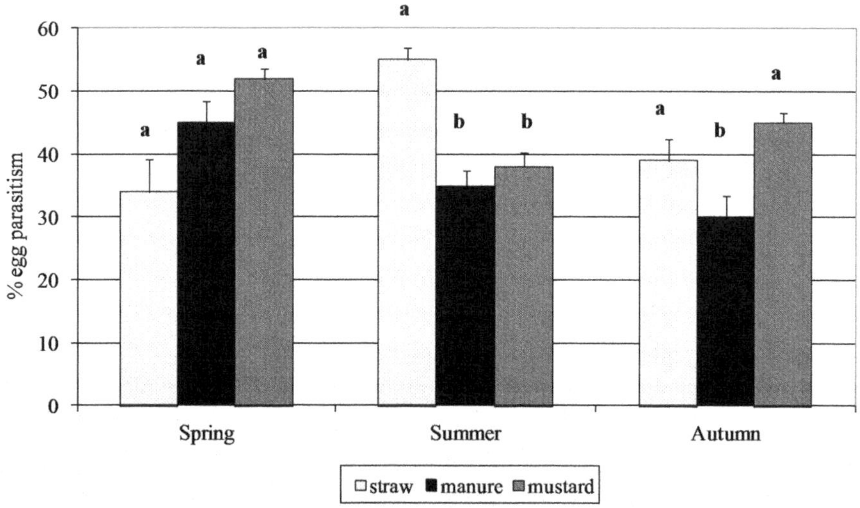

Fig. 16.7 Infection of *Heterodera schachtii* eggs with fungi from plots with sugar beet in a 2-year rotation (average for 1 year) (n = 4 replicates). Different letters above bars indicate significant differences between treatments ($P < 0.05$)

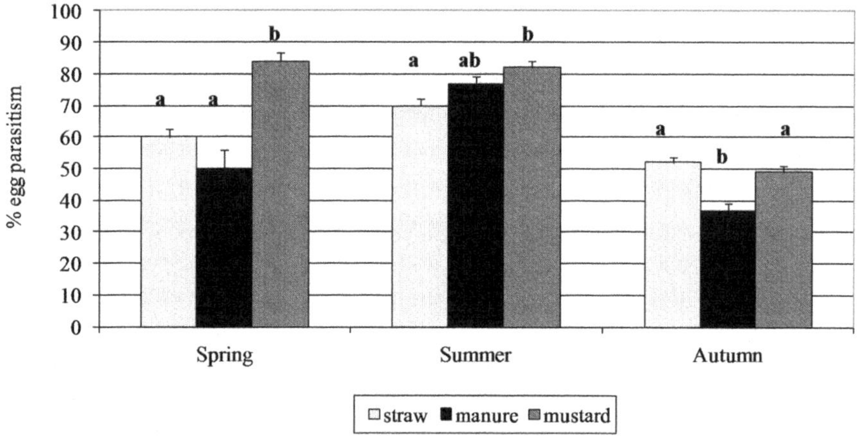

Fig. 16.8 Infection of *Heterodera schachtii* eggs with fungi from plots with sugar beet in a 3-year rotation (average for 1 year), (n = 4 replicates). Different letters above bars indicate significant differences between treatments ($P < 0.05$)

Sixteen species of fungi (Table 16.2) were identified from the eggs of the beet cyst nematode. A greater diversity of species was isolated from fields where the 3-year sugar beet rotation was used than from fields with the 2-year rotation. The nematophagous fungus *P. chlamydosporia* played the most important role in reduc-

Table 16.2 Species of fungi isolated from eggs and percent (%) of parasitized eggs inside cysts of *Heterodera schachtii* according to season

	% parasitized eggs inside cysts (40 cysts examined)		
Species of fungus	Spring	Summer	Autumn
Beauveria bassiana	0	0	2.5
Cladosporium sp.	2.5	0	0
Cylindrocarpon destructans (= *Ilyonectria destructans*)	10–40	0	0
Fusarium oxysporum	16.7	3	1–10
Fusarium solani	0	10	0
Fusarium sp.	15–45	25	25
Gliocladium roseum (= *Clonostachis rosea*)	3.2–15.8	34.8–35	7.6
Gliocladium sp.	15	2–5	24
Paecilomyces lilacinus (= *Purpureocillium lilacinum*)	1–13	0	20
Paecilomyces sp.	0	5	0
Pochonia chlamydosporia	30	30–60.5	15
Scopulariopsis sp.	0.2	28.3	0
Trichoderma sp.	0	0	10
Lecanicillium lecanii	0	3.5	0
Lecanicillium sp.	5	0	15

ing populations of *H. schachtii* and was found to infect more than 60% of nematode eggs in the samples. *Pochonia chlamydosporia* occurred in all of the fields and within each fertiliser treatment.

The results highlighted the possibility of using domestic strains of the nematophagous fungus *P. chlamydosporia* in controlling *H. schachtii* under field conditions. In the development of a bio-management strategy in integrated pest control on fields with sugar beet infected with *H. schachtii*, we can use organic fertilisers and plant rotations which can increase the efficacy of the parasitic fungi and can be important factors in reducing contamination of the soil environment. Several years of observations revealed a trend towards increased incidence of egg parasitism by fungi where the straw fertiliser was used. This occurred mainly on plots when sugar beet was cultivated in 3-year rotations. It is known that straw fertiliser provides economic benefits to farms. Straw has a positive effect on organic matter and nutrient balance in soil, microfauna and microflora development. It appears that straw properties favoured soil fungi development and can be useful for its efficacy in controlling nematode populations under field conditions. Such activity is an important component in the general principles of IPM because enhancement of important beneficial organisms includes nematophagous fungi (Sosnowska and Banaszak 2000).

16.4.3.2 Polish Strains of *Pochonia chlamydosporia* as Biological Control Agents for Root-Knot Nematodes

Pochonia chlamydosporia strains differ and show different virulence towards plant-parasitic nematodes. They are not effective against *Meloidogyne arenaria*, but they are most effective against *M. hapla* and *M. incognita* at 25 °C. At this temperature, the most effective against *M. hapla* was strain Vc-1 (67% eggs infected), and this was also true for *M. incognita* (77% eggs infected). At 15 °C, the most effective strain against eggs of *M. hapla* was also Vc-1 (50%), while at 20 °C strains Vc-1 and Vc-3 showed good effectiveness (45% and 49%) (Table 16.3). Growth and sporulation of strains are also dependent on temperature. The strains did not grow at temperatures of 5 °C and 10 °C. Domestic strains of *P. chlamydosporia* started to grow at 15 °C. All strains produced a similar number of chlamydospores at 15 °C and 25 °C. The differences between strains was found at 20 °C as Vc-1 and Vc-3 produced the greatest number of chlamydospores, while Vc-2 the fewest. All strains of *P. chlamydosporia* failed to grow at 30 °C and did not produce chlamydospores (Table 16.4). The optimal temperature for growth was 25 °C for all strains.

There were different criteria for strain selectivity based on sporulation intensity, colonisation and infestation of rhizosphere that allowed the selection of the most suitable strain for usage either in field conditions or in glasshouse production. The

Table 16.3 Effect of Polish strains of *Pochonia chlamydosporia* on egg parasitism of *Meloidogyne* spp. under laboratory conditions

	Percent (%) of infected eggs of *Meloidogyne* spp. according to temperature				
	M. arenaria	*M. hapla*			*M. incognita*
		% egg infection			
Strain	25 °C	15 °C	20 °C	25 °C	25 °C
Vc-1	$45^a \pm 7.8$	$50^a \pm 4.2$	$45^b \pm 3.2$	$67^a \pm 2.1$	$77^a \pm 3.1$
Vc-2	$50^a \pm 4.8$	$27^b \pm 4.3$	$30^a \pm 0.9$	$51^b \pm 1.7$	$68^b \pm 2.1$
Vc-3	$33^b \pm 6.9$	$37^c \pm 2.3$	$49^b \pm 5.0$	$56^b \pm 5.7$	$76^a \pm 4.2$

[abc]repetition of at least one letter inside column indicates lack of significant difference between isolates (Tukey's test, $P < 0.0001$), n = 3 replicates, ± standard error

Table 16.4 Effect of temperature on chlamydospore production by Polish strains of *Pochonia chlamydosporia*

	Number of chlamydospores × 10^4/colony					
Treatment	5 °C	10 °C	15 °C	20 °C	25 °C	30 °C
Vc-1	0	0	$37.3^a \pm 1.76$	$44.7^a \pm 5.21$	$35.3^a \pm 10.35$	0
Vc-2	0	0	$32.0^a \pm 7.02$	$21.3^b \pm 5.70$	$34.0^a \pm 8.08$	0
Vc-3	0	0	$38.7^a \pm 11.68$	$59.3^a \pm 7.86$	$21.3^a \pm 5.70$	0

[abc]repetition of at least one letter inside column indicates lack of significant difference between isolates (Tukey's test, $P < 0.05$), n = 3 replicates, ± standard error

results from this research appear to support the use of the domestic strain Vc-1 *P. chlamydosporia* as the best potential BCA for control of *Meloidogyne* species. This strain produced the highest number of chlamydospores at 20 °C and 25 °C, colonized 80% of the root surface without damage to the plants, and produced the highest efficacy in reducing the number of root-knot nematode eggs. It can be recommended for *M. hapla* control under field conditions because the preferred temperatures were lower than 15 °C, temperatures that are often found in Polish conditions.

16.5 Concluding Remarks

The variation within *P. chlamydosporia* indicates that careful strain selection is essential for the development of the fungus for biological control (Sosnowska et al. 2001). The results presented show that nematophagous fungi, including *P. chlamydosporia*, may play a significant role in the control of plant-parasitic nematodes in climatic conditions such as those found in Poland. However, they are only part of the strategy to solve the problem. There is still a need to conduct more research as some aspects, such as the method of investigating the impact of fungi on egg parasitism in egg masses of root-knot nematodes, the influence of manure on fungus sporulation, the composition of fungi in the nematode population, and identification of strains with the best efficacy. Criteria commonly applied for evaluation and selectivity of strains require more research as they do not always provide satisfactory information on strain behaviour, the results obtained under laboratory conditions not always correlating with those obtained under field conditions.

16.6 Future Perspectives

There is no doubt that *Pochonia chlamydosporia* has potential for the management of endoparasitic nematode pests of temperate field crops. In this chapter, we have provided observations and experimental work from Norway, Poland and the UK to support the application of *P. chlamydosporia* to cyst and root-knot nematodes. The feasibility of *P. chlamydosporia* being utilised on a field scale depends on the following factors:

- ***Availability of nematicides:*** If nematicides become unavailable due to withdrawal or restricted use as a result of EU pesticide legislation (Directive (EC) No.1107/2009) then crop managers will require alternative pest management strategies.
- ***Cost*:** A formulated product containing *P. chlamydosporia* would need to be no more expensive than the price of nematicides. Tobin (2008) calculated that the cost of one particular raw substrate (rice) alone would be in excess of €380/ton.

- ***Ease of use:*** Previous work has mainly focused on the use of chlamydospores on colonised substrates, which may require high field rates of up to 5 tonnes/ha. Researchers in Italy have also produced an oil emulsion formulation that contains chlamydospores, conidia and hyphae of *P. chlamydosporia* (Manzanilla-López et al. 2013). This type of formulation can be applied using trickle-tape irrigation systems.
- ***Compatibility with other crop protection inputs:*** In section 15.2.3.3 we discussed the compatibility of *P. chlamydosporia* with the fungicide azoxystrobin. Further research may be required to assess compatibility with other crop protection inputs.
- ***Selection of appropriate strains:*** See Chaps. 12 and 15 for more details

Management of plant-parasitic nematodes is best achieved using an IPM plan. *Pochonia chlamydosporia* has a role in the longer term management of these pests through its ability to reduce population densities in soil. In order to preserve yield it is important that crop managers and technical advisors consider the use of *P. chlamydosporia* alongside other compatible pest management strategies, such as resistant/tolerant cultivars and nematicide inputs.

References

Bye, S. A., Aarstad, P. A., Løvberget, A., et al. (2015). Jordbruk og miljø Tilstand og utvikling 2014, Statistisk sentralbyrå, Statistics Norway, Rapporter 2015/18, Oslo-Kongsvinger, pp. 1–146.

CABI. (2015). *Meloidogyne fallax* (false Columbia root-knot nematode). [Online] Available at: http://www.cabi.org/isc/datasheet/33241 . Accessed 30 Mar 2016.

CABI. (2016). *Meloidogyne hapla* (root knot nematode). [Online] Available at: http://www.cabi.org/isc/datasheet/33244. Accessed 30 Mar 2016.

Crump, D. H. (1987a). A method for assessing the natural control of cyst nematode populations. *Nematologica, 33*, 232–243.

Crump, D. H. (1987b). Effect of time of sampling method of isolation and age of nematode on the species of fungi isolated from females of *Heterodera schachtii* and *H. avenae*. *Revue Nématol, 10*, 369–373.

Crump, D. H., & Flynn, C. A. (1995). Isolation and screening of fungi for the biological control of potato cyst nematodes. *Nematologica, 41*, 628–638.

Crump, D. H., & Kerry, B. R. (1977). Maturation of females of the cereal cyst-nematode on oat roots and infection by an Entomophthora-like fungus in observation chambers. *Nematologica, 23*, 398–402.

Dackman, C., & Nordbring-Hertz, B. (1985). Fungal parasites of the cereal cyst nematode *Heteroderaavenae* in southern Sweden. *Journal of Nematology, 17*, 50–55.

Dackman, C., Chet, I., & Nordbring-Hertz, B. (1989). Fungal parasitism of the cyst nematode *Heterodera schachtii*: Infection and enzymatic activity. *FEMS Microbiology Ecology, 62*, 201–208.

Davies, K. G., De Leij, F. A. A. M., & Kerry, B. R. (1991). Microbial agents for the biological control of plant parasitic nematodes in tropical agriculture. *Tropical Pest Management, 37*, 303–320.

Defra. (2015). *Agriculture in the United Kingdom 2014*. [Online] Available at: https://www.gov.uk/government/statistics/agriculture-in-the-unitedkingdom-2014 . Accessed 30 Mar 2016.

Domsch, K. H., Gams, W., & Anderson, T.-H. (1980). *Compendium of soil fungi* (pp. 525–536). London: Academic.
Fera. (2007). *Meloidogyne minor*. [Online] Available at: https://secure.fera.defra.gov.uk/phiw/riskRegister/plant-health/documents/meloidogyneminor.pdf. Accessed 30 Mar 2016.
Gams, W. (1988). A contribution to the knowledge of nematophagous species of *Verticillium*. *Netherlands Journal of Plant Pathology, 94*, 123–148.
Giné, A., Carrasquilla, M., Martínez-Alsonso, et al. (2016). Characterization of soil suppressiveness to root-knot nematodes in organic horticulture in plastic greenhouse. *Front Plant Science*. doi:10.3398/fpls.2016.00164.
Holgado, R., & Crump, D. H. (2003). First record on the occurrence of nematophagous fungi parasitizing cyst nematodes in Norway. *International Journal of Nematology, 13*, 65–71.
Holgado, R., & Magnusson, C. (2010). Management of potato cyst nematodes (*Globodera* spp.) populations under Norwegian conditions. In P. P. J. Haydock (Ed.), *Aspects of applied biology, 3rd symposium on potato cyst nematodes 103* (pp. 87–92). Warwick: Association of Applied Biologists.
Holgado, R., Støen, M., Magnusson, C., et al. (2003). The occurrence and hosts of cereal cyst nematodes (*Heterodera* spp.) in Norway. *International Journal of Nematology, 13*, 1–19.
Holgado, R., Andersson, S., Rowe, J., et al. (2009). Management strategies for cereal cyst nematodes *Heterodera* spp., in Norway. In I. T. Ryley, J. M. Nicol, & A. A. Dababat (Eds.), *Cereal cyst nematodes: Status, research and outlook* (pp. 154–159). Ankara: CIMMYT.
Holgado, R., Niere, B., Forbord, J. O., et al. (2010). *Resultater fra pilotprosjekt om potetcystenematoder Bioforsk FOKUS, 5*, 148–149.
Holgado, R., Magnusson, C., Rasmussen, I., et al. (2011). Kartleggingsprogrammet for Potetcystenematoder (*Globodera rostochiensis* og *G. pallida*) i 2010. *Rapport, 6*(143), 25.
Holgado, R., Magnusson, C., Rasmussen, I., et al. (2012). Kartleggingsprogrammet for Potetcystenematoder (*Globodera rostochiensis* og *G. pallida*) i 2011. *Bioforsk Rapport, 7*(53), 12.
Holgado, R., Magnusson, C., Rasmussen, I., et al. (2013). Kartleggingsprogrammet for Potetcystenematoder (*Globodera rostochiensis* og *G. pallida*) i 2012. *Bioforsk Rapport, 8*(63), 30.
Irving, F., & Kerry, B. R. (1986). Variation between strains of the nematophagous fungus *Verticillium chlamydosporium* Goddard II. Factors affecting parasitism of cyst nematode eggs. *Nematologica, 32*, 474–485.
Jacobs, H., Gray, S. N., & Crump, D. H. (2003). Interactions between nematophagous fungi and consequences for their potential as biological agents for the control of potato cyst nematodes. *Mycological Research, 107*, 47–56.
Jaffee, B. A. (1992). Population biology and biological control of nematodes. *Canadian Journal of Microbiology, 38*, 359–364.
Kerry, B. R. (1981). Fungal parasites: A weapon against cyst nematodes. *Journal of Nematology, 12*, 253–259.
Kerry, B. R. (1989). Fungi as biological control agents for plant parasitic nematodes. In J. M. Whipps & R. D. Lumsden (Eds.), *Biotechnology of Fungi for improving plant growth* (pp. 153–170). Cambridge: Cambridge University Press.
Kerry, B. R. (1990). An assessment of progress toward microbial control of plant parasitic nematodes. *Journal of Nematology, 22*(Suppl), 621–631.
Kerry, B. R. (1991). Methods for studying the growth and survival of the nematophagous fungus, *Verticillium chlamydosporium* Goddard, in soil. In B. R. Kerry, D. H. Crump (Eds.). *Methods for studying nematophagous fungi*. IOBC/WPRS Bulletin XIV/2, pp. 34–38.
Kerry, B. R. (2000). Rhizosphere interactions and the exploitation of microbial agents for the biological control of plant parasitic nematodes. *Annual Review of Phytopathology, 38*, 423–441.
Kerry, B. R., & Crump, D. H. (1977). Observations on fungal parasites of females and eggs of the cereal cyst nematodes, *Heterodera avenae* and other cyst nematodes. *Nematologica, 23*, 193–201.

Kerry, B. R., & Crump, D. H. (1980). Two fungi parasitic on females of cyst nematodes (*Heterodera* spp.) *Transactions of the British Mycological Society, 74*, 119–125.
Kerry, B. R., & Crump, D. H. (1998). The dynamics of the decline of the cereal cyst nematode, *Heterodera avenae*, in four soils under intensive cereal production. *Fundamental and Applied Nematology, 21*, 617–625.
Kerry, B. R., & Hirsch, P. R. (2011). Ecology of *Pochonia chlamydosporia* in the rhizosphere at the population, whole organisms and molecular scales. In K. Davies & Y. Spiegel (Eds.), *Biological control of plant-parasitic nematodes: Building coherence between microbial Ecology and molecular mechanisms, Progress in Biological Control* (Vol. 11, pp. 71–182). Dordrecht: Springer.
Kerry, B. R., Crump, D. H., & Mullen, L. A. (1982). Natural control of the cereal cyst nematode *Heterodera avenae* Woll by soil fungi at three sites. *Crop Protection, 1*, 99–109.
Kerry, B. R., Kirkwood, I. A., De Leij, F. A. A. M., et al. (1993). Growth and survival of *Verticillium chlamydosporium* Goddard, a parasite of nematodes in soil. *Biocontrol Science and Technology, 3*, 355–365.
Kerry, B. R., Davies, K. G., Esteves, .I (2008). Managing potato cyst nematode through maximising natural decline and population suppression. DEFRA LINK Project LK0966 – Final Project Report, 71pp.
Maciá-Vicente, J. G., Rosso, L. C., Ciancio, A., et al. (2009). Colonisation of barley roots by endophytic *Fusarium equiseti* and *Pochonia* chlamydosporia: Effects on plant growth and disease. *The Annals of Applied Biology, 155*, 391–401.
Magnusson, C., Holgado, R., Rasmussen, I., et al. (2011). Kartleggingsprogrammet for potetcystenematoder (*Globodera rostochiensis* og *G. pallida*) i 2009. *Bioforsk Rapport, 6*(159), 25.
Manzanilla-López, R. H., Esteves, I., Powers, S. J., et al. (2011). Effects of crop plants on abundance of *Pochonia chlamydosporia* and other fungal parasites of root-knot and potato cyst nematodes. *The Annals of Applied Biology, 159*, 118–129.
Manzanilla-López, R. H., Esteves, I., Finetti-Sialer, M. M., et al. (2013). *Pochonia chlamydosporia*: Advances and challenges to improve its performance as a biological control agent of sedentary endo-parasitic nematodes. *Journal of Nematology, 45*, 1–7.
Mitchinson, S. M. (2009). *New cyst nematode threats to cereals in Southern Britain*. PhD Thesis, Reading University, UK, p. 272.
Morgan-Jones, G., Rodriguez-Kabana, R., & Jatala, P. (1986). Fungi associated with cysts of potato cyst nematodes in Peru. *Nematropica, 16*, 21–31.
Ngala, B. M., Haydock, P. P. J., Woods, S., et al. (2015). Biofumigation with *Brassica juncea*, *Raphanus sativus* and *Eruca sativa* for the management of field populations of the potato cyst nematode *Globodera pallida*. *Pest Management Science, 71*, 759–769.
Rodriguez-Kabana, R., & Morgan Jones, G. (1988). Potential for nematode control by microfloras endemic in the tropics. *Journal of Nematology, 20*, 191–203.
Shepherd, A. M. (1986). Extraction and estimation of cyst nematodes. In J. F. Southey (Ed.), *HMSO laboratory methods for work with plant and soil nematodes, Reference Book 402* (pp. 31–49). London: Ministry of Agriculture, Fisheries and Food.
Sosnowska, D. (1996). Fungi occurring on sugar beet nematode (*Heterodera schachtii* Schmidt) in Wielkopolska region. In P. H. Smits (Ed.). *Insect pathogens and insect parasitic nematodes* (IOBC WPRS Bulletin 19, pp. 204–207).
Sosnowska, D., & Banaszak, H. (2000). The effect of organic fertilizers on fungi parasitization of beet cyst nematode (*Heterodera schachtii* Schmidt) eggs in sugar beet cultivated in a three years rotation. *Journal of Plant Protection Research, 40*, 73–79.
Sosnowska, D., Mauchline, T. H., Bourne, J. M., et al. (2001). Screening of Polish, other European and tropical isolates of *Verticillium chlamydosporium* to assess their potential as biological control agents of root-knot nematodes. *Journal of Plant Protection Research, 41*, 77–84.
Stirling, G. R., & Smith, L. R. (1998). Field tests of formulated products containing either *Verticillium chlamydosporium* or *Arthrobotrys dactyloides* for biological control of root-knot nematodes. *Biological Control, 11*, 231–239.

Tobin, J. D. (2008). The biological control of potato cyst nematodes (*Globodera pallida* and *G. rostochiensis*) using the nematophagous fungus *Pochonia chlamydosporia*. PhD Thesis, Harper Adams University, UK.

Tobin, J. D., Haydock, P. P. J., Hare, M. C., et al. (2008a). Effect of the fungus *Pochonia chlamydosporia* and fosthiazate on the multiplication rate of potato cyst nematodes (*Globodera pallida* and *G. rostochiensis*) in potato crops grown under UK field conditions. *Biological Control, 46*, 194–201.

Tobin, J. D., Haydock, P. P. J., Hare, M. C., et al. (2008b). The compatibility of the fungicide azoxystrobin with *Pochonia chlamydosporia*, a biological control agent for potato cyst nematodes (*Globodera* spp.) *The Annals of Applied Biology, 152*, 301–305.

Tribe, H. T. (1979). Extent of disease in populations of *Heterodera*, with especial reference to *H. schachtii*. *The Annals of Applied Biology, 92*, 61–72.

Trudgill, D. L., Evans, K., & Faulkner, G. (1972). A fluidising column for extracting nematodes from soil. *Nematologica, 18*, 469–475.

Verdejo-Lucas, S., Sorribas, F. J., Ornat, C., et al. (2003). Evaluating *Pochonia chlamydosporia* in a double-cropping system of lettuce and tomato in plastic houses infected with *Meloidogyne javanica*. *Plant Pathology, 52*, 521–528.

Viaene, N., Coyne, D. L., & Davies, K. G. (2013). Biological and cultural management. In R. N. Perry & M. Moens (Eds.), *Plant Nematology* (2nd ed., pp. 346–369). Wallingford: CAB International.

Wesemael, W. M. L., Anthoine, G., Griffin, D., et al. (2014). Quarantine nematodes in potato: Practical solutions using molecular tools. *Potato Research, 57*, 365–366.

Williams, J. C. (2011). New EU pesticide legislation – The view of a manufacturer. *Aspects of Applied Biology – Crop Protection in Southern Britain, 106*, 269–274.

Zavala-Gonzalez, E. A., Escudero, N., Lopez-Moya, F., et al. (2015). Some isolates of the nematophagous fungus *Pochonia chlamydosporia* promote root growth and reduce flowering time of tomato. *The Annals of Applied Biology, 166*, 472–483. doi:10.1111/aab.12199.

Chapter 17
Pochonia chlamydosporia: A Promising Biotechnological Tool Against Parasitic Nematodes and Geohelminths

Pedro Mendoza de Gives and Fábio Ribeiro Braga

Abstract A number of parasitic diseases seriously affect animal/crop production worldwide. Traditionally, the use of chemical anthelmintic drugs or pesticides is considered the commonest method of controlling such parasites. However, this has triggered an imminent risk to public health due to the continuous exposure of people to those chemical products. In addition, parasites, after being in continuous contact with the pesticide or anthelmintic chemical molecules for long periods, eventually develop resistance, surviving dosages that used to kill them. During the last decades, the use of environmentally friendly alternatives has been extensively investigated in the search for healthier animal or plant products for human consumption. Likewise, the use of beneficial microorganisms as natural control agents is gaining a very good reputation over chemical anthelmintic drugs for cattle and sheep or for pests and plant pathogens affecting important crops worldwide. Nematophagous fungi are natural nematode antagonists that offer very good hopes for the control of animal and plant-parasitic nematodes. In particular, the fungus *Pochonia chlamydosporia* has shown an enormous potential to control a number of genera/species of plant-parasitic nematodes of economic importance. The present chapter provides a general view of the potential use of this promising biotechnological tool against plant- and animal-parasitic nematodes.

P. Mendoza de Gives (✉)
National Centre for Disciplinary Research in Veterinary Parasitology (CENID-PAVET-INIFAP-MEXICO), Jiutepec, MOR, Mexico
e-mail: pedromdgives@yahoo.com

F.R. Braga
Department of Veterinary, Federal University of Viçosa,
University Campus s/n, P.O. Box , 3657000, Viçosa, MG, Brazil
e-mail: fabioribeirobraga@hotmail.com

R.H. Manzanilla-López, L.V. Lopez-Llorca (eds.), *Perspectives in Sustainable Nematode Management Through* Pochonia chlamydosporia *Applications for Root and Rhizosphere Health*, Sustainability in Plant and Crop Protection,
DOI 10.1007/978-3-319-59224-4_17

17.1 Introduction

The soil biota is a micro-world of organisms of different taxa and species that includes bacteria, fungi, protozoa, nematodes, earthworms, and arthropods, all sharing the same microenvironment (Bhattarai et al. 2015). Due to their close association in the soil they have developed a number of biological associations. Microorganisms living in soil need to be nourished in order to perform all their biological activities and so there are endless biological struggles taking place in order to survive. Within the whole soil microbiota, nematodes and micro-fungi occupy a very important place due not only due to their enormous populations but also to their important ecological roles in the rhizosphere. Nematodes in soil play an important role in bio-degrading organic matter and in trophic chains. Free-living nematodes, by feeding on other organisms – including plants, fungi, bacteria, micro-arthropods, and other nematodes – participate in nitrogen re-cycling in nature (Ferris 2010). On the other hand, a group of micro-fungi called 'nematophagous fungi' also play an important role as one of the main bio-control agents of gastrointestinal parasitic nematodes in domestic animals (Liu et al. 2015) – they form trapping devices to capture nematodes and feed on them (Mendoza de Gives 2011). Nematophagous fungi are also able to produce substances or metabolites with nematicidal activity (Li et al. 2007; see also Chap. 7). Other type of nematophagous fungi, i.e., *Pochonia chlamydosporia*, envelop nematode eggs with hyphae and develop appressoria that penetrate the egg shell and then invade the egg, eventually feeding on the embryo inside (Trifonova and Karadjova 2003). The present chapter presents a general view of one of the most important species of nematophagous fungi, *P. chlamydosporia*, and its role in the control of plant- and animal-parasitic nematodes and also presents new biological activities attributed to this important microorganism.

17.2 Nematodes and the Soil Environment

Nematodes are represented in the soil environment by large populations of individuals of different genera/species living either as free-living organisms or as parasites of a wide range of hosts; including plants, animals and human beings (Mendoza de Gives and Alatorre 2011). Free-living nematodes have an important role in recycling nitrogen in soil and other bio-materials i.e., decaying plant material and animal faeces. Plant-parasitic nematodes (PPN) are adapted to a parasitic life into the plant tissues. They cause diseases and are also able to introduce other plant pathogens into the plant. As a consequence, enormous economic losses in yield are recorded every year in major world crops (Singh and Kumar 2013; Mehl and Kleczeuski 2015). Similarly, other groups of nematodes are adapted to live as parasites of animals where they complete part of their life cycle, causing varying levels of disease that diminish or compromise the health and life of animals and which may result in important economic losses in the livestock industry (Qamar et al. 2015).

17.3 Control Measurements Against Important Nematodes in Agriculture and the Livestock Industry

Different strategies against parasites of plants or animals have been proposed mainly through crop rotation or management of grazing animals. However, the commonest method to control nematode parasites of plants or animals is mainly based on the use of chemical drugs or pesticides, which help to reduce the damage caused. However, there are many problems in the use of this control measure, including the imminent threat of resistance development in the parasites and the potential risk posed by chemical residues to public health as they can remain in products and by-products of plant and animal origin destined for human consumption (Cooper et al. 2012). Additionally, some chemically active molecules are eliminated in the faeces, thereby contaminating the soil and affecting non-target, possibly beneficial, organisms (Horvat et al. 2012).

17.4 Nematophagous Fungi

Nematophagous fungi are a group of soil micro-fungi that are considered as one of the main natural enemies of nematodes in soil. They are saprophytes but they can easily become either parasites or predators of nematodes in soil (Mendoza de Gives 2011). There are different groups of nematophagous fungi, these being grouped together depending on their mode of action against nematodes. For example, one group of fungi are able to produce trapping devices from their mycelia, such structures being designed to capture, penetrate and kill nematodes in soil and eventually to feed on the remains. This group is referred to as the 'nematode-trapping fungi' (Casas-Flores and Herrera-Estrella 2015). Another group possesses small spores that are ingested by nematodes and germinate in the nematode intestine from where they destroy the host. This group is referred to as the 'endozoic fungi' (Barron 1977). Within this group there are fungi that possess adhesive spores which adhere to the nematode cuticle before germinating, penetrating the host body and consuming the contents. A third group of nematophagous fungi includes those that develop structures which target the nematode egg shell, dissolving their way through and finally feeding on the nematode embryo inside. This group of fungi are called the 'nematode egg-parasitic fungi' (Mehl and Kleczeuski 2015).

17.5 *Pochonia chlamydosporia* (Goddard) Zare and W. Gams (= *Verticillium chlamydosporium*)

Pochonia chlamydosporia is a nematophagous fungus belonging to the family *Clavicipitaceae* (*Ascomycota*: *Pezizomycotina*: *Sordariomycetes*: *Hypocreales*) (Sheng et al. 2015) (see Chap. 2 for systematics). This species has been classified as a saprophytic organism but is also a facultative parasite of nematode eggs, mainly of PPN eggs; i.e., the root-knot nematode *Meloidogyne* spp., the false root-knot nematode *Nacobbus* spp., and the cyst nematodes i.e., *Heterodera* spp. and *Globodera* spp. (Esteves et al. 2009). A number of research works have been published on the potential use of *P. chlamydosporia* in the control of root-knot nematodes (RKN).

17.6 Experiences with *Pochonia chlamydosporia* in the Control of Plant-Parasitic Nematodes

Many papers in the literature have reported upon interesting experiences in many countries on the use of *P. chlamydosporia* in the control of nematodes affecting different crops (see Chaps. 14 and 15). The parasitic activity of a number of *P. chlamydosporia* isolates has been demonstrated in *Nacobbus aberrans sensu lato* where 70–80% egg infection can occur (Franco-Navarro et al. 2008; see also Chap. 15). In glasshouse trials, *P. chlamydosporia* has been shown to significantly reduce the numbers of RKN egg masses and nematode eggs on tomato roots by almost 50% (Yang et al. 2012). Recent studies have shown that, in addition to their parasitic activity against nematode eggs, some isolates of *P. chlamydosporia* are root endophytes, promoting growth in some plants (see Chap. 14), for example, by increasing the number of secondary roots, the total weight of tomato seedlings and the mature fruit weight in tomato plants (Zavala-Gonzalez et al. 2015). Additionally, new fungal polyketide metabolites, identified as Benzenediol lactones possessing a macrolide core structure fused into a resorcinol aromatic ring, are being identified as produced by *Pochonia* sp. and other genera of fungi including *Aigialos*, *Cochliobolus*, *Curvularia*, *Fusarium*, *Humicola*, *Lasiodiplodia*, and *Penicillium*. These metabolites are reported to have various biological activities i.e., cytotoxicity, nematicidal properties, inhibition of various kinases, receptor agonists, anti-inflammatory activities, heat shock response, and immune system modulatory activities, etc. (Shen et al. 2015) – even an anti-cancer activity has being proposed (Wang et al. 2013) (See Chap. 7 on secondary metabolites).

17.7 Animal-Parasitic Nematodes and Other Geohelminths

Geohelminths or soil transmitted worms, including some animal-parasitic nematodes, are cosmopolitan parasites that depend on multiple factors to exist in the soil environment in order to complete part of their life cycle (Araújo et al. 2009). Soil containing infective geohelminth eggs is the main source of infection in both human and animal gastrointestinal helminthiasis. Studies on the natural processes involved in the destruction of geohelminth eggs are still in their initial stages, but nematophagous fungi represent an alternative strategy to be used, in combination with other prophylactic measures such as public education concerning zoonoses, to control

Table 17.1 List of geohelminth species that have been screened under *in vitro* conditions for an ovicidal effect by *Pochonia chlamydoporia*. Geohelminth eggs differ in size, egg shell thickness and embryo features

Species	Egg length (μm)	Egg width (μm)	Egg shell features	References
Ascaris lumbricoides	45–75	35–60	Mammillated. Four layers thick (3–4 μm)	Lysek, Sterba (1991) and Braga et al. (2007)
Ascaris suum	50–70	40–60	Similar to *A. lumbricoides*	Ferreira et al. (2011)
Austroxyuris finlaysoni	–	–	–	Braga et al. (2010c)
Dipylidium caninum	31–50	27–40	Oncosphere, egg capsules	Araujo et al. (2009)
Echinostoma paraensei	80–135	55–80	Operculate	Lelis et al. (2014)
Fasciola hepatica	130–150	60–90	Operculate	Braga et al. (2008a)
Oxyuris equi	80–90	42	Thick	Braga et al. (2010a)
Parascaris equorum	90–100		Thick outer egg shell	De Carbalho et al. (2013)
Schistosoma mansoni	110–170	40–70	Prominent lateral spine	Braga et al. (2008b)
Small strongyles (cyathostomin eggs)	–	–	–	Braga et al. (2010b)
Taenia taeniaeformis	37	–	Three layered egg shell	Braga et al. (2009)
Taenia saginata	31	34	Thick egg shell	Araújo et al. (2009)
Trichuris vulpis	79	38		Silva et al. (2010)
Toxocara canis	72	85	Thick, coarsely pitted surface	Frassy et al. (2010); Maciel et al. (2012); Araujo et al. (2013)
Toxocara vitulorum	75	90	Thick, coarsely pitted surface	Braga et al. (2010d)

species of epidemiological importance. Nematophagous fungi that control geohelminths include predators (e.g., *Monacrosporium thaumasium* [= *Arthrobotrys thaumasia*], *Duddingtonia flagrans*), endoparasites and opportunists (Araújo et al. 2009). The effectiveness of nematophagous fungi against eggs and infective larvae of gastrointestinal helminths of domestic animals has been documented (Mota et al. 2003; Braga et al. 2008a, b, 2010a, b, c, d). Lysek (1976) established a qualitative method for classifying ovicidal effects, which was initially proposed for the activity of *P. chlamydosporia* against *Ascaris lumbricoides* eggs. The three major types include: i) physiological type or biochemical effect without morphological damage to egg shell; ii) lytic biochemical type, with progressive morphological, changes of the egg shell and damage to the embryo; and iii) lytic and morphological effects with hyphal penetration and colonization of eggs resulting in embryo death. The three major types of ovicidal effect have been reported *in vitro* experiments done with eggs of different geohelminths (see Table 17.1, Fig. 17.1). The controlling

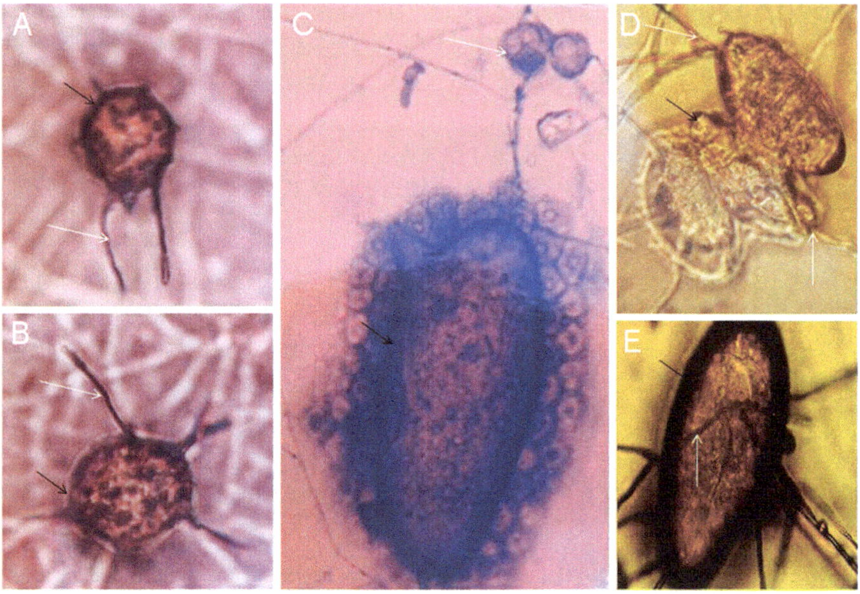

Fig. 17.1 Hyphae and chlamydospores of *Pochonia chlamydosporia* (*white arrows*) colonizing geohelminth eggs (*black arrows*). A, B: *Taenia* sp.; C: *Shistosoma mansoni*; D, E: *Fasciola hepatica* eggs 10 days after being exposed to the fungus (Courtesy F.R. Braga)

action of nematophagous fungi can be used in targeting the faeces environment to combat the eggs and free-living larval stages of geohelminths, including animal-parasitic nematodes.

17.7.1 Assessing Nematicidal Activity Against Animal-Parasitic Nematodes

The life span of PPN is very different to that of animal-parasitic nematodes. Plant-parasitic nematode eggs can remain viable for longer periods in the soil and rhizosphere, where they are protected within root debris, egg masses and/or cysts, than the eggs of animal-parasitic nematodes. Larval hatching of animal-parasitic nematodes basically depends on micro-environmental conditions, mainly temperature and humidity, prevailing within the faeces after being discharged into the environment by the animal. However, the period of time that the eggs spend outside their animal host has a positive implication in the use of nematode-trapping fungi for controlling the parasitic nematodes of ruminants, since the time for spore germination and fungal development *in situ* is short and fast enough to allow the fungus to colonise the dung and capture most of the recently hatched larvae. In nematode egg-parasitic fungi, such as *P. chlamydosporia*, the time taken to colonise faecal or dung material is slightly slower than for nematode-trapping fungi and therefore there is insufficient time to parasitize the eggs of ruminant-parasitic nematodes which, depending on micro-environmental conditions, can hatch in just a few days (Ashad et al. 2011). This colonization time difference gives the parasitic nematode larvae of ruminants a chance to escape from fungal activity and thus avoid being parasitized. In a study by Silva et al. (2011) no activity of *P. chlamydosporia* was found against *Haemonchus contortus* infective larvae. Nevertheless, some studies have shown that some *P. chlamydosporia* isolates have activity against nematode larvae of other nematode genera/species and even other groups of helminths (Araujo et al. 2013).

A number of studies have shown that *P. chlamydosporia* is able to produce mycelia that grow to surround the egg shell of different animal-parasitic nematodes and form appressoria to penetrate the egg shell, thus allowing the fungus to enter and feed on the embryo or egg contents. For example, Ferreira et al. (2011), reported an ovicidal activity of some *P. chlamydosporia* isolates on *Ascaris suum* eggs, a parasite of pigs. On the other hand, an ovicidal effect of *P. chlamydosporia* against *Taenia taeniformis* (a metacestode parasite of cats and rats) was reported by Braga et al. (2009). Similarly, an *in vitro* activity of *P. chlamydosporia* and other nematophagous fungi against the eggs of *Fasciola hepatica* (a trematode) was reported (Braga et al. 2008a). There are also some reports concerning the effect of *P. chlamydosporia* against parasites of equines. Braga et al. (2009) evaluated the viability and the ovicidal activity of some *P. chlamydosporia* isolates after passage through the gastrointestinal tract of horses and found that one isolate was able to survive and retain its activity against *Oxyuris equi* (Braga et al. 2010a). The antagonistic activity of *P. chlamydosporia* was also reported on *Toxocara canis* eggs (Hiura et al. 2015). Similarly, an *in vitro* ovicidal activity of *P. chlamydosporia* and other fungi was shown against *Trichuris vulpis* eggs (a nematode parasite of dogs) (Silva et al. 2010).

17.7.2 **Pochonia chlamydosporia** ***Mode of Action on Nematode Eggs***

Pochonia chlamydosporia and its mode of action have been studied for many years. Work in the UK demonstrated the presence of enzymes with chitinase and elastase activity to be associated with the lethal activity of *P. chlamydosporia* against nematode eggs (Ward et al. 2012; Braga and Jackson 2014). A crude enzymatic extraction of this fungus showed ovicidal activity against cyathostomin eggs (Braga et al. 2010b). The ovicidal activity of *P. chlamydosporia* has been demonstrated against nematodes affecting other animal species i.e., *Ascaridia galli* and *Toxocara canis*, parasites of birds and dogs, respectively (Thapa et al. 2015). Likewise, the activity of *P. chlamydosporia* as a biocontrol agent of *F. hepatica* was reported in Brazil (Braga et al. 2008a). Pellets containing *P. chlamydosporia* chlamydospores were supplied to cattle infected with *F. hepatica* and a reduction in the number of parasite eggs was observed in the animals receiving the fungal preparation (Dias et al. 2012).

Treated animals also showed a higher weight gain than the control group without the fungus (Dias et al. 2013). On the other hand, a gene called *pcchi44* associated with the production of a chitinase enzyme has being considered as a virulence factor against nematodes (Mi et al. 2010).

17.7.3 Experimental Assays

The following method has been used to test the ovicidal capacity of many geohelminth species (Araújo et al. 2009). In summary, isolates are kept in tubes containing 2% CMA in the dark at 4 °C for 10 days. Culture plugs (4 mm diameter) are taken from fungal colonies, plated onto 9-cm diameter Petri-dishes containing 20 ml 2% WA and stored in the dark at 25 °C for 10 days. One thousand eggs are placed on the surface of the Petri-dishes containing 2% WA and the fungal isolates. After 5, 10 and 15 days, 100 eggs are removed from the plates. Using a spatula, the eggs are placed on the surface of slides containing a drop of 1% Amam blue. Eggs are examined under a light microscope (40 × magnification), SEM and evaluated according to Lysek et al. (1982). Variations include plating the colonised plugs onto 2% PDA instead of 2% WA and storing in the dark at 25 °C for 10 days. Subsequently, colonized plugs are placed onto 2% WA fresh cultures for 10 days (Araujo et al. 2009; Silva et al. 2010). Instead of fungus-colonized plugs, chlamydospores have been used as inoculum (Araujo et al. 2013).

The passage and survival of nematophagous fungi through the gastrointestinal tract has been assessed using mycelia in sodium alginate pellets. The following protocol is taken from Braga et al. (2010a). Fungal mycelia are obtained by transferring mycelia-colonized plugs (2% CMA) into 150 ml potato-dextrose medium, pH 6.5 and incubated at 120 rpm in the dark at 26 °C for 10 days. Mycelia are then removed, pelletized in sodium alginate and given to de-wormed animals. Faecal samples are collected at different intervals (98–72 h) after pellet administration and placed in a Petri-dish containing 2% WA. Using this method, *P. chlamydosporia* (isolate VC4) sodium alginate pellet survival was assessed *in vitro* after passage through the gastrointestinal tract of dewormed horses against *Oxyuris equi* eggs. The isolate remained viable and maintained its ovicidal activity. Two percent WA is a nutritionally poor media and therefore the addition of faecal material to the Petri-dishes was crucial for conidia production and destruction of *O. equi* eggs by the fungus. Anderson et al. evaluated the effect of *P. chlamydosporia* isolate VC1 sodium alginate pellets on *F. hepatica* eggs after passing through the intestinal tract of cattle. The ovicidal effect was observed 7 days after interaction. The fungus showed activity on *F. hepatica* eggs after 35 days of incubation in the cattle stool samples. The formulation efficacy after 96 h suggests that this fungus, due to the possibility of being applied to animals under *in vivo* conditions, could be applied at intervals of twice a week. The presence of *P. chlamydosporia* in the environment could also enable the reduction of the freshwater mollusc infection (an intermediate host of *F. hepatica*).

17.8 Conclusions and Future Perspectives

There are excellent prospects for using nematophagous fungi in the control of gastrointestinal nematodiasis of domestic animals. The fungus has been used under laboratory conditions and has shown ovicidal activity against parasitic helminths of humans, cattle and other domestic animals. So far, however, the most effective activity has been observed in fresh faeces (Braga et al. 2010a). Nematode animal parasites develop through egg to larval stages before migrating to the pasture as infective larvae or, as in some helminth species, the infective stage is developed inside the egg (Larsen 1999; Mota et al. 2003). Therefore, to be considered as an efficient BCA, nematophagous fungi must have the ability to survive the passage through the gastrointestinal tract of domestic animals and then germinate in the faeces. *Pochonia chlamydosporia* (VC4) survived the passage through the gastrointestinal tract of horses, germinated in the faeces and produced chlamydospores. Parasite resistance to one or more anthelmintics is widespread, particularly in domestic animals, leaving many countries without adequate means of controlling gastrointestinal nematode parasitism. Singh et al. (2010) observed that treatment with benzimidazoles did not inhibit the proliferation of *P. chlamydosporia* after interaction with the anthelmintic. Therefore, the biological control of gastrointestinal nematodiasis using ovicidal nematophagous fungi could represent an important tool in combination with chemical control (Ferreira et al. 2011).

Helminth eggs have some physiological and morphological features that may be responsible for the differential action of *P. chlamydosporia* on different species, including age of the eggs. Immature eggs of *T. canis* are more susceptible to infection than mature eggs (Maciel et al. 2012). *Fasciola hepatica* eggs are large (150 μm diameter), *A. suum* eggs are 75 μm in diameter and present a relative thicker shell. The egg diameter is 37 μm in *T. taeniformis* and the egg capsule is generally thinner in taenids. *Pochonia chlamydosporia* (isolates VC1 and VC4) showed ovicidal effect on *D. caninum* egg capsules. The egg capsule is formed by quantities of mucopolysaccharide-positive vitelline material inside the uterus. This material cements the capsules to form clusters of eggs. The structure of the capsule is formed by polysaccharides or glycoproteins. The egg capsules measure 120–200 μm, and the eggs 26–50 μm (Araujo et al. 2009). The penetration mechanism of ovicidal fungi when parasitizing eggs is still not clear but enzymatic activity may be one of the main factors in the process of attack and penetration (Braga et al. 2009). Differences in the interaction of *P. chlamydosporia* with eggs of some helminth genera have been attributed to egg size, and egg shell composition. Eggs of *A. lumbricoides* and *A. suum* have a thick chitin-protein capsule (Braga et al. 2010b). Nematophagous fungi secrete extracellular enzymes of the protease type that have an important role in the infection and destruction (ovicidal activity) of nematode eggs (Braga et al. 2010c). The effect of *P. chlamydosporia* isolate VC4 crude extract on the hatching of plated cyathostomin eggs (small strongyles) and in faeces samples resulted in a 67% reduction in the number of L3 recovered from the coprocultures. *Pochonia chlamydosporia* isolate VC4 was cultured in Erlenmeyer flasks

(250 ml) containing 50 ml of liquid minimal medium (0.3 g l/1 NaCl, 0.3 g l/1 $MgSO_4.7H_2O$, 0.3 l/1 K_2HPO_4 and 0.2 g l/1 yeast extract) supplemented with 0.2% gelatine for production of the crude enzymatic extract (Esteves et al. 2009; Braga et al. 2010c).

In conclusion, the fungus *P. chlamydosporia* has been shown to have an important potential in the control of diseases caused by PPN affecting economically important crops. The knowledge of modern technologies of fungal mass production, as well as epidemiological and ecological aspects of the fungi and PPN, are crucial elements to achieve the best bio-technological use of *P. chlamydosporia* in the control of RKN. A replacement (at least partially) of chemical pesticides against root galling nematodes by *P. chlamydosporia* is expected and hopefully we will take advantage of this and other beneficial microorganisms in the control of PPN in the near future. In the case of ruminant parasitic nematodes, due to the fast development of eggs in faeces (sometimes in hours) and the rapid hatching nematode larvae, *P. chlamydosporia* has only a brief window of opportunity to grow on the nematode egg-shell before larval emergence take place. However, more *P. chlamydosporia* strains must be assessed in order to find a candidate for controlling parasitic nematodes of ruminants. On the other hand, extracellular fungal products with a number of medical properties are being discovered, including nematicidal activity and, perhaps, some secondary metabolites produced by *P. chlamydosporia* could, in future, have an important role in the control of the important parasitic diseases affecting cattle and sheep worldwide. More surveys on nematophagous fungi as natural inhabitants of soil in public parks should also be encouraged in order to obtain native isolates with potential as biological control agents of geohelminths (see Gortari et al. 2007).

References

Araujo, J. M., Araújo, J. V., Braga, F. R., et al. (2009). Activity of the nematophagous fungi *Pochonia chlamydosporia, Duddingtonia flagrans* and *Monacrosporium thaumasium* on egg capsules of *Dipylidium caninum. Veterinary Parasitology, 166*, 86–89.

Araujo, J. M., Araújo, J. V., Braga, F. R., et al. (2013). Predatory activity of chlamydospores of the fungus *Pochonia chlamydosporia* on *Toxocara canis* eggs under laboratory conditions. *Revista Brasileira de Parasitologia Veterinária, 22*, 171–174.

Ashad, A. F., Anisuzzaman, B. N., Dey, R. A., et al. (2011). Factors affecting the development and hatching of eggs and the survival of infective larvae of *Haemonchus contortus* in laboratory conditions. *Progressive Agriculture, 22*, 75–83.

Barron G. L. (1977). *The Nematode-Destroying Fungi. Topics in Microbiology No. 1*. Department of Environmental Biology, University of Guelph, Canada. Canadian Biological Publications.

Bhattarai, A., Bhattarai, B., & Pandey, S. (2015). Variation of soil microbial population in different soil horizons. *Journal of Microbiology & Experimentation, 2*(2), 00044. doi:10.15406/jmen.2015.02.00044.

Braga, F. R., Araújo, J. V., Campos, A. K., et al. (2007). Observação in vitro da ãçao dos isolados fúngicos *Duddingtonia flagrans, Monacrosporium thaumasium* e Verticillium chlamydosporium sobre ovos de *Ascaris lumbricoides* (Linnaeus, 1758). *Revista da Sociedade Brasileira de Medicina Tropical, 40*, 356–358.

Braga, F. R., Araújo, J. V., Campos, A. K., et al. (2008a). *In vitro* evaluation of the action of the nematophagous fungi *Duddingtonia flagrans, Monacrosporium sinense* and *Pochonia chlamydosporia* on *Fasciola hepatica* eggs. *World Journal of Microbiology and Biotechnology, 24*, 1559–1564.

Braga, F. R., Araújo, J. V., Campos, A. K., et al. (2008b). *In vitro* evaluation of the action of the nematophagous fungi *Duddingtonia flagrans, Monacrosporium sinense* and *Pochonia chlamydosporia* on *Schistosoma mansoni* eggs. *World Journal of Microbiology and Biotechnology, 24*, 2713–2716.

Braga, F. R., Araújo, J. V., Carvalho, R. O., et al. (2009). Ovicidal effect of nematophagous fungi on *Taenia taeniaeformis* eggs. *World Journal of Microbiology and Biotechnology, 25*, 533–535.

Braga, F. R., Araújo, J. V., Ricardo Silva, A., et al. (2010a). Viability of the nematophagous fungus *Pochonia chlamydosporia* after passage through the gastrointestinal tract of horses. *Veterinary Parasitology, 168*, 264–268.

Braga, F. R., Araújo, J. V., Carvalho, O. R., et al. (2010b). Ovicidal action of a crude enzymatic extract of the fungus *Pochonia chlamydosporia* against cyathostomin eggs. *Veterinary Parasitology, 172*, 264–268.

Braga, F. R., Araújo, J. V., Silva, A. R., et al. (2010c). *Duddingtonia flagrans, Monacrosporium thaumasium* and *Pochonia chlamydosporia* as possible biological control agents of *Oxyuris equi* and *Austroxyuris finlaysoni*. *Journal of Helminthology, 84*, 21–25.

Braga, F. R., Ferreira, S. R., Araújo, J. V., et al. (2010d). Predatory activity of *Pochonia chlamydosporia* fungus on *Toxocara* (syn. *Neoascaris*) *vitulorum* eggs. *Tropical Animal Health and Production, 42*, 309–314.

Braga, F. R., & Araújo, J. V. (2014). Nematophagous fungi for biological control of gastrointestinal nematodes in domestic animals. *Applied Microbiology and Biotechnology, 98*, 71–82.

Casas-Flores, S., & Herrera-Estrella, A. (2015). Antagonism of plant parasitic nematodes by fungi. In C. Kubicek & I. Druzhinina (Eds.), *Environmental and microbial relationships* (pp. 43–66). Heidelberg: Springer.

Cooper, K. M., Whelan, M., Kennedy, D. G., et al. (2012). Anthelmintic drug residues in beef: UPLC-MS/MS method validation, European retail beef survey, and associated exposure and risk assessments, food additives and contaminants: Part A: Chemistry, analysis, Control, exposure and amp. *Risk Assessment, 29*, 746–760.

de Carvalho, L. M., Braga, F. R., Domingues, R. R., et al. (2013). Interaction of the nematophagous fungus *Pochonia chlamydosporia* and *Parascaris equorum* eggs in different culture media. *Journal of Basic Microbiology, 54*(Suppl 1), S109–S114. doi: 10.1002.

Dias, A. S., Araújo, J. V., Braga, F. R., et al. (2012). Biological control of *Fasciola hepatica* eggs with the *Pochonia chlamydosporia* fungus after passing through the cattle intestinal tract. *Parasitology Research, 110*, 663–667.

Dias, S. A., Araújo, B. R., Pippin, C. A., et al. (2013). *Pochonia chlamydosporia* in the biological control of *Fasciola hepatica* in cattle in southern Brazil. *Parasitology Research, 112*, 2131–2136.

Esteves, I., Peteira, B., Atkins, S. D., et al. (2009). Production of extracellular enzymes by different isolates of *Pochonia chlamydosporia*. *Mycological Research, 113*, 867–876.

Ferreira, S. R., Araújo, J. V., Braga, F. R., et al. (2011). Ovicidal activity of seven *Pochonia chlamydosporia* fungal isolates on *Ascaris suum* eggs. *Tropical Animal Health and Production, 43*, 639–642.

Ferris, H. (2010). Contribution of nematodes to the structure and function of the soil food web. *Journal of Nematology, 42*, 63–67.

Franco-Navarro, F., Vilchis-Martínez, K., & Miranda-Damián, J. (2008). Nuevos registros de *Pochonia chlamydosporia*: aislamiento, colonización de raíces y parasitismo de huevos de *Nacobbus aberrans*. *Nematropica, 39*, 133–142.

Frassy, L. N., Braga, F. R., Silva, A. R., et al. (2010). Destruction of *Toxocara canis* eggs by the nematophagous fungus *Pochonia chlamydosporia*. *Revista da Sociedade Brasileira de Medicina Tropical, 43*, 102–104.

Gortari, C., Cazau, C., & Hours, R. (2007). Nematophagous fungi of *Toxocara canis* in a public park of La Plata, Argentina. *Revista Iberoamericana de Micología, 24*, 24–28.
Hiura, E., Lopez, A. C. G., Da Paz, J. S., et al. (2015). Fungi predatory activity on embryonated *Toxocara canis* eggs inoculated in domestic chickens (*Gallus gallus domesticus*) and destruction of second stage larvae. *Parasitology Research, 114*, 3301–3308.
Horvat, A. J. M., Babic, S., Pavlovic, D. M., et al. (2012). Analysis, occurrence and fate of anthelmintics and their transformation products in the environment. *Trends in Analytical Chemistry, 31*, 61–84.
Larsen, M. (1999). Biological control of helminths. *International Journal for Parasitology, 29*, 139–146.
Lelis, R. T., Braga, F. R., de Carvalho, L. M., et al. (2014). Effect of the fungus *Pochonia chlamydosporia* on *Echinostoma paraensei* (Trematoda: Echinostomatidae). *Acta Tropica, 139*, 88–92.
Li, G., Zhang, K., Xu, J., et al. (2007). Nematicidal substances from fungi. *Recent Patents on Biotechnology, 1*, 212–1233.
Liu, W., Han, Y., Wang, B. B., et al. (2015). Isolation, identification, and characterization of the nematophagous fungus *Monacrosporium salinum* from China. *Journal of Basic Microbiology, 55*, 992–1001.
Lysek, H. (1976). Classification of ovicide fungi according to type of ovidicity. *Acta Universal Palak Olomue, 76*, 9–13.
Lysek, H., & Sterba, J. (1991). Colonization of *Ascaris lumbricoides* eggs by the fungus *Verticillium chlamydosporium* Goddard. *Folia Parasitologica, 38*, 255–259.
Lysek, H., Fassatiová, P. N. C., et al. (1982). Ovicidal fungi in soils of Cuba. *Folia Parasitologica, 29*, 265–270.
Maciel, S. A., Freitas, G. L., Figueiredo, D. L., et al. (2012). Antagonistic activity of the fungus *Pochonia chlamydosporia* on mature and immature *Toxocara canis* eggs. *Parasitology, 139*, 1074–1085.
Mehl L. H., Kleczeuski N. (2015). Disease and Nematode Management in Field Crops. Disease and Nematode Management in Field Crops: Corn and Sorghum 3–1. https://pubs.ext.vt.edu/456/456-016/Section03-Diseases-and-Nematodes-1-full.pdf
Mendoza de Gives, P. (2011). Carnivorous fungi: The cruelest executioners of nematodes in the soil. *Mushroom Journal, 1*, 31–36.
Mendoza de Gives, P., & Alatorre, R. (2011). Nematodos: Un mundo de gusanos. *Ciencia y Desarrollo, 237*, 40–44.
Mi, Q. L., Yang, J. K., Ye, F. P., et al. (2010). Cloning and over expression of *Pochonia chlamydosporia* chitinase gene *pcchi44*, a potential virulence factor in infection against nematodes. *Process Biochemistry, 45*, 810–814.
Mota, M. A., Campos, A. K., & Araújo, J. V. (2003). Controle biológico de helmintos parásitos de animais: atual e perspectivas futuras. *Pesquisa Veterinária Brasileira, 23*, 93–100.
Qamar, F. M., Maqbool, A., & Ahmad, N. (2015). Economic losses due to haemonchosis in sheep and goats. *Science International, 23*, 321–324.
Shen, W., Mao, H., Huang, Q., et al. (2015). Benzenediol lactones: A class of fungal metabolites with diverse structural features and biological activities. *European Journal of Medicinal Chemistry, 97*, 747–777.
Silva, A. R., Araújo, J. V., Braga, R. F., et al. (2010). In vitro ovicidal activity of the nematophagous fungi *Duddingtonia flagrans*, *Monacrosporium thaumasium* and *Pochonia chlamydosporia* on *Trichuris vulpis* eggs. *Veterinary Parasitology, 172*, 76–79.
Silva, A. R., Araújo, J. V., Braga, F. R., et al. (2011). Comparative analysis of destruction of the infective forms of *Trichuris trichiura* and *Haemonchus contortus* by nematophagous fungi *Pochonia chlamydosporia*; *Duddingtonia flagrans* and *Monacrosporium thaumasium* by scanning electron microscopy. *Veterinary Microbiology, 147*, 214–219.
Singh, R., & Kumar, U. (2013). Assessment of nematode distribution and yield losses in vegetable crops of western Uttar Pradesh in India. *International Journal of Science and Research, 4*, 2812–2816.

Singh, R. K., Sanyal, P. K., Patel, N. K., et al. (2010). Fungus-benzimidazole interactions: A prerequisite to deploying egg-parasitic fungi *Paecilomyces lilacinus* and *Verticillium chlamydosporium* as biological control agents against fasciolosis and amphistomiasis in ruminant livestock. *Journal of Helminthology, 84*, 123–131.

Thapa, S., Meyling, N. V., Katakam, K. K., et al. (2015). A method to evaluate relative ovicidal effects of soil microfungi on thick-shelled eggs of animal-parasitic nematodes. *Biocontrol Science and Technology, 25*, 756–767.

Trifonova, Z., & Karadjova, J. (2003). Fungal parasitism of the cyst and eggs of the *Globodera rostochiensis*. *The Journal of Agricultural Science, 48*, 103–110.

Wang, Y., Sun, L., Yi, S., Huang, Y., Lenaghan, S. C., & Zhang, M. (2013). Naturally occurring nanoparticles from arthrobotrys oligospora as a potential immunostimulatory and antitumor agent. *Advanced Functional Materials, 23*, 2175–2184. doi:10.1002/adfm.201202619.

Ward, E., Kerry, B. R., Manzanilla-López, R. H., et al. (2012). The *Pochonia chlamydosporia* serine protease gene *vcp1* is subject to regulation by carbon, nitrogen and pH: Implications for nematode biocontrol. *PloS One, 7*(4), e35657. doi:10.1371/journal.pone.0035657.

Yang, I., Lofreedo, A., Borneman, J., et al. (2012). Biocontrol efficacy among strains of *Pochonia chlamydosporia* obtained from a root-knot nematode suppressive soil. *Journal of Nematology, 44*, 67–71.

Zavala-Gonzalez, E. A., Escudero, N., Lopez-Moya, F., et al. (2015). Some isolates of the nematophagous fungus *Pochonia chlamydosporia* promote root growth and reduce flowering time of tomato. *The Annals of Applied Biology, 166*, 472–483.

Part VI
Future Perspectives

Chapter 18
Future Perspectives

Rosa H. Manzanilla-López and Luis V. Lopez-Llorca

Abstract Research on *Pochonia chlamydosporia* biology, diversity and ecology, especially multitrophic interactions, has improved our knowledge of the fungus. This has been applied to enhance the effectiveness of fungal strains and develop *Pochonia*-compatible Integrated Pest Management strategies. Several *Pochonia*-based commercial bionematicides are currently available in the biopesticide market. However, control success and adoption of the current *P. chlamydosporia*-based products relies upon sustainable production, proper use and maintenance of isolates/strains, as well as the development of more efficient formulations and application methods. This chapter highlights some of the requirements that are necessary to enhance our understanding of those mechanisms that underpin the potential of the fungus in controlling nematodes and enhancing crop performance. Furthermore, exploration of *P. chlamydosporia* multitrophic phases could pave the way to identify new applications, a good example being the use of *P. chlamydosporia* secondary metabolites in human and veterinary medicine.

R.H. Manzanilla-López (✉)
Centro de Desarrollo de Productos Bióticos (Visiting Professor),
Yautepec, Morelos, C.P. 62731, Mexico
e-mail: rosa.manzanilla@gmail.com

L.V. Lopez-Llorca
Laboratory of Plant Pathology, Multidisciplinary Institute for Environmental Studies (MIES)
Ramón Margalef, Department of Marine Sciences and Applied Biology,
University of Alicante, Alicante, Spain

Department of Agri-Food Engineering and Biotechnology, Universitat Politècnica de
Catalunya, Barcelona, Spain
e-mail: luislopezllorca1@gmail.com

R.H. Manzanilla-López, L.V. Lopez-Llorca (eds.), *Perspectives in Sustainable Nematode Management Through* Pochonia chlamydosporia *Applications for Root and Rhizosphere Health*, Sustainability in Plant and Crop Protection,
DOI 10.1007/978-3-319-59224-4_18

18.1 Introduction

The development and use of *P. chlamydosporia* as a biocontrol agent (BCA) of plant-parasitic nematodes (PPN) has been illustrated through the chapters of this volume. Practical experiences support the use of the fungus for long term management of PPN (Chaps. 15 and 16) although there is a need for a deeper knowledge of its biology and ecology in order to provide the minimum initial conditions required to establish and enhance regulation of nematode populations in a sustainable manner.

The effectiveness of biological control (BC) of PPN has often been questioned due to inconsistent results. Although nematode biological control could be considered risky by some farmers it may appeal to others, especially if the use of nematicides becomes restricted or more expensive. Under these conditions, biological control is likely to be one of the few remaining options to control nematodes. However, whatever the reasons to choose BC, the BCA-based products should have a consistent performance and be cost-effective.

An incomplete knowledge of the environmental parameters affecting transmission, lack of field and farm-scale evaluation and quality control problems in production, formulation and BCA application continue to be among the limitations to uptake fungal control agents (Judith Pell pers. comm.). Other major technical constraints to use BCA include production and transport costs of antagonists to the required location. Another limitation, from an ecological perspective, is represented by the buffering activities of soil microbial communities operating against introduced BCAs. Despite these constraints, BCAs have several advantages over chemical pesticides. The most convenient is that BCAs are non-toxic products. This allows their use and application in nematode infected fields without the need to observe strict pre-harvest and re-entry intervals or enforce personal protective equipment regulations. Biological control agents can also be compatible with organic farming, provide protection to sown crops through field application, and may have a lasting control effect. In addition to acting as parasites, some BCA also behave as endophytes and growth promoters that help increase crop yield. Furthermore, effectiveness of the BCA can be enhanced through manipulation of the environment (Stirling 1991, 2014).

Biological control is complex and farming systems have a marked effect on the way a given BCA is used and applied. Growers operating in the agricultural systems from developed countries require formulated products with a good shelf-life, low application dosage and should preferably be delivered as seed treatments, in planting furrows or transplant media to suppress nematode populations (Kerry 1997; Timper 2014). By contrast, in the subsistence farming systems of developing countries where quality control is minimal or absent, large application rates of a BCA such as *Pochonia* (30 kg – 5 t/ha) can be used as long as the BCA is low-cost and produced locally (Kerry 1997). The need for seed dressing, root dressing or seedbed application methods to protect the growing seedling from early damage by PPN has also been pointed out in Chap. 15. Although BCA seedbed application has been tested (Gowen 2002), it is an area that needs further exploration in *Pochonia*.

18.2 Biopesticides Market

Economics undoubtedly plays a major role in developing BCA products for nematode control. Biopesticides have become the preferred control option for food value chains. Consequently, the biopesticide market (BM) has been growing at a faster rate than that of the synthetic pesticides (OSU 2012; IPMnet News 2012; Markets and Markets 2016; [http://www.marketsandmarkets.com]). It has been estimated that the BM was over US$2.0 billion in 2015; the largest market for biopesticides being the USA (36%) followed by Europe, Asia-Pacific, and Latin America. However, BCA products for nematode control remain few in number and have a more restricted market than nematicides (Table 18.1).

Pochonia-based products are currently available in the BM of China, Italy, Kenya, Nicaragua and Brazil. Dudutech K. Ltd. (Kenya) has produced 40-100 kg/week at industrial scale costs of £10.00/kg to treat one hectare (Gowen 2005). BiotorLabs Production Plant in Matagalpa (Nicaragua) has the capacity to produce 3000 kg/month (36,000 kg/year) of formulated KlamiC®, a quantity sufficient to treat 12,000 ha of diverse crops. The Rizoflora Pilot Production Plant in Viçosa (Brazil) is capable of producing 30,000 kg of Rizotec®/year (Chap. 15). Formulation, application methods and rates have been targeted and improved to meet market standards. Rizotec® is available as a wettable powder formulation of 5.2×10^7 chlamydospores/g of isolate Pc-10 of *P. chlamydosporia* var. *chlamydosporia* to be applied at a rate of 3 kg/ha (Bontempo et al. 2014). The bionematicide KlamiC®, obtained from the fungus strain SD 187 IMI of *P. chlamydosporia* var. *catenulata*, is available as a powdered colonized rice substratum, with 1.6×10^7 chlamydospores/g (Montes de Oca et al. 2009; Chap. 15).

More often than not, biological control results from multiple interactions between more than one microorganism (Timper 2014; Stirling 2014). Alternative plant defence methods against soil-borne pathogens will increasingly require more innovative approaches aimed at improved exploitation of natural microbe communities, including soil biological antagonists, in a more holistic and environment-friendly manner (Sellitto et al. 2016). Both, Rizotec® and KlamiC®, products have been tested alone or in combination with other bacteria- and nematophagous fungi based products, either singly or as a BCA cocktail mix formulation. Some of the BCA tested under field conditions, either alone or in combination with *P. chlamydosporia* Pc-10 isolate, include the bacteria *Bacillus subtilis*, *Bacillus* sp., nematophagous trapping fungi (*Arthrobotrys* spp.), and the nematophagous fungus *Paecilomyces lilacinus* (= *Purpureocillium lilacinum*) (Bontempo et al. 2014). *Bacillus* spp. (e.g., *Bacillus subtilis* and *B. lenchiniformis*) adversely affect root finding behaviour by juvenile nematodes, thereby causing their death through starvation. Commercial preparations including nematode trapping fungi have been tested to control cyst

Table 18.1 Commercial biological products for managing plant-parasitic nematodes after Chen and Dickson (2012)

BCA	Type of antagonist	RKN (+)	Other PPN	Countries where the products are marketed
Bacillus spp.	Antibiotic bacterium	+		Brazil[a], Germany, India, Israel, South Africa, USA
Mycorrhizal fungi	Endophytes	+		Italy[a], USA
Myrothecium verrucaria	Antibiotics produced by the fungus	+		USA
Purpureocillium lilacinum	Egg-parasitic fungus	+		Australia, Brazil[a], China, Germany, India, Kenya, Singapore, South Africa, USA
Pasteuria penetrans	Obligate parasite of second-stage juvenile	+		USA[b]
Pasteuria nishizawae Pn1	Obligate parasite of second-stage juvenile		*Heterodera glycines*	Canada
Pochonia chlamydosporia	Egg-parasitic fungus	+		Cuba, China, Kenya, Portugal, Italy[a]
Trichoderma spp.	Toxins, proteolytic enzymes	+		Brazil[a], India, Israel, Kenya, Sweden, New Zealand
Pseudomonas fluorescens	Antibiotic bacterium	+	PCN	India
Abamectin	Toxic metabolites produced by bacterium	+		Worldwide
Nematode trapping fungi	Parasitism of second-stage juvenile and eggs	+		Brazil[a], USA

BCA Biological control agent, *RKN* Root-knot nematodes, *PCN* Potato cyst nematodes
[a]BCA also included as part of cocktail mix formulation
[b]Syngenta acquired Pasteuria Bioscience Inc. in 2012

nematodes and RKN, although they have limitations as trapping activity must be sustained for long periods to give adequate control (Stirling 1991). The fungi *Purpureocillium lilacinum*, *P. chlamydosporia* and *Trichoderma harzianum* are not only rhizosphere competent and facultative parasites of nematode eggs, but are also plant endophytes. The multitrophic potential of these fungi may be further explored to broaden their use as plant growth enhancers and to promote plant defences to attack by other soil-borne pathogens. In this respect, *P. chlamydosporia* can reduce take-all (*Gaeumannomyces graminis*) in cereals (Monfort et al. 2005; Maciá-Vicente et al. 2008).

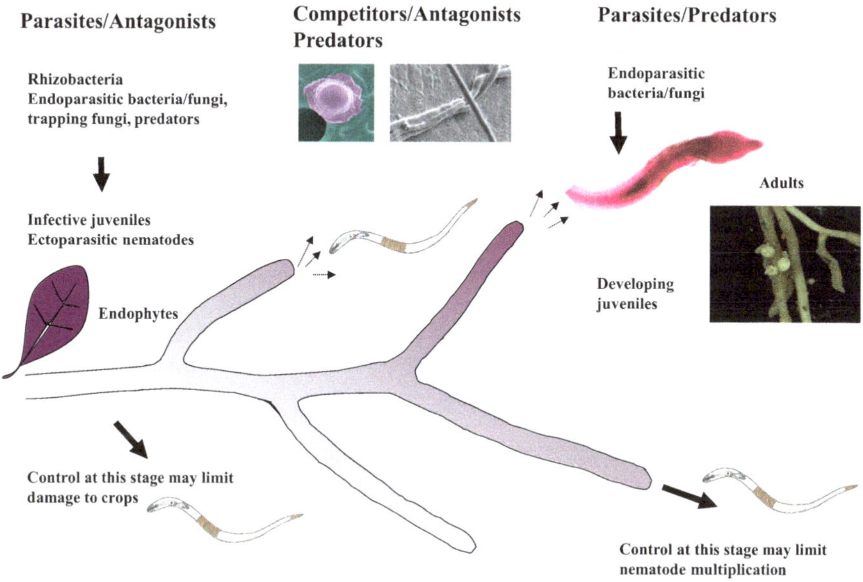

Fig. 18.1 The interactions of parasites, competitors, antagonists and predators, and the different types of control for different stages of plant-parasitic nematodes (After Brian R. Kerry and P.R. Hirsch)

The BCA cocktail mix formulation of some products may, depending on the BCAs included and their individual mode of action, attempt to mimic the effects of the interactions of parasites, competitors, antagonists and predators present in the soil and rhizosphere, and the different types of control that they exert on different stages of plant-parasitic nematodes throughout the crop and nematode life cycles (Fig. 18.1). Therefore, targeting different development stages of the nematode(s) with different BCA can be achieved with a mix of BCA contained in a single product. Profix Max™ is formulated as a mix of bacteria (*Bacillus* spp.) and predatory fungi such as *Arthrobotrys oligospora*, *A. musiformis* and *Monacrosporium robustum*, to target RKN and *Rotylenchulus reniformis*. Pochar®, a *Pochonia chlamydosporia*-based commercial product, was developed by Microspore (Larino, Italy). The product includes the BCAs *A. oligospora*, *Glomus* sp. and *Bacillus* spp. Pochar® can be applied via irrigation and has been tested in field trials against RKN on potato. Furthermore, the growth promotion effect of the *P. chlamydosporia* isolate (DMS 26985) of Pochar® has also been tested on its own (with no RKN) on tomato. The application method of Pochar® with soil or plant leaf treatments followed by irrigation was shown to be effective in the management of RKN in potato, and the isolate DMS 26985 sustained tomato plant growth and gave higher fruit weight. Such field applications showed the potential of this product to manage RKN in integrated or organic farming (Sellitto et al. 2016).

It may be too early to predict both the long term niche for *Pochonia*-based products in the biomarket and how well they can compete with other nematophagous-based products (Stirling 1991), but their potential role is promising. Recently, in addition to its nematophagous activity, some other beneficial potential effects of *P. chlamydosporia* to the plant have been demonstrated. In the absence of nematodes, *Arabidopsis thaliana* (model) plants inoculated with the fungus reduced their flowering time. The fungus also stimulated plant growth and gave a higher yield in terms of seed production per plant (Zavala-Gonzalez et al. 2017). This has also been demonstrated in tomato for some *P. chlamydosporia* strains (Zavala-Gonzalez et al. 2015).

However, before choosing to apply one or more BCA-based products simultaneously, other factors such as egg age and pre-colonization of cysts by other fungi (Chen and Chen 2003), specific mode of action of single BCA on nematodes (i.e., eggs or second-stage juveniles) and the antagonistic effect exerted on *P. chlamydosporia* by native nematophagous fungi and bacteria should also be considered (Chapts. 4 and 15).

18.3 Biological Control Approaches

Biological control agents are not a complete replacement for pesticides and to be effective BCA should be integrated with other IPM strategies. Although pesticide application is likely to affect the natural enemies of nematodes, both nematicides and naturally occurring opportunistic fungi can provide an acceptable overall control as part of an integrated management programme – antagonistic fungi provide adequate control of the second generation of nematodes while nematicides control the first (Crump and Kerry 1987; Stirling 1991). In situations where naturally occurring parasites and predators of nematodes exert a significant regulatory effect on nematode populations, it may be necessary to modify the selection of a pesticide and its use and compatibility with existing or introduced BCA in order to retain BCA regulatory activity (Chaps. 15 and 16). In an integrated soil biology management scenario, soil food web regulatory processes determine the size, composition and activity of the soil biological community. This 'biological buffering' or the balance of nature, affects all organisms, including nematode pests and is the basis of biological control (Stirling 2014).

Conservation biological control uses tactics and approaches involving the manipulation of the environment. Unfortunately, biological control is slow to establish in soil and can be difficult to manipulate (Kerry 1991). Habitat manipulations may entail the induction or enhancement of favourable conditions for the BCA that are either lacking in the target niche or are present at levels inadequate to enhance BCA survival and effectiveness (Barbosa 1998). An inundative (i.e., release of large quantities of parasites or predators) or a conservation approach to biological control can be adopted within a more holistic approach to IPM and still be capable of delivering a sustained crop yield while preserving soil health (Stirling 2014). The latter approach demands in depth information of soil quality and soil health and requires dealing with the soil as a living, rather than an inert, system (Stirling 2014). Each

BCA is a living organism that requires suitable environmental conditions to thrive and be effective. For these reasons a single BCA cannot be expected to work in all environments and under all conditions. Some nematophagous fungi species, including *P. chlamydosporia* strains, have developed a degree of specialization towards nematodes (host preference) but they also have a saprophytic component. Ideally, selection of BCA strains should be based on host preference and BCA management 'tailored' according to crop and a 'soil health diagnostic profile'. The soil health profile is reflected in physicochemical analyses and crop fertilization requirements which may also affect the nematode, microbial interactions and BCA performance (see Bontempo et al. 2014). Therefore, additional compatibility tests should be carried out, such tests being carefully designed and based on solid data derived from interaction experiments between *Pochonia*-based bionematicides and other commercially available biopesticides/bionematicides formulations and agrochemical/botanical products (Chap. 15). Soil receptivity tests to BCA, primary devised for *Trichoderma* spp., have also been used to analyse *Pochonia* spp. strains (Monfort et al. 2006).

18.4 Agricultural Practices and Integrated Pest Management

Although *P. chlamydosporia* has been found in soils suppressive to nematodes, its presence does not necessarily reduce nematode populations, as discussed in Chap. 16. This highlights the need to carry out more long term studies and field trials to assess IPM strategies such as the effect of crop rotations and other potentially compatible practices on *Pochonia*-based products, isolates and metabolites. The multitrophic phases so far identified in *P. chlamydosporia* show that the fungus can act as a saprophyte, pathogen of invertebrates (insect, nematodes and molluscs), hyperparasite of soil-borne fungi, plant mutualist, and ecosystem service provider (= biofertilizer) by influencing nitrogen movement from the soil into the plant (Chaps. 2 and 3).

The cropping sequencing in a rotation can have different effects on the nematode, *P. chlamydosporia*, and other facultative nematode parasites and predators (Stirling 1991; Manzanilla-López et al. 2011; Timper 2014). Plant and environment have a profound influence on the *Pochonia*-nematode interaction as discussed in Chap. 4. Temperature, pH, soil type, soil microbiota and roots can enhance or reduce the parasitism of the nematode by the fungus. Plant root distribution patterns, composition of root exudates and organic matter C:N ratio input vary from crop to crop. The method and timing of application of *Pochonia* can also affect the establishment of the fungus in the soil and control of nematodes (Chaps. 11 and 12).

The importance of a better understanding of the effect of agricultural practices on soil quality, soil biodiversity, multitrophic behaviour of the fungus, and their interactions will benefit decision making in IPM/ or integrated soil biology management. Farming systems can enhance general nematode suppression in soil by increasing organic matter via minimum tillage (Stirling 2014; Timper 2014). Conservation tillage systems include technologies that offer numerous benefits (http://www.ctic.purdue.edu/resourcedisplay/293/) as part of an integrated nutrient and pest management strategy (http://www.unep.or.jp/ietc/Publications/TechPublications/TechPub-8a/tillage.asp). However, incorrect application of some of these technologies can result in environments that encourage pests and weeds, thereby necessitating additional management strategies to overcome these induced problems. Some information is available about the effect of organic matter on the performance of *P. chlamydosporia* but little is known regarding the effect of conservation tillage on performance of the fungus.

18.5 Achieving Sustainability

True sustainability can be achieved through IPM technologies when combined in a balanced fashion (Naranjo et al. 2015). Cultivation in its various forms entails some abrasive effects on the soil biota, the loss of soil organic matter, and the exposure of organisms to heating and drying by the sun and wind (Sánchez-Moreno et al. 2009; Bontempo et al. 2014). Reduced or conservation tillage, when feasible, may therefore help conserve antagonists of nematodes to a greater extent than occurs under conventional practices (Stirling 2014).

Biological control is considered environmentally-friendly (Naranjo et al. 2015). However, it is important to be aware that farmers will use BC not simply because it reduces pesticide applications, but because it works better (http://www.ecpa.eu/dashboard/life-cycle/ipm). Effective biocontrol may be the result of a number of different and complex mechanisms working synergistically to achieve disease or pest control (Naranjo et al. 2015). Understanding the different ecosystem services (e.g., decomposition, nutrient cycling, suppression of noxious and pathogenic organisms) that can be supplied by BC and BCA help to broaden BC utility and their value as perceived by farmers and other stakeholders (Naranjo et al. 2015).

18.6 Towards Innovative Approaches

The discovery of induced resistance pathways in plants has opened the possibility to chemically activate one or more such pathways for the development of innovative strategies for crop protection (Chap. 14). The potential of priming or sensitization of a cell or organism for enhanced defence by beneficial microbes for plant protection has been recognized (Conrath et al. 2015). The *Trichoderma*-fertile-based TrichoPlus™ protects crops from damping-off disease and promotes plant growth. Furthermore, synthetic agrochemicals, beneficial microbes and/or natural priming activators can be complementary in their use (Conrath et al. 2015).

The development of a sustainable, natural and effective strategy of nematode control may rely, for example, on the activation of the Jasmonic Acid pathway by plant defence activators that reduce RKN penetration and promote egg parasitism by *P. chlamydosporia* as discussed in Chap. 14 by dos Santos et al. The putative modulation of the tritrophic interaction (host plant, nematode and *P. chlamydosporia*) by the activation of biochemical defence pathways may lead to a better understanding of how the organisms interact at this level and how these interactions can be further exploited to promote nematode control. In this respect, root endophytism of *P. chlamydosporia* is modulated by jasmonate signalling (Zavala-Gonzalez et al. 2017).

18.7 *Pochonia* Secondary Metabolites

The classification, occurrence, biological activities and functions of a large number of secondary metabolites from *P. chlamydosporia* and other related species are revised by Niu in Chap. 7. *Pochonia chlamydosporia* metabolites such as pyranones, phomalactone and aurovertins show nematicidal activity. Secondary metabolite production from *P. chlamydosporia* strains is an area justifying further exploration as fungal metabolites may also vary across spatial scales varying from the geographical to field level. A better understanding of secondary metabolites and their natural function in *P. chlamydosporia*, together with a knowledge of their biosynthesis, would aid in selection and development of highly effective BCA.

The metabolites of *P. chlamydosporia* have also attracted attention for their molecular architectures, antibacterial, antifungal, antioxidative, anti-malarial, anti-virus, and anti-tumour activities. However, as pointed by Larriba et al. (2014): "despite the potential of pochonins in biotechnology and pharmacology, only two genes encoding polyketide synthases of *P. chlamydosporia* have been cloned and characterised." This demonstrates the need to continue exploring the potential of *P. chlamydosporia* secondary metabolites by taking advantage of new technologies such as *-omics*.

18.8 Molecular and *-omics* Approaches

Current molecular identification of *P. chlamydosporia* varieties, as well as that of former species included in the genus *Pochonia* (Chap. 2), relies increasingly on different sets of molecular markers. Classification is mainly based on the phylogenetic analysis of several marker gene sequences such as those of *β-tubulin*, ITS, *nrSSU*, *nrLSU*, *rpb1*, *rpb2*, and *EF1-α*. *Pochonia* taxonomy and phylogenetics are both complex and plastic and are under constant revision as new molecular information becomes available (Chap. 2).

The diversity of *P. chlamydoporia* populations has been revealed using molecular fingerprinting techniques. Quantification (abundance) of the fungus in soil and roots can be done using qPCR alone or in combination with other classic microbiological techniques (Escudero and Lopez-Llorca 2012). However, more research is still needed to develop epidemiological dose-response effect models for the fungus (Chap. 13). In other disciplines, such as plant pathology, genomics has speeded up the development of pathogen detection methods and improved disease epidemiology studies (Sundin et al. 2016). The availability of whole genome resources coupled with next generation sequencing (NGS) technologies has stimulated genomics-based approaches to improve disease resistance in crops and increased the pace at which whole plant and pathogen genomes are becoming available (Klosterman et al. 2016). Engineering fungi (including *Pochonia*) for increased virulence, either by expressing transgene-encoded toxin regulators or by elevated production of native metabolite virulence factors, is expected to provide more attractive biocontrol agents (Chap. 6). New procedures based on CRISPR-Cas technology are now available for fungal pathogens (Arazoe et al. 2015) and could be used in future studies for genome editing of *P. chlamydosporia.*

The study of saprophytic/parasitic phases of the fungus using molecular tools (Chaps. 5 and 13) is an important step in understanding the fungus-plant-nematode tritrophic interaction. Molecular advances and genome sequencing of *P. chlamydosporia* have opened new avenues for studying the molecular mechanisms of differentiation, pathogenesis and diagnostics of the fungus (Chaps. 2 and 13). The next step is to couple these studies with, or take advantage of data from, fully sequenced genomes of major PPN pests (e.g., *Globodera rostochiensis*, *Radopholus similis*, *Meloidogyne incognita*, *M. hapla*, *Rotylenchulus reniformis*), model nematodes such as *Caenorhabditis elegans* or *Pristionchus pacificus*, plant hosts (e.g., *Solanum lycopersicum*, *S. tuberosum*), and model plants (*Arabidopsis thaliana*, *Lotus japonicus*). The benefits of functional genomics research should be extended to animal-parasitic nematodes (Chap. 17) for which several genomes are now available (e.g., *Ascaris suum*, *Haemonchus contortus*, *Trichuris suis*).

Fungal proteomics research is steadily growing as a result of an increase in the number of sequenced and well annotated fungal genomes (Chap. 6). A complete *P. chlamydosporia* genome sequence could help with protein identification of fungal secretomes under various conditions (Chaps. 6, 8, 10). Mitogenome analysis of *P. chlamydosporia* will also help to improve understanding of the evolution of this fungus (Chap. 9).

18.9 Mining and Harnessing *-omics* Approaches

The knowledge as to how fungi infect/colonize their hosts is improving thanks to the information available from genome sequences of fungi with different life styles, especially pathogens and mutualists (Tunlid and Talbot 2002). Molecular phylogeny has shown that the parasitic and symbiotic fungus life styles have evolved repeatedly within the fungal kingdom. At the genomic level, this may account for adaptations involving the presence of novel genes, differences in gene expression, gene loss and duplication (Tunlid and Talbot 2002). Ideally, more isolates of *P. chlamydosporia* from various ecological backgrounds should be sequenced to explore the mechanisms acting in the parasitic/symbiotic phases of the fungus.

Complete genome sequences and other genomic approaches offer a starting point for generating new hypotheses on the mechanisms involved in the evolution of parasitism, pathogenesis and symbiosis, and which allow a fungus to grow and proliferate within another living organism (Tunlid and Talbot 2002). According to Sundin et al. (2016) genomics has affected traditionally hypothesis-driven science, in the sense that genome sequence or transcriptomic data may be first obtained and then generate novel research questions based on that molecular data. This hypothesis-driven approach can also be applied to *Pochonia*.

Mining database information from genomics and other *-omics* data may help to manipulate/engineer genomes of both host and parasite to develop novel gene-based control strategies (Finetti-Sialer and Manzanilla-López 2011). The differential gene expression induced by *P. chlamydosporia* in the presence of nematodes or other pathogens may also uncover novel pest and disease management strategies. Fungus transcriptomic analyses may also support the production of industrial and commercial bioformulations for plant protection, through the induction of endogenous plant defence mechanisms (Chap. 5). *Pochonia* '*-omics*' research should also take advantage of NGS technology that has made possible the metagenomic analysis of plant-associated microbial communities, and the emerging field of plant microbiome- or phytobiome-research, which may help in elucidating the interactions between plants and their phytobiomes (Klosterman et al. 2016). Furthermore, it is likely that the construction of genome-edited plants with durable resistance and the discovery of pathogen targets for chemical inhibition can be achieved through genomics (Sundin et al. 2016). Phytobiomes could, at least in theory, be engineered as biopesticides to control specific pathogens by enhancing some of their microbe members (Klosterman et al. 2016).

18.10 Applied Research

A current key challenge that lies ahead is how to integrate the massive datasets continuously generated through molecular and *-omics* studies in order to apply them towards practical benefits for crop production and pest management. The availability of genome sequences has made possible the massive analysis of expression levels of large sets of genes in different species of fungal pathogens and symbionts (Tunlid and Talbot 2002; Westermann et al. 2012). A similar approach may help in the screening/selection process of *Pochonia* strains with different nematode host preferences and rhizosphere colonization abilities. It may also be extended to endophytic and growth promoter fungal strains. It is expected that the structural and functional analysis of the *P. chlamydosporia* genome will provide a platform for understanding, and taking advantage of, the molecular mechanisms involved in the multitrophic lifestyle of this fungus (Larriba et al. 2014). Enhancing biological control organisms through phenotype selection, mutagenesis, or genetic engineering is an emerging field (Timper 2014). The role of hydrolytic enzymes/secondary metabolites of *P. chlamydosporia* in the infection process is also a subject for genetic engineering.

Priming or sensitization of a cell or organism for enhanced defence is yet another area that demands further exploration (Chap. 14). Primed plants respond to very low levels of a stimulus and show more rapid and robust activation of defence responses when challenged by pathogens, insects, or abiotic stress, and this is frequently associated with local and systemic immunity and stress tolerance as shown in other fungi such as *Trichoderma asperellum* that incites a protective effect on plants through activation of the *TIPK* gene (Shorest et al. 2006).

Larriba et al. (2014) showed the existence of a close relationship between *P. chlamydosporia*, entomopathogenic, mycoparasitic (*Trichoderma* spp.) and endophytic fungi (*Epicloë festucae*). The wide array of hydrolytic enzymes and transporters encoded by *P. chlamydosporia* are likely to support its multitrophic behaviour (pathogenic, endophytic and saprophytic). It is expected that genomic-level information will help enhance the diverse capabilities of the fungus, not only as a BCA of PPN, but as a growth-promoting agent of crop plants (Larriba et al. 2014) and also acting as a priming agent against other soil-borne pathogens.

18.11 In Search of the Holy Grail

The search for new isolates with potential as BCA of nematodes is expected to continue (Chaps. 11, 12, 15, and 16). However, this can be an expensive and time consuming process, even when assisted by molecular technologies and NGS. There are already many isolates of the fungus deposited in fungal collections. Their characterization and potential remains to be explored by taking advantage of high-throughput technologies such as robotics. The collections of the fungus (at one stage there were 473 isolates of the fungus in the *Pochonia* collection of Rothamsted Research) represent a valuable resource that can be used for research purposes through inter-institutional transfer agreements. Other nematode pests, such as the reniform nematode, a species that causes US$130 million of annual losses to the USA cotton industry and has supplanted *M. incognita* as the major nematode pest of cotton in Mississippi, Louisiana and Alabama (Robinson 2007; Nyaku et al. 2014), have been recorded as *Pochonia* hosts and deserve detailed exploration as potential BC candidates. Another nematode that could be intensively screened for *P. chlamydosporia* parasitism is *Tylenchulus semipenetrans*, the citrus nematode.

18.12 Concluding Remarks

As pointed out at the beginning of this book (Chap. 1), the foundations of *Pochonia* research were largely laid by collaboration between diverse international research groups, both molecular and non-molecular in focus. *Pochonia clamydosporia* laboratory and field applied research should strive towards a more holistic, inter- and trans-disciplinary approach. This will lead to a new paradigm of sustainable ideas/solutions for nematode management and biotechnological products (e.g., medicinal drugs) based on the fungus. Education and knowledge transfer should also be included since education and detailed BC surveys can help to target stakeholder-perceived value of BCA that may help identify the economic value of the diverse ecosystem services provided by BCA (Naranjo et al. 2015) such as *P. chlamydosporia*.

To promote safe and adequate use of *Pochonia* BCAs within an IPM, farmer workshops can be implemented during crop protection meetings. This and other related activities may help to build up contacts with governmental bodies, research organizations and decision makers, with the aim of increasing stakeholder appreciation of BCA (Naranjo et al. 2015). Farmer participatory research programmes to integrate *P. chlamydosporia* and *Pasteuria penetrans* for RKN management has been pioneered in Kenya (Gowen 2002, 2005) and Cuba. A similar approach can be used with small- and medium-scale farmers in countries where the fungus is being produced, either at small scale by publicly funded research laboratories or at a large scale by private industry (Chap. 1).

Currently the biopesticide industry has an estimated value of US$1.6 billion with an expected annual growth rate of 16% through 2019 (Agrow Agribusiness Intelligence 2016). Worldwide, the trend for mergers, acquisitions and deals between biopesticide and 'biorational' products companies is a reality where large transnational agro-technology companies can acquire small biopesticide companies and start-ups. In such a scenario, publicly funded institutions, included spin-offs or start-ups, should ensure that the benefits of BCA research will continue to trickle down to small farmers in an affordable manner, especially in countries with an emerging economy, and regardless of the increasingly prevalent commercially oriented business attitudes.

An important issue addressed in Chap. 15 referred to the need to ensure a safe and proper use of *Pochonia*-based products in countries where phytosanitation official norms are non-existent or are currently waiting to be updated to include nematophagous fungi. Biological control agents may be categorized as relatively 'high risk' organisms (Stirling 1991) and extensive toxicological testing and risk analysis assessment are likely to be required by governments before permission is granted to use nematophagous and other BCA from foreign countries (see Wabule et al. 2004 and Chap. 15). Development of biopesticide registration and risk assessment guidelines can be facilitated through amendments to existing legislation for conventional pesticides and the development of a harmonised set of guidelines for BCA use and movement within and outside countries by using models from the European Union

(REBECA http://www.rebeca-net.de/?p=999), the Organization for Economic Co-operation and Development, or the US-Environmental Protection Agency. Heavier regulatory burdens can be avoided by providing the minimum necessary information on the safety and efficacy of the biopesticide (Cherry 2006; Bigler et al. 2005).

Despite the fact that BC is a naturally occurring phenomenon in agroecosystems, it will only succeed if the soil biological community is nurtured rather than depleted by the practices used to grow crops (see Barbosa 1998; Sánchez-Moreno et al. 2009; Stirling 2014; Timper 2014). As pointed out by Stirling (2014) "Wherever agriculture is practiced the requisite components of the soil food web will already be present and adapted to local conditions. The challenge is to redesign management practices to enhance their activity".

In a world scenario of increasing demand for food and fibre, crop protection from pest losses will require sound ecologically-based and economically viable IPM practices based on innovative solutions (Naranjo et al. 2015). Biological control and BCA such as *Pochonia* have the potential to meet that challenge.

References

Agrow Agribusiness Intelligence. (2016). *Biologicals 2016* (an analysis of corporate, products and regulatory news in 2015/2016). https://www.agra-net.com/agra/agrow/supplements-reports/supplements/article518033.ece/BINARY/Agrow_Biopesticide_2016-supplement.pdf. pp 14, 17–19.

Arazoe, T., Miyoshi, K., Yamato, T., et al. (2015). Tailor-made CRISPR/Cas system for highly efficient targeted gene replacement in the rice blast fungus. *Biotechnology and Bioengineering, 112*, 2543–2549. doi:10.1002/bit.25662.

Barbosa, P. (1998). Agroecosystems and conservation biological control. In P. Barbosa (Ed.), *Conservative biological control* (pp. 39–54). San Diego: Academic.

Bigler, F., Looman, S. A., & van Lenteren, J. (2005). *Harmonization of the regulation of invertebrate biological control agents in Europe*. Second international symposium on biological control of arthropods. Davos, Switzerland, pp. 692–700.

Bontempo, A. F., Fernandes, R. H., Lopes, J., et al. (2014). *Pochonia chlamydosporia* controls *Meloidogyne incognita* on carrot. *Australasian Plant Pathology, 43*, 421–424.

Chen, S. Y., & Chen, F. J. (2003). Fungal parasitism of *Heterodera glycines* eggs as influenced by egg age and pre-colonization of cysts by other fungi. *Journal of Nematology, 35*, 271–277.

Chen, S., & Dickson, D. W. (2012). Biological control. In R. H. Manzanilla-López & N. Marbán-Mendoza (Eds.), *Practical plant nematology* (pp. 761–811). Mexico: BBA Biblioteca básica de agricultura, Editorial del Colegio de Postgraduados.

Cherry, A. (2006). *Crop protection programme. Development of biopesticide registration and risk assessment guidelines for Ghana*. R8430 (ZA 0659). Final technical report. http://www.fao.org/docs/eims/upload/agrotech/2003/r8430_ftr.pdf

Conrath, U., Beckers, G. J. M., Caspar, J. G., et al. (2015). Priming for enhanced defense. *Annual Review of Phytopathology, 53*, 97–119.

Crump, D. H., & Kerry, B. R. (1987). Studies on the population dynamics and fungal parasitism of *Heterodera schachtii* in soil from a sugar beet monoculture. *Crop Protection, 6*, 49–55.

Escudero, N., & Lopez-Llorca, L. V. (2012). Effects of plant growth and root-knot nematode infection of an endophytic GFP transformant on the nematophagous fungus *Pochonia chlamydosporia*. *Symbiosis, 57*, 33–42.

Finetti-Sialer, M. M., & Manzanilla-López, R. H. (2011). Exploiting "-omics" and molecular approaches in plant nematology research. In R. Rodríguez-Herrera, C. N. Aguilar, J. K. Simpson-Williamson, et al. (Eds.), *Phytopathology in the omics era, Transworld research network* (pp. 39–68). Thiruvananthapuram: Research Signpost.
Gowen, S. R. (2002). *Integrated management of root-knot nematodes on vegetables in Kenya R 7472 (Za 0324).* Final technical report (1 October 1999–30 September 2002). Reading, UK, 52 pp. DFID R7472 Crop Protection Programme.
Gowen, S. R. (2005). *Promotion of sustainable approaches for the management of root-knot nematodes of vegetables in Kenya (R8296 (ZA 0568)).* Crop protection programme. Final Report, DFID, 39 pp.
Kerry, B. R. (1991). Methods for studying the growth and survival of the nematophagous fungus, *Verticillium chlamydosporium* Goddard, in soil. *IOBC/WPRS Bulletin, 14*, 34–38.
Kerry, B. R. (1997). *Biological control of nematodes: Prospects and opportunities.* FAO Corporate Document Repository. Plant nematode problems and their control in the Near East Region. FAO Plant Production and Protection Paper, 144.
Klosterman, S. J., Rollins, J. R., Sudarshana, M. R., & Vinatzer, B. A. (2016). Disease management in the genomics era–Summaries of focus issue papers. *Phytopathology, 106*, 1068–1070.
Larriba, E., Jaime, M. D. L. A., Carbonell-Caballero, J., et al. (2014). Sequencing and functional analysis of the genome of a nematode egg-parasitic fungus, *Pochonia chlamydosporia. Fungal Genetics and Biology, 65*, 69–80.
Maciá-Vicente, J. G., Jansson, H., Mendgen, K., et al. (2008). Colonization of barley roots by endophytic fungi and their reduction of take-all caused by *Gaeumannomyces graminis* var. *tritici. Canadian Journal of Microbiology, 54*, 600–609.
Manzanilla-López, R. H., Esteves, I., Powers, S. J., et al. (2011). Effect of crop plants on abundance of *Pochonia chlamydosporia* and other fungal parasites of root-knot and potato cyst nematodes. *The Annals of Applied Biology, 159*, 118–129.
Markets and Markets. (2016). Biopesticides market by type (bioinsecticides, biofungicides, bioherbicides, and bionematicides), origin (beneficial insects, microbials, plant-incorporated protectants, and biochemicals), mode of application, formulation, & crop type – Global Forecast to 2022. November 2016. Report Code: AGI 2716. http://www.marketsandmarkets.com/Market-Reports/biopesticides-267.html
Monfort, E., Lopez-Llorca, L. V., Jansson, H.-B., et al. (2005). Colonisation of seminal roots of wheat and barley by egg-parasitic nematophagous fungi and their effects on *Gaeumannomyces graminis* var. *tritici* and development of root-rot. *Soil Biology and Biochemistry, 37*, 129–1235.
Monfort, E., Lopez-Llorca, L. V., Jansson, H. B., et al. (2006). *In vitro* soil receptivity assays to egg parasitic nematophagous fungi. *Mycological Progress, 5*, 18–23.
Montes de Oca, N., Arévalos, J., Nuñez, A., et al. (2009). Klamic: Experiencia técnica-productiva. *Revista de Protección Vegetal, 24*, 62–65.
Naranjo, S. E., Ellsworth, P. C., & Frisvold, G. B. (2015). Economic value of biological control in integrated pest management of managed plant systems. *Annual Review of Entomology, 60*, 621–645.
Nyaku, S. T., Sripathi, V. R., Kantety, R. V., et al. (2014). Characterization of the reniform nematode genome by shotgun sequencing. *Genome, 57*, 209–221.
OSU Oregon State University. (2012). Integrated plant protection center. *IPMnet News*, 192 January 2012, issue no. 192. http://www.ipmnet.org/IPMNews/2012/news192.html
Robinson, A. F. (2007). Reniform in U.S. Cotton: When, where, why, and some remedies. *Annual Review of Phytopathology, 45*, 263–288.
Sánchez-Moreno, S., Nicola, N. L., Ferris, H., et al. (2009). Effects of agricultural management on nematode–mite assemblages: Soil food web indices as predictors of mite community composition. *Applied Soil Ecology, 41*, 107–117.
Sellitto, V. M., Curto, G., Dallavalle, E., et al. (2016). Effect of *Pochonia chlamydosporia*-based formulates on the regulation of root-knot nematodes and plant growth response. *Frontiers in Life Science.* https://doi.org/10.1080/21553769.2016.1193827.

Shorest, M., Gal-On, A., Leibamn, D., et al. (2006). Characterization of a mitogen-activated protein kinase gene from cucumber required for *Trichoderma*-conferred plant resistance. *Plant Physiology, 142*, 1169–1179.

Stirling, G. R. (1991). *Biological control of plant parasitic nematodes: Progress, problems and prospects*. Wallingford: CABI Publishing.

Stirling, G. R. (2014). *Biological control of plant-parasitic nematodes: Soil ecosystem management in sustainable agriculture* (2nd ed.). Croydon: CABI.

Sundin, G. W., Wang, N., Charkowski, A. O., et al. (2016). Perspectives on the transition from bacterial phytopathogen genomics studies to applications enhancing disease management: From promise to practice. *Phytopathology Perspectives, 106*, 1071–1082.

Timper, P. (2014). Conserving and enhancing biological control of nematodes. *Journal of Nematology, 46*, 75–89.

Tunlid, A., & Talbot, N. J. (2002). Genomics of parasitic and symbiotic fungi. *Current Opinion in Microbiology, 5*, 513–519.

Wabule, M. N., Ngaruiga, P. N., Kimmins, F. K. et al. (2004). *Registration for biological control agents in Kenya*. Proceedings of the pest control products Board/Kenya/Agricultural Research Institute/Department for international development crop protection programme workshop, Nakuru, Kenya, 14–16 May 2003. KARI/PCPB, Nairobi, Kenya, and Natural Resources International Ltd., Aylesford, UK.

Westermann, A. J., Gorski, S. A., & Vogel, J. (2012). Dual RNA-seq of pathogen and host. *Nature Reviews Microbiology, 10*, 618–630. http://dx.doi.org/10.1038/nrmicro2852.

Zavala-Gonzalez, E. A., Escudero, N., Lopez-Moya, F., et al. (2015). Some isolates of the nematophagous fungus *Pochonia chlamydosporia* promote root growth and reduce flowering time of tomato. *The Annals of Applied Biology, 166*, 472–483.

Zavala-Gonzalez, E. A., Rodríguez-Cazorla, E., Escudero, N., et al. (2017). *Arabidopsis thaliana* root colonization by the nematophagous fungus *Pochonia chlamydosporia* is modulated by jasmonate signalling and leads to accelerated flowering and improved yield. *The New Phytologist, 213*, 351–364.

Index

R.H. Manzanilla-López, L.V. Lopez-Llorca (eds.), *Perspectives in Sustainable Nematode Management Through* Pochonia chlamydosporia *Applications for Root and Rhizosphere Health*, Sustainability in Plant and Crop Protection,
DOI 10.1007/978-3-319-59224-4

Q

R

Printed by Printforce, the Netherlands